Methylotrophs and Methylotroph Communities

https://doi.org/10.21775/9781912530045

Edited by

Ludmila Chistoserdova

Department of Chemical Engineering
University of Washington
Seattle
WA
USA

Caister Academic Press

Caister Academic Press
Norfolk, UK

www.caister.com

British Library Cataloguing-in-Publication Data
A catalogue record for this book is available from the British Library

ISBN: 978-1-912530-04-5 (hardback)

ISBN: 978-1-912530-05-2 (ebook)

Cover design adapted with permission from Dr Kelly C. Wrighton, Fort Collins, CO, USA

Contents

Preface

This volume is a compilation of articles that present the recent data and the novel outlooks at the diverse areas of the methylotrophy field, encompassing the fundamentals of this research field, such as updates on the biochemistry, physiology and systematics of methylotrophs, as well as the newly emerging areas, such as laboratory evolution of methylotrophs, enumeration of methylotrophs in novel environments such as clouds, and experimenting with synthetic methylotrophs and methylotroph communities. In general, methylotrophy is a rather small field within the broader field of microbiology. While over one hundred years of age, the field has been experiencing a remarkable rejuvenation, partly fuelled by the omics capabilities, but also defined by the bold new questions that the established as well as the newly emerging methylotrophysts want to approach. The young and upcoming methylotrophy researchers were especially welcomed to contribute to the volume. Focused articles featuring results from specific research areas/teams have also been encouraged.

In Chapter 1, Semrau and DiSpirito provide a review of the current knowledge on the aerobic methanotrophy and describe the potential of these organisms for environmental, industrial and medical applications. In Chapter 2, Knief provides an extensive overview of methane-producing and methane-consuming microbes from diverse environmental niches, along with a comprehensive update on the classification of microorganisms involved in the global methane cycle. In Chapter 3, Smith and Wrighton provide a metagenomic insight into methanotrophy, through the analysis of the currently available sequence datasets, originating from a variety of environments. In Chapter 4, Khmelenina and colleagues present recent insights from comparative analyses of properties of some key enzymes in methylotrophy, and discuss the recent adjustments to understanding of key metabolic pathways in aerobic methanotrophy. In Chapter 5, Skovran and colleagues elaborate on the newly emerged role of lanthanides in methylotrophy. They present data on the complex regulatory networks involved in balancing the roles of alternative methanol dehydrogenases, and they review the potential of lanthanide enzymes for biotechnological applications. In Chapter 6, Czarnecki and Bartosik provide detailed overview of methylotrophy in species of the genus *Paracoccus*, revealing great variability of C1 metabolism in these species and discussing potential evolutionary mechanisms for such metabolic flexibility. In Chapter 7, Mausz and Chen summarize the latest developments in analytical methods for quantifying methylated amines in marine waters and sediments, and they discuss the metabolic pathways leading to the formation and the degradation of methylated amines, with a special focus on the novel biochemistry and structural biology of the enzymes for these transformations. In Chapter 8, Bringel and colleagues present the state of the art of the knowledge on microbial degradation of chloromethane, an important atmospheric pollutant, uncovering the many unknowns that still exist in both understanding the pathways for chloromethane degradation and the microorganisms/communities contributing to the global chloromethane cycle. In Chapter 9, Schäfer and Eyice provide an update on the diversity of microorganisms involved in the production and the degradation of methanethiol, an important methylated intermediate in the global sulfur cycle. In Chapter 10, Yu and colleagues present the recent data that support the communal function in aerobic methane oxidation, including insights from manipulation of synthetic methanotrophic communities. They further elaborate on the role of lanthanide-dependent alcohol dehydrogenases in methylotrophy, and on the multiple factors that regulate these enzymes. In Chapter 11, Yurimoto and Sakai describe the physiology of methylotrophic yeasts that proliferate and survive on plant leaves, feeding on the excreted methanol. They focus on the specific adaptations for the life in the phyllosphere, which include the specific mechanisms for sensing methanol. In Chapter 12, Puri introduces an emerging subfield of methylotrophy, dealing with the potentially methylotroph-specific specialized (secondary) metabolites. The author examines the predicted biosynthetic potential of several methylotrophs, and reviews some of the specialized metabolites that have been recently characterized from these organisms. In Chapter 13, Zheng and colleagues review the recent progress towards engineering a model methylotroph, *Methylobacterium extorquens*, towards industrial production of several platform chemicals from methanol, as well as the recent progress in engineering synthetic methylotrophs based on the established industrial strains of bacteria and yeasts. In Chapter 14, Claassens and colleagues discuss the use of a step-wise, modular engineering approach for synthetic implementation of pathways for assimilation of carbon from methanol and formate, this approach providing opportunities for identification and resolution of metabolic barriers hampering pathway performance, these being essential for rewiring microbial metabolism towards the desired growth phenotypes and sustainable bioproduction. Finally, in Chapter 15, Marx describes his personal account of the development of *Methylobacterium* as a model system for experimental evolution, designed to study questions at the intersection of metabolism and evolution, a 15-year long journey. As such, the volume covers a remarkable diversity of the topics relevant to methylotrophy, and thus well represents the current state of the field. It should be useful for a wide audience, both researchers familiar with the concepts and the details of methylotrophy metabolism, and the newcomers to the field. I want to thank all the authors for their contributions. I thoroughly enjoyed reading each chapter, and I hope the future reader finds the volume as enjoyable as informative.

Methanotrophy – Environmental, Industrial and Medical Applications

1

Jeremy D. Semrau[1]* and Alan A. DiSpirito[2]

[1]Department of Civil and Environmental Engineering, University of Michigan, Ann Arbor, MI, USA.
[2]Roy J. Carver Department of Biochemistry, Biophysics and Molecular Biology, Iowa State University, Ames, IA, USA.

*Correspondence: jsemrau@umich.edu

https://doi.org/10.21775/9781912530045.01

Abstract

Aerobic methanotrophs are an intriguing group of microbes with the singular ability to consume methane as their sole source of carbon and energy. As such, methanotrophs are receiving increased attention to control methane emissions to limit future climate change. Methanotrophs have a wide range of other applications, including pollutant remediation and methane valorization (e.g. conversion of methane to protein, bioplastics, and biodiesel amongst other products). Methanotrophs also produce a novel copper-binding compound – methanobactin – that has significant potential for the treatment of copper-related human pathologies. Here we provide an overview of aerobic methanotrophy, describe current and future applications of these unique microbes, as well as discuss various strategies one can consider to better realize the opportunities these microbes present.

Introduction

Aerobic methanotrophs, or methane-oxidizing bacteria, are a group of microbes with great environmental and industrial importance. For example, methanotrophs are well known to play a key role in controlling the net emission of methane from soils, a potent greenhouse gas with a global warming potential ≈34 times that of carbon dioxide over a 100-year time frame (Myhre *et al.*, 2013). In fact, it is estimated that as much as 90% of methane generated in anaerobic soils via methanogenesis may be removed via methanotrophy (Chowdhury and Dick, 2013). Further, methanotrophs oxidize methane under ambient temperatures and pressures, and thus are attractive platforms for the valorization of methane to products such as single-cell protein, bioplastics, and biofuels (Semrau *et al.*, 2010; Khmelenina *et al.*, 2015; Strong *et al.*, 2015, 2016). Interest in commercial application of methanotrophy has dramatically accelerated in recent years as methane prices have become quite low, with the industrial price of natural gas dropping from \$13.06 per 1000 ft^3 in July 2008 to \$3.92 per 1000 ft^3 in November 2017 (United States Energy Information Administration, 2018). Herein we provide an overview of aerobic methanotrophy followed by a discussion of current and developing applications.

Overview of methanotrophic diversity

Aerobic methanotrophs are distinguished from other microorganisms by their ability to utilize methane as their sole carbon and energy source, yet are phylogenetically and physiologically quite diverse. Although most characterized aerobic methanotrophs can be considered mesophilic, i.e. pH and temperature optima around 7 and 30°C, respectively, there are many examples of aerobic methanotrophs living under more 'extreme' conditions. That is, acidophilic (growth at pH <3), alkaliphilic (growth at pH >9) thermophilic (growth above 50 °C), psychrophilic (growth below 15°C) as well as halophilic (growth at salt concentrations >1 M) methanotrophs have been isolated (Op den Camp *et al.*, 2009; Semrau *et al.*, 2010; Knief, 2015).

Perhaps not surprisingly then, aerobic methanotrophs exhibit very broad phylogenetic diversity, with extant strains grouping with the Proteobacteria, Verrucomicrobia, and NC10 phyla. More precisely, 20 genera have been characterized in the *γ-Proteobacteria* class, six genera in the *α-Proteobacteria* class, two genera in the Verrucomicrobia phylum and one in the NC10 phylum (Table 1.1). It is quite likely that as more researchers focus on methanotrophy, the phylogenetic and physiological diversity of these intriguing microbes will greatly expand. In this review, however, we will focus on the *Proteobacteria* methanotrophs as these microbes are ubiquitous, play critical roles in mitigating greenhouse gas emissions *in situ*, and we have a greater depth of knowledge of the genetics, biochemistry and metabolism of these strains. We hasten to stress, however, that undoubtedly other methanotrophs possess intriguing properties with great environmental and industrial relevance.

A key enzyme in aerobic methanotrophy – methane monooxygenase

As noted above, methanotrophs are noted by their ability to utilize methane as their sole source of carbon and energy. Doing so, however, requires a remarkable enzyme that converts methane to methanol, the methane monooxygenase. Cleaving the C–H bond in methane is inherently challenging given the high bond dissociation energy of the C–H bond in methane – 104 kcal/mol. What is perhaps even more remarkable is that aerobic methanotrophs have not one, but two forms of methane monooxygenase. One form of the enzyme,

Table 1.1 General phylogeny of aerobic methanotrophs[a]

Phylogeny	Reference
Phylum	
Gammaproteobacteria	
Family	
Methylococcaceae	
Genera	
Clonothrix[b]	Vigliotta *et al.* (2007)
Methylobacter	Whittenbury *et al.* (1970)
Methylocaldum	Bodrossy *et al.* (1997)
Methylococcus	Foster and Davis (1966)
Methylogaea	Geymonat *et al.* (2011)
Methyloglobulus	Deutzmann *et al.* (2014)
Methylomagnum	Khalifa *et al.* (2015)
Methylomarinum	Hirayama *et al.* (2013)
Methylomicrobium	Bowman *et al.* (1995)
Methylomonas	Brown *et al.* (1964)
Methyloparacoccus	Hoefman *et al.* (2014)
Methyloprofundus	Tavormina *et al.* (2015)
Methylosarcina	Wise *et al.* (2001)
Methylosoma	Rahalkar *et al.* (2007)
Methylosphaera	Bowman *et al.* (1997)
Methylovulum	Iguchi *et al.* (2011a)

Table 1.1 Continued

Phylogeny	Reference
Family	
Methylothermaceae	
Genera	
Methylohalobius	Heyer *et al*. (2005)
Methylomarinovum	Hirayama *et al*. (2014)
Methylothermus	Bodrossy *et al*. (1999)
Family	
Crenotrichaceae	
Genera	
Crenothrix[b]	Stoecker *et al*. (2006)
Phylum	
Alphaproteobacteria	
Family	
Methylocystaceae	
Genera	
Methylocystis	Whittenbury *et al*. (1970)
Methylosinus	Whittenbury *et al*. (1970)
Family	
Beijerinkiaceae	
Genera	
Methylocella	Dedysh *et al*. (2000)
Methylocapsa	Dedysh *et al*. (2002)
Methyloferula	Vorobev *et al*. (2011)
Methyloaffinis[b]	Pratscher *et al*. (2018)
Phylum	
Verrucomicrobia	
Family	
Methylacidiphilaceae	
Genera	
Methylacidiphilum	Op den Camp *et al*. (2009)
Family	
Unclassified[c]	
Genera	
Methylacidimicrobium	van Teeseling *et al*. (2014)
Phylum	
NC10	
Family	
Unclassified	
Genera	
Methylomirabilis[b]	Ettwig *et al*. (2010)

[a]For the sake of brevity, methanotrophic species are not listed. For a thorough list/description of validated aerobic methanotrophic species, the reader is directed to (Knief, 2015); [b]No type strains have been isolated/purified, thus these genera should be considered Candidatus; [c]Using a cut-off of 86.5% to distinguish bacteria of different families (Yarza *et al*., 2014), *Methylacidimicrobium* can be considered to be a member of the *Methylacidiphilaceae* family as its 16 rRNA sequence is 89–90% identical to that of *Methylacidiphilum* (van Teeseling *et al*., 2014).

the particulate methane monooxygenase (pMMO) is found in most known methanotrophs (Table 1.2) and is located in the cytoplasmic membrane (Stanley *et al.*, 1983; Dalton *et al.*, 1984; Zahn and Dispirito, 1996; Murrell *et al.*, 2000; Basu *et al.*, 2003; Choi *et al.*, 2003; Lieberman *et al.*, 2003;

Table 1.2 Occurrence of genes encoding for subunits of pMMO, sMMO, or pXMO as well methanobactin biosynthesis in the genomes of select methanotrophs. pMMO – *pmoA;* sMMO – *mmoX*; pXMO – *pxmA*; methanobactin – *mbnB/C*

Strain	Class	*pmoA*	*mmoX*	*pxmA*	*mbnB/C*
Methylobacter tundripaludum SV96	Gammaproteobacteria	Yes	No	Yes	No
Methylobacter marinus A45	Gammaproteobacteria	Yes	No	Yes	No
Methylobacter tundripaludum 21/22	Gammaproteobacteria	Yes	No	Yes	No
Methylobacter luteus IMV-B-3098	Gammaproteobacteria	Yes	No	Yes	No
Methylobacter tundripaludum 31/32	Gammaproteobacteria	Yes	No	Yes	No
Methylobacter whittenburyi ACM	Gammaproteobacteria	Yes	No	Yes	No
Methylobacter sp. BBA5.1	Gammaproteobacteria	Yes	No	Yes	No
Methylocaldum szegediense O-12	Gammaproteobacteria	Yes	No	No	No
Methylococcus capsulatus str. Bath	Gammaproteobacteria	Yes	Yes	No	No
Methyloglobulus morosus KoM1	Gammaproteobacteria	Yes	No	Yes	No
Methylohalobius crimeensis 10Ki	Gammaproteobacteria	Yes	No	No	No
Methylomarinum vadi strain IT-4	Gammaproteobacteria	Yes	No	No	No
Methylomicrobium alcaliphilum	Gammaproteobacteria	Yes	No	No	No
Methylomicrobium album BG8	Gammaproteobacteria	Yes	No	Yes	No
Methylomicrobium buryatense 5G	Gammaproteobacteria	Yes	Yes	No	No
Methylomicrobium agile ATCC 35068	Gammaproteobacteria	Yes	No	Yes	No
Methylomonas methanica MC09	Gammaproteobacteria	Yes	Yes	No	No
Methylomonas sp. MK1	Gammaproteobacteria	Yes	Yes	Yes	No
Methylomonas sp. 11b	Gammaproteobacteria	Yes	Yes	Yes	No
Methylomonas sp. LW13	Gammaproteobacteria	Yes	Yes	Yes	No
Methylomonas denitrificans FJG1	Gammaproteobacteria	Yes	No	Yes	No
Methylosarcina fibrata AML-C10	Gammaproteobacteria	Yes	No	No	No
Methylosarcina lacus LW14	Gammaproteobacteria	Yes	No	No	No
Methylovulum miyakonense HT12	Gammaproteobacteria	Yes	Yes	No	No
Methylocapsa acidiphila B2	Alphaproteobacteria	Yes	No	No	No
Methylocapsa aurea KYG	Alphaproteobacteria	Yes	No	No	No
Methylocella silvestris BL2	Alphaproteobacteria	No	Yes	No	No
Methylocystis sp. SB2	Alphaproteobacteria	Yes	No	Yes	Yes
Methylocystis sp. SC2	Alphaproteobacteria	Yes	No	No	Yes
Methylocystis rosea SV97	Alphaproteobacteria	Yes	No	Yes	Yes
Methylocystis sp. ATCC 49242 Rockwell	Alphaproteobacteria	Yes	No	No	No
Methylocystis parvus strain OBBP	Alphaproteobacteria	Yes	No	No	Yes
Methyloferula stellata AR4	Alphaproteobacteria	No	Yes	No	No
Methylosinus sp. LW4	Alphaproteobacteria	Yes	Yes	No	Yes
Methylosinus trichosporium OB3b	Alphaproteobacteria	Yes	Yes	No	Yes
Methylosinus sp. LW3	Alphaproteobacteria	Yes	Yes	No	Yes
Methylosinus sp. PW1	Alphaproteobacteria	Yes	Yes	No	Yes
Methylocystis sp. LW5	Alphaproteobacteria	Yes	Yes	No	Yes

Lieberman and Rosenzweig, 2005). Another form, the soluble methane monooxygenase (sMMO), is found in some methanotrophs (Table 1.2) and is located in the cytoplasm (Stirling *et al.*, 1979; Stanley *et al.*, 1983; Dalton *et al.*, 1984; Fox *et al.*, 1989; Stainthorpe *et al.*, 1989, 1990a,b; Lipscomb, 1994; Wallar and Lipscomb, 1996). The sMMO has relatively high turnover but poor affinity for methane. sMMO also has a broad substrate range and as a result has significant potential in biocatalysis (Colby *et al.*, 1977; Tonge *et al.*, 1977; Trotsenko and Murrell, 2008; Semrau *et al.*, 2010; Semrau, 2011; Kalyuzhnaya *et al.*, 2015). pMMO, conversely, has relatively low turnover but greater affinity for methane, and so has greater significance for the removal of atmospheric methane (Lee *et al.*, 2006). Interestingly, methanotrophic growth via pMMO is energetically more efficient than growth via sMMO (Trotsenko and Murrell, 2008; Kalyuzhnaya *et al.*, 2015). The operons for both sMMO (*mmo*) and pMMO (*pmo*) are shown in Fig. 1.1, with most genes known to encode for polypeptides of either MMO. The only exceptions are *mmoR* and *mmoG*, known to encode for a regulatory element of the *mmo* operon and a GroEL homologue, respectively (Stafford *et al.*, 2003), and *mmoD*, that plays a critical role in controlling the relative expression of the two forms of MMO for those methanotrophs that can express both sMMO and pMMO (Semrau *et al.*, 2013; Yan *et al.*, 2016).

In addition to the canonical forms of MMO, there is evidence for a third form, or pXMO

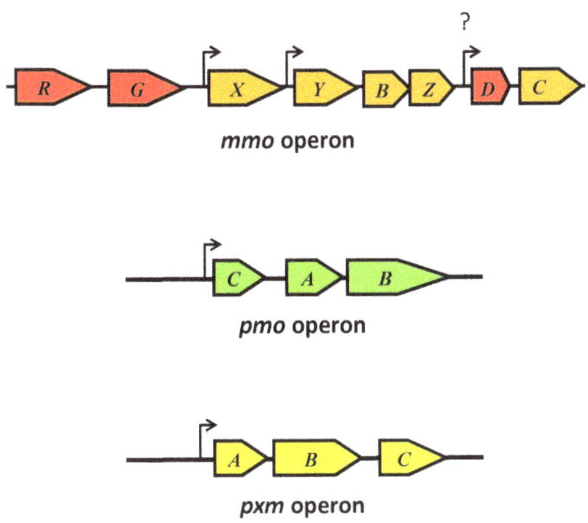

Figure 1.1 *mmo*, *pmo*, and *pxm* operons. Known regulatory genes shown in red.

(Tavormina *et al.*, 2011). This third MMO appears to be a divergent form of pMMO, and has a different gene organization, i.e. *pxmABC* versus *pmoCAB* (Fig. 1.1). Many (but not all) γ-Proteobacteria methanotrophs have been found to harbour these genes, but to date only a small number of α-Proteobacteria methanotrophs appear to have the *pxm* operon, e.g. *Methylocystis rosea* SV97 and *Methylocystis* strain SB2 (Table 1.2 (Ghashghavi *et al.*, 2017; Gu, 2017)). It is unclear what this significance of pXMO may be. Although expression is evident (Tavormina *et al.*, 2011; Kits *et al.*, 2015), it is typically quite low and its role is still undecided in methanotrophic metabolism. It may be, as speculated by Tavormina *et al.* (2011), that pXMO promotes methanotrophic fitness by extending the range of substrates these methanotrophs can utilize for growth and/or serving to remove various toxins. Alternatively, it has been suggested that expression of *pxm* genes increase under hypoxia and in the presence of nitrate. As a result, pXMO may play a role in helping methanotrophs deal with oxygen limitation (Kits *et al.*, 2015). There is still much to be learned, however, as what environmental conditions induce expression of *pxm* genes and if these are in any way coordinated with expression of other MMO genes.

Methanotrophic response to copper

Aerobic methanotrophs are sensitive to copper, and it is a key factor regulating expression of genes encoding for polypeptides of sMMO and pMMO (Stanley *et al.*, 1983; Dalton *et al.*, 1984; Prior and Dalton, 1985a,b; Choi *et al.*, 2003; Han and Semrau, 2004; Semrau *et al.*, 2010, 2013). *mmoX* is expressed only under low-copper conditions while *pmoA* expression increases substantially when copper is increased. Further, no activity is found in the soluble (cytoplasmic) fraction above 1 μM copper (i.e. no sMMO activity), and as the copper concentration increases, MMO activity in the membranes and whole-cells increases substantially (i.e. increased pMMO activity, Fig. 1.2). Such findings can be explained as sMMO is known to have a diiron complex in its active site, while copper occupies at least two of three or one of two metal centres found in purified pMMO (Lipscomb, 1994; Lieberman *et al.*, 2003; Balasubramanian *et al.*, 2010; Semrau *et al.*, 2010). Interestingly, addition of copper can also affect cell composition.

That is, when *Methylococcus capsulatus* Bath is grown in the absence of copper (and expressing sMMO), a substantial fraction of cell carbon is found as polyhydroxybutyrate (Choi *et al.*, 2003). As copper is added, however, the distribution of cell carbon changes with polyhydroxybutyrate content decreasing and intracytoplasmic membrane (lipid) content increasing. As such, one may be able to use copper to control cell composition and increase either the production of a bioplastic precursor (polyhydroxybutyrate) or a biodiesel precursor (lipid) as described later in this chapter.

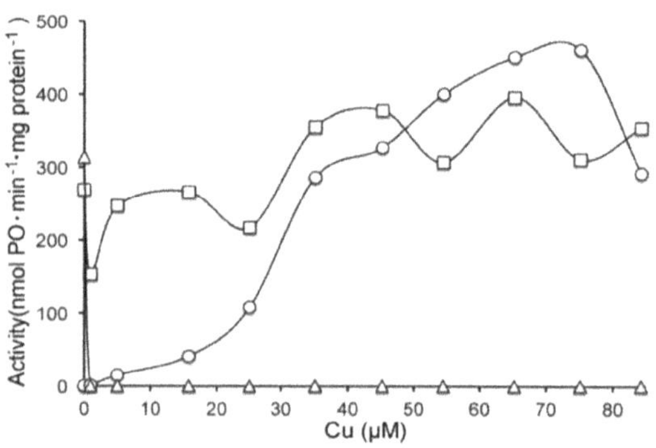

Figure 1.2 MMO activity in *M. capsulatus* Bath as a function of copper. (□): whole cells, (△): soluble fraction, (○): membrane fraction (DiSpirito, unpublished data). PO = propylene oxide.

Copper uptake in methanotrophs

Clearly methanotrophs sense and respond to copper in their environment, but it was only recently that the mechanism(s) used by methanotrophs to do so was discovered. Some aerobic methanotrophs utilize an intriguing copper-binding compound analogous to a siderophore, dubbed a chalkophore ('chalko' is Greek for copper). The first structurally characterized chalkophore, methanobactin, was from the *α-Proteobacteria, Methylosinus trichosporium* OB3b (Kim *et al.*, 2004) and is a modified polypeptide with two oxazolone rings with thioamide groups that are responsible for copper binding (Fig. 1.3). Subsequently methanobactin has been isolated and purified from a variety of methanotrophs and is found to have several general characteristics, including being a small (less than 1300 Da) modified polypeptide chain with two heterocyclic rings, either an imidazole, oxazolone or pyrazinedione ring, each with an associated thioamide group that together are responsible for copper binding (Kim *et al.*, 2004; Choi *et al.*, 2006b; Behling *et al.*, 2008; Krentz *et al.*, 2010; El Ghazouani *et al.*, 2011, 2012; Bandow *et al.*, 2012; Vorobev *et al.*, 2013; DiSpirito *et al.*, 2016; Kenney *et al.*, 2016). Methanobactin

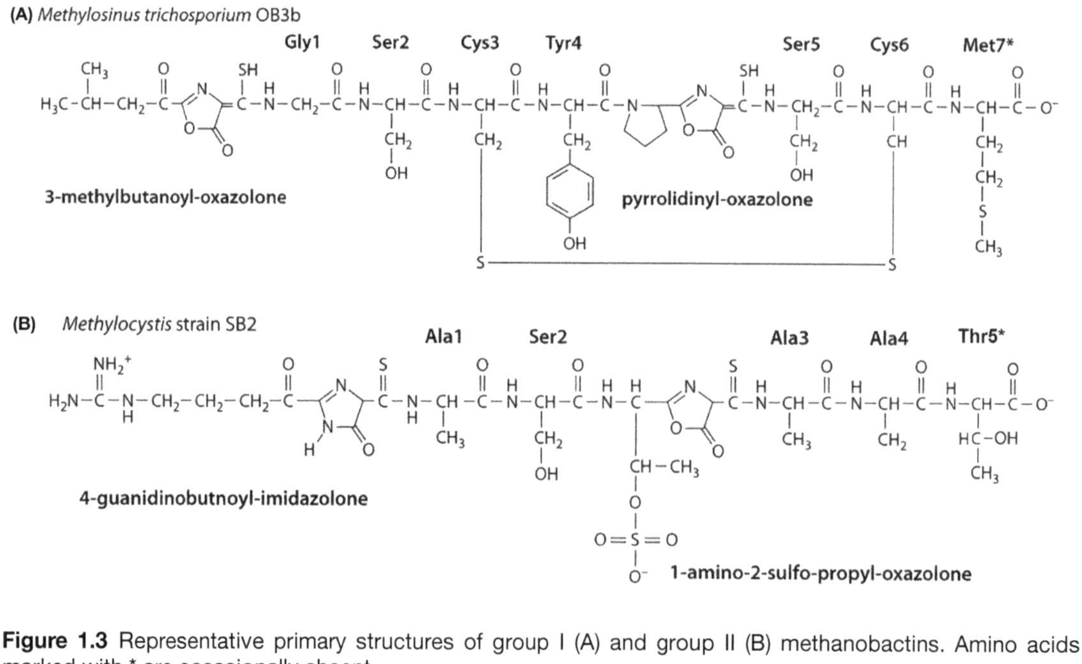

Figure 1.3 Representative primary structures of group I (A) and group II (B) methanobactins. Amino acids marked with * are occasionally absent.

has an extremely high copper-binding constant (Kim *et al.*, 2004; Behling *et al.*, 2008; Krentz *et al.*, 2010; El Ghazouani *et al.*, 2011, 2012; Bandow *et al.*, 2012; Vorobev *et al.*, 2013). Methanobactins can be divided into two general groups on the basis of the types of heterocyclic rings present as well as the presence/absence of a sulfate group (Fig. 1.3). Specifically, Group I methanobactins have two oxazolone rings while Group II methanobactins have a C-terminal oxazolone ring, with the other ring being either imidazolone or a pyrazinedione. Group I methanobactins also contain Cys residues in the mature peptide, while those in Group II do not. The N-terminal amino acid in Group I methanobactin is deaminated, while N-terminal deamination is not observed in Group II methanobactins (Gu *et al.*, 2017). Finally, Group II methanobactins have an associated sulfate group that has not been found to date in Group I methanobactins (Krentz *et al.*, 2010; El Ghazouani *et al.*, 2012; DiSpirito *et al.*, 2016).

Biochemical analyses indicated that methanobactin could be formed from a polypeptide precursor with the heterocyclic rings derived from various amino acids, primarily cysteines (Krentz *et al.*, 2010). Interrogation of available methanotrophic genomes found one possible candidate gene, *mbnA*. Deletion of *mbnA* in *M. trichosporium* OB3b showed that it indeed encodes for the precursor of methanobactin and that it is part of an operon (Fig. 1.4; Semrau *et al.*, 2013), with many genes of unknown function (involved in ring formation), an extrusion protein (that may serve to secrete methanobactin), as well as an aminotransferase recently shown to de-aminate the N-terminal leucine of methanobactin from *M. trichosporium* OB3b, required for subsequent formation and/or stability of the 3-methylbutanoyl-oxazolone moiety (Gu *et al.*, 2017). Downstream of *mbnN* (encoding for the aminotransferase) is another gene cluster that encodes for a di-haem cytochrome peroxidase and its partner protein that may be involved in methanobactin synthesis (Semrau *et al.*, 2018). Upstream of *mbnA* is another gene cluster under the control of a separate promoter with genes encoding for a TonB-dependent transporter (*mbnT*) responsible for methanobactin uptake (Gu *et al.*, 2016), as well as a putative membrane sensor (*mbnR*) and an extracytoplasmic function sigma factor (*mbnI*). A similar system is frequently found in siderophore synthesis where an outer membrane transporter binds a ferri-siderophore, transmitting a signal to a membrane sensor that then activates an extracytoplasmic function sigma factor. This ultimately induces expression of genes required for siderophore synthesis, as well as in some cases genes unrelated to siderophore production or uptake, e.g. genes encoding for exotoxins and proteases (Crosa, 1997; Lamont *et al.*, 2002; Visca *et al.*, 2002; Mahren and Braun, 2003; Braun *et al.*, 2006; Grosse *et al.*, 2007; Brooks and Buchanan, 2008). MbnI may thus be involved in regulating expression of the methanobactin gene cluster and/or of other gene clusters, e.g. the *mmo* and *pmo* operons and perhaps others as yet unknown. Interestingly, interrogation of available microbial genomes indicated that both methanotrophs and non-methanotrophs have similar methanobactin gene clusters as that found in *M. trichosporium* OB3b, suggesting that methanobactin-like compounds may be widespread in nature (Haft *et al.*, 2013; Kenney and Rosenzweig, 2013; Semrau *et al.*, 2013). To date, however, *mbn* genes have been only found in methanotrophs grouping in the α-Proteobacteria, suggesting that γ-Proteobacteria methanotrophs (that also require copper) utilize a different system for copper

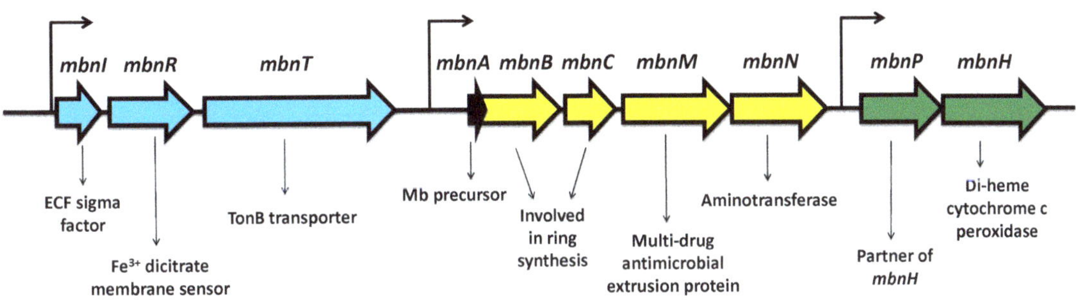

Figure 1.4 Methanobactin gene cluster from *M. trichosporium* OB3b.

sequestration. For example, the γ-Proteobacteria methanotroph, *M. capsulatus* Bath, appears to rely on a surface-bound protein with a tryptophan converted to a kynurenine (MopE) and a secreted form of it (MopE*) for copper binding (Fjellbirkeland *et al.*, 1997; Karlsen *et al.*, 2003; Helland *et al.*, 2008; Ve *et al.*, 2012), although there is some evidence of a methanobactin-like compound in this methanotroph that has not been extensively characterized (Zahn and Dispirito, 1996; Choi *et al.*, 2005). Interestingly, many γ-Proteobacteria methanotrophs have TonB-dependent transporters similar to MbnT, and it may be that some γ-Proteobacteria methanotrophs act as 'cheaters' where they utilize methanobactin made by other methanotrophs to satisfy copper requirements.

Environmental applications/ implications of methanotrophy

Organic pollutant degradation

Given the ubiquity of methanotrophy in terrestrial, freshwater and marine environments, there has been interest for decades to harness the activity of these microbes to address a number of environmental issues, initially for pollutant degradation. The utility of methanotrophs for bioremediation, particularly for organic pollutants is enhanced as both pMMO and sMMO are relatively non-specific and will oxidize a number of compounds in addition to methane. Early work showed that sMMO can degrade a wide range of organic compounds, i.e. alkanes up to C_8, as well as ethers, cyclic alkanes, and aromatic hydrocarbons (Colby *et al.*, 1977; Burrows *et al.*, 1984). pMMO can oxidize alkanes up to C_5, but not aromatic compounds (Burrows *et al.*, 1984). Further, both forms of MMO can attack ethenes, especially chlorinated ethenes that are commonly found at hazardous waste sites (Zahniser, 2015). Alkanes are general converted to the corresponding alcohol (with the hydroxyl group primarily at the terminal carbon), while aromatic hydroxylation occurs at the meta position (Strong *et al.*, 2017). Alkenes are oxygenated across the carbon-carbon double bond, with the subsequent epoxides commonly spontaneously undergoing a variety of abiotic oxidation and hydrolysis reactions (e.g. Fox *et al.*, 1990; Lontoh *et al.*, 2000). The reader is directed to Jiang *et al.* (2010), who provide an excellent summary of organic compounds attacked by both sMMO and pMMO, as well as the products of such transformation. Finally, it should be noted that not only does sMMO attack more compounds than pMMO, it turns over these compounds more quickly.

It should be stressed, however, that degradation of these compounds by either form of MMO is 'co-metabolic', i.e. the continued degradation can occur only 'in the obligate presence of a growth substrate or another transformable compound' (Dalton and Stirling, 1982) as these pollutants typically can serve neither as a carbon nor energy source to methanotrophs. Rather, transformation of these compounds requires reducing equivalents, whose consumption can limit methanotrophic growth and activity. In addition, toxic products are commonly formed from these co-metabolic reactions (e.g. Fox *et al.*, 1990; Lontoh *et al.*, 2000; Jiang *et al.*, 2010; Semrau, 2011). As a result, although use of sMMO-expressing methanotrophs for the biodegradation of organic pollutants may appear attractive given these microbes can attack a wider range of compounds and degrade them more quickly than pMMO-expressing methanotrophs, it may actually be counter-productive. That is, methanotrophs expressing pMMO, although attacking fewer co-metabolic pollutants and degrading them slower, more easily turnover methane in their presence, thus allowing for greater overall activity and growth that then leads to greater overall degradation of select pollutants. In other words, a pMMO-expressing methanotroph, or 'tortoise' may actually be better than an sMMO-expressing methanotroph, or 'hare' for bioremediation of polluted sites (Lee *et al.*, 2006). Indeed, studies have found that pMMO-expressing methanotrophs are responsible for pollutant degradation *in situ* and in enrichment cultures (Shukla *et al.*, 2009; Paszczynski *et al.*, 2011). The take-home message is that effective use of methanotrophs for pollutant degradation is not as simple as it might first seem. Rather, it requires one to consider what is to be degraded and how effective turnover of methane can be maintained to ensure that the appropriate conditions (i.e. pMMO- vs sMMO-promoting conditions) are provided to facilitate methanotrophic growth that then leads to greater pollutant removal.

Control of methane emissions

Methane is a potent greenhouse gas, with a global warming potential 84 times that of carbon dioxide over a 20-year time frame (Myhre *et al.*, 2013). The current atmospheric methane concentration of 1.8 ppm_v (or 154 ppm_v in CO_2 equivalents) is expected to double by 2100 (Saunois *et al.*, 2016). Moreover, the radiative forcing incurred per unit increase in concentration is larger for methane than for carbon dioxide (Lashof and Ahuja, 1990). Of particular concern are emissions of methane from permafrost and thermokarst lakes in warming Arctic regions (Nzotungicimpaye and Zickfeld, 2017). The positive feedback loops resulting from such releases are projected to discharge 30 to 63 gigatons of permafrost carbon to the atmosphere in the next three decades and 232 to 380 gigatons by 2100 (Schuur and Abbott, 2011). Thus, both prevention of methane emissions from warming arctic regions as well as from agriculture systems (e.g. rice paddies) is a topic of great importance, as are strategies for removal of methane from the atmosphere. Pursuing such a two-pronged approach could have a significant effect in the short term on reducing future global warming.

Currently, the primary sink of methane after entering the atmosphere is reaction with the hydroxyl radical (OH·), with initial estimates that this mechanism is responsible for ≈90% of atmospheric methane removal (Wuebbles and Hayhoe, 2002). Subsequent studies suggest this sink may be overestimated, indicating that other sinks of atmospheric methane exist (Wang *et al.*, 2008). Although microbial consumption of methane (i.e. methanotrophy) is well known to occur in the subsurface where methane concentrations are high (Hanson and Hanson, 1996), it was initially widely believed that biological removal of atmospheric methane was unlikely given the relatively poor affinity cultured methanotrophs had for methane (on the order of 1000–10,000 ppm_v). It had been speculated, however, that methanotrophs capable of oxidizing methane at atmospheric levels could exist (Bender and Conrad, 1992). It is now known that biological removal of atmospheric methane can and does occur, and that some environments, e.g. upland forest soils, act as sinks of atmospheric methane through methanotrophic activity (Bull *et al.*, 2000; Knief *et al.*, 2003; Knief and Dunfield, 2005). More recently it was shown that some methanotrophs, e.g. *Methylocystis* strain SC2, can express alternative forms of pMMO, with one isoform having an affinity for methane of ≈80 ppm_v (Baani and Liesack, 2008). Further, this methanotroph, when expressing this isoform could grow when at least 10 ppm_v methane was added and 1.75 ppm_v methane was sufficient to maintain cells.

With such information, it is now apparent that methanotrophs can oxidize methane at concentrations found in the atmosphere. Intriguingly, analyses of a forest soil exhibiting atmospheric methane uptake indicated that a member of a hitherto unknown genus in the *Beijerinckiaceae* family, with the proposed name of '*Methyloaffinis lahnbergensis*', can also carry out atmospheric methane oxidation (Pratscher *et al.*, 2018). Using the 16S rRNA and *pmoA* sequence from this genome to query available genomic databases finds that this microbe is present in a wide range of environments, including Arctic/permafrost soil, high elevation crusts, alpine grasslands as well as cave biofilms, suggesting that microbes able to take up atmospheric concentrations of methane may be more widespread than initially believed (Pratscher *et al.*, 2018).

What is equally exciting is that some locations widely believed to act primarily as methane sources (e.g. rice paddies and other agricultural soils) can, when properly stimulated, remove atmospheric methane (Ho *et al.*, 2015; Cai *et al.*, 2016). That is, in these situations, more 'conventional' methanotrophs, i.e. those that do not do not align phylogenetically with known high affinity methane oxidizers, appear to be responsible for atmospheric methane consumption, but that such consumption was dependent on periods of methane enrichment alternated with no provision of methane. Such patterns are likely to occur *in situ*, and thus it appears that different methanotrophs have developed different strategies to survive when methane approaches atmospheric concentrations. Given these findings, it may be possible to harness the activity of methane-oxidizing bacteria to not only remove methane from 'hotspots', i.e. areas where ambient concentrations are greater than 500 ppm (e.g. above landfills and in concentrated animal feeding operations; Yoon *et al.*, 2009; Ganendra *et al.*, 2015) but also directly from the atmosphere. Such strategies should be more fully examined to determine if methanotrophs can be appropriately stimulated to remove methane from a variety of environments (including the open atmosphere) such that a net reduction of greenhouse gas emissions occurs. That is, life cycle assessments of any proposed strategy should be performed to insure that it does not indirectly lead to net greenhouse gas loading to the atmosphere, e.g. increased carbon dioxide emissions from resource and power consumption required to stimulate methanotrophic activity.

Metal detoxification

As described above, methanotrophs are sensitive to copper, and have effective means of sequestering copper from the environment. Methanotrophs, however, have been extensively shown to bind and detoxify many other metals, including toxic forms of chromium and selenium (Al Hasin *et al.*, 2010; Lai *et al.*, 2016; Eswayah *et al.*, 2017). These studies clearly show there is significant potential for the use of methanotrophy for the bioremediation of sites contaminated with these metals, but much remains to be learned as how these metals are bound, taken up and transformed. That is, although one may suspect that methanobactin may be involved in transforming these metals, it should be noted that in these studies, the methanotrophic type strain *Methylcoccus capsulatus* Bath was used, and this methanotroph does not have the operon for methanobactin. It may be that MopE and/or MopE* can bind metals such as chromium and selenium, and facilitate delivery to *M. capsulatus* Bath. Alternatively, uptake may be facilitated by some other active mechanism or it may be due to non-specific sorption onto the cell surface, e.g. possibly sulfhydryl groups (Eswayah *et al.*, 2017). In any event, detoxification is dependent on the availability of methane that likely serves as an electron source for metal reduction (Al Hasin *et al.*, 2010; Eswayah *et al.*, 2017).

Although methanobactin is not involved in metal detoxification by *M. capsulatus* Bath, methanobactin has been shown to bind and detoxify mercury. Specifically, both Group I and II methanobactins will bind Hg[II], even in the presence of copper due to the initial rapid binding of Hg[II] (Vorobev *et al.*, 2013; Baral *et al.*, 2014). Mercury binding by methanobactin also reduced Hg[II] toxicity to methanotrophs unable to produce methanobactin, suggesting that methanotrophic production of methanobactin *in situ* may provide protection to the broader microbial community from mercury toxicity (Vorobev *et al.*, 2013).

Interestingly, not only will methanobactin bind inorganic mercury, it can also bind the much more toxic methylmercury (Baral *et al.*, 2014). Interestingly, methylmercury uptake seems to be a general phenomenon exhibited by methanotrophs, including those that do not possess the methanobactin gene cluster (e.g. *M. capsulatus* Bath; Lu *et al.*, 2017). It is unknown how these methanotrophs sequester methylmercury, but again MopE or MopE* may play a role. Interestingly, however, only those methanotrophs capable of expressing methanobactin were found to not only sequester, but also degrade methylmercury (Lu *et al.*, 2017). Such a finding is remarkable as in these methanotrophs, *merB*, encoding for the organomercurial lyase, is not in their genomes, indicating that some other mechanism demethylates methylmercury. Although one may suspect that methanobactin may be responsible for both methylmercury uptake and degradation, assays using purified methanobactin found no significant methylmercury degradation, suggesting that methanobactin works in concert with (as yet unknown) proteins to facilitate methylmercury degradation. Whatever the process whereby methylmercury is degraded by methanotrophs expressing methanobactin, such findings may have significant environmental relevance as in these assays, low (nanomolar) levels of methylmercury were used, similar to that found in the environment (Barkay and Wagner-Döbler, 2005; Lu *et al.*, 2016).

Methanotrophic sequestration of copper and its impact on activity of other microbes

Although methanobactin will adventitiously bind a variety of metals in addition to copper (i.e. mercury), most metals are bound with much weaker affinity (Choi *et al.*, 2006a). Thus, the predominant (but clearly not the only possible) function of methanobactin is to collect copper to support methanotrophic physiology. Sequestration of copper by methanobactin could then prevent other microbes from collecting copper as they may not have the mechanism required for uptake of copper–methanobactin complexes and/or be able to liberate copper from methanobactin. It should be noted that although copper can be toxic due to its high redox activity and binding to iron–sulfur cluster sites (Macomber and Imlay, 2009; Pham *et al.*, 2013), copper is required as a trace nutrient, and respiration by some microbes is strongly dependent on copper availability. For example, complete denitrification of nitrate to dinitrogen requires copper for the activity of the nitrous oxide reductase (NosZ (Brown *et al.*, 2000)). When several denitrifying bacteria were incubated axenically, little and transient amounts of N_2O were observed. When these same cultures were incubated either in the presence of methanobactin or active cultures of *M. trichosporium* OB3b, substantial amounts of N_2O were observed, indicating that methanobactin, through binding of copper, limited the activity of NosZ (Chang *et al.*, 2018). In support of this conclusion, when denitrifying bacteria were incubated in the presence of a mutant of *M. trichosporium* OB3b defective in the production of methanobactin, little N_2O was observed. It should be stressed that microbes that occupy niches similar to methanotrophs, e.g. ammonia-oxidizing bacteria and archaea as well as the methanethiol degrading methylotroph, *Hyphomicrobium* sp. VS, also have strong copper requirements (Ensign *et al.*, 1993; Juliette *et al.*, 1995; Walker *et al.*, 2010; Amin *et al.*, 2013; Jacquot *et al.*, 2014; Eyice *et al.*, 2018). It would be of great value to determine if copper competition between methanotrophs and these microbes also occurs and, if so, what the impact may be.

Medical applications of methanotrophy

Treatment of Wilson disease

Potential use of methanobactin in treating other copper-related pathologies

Methanobactin may also be useful for treating copper-associated pathologies such as Alzheimer's disease and some cancers (Folkman and Shing, 1992; Pan *et al.*, 2002; Brady *et al.*, 2014; Garber, 2015). Alzheimer's disease is associated with the presence of excess misfolded proteins resulting in neuronal-damaging plaques (Gaggelli *et al.*, 2006; Ballard *et al.*, 2011; Davies *et al.*, 2014; Gamez and Caballero, 2015). Copper increases the severity of Alzheimer's disease by not only increasing the production of misfolded proteins, but also by inhibiting export of these proteins from brain tissue, collectively leading to greater plaque formation (Singh *et al.*, 2013). It is tempting to speculate that methanobactin, given its high affinity for copper, may be useful in treating Alzheimer's disease through the removal and/or prevention of copper partitioning to brain tissue. Although such applications are certainly intriguing, it should be stressed that much more investigation is needed to verify the utility of methanobactin for treating Alzheimer's disease.

Active oncogenesis of a variety of cancers is also influenced by copper availability. That is, copper is required for serine/threonine-protein kinase B-Raf (BRAF)-driven tumorigenesis. If copper uptake is disrupted through the addition of tetrathiolmolybdate or mutations in the copper binding site of BRAF, tumorigenesis is inhibited (Brady *et al.*, 2014). As such, inhibiting kinase activity with methanobactin, a much more potent copper chelator, merits consideration for the treatment of a variety of cancers.

Industrial applications of methanotrophy

As noted earlier, methane prices have decreased significantly in recent years, largely due to the development of technologies to extract methane from previously inaccessible locations, e.g. hydraulic fracturing of shales coupled with directional drilling. This has fostered a great deal of interest in the use of methanotrophy to valorize methane (a relatively inexpensive substrate) to diverse products including protein (for use as a feed supplement) liquid fuels (for use in the transportation sector), plastic precursors (for use in packaging and biomedical applications), osmo-protectants (for use in cosmetics), and potentially as human health supplements. Here we summarize current products and strategies of methane 'bio-valorization' as there have been recently many excellent reviews on the valorization of methane via methanotrophy (e.g. Chistoserdova and Kalyuzhnaya, 2018; Fei *et al.*, 2014; Kalyuzhnaya *et al.*, 2015; Khmelenina *et al.*, 2015; Strong *et al.*, 2015, 2016, 2017; Kalyuzhnaya, 2016). The reader is strongly encouraged to peruse these reviews for more information.

Production of single-cell protein

The initial industrial application of methanotrophy for product development was the conversion of methane to protein. Commercial production first started with the company now known as UniBio (www.unibio.dk), producing 'uniprotein' using a continuous fermentation process cultivating a moderate thermophilic methanotroph, *Methylococcus capsulatus* Bath. The resulting product has been used as a feed supplement for a number of animals, including chickens, various fish species, dogs and mink (Øverland *et al.*, 2010a,b; Romarheim *et al.*, 2011; Anvar *et al.*, 2014). It should be kept in mind that axenic cultures were not possible, with the system repeatedly invaded by an *Aneurinibacillus* species, a *Brevibacillus agri* strain, and a *Ralstonia* species (Bothe *et al.*, 2002). These contaminants, however, are of limited concern as they do not produce enterotoxins. Rather their presence seems beneficial as they appear to consume metabolites that otherwise may have inhibited methanotrophic growth. Interest in this area is increasing, with a second company, Calysta, also using a methanotrophic platform for the production of single-cell protein (www.calysta.com).

Plastic production

Another current commercial application of methanotrophy is the production of polyhdroxyalkanoates for bioplastics. In particular, methanotrophs that rely on the serine cycle for carbon assimilation (primarily α-Proteobacteria methanotrophs, but some γ-Proteobacteria methanotrophs appear to have components of the serine cycle (Ward *et al.*, 2004; Semrau *et al.*, 2010)) are targeted for production of polyhdroxyalkanoates. The serine cycle can be conjoined with the ethylmalony-CoA pathway where the intermediate 3-hydroxybutryl-CoA can be converted into polyhydroxybutyrate. A great deal of research has shown that the production of polyhydroxyalkanoates is dependent on growth conditions, e.g. copper-limitation in *M. capsulatus* Bath (Choi *et al.*, 2003) or nitrogen (as ammonia) limitation in *Methylocystis parvus* OBBP and *Methylosinus trichosporium* OB3b (Asenjo and Suk, 1986; Rostkowski *et al.*, 2013). There is also some effort under way to generate modified polyhydroxyalkanoates through the addition of valerate or *n*-pentanol (Myung *et al.*, 2015; Cal *et al.*, 2016). A significant reason methanotrophy is attractive for the conversion of methane to polyhydroxyalkanoates is the naturally high conversion rate, i.e. up to 68% of dry biomass can be polyhydroxyalkanoates (Asenjo and Suk, 1986) although yields on the order of 50% or less are more common (Pieja *et al.*, 2011; Chidambarampadmavathy *et al.*, 2015; Myung *et al.*, 2015; Zhang *et al.*, 2017). As a result, several companies are pursuing large-scale bio-production of polyhydroxyalkanoates from methane, perhaps most notably NewLight Technologies (https://www.newlight.com/) and Mango Materials (http://mangomaterials.com/).

Biodiesel production

Methanotrophs biomass naturally has a high fraction as lipids that make up the internal cytoplasmic membranes, upwards of 20% (Khmelenina *et al.*, 2015). These membranes, either as discs stacked throughput the cell for γ-Proteobacteria or along the periphery of the cell for α-Proteobacteria, are common features as they are used to house the pMMO, and thus their production can be expected to increase with increasing copper as this increases pMMO synthesis. As such, methanotrophs could be used for the production of biodiesel from these lipids. These lipids, however, are polar phospholipids mainly composed of phosphatidylglycerol, phosphoditylcholine and phosphatidylethanolamine, with $C_{16:1}$ fatty acids for γ-Proteobacteria and $C_{18:1}$ fatty acids for α-Proteobacteria. Although the size of these fatty acids is appropriate for biodiesel production, the high heteroatom content (i.e. N and P) make downstream processing more challenging than use of nonpolar lipids (Fei *et al.*, 2014; Khmelenina *et al.*, 2015; Strong *et al.*, 2015; Strong *et al.*, 2016). Further, it has been estimated that lipid content would have to be increased to ≈35% to be economically viable (Khmelenina *et al.*, 2015). Nonetheless, the use of a non-agricultural source (methane) for biodiesel production has attraction as it avoids the food versus fuel dilemma of using agricultural sources. Finally, as noted by others (Strong *et al.*, 2015, 2016), the production of polar lipids via methanotrophy may have medical benefit as some studies show that it may lower LDL:HDL cholesterol levels (Müller *et al.*, 2004). Thus, this could be a particularly lucrative application of methanotrophy.

Ectoine biosynthesis

Some methanotrophs are extremely halophilic, and as such, require osmotic stabilizers or compatible solutes. One such stabilizer is the cyclic imino acid, ectoine, that is currently biologically produced at scale using *Halomonas elongate* DSM 2581^T (Schwibbert *et al.*, 2011). This compound, with the ability to stabilize proteins, nucleic acids and nucleic acid–protein complexes has use in the cosmetic industry as a moisturizer, and some potential medical applications also exist (Pastor *et al.*, 2010). Some halophilic methanotrophs, i.e. various *Methylomicrobium* species also produce ectoine as an osmotic stabilizer (Khmelenina *et al.*, 2000, 2015; Reshetnikov *et al.*, 2006, 2011a,b). As stressed by Strong *et al* (2016), high value applications of methanotrophy for production of lipids for medical use and ectoine for cosmetics may be required to make methanotrophic creation of other products such as fuels and plastics economically viable. That is, the simultaneous realization of multiple products from methanotrophic platforms may be necessary to ensure that methane valorization can be attractive and is pursued more vigorously in the future for many different products. Doing so will make it more likely that we can achieve the more important and broader goals of creating more sustainable production platforms while also reducing net emissions of greenhouse gases like methane.

Current/future strategies for the enhancement of methane valorization

Obviously, there is great, yet unrealized potential for the use of methanotrophy to convert methane to various valuable products. These opportunities in many (but not all) cases have yet to be commercialized for a variety of reasons, including slow methanotrophic growth, limited product yield, and low rates of production. These difficulties are compounded by the fact that methane is only sparingly soluble in water, with mass transfer limitation a major obstacle to scaling up production. One must also realize the methane–air mixtures can be explosive (i.e. the lower and higher explosive limits of methane–oxygen are 5–15% (Zlochower and Green, 2009)). Nonetheless, there are ingenious engineering strategies to circumvent many of these problems via novel reactor/gas transfer designs that are reviewed elsewhere (Fei *et al.*, 2014; Strong *et al.*, 2015, 2016). Here we will focus on efforts to manipulate methanotrophic metabolism to enhance methane valorization.

Manipulation of growth conditions

The simplest strategy for enhancing methanotrophic platforms for methane valorization is the use of selective growth conditions. Such strategies have been frequently employed, e.g. varying copper to control expression of alternative forms of MMO, enhanced polyhydroxyalkanoate production under nitrogen limitation etc. Such efforts, although relatively simple in concept, could in practice be challenging given the wide range of parameters one may wish to vary to direct methane-derived carbon to various products. One promising strategy to advance screening of various conditions would be the use of microfluidic systems where many parameters could be varied simultaneously. Although this approach has some attraction, the provision of methane and oxygen may be problematic in such systems, and a simple, yet robust screening method must be created to quantify levels of product(s) of interest easily and quickly. To the best of our knowledge, such efforts have not been discussed in the archival literature, but warrant further consideration to not only screen extant methanotrophic species for enhanced product creation, but also to isolate and characterize novel methanotrophs from different environments.

Metabolic engineering

One emerging strategy to enhance the utility of methanotrophy for methane valorization is the use of metabolic engineering to remove bottlenecks and/or engineer shunts to facilitate production of specific compounds. This is increasingly attractive, especially as genetic tools have been developed for many methanotrophs (Kalyuzhnaya *et al.*, 2015), as well as increasingly more methanotrophic genomes have been sequenced. Initial efforts have shown that methanotrophs can be engineered for the production of lactic acid and carotenoids, as well as increased production of fatty acids (Ye *et al.*, 2007; Henard *et al.*, 2016; Demidenko *et al.*, 2017). Although titres to date are low, these studies clearly show proof-of-concept and are well worth pursuing further.

Use of mixed microbial communities for methane valorization

One very intriguing future research direction is not the use of axenic cultures for methane valorization, but the deliberate construction of mixed communities. That is, although axenic cultures are critical for elucidating pathways and mechanisms of substrate transformation, artificial or synthetic communities may perform better for specific applications. That is, mixed cultures of methanotrophs with heterotrophs that remove toxic metabolites may be needed for the long-term stable activity of methanotrophs [as was concluded for the production of single cell protein (Bothe *et al.*, 2002)].

On the other hand, it may be infeasible to engineer a methanotroph to carry out the transformation of methane to any specific product as the amount of effort may simply be too great for any one species. Rather, it may be more profitable to use a 'division of labour', where different microbes perform different steps in some metabolic pathway (Tsoi *et al.*, 2018). In this case, methanotrophs could be engineered to convert methane to some intermediate that would then be further transformed by another microbe – perhaps an engineered heterotroph – to further the conversion of this intermediate to the final product. One could then envisage even grander interactions between community members, i.e. where the heterotroph, in addition to removing some of the metabolic burden from methanotrophs for methane conversion may also provide some trace nutrient to the methanotroph that stimulates the overall population. Such an idea is certainly intriguing, and the idea of microbial interdependence an important concept in microbial ecology (Hug and Co, 2018). Indeed, it has been found that methanotrophs thrive better in the presence of heterotrophs (Ho *et al.*, 2014), possibly through provision of cobalamin from the heterotrophs (Iguchi *et al.*, 2011b). As such, in developing synthetic microbial communities for methane valorization, one should consider engagement between various partners in all directions. To be sure, this is a complex problem, but advances in modelling strategies could provide insight into how best to foster specific microbial interactions (e.g. Larsen *et al.*, 2012; Xiao *et al.*, 2017).

Conclusions and suggestions for future work

As a field, methanotrophy has made significant breakthroughs for environmental applications such as pollutant degradation, and is on the cusp of making equally, if not more significant contributions for the control of greenhouse gas emissions as well as the valorization of methane. There is much that remains to be learned, however, to ensure that optimal strategies are developed. Future work should examine in more detail how to best enhance methane consumption *in situ* while minimizing impact on production of other greenhouse gases such as nitrous oxide, given that methanotrophs can 'starve' other microbes for copper. Along these lines, the exact mechanism whereby different metals are detoxified, as well as how to best manipulate methanotrophic activity *in situ* for enhanced pollutant degradation needs to be more fully explored. For methane valorization, there is still much more information to gather as how best to apply metabolic engineering/synthetic microbial systems to shunt methane-derived carbon effectively to specific products. There is also a great deal of work to be done elucidating how different forms of methanobactin are synthesized, and how one can modify either precursor polypeptide or biosynthetic enzymes to alter the size and composition of mature methanobactin. It is hoped that overviews such as this stimulate more interest in the study of methanotrophy, an intriguing group of microbes.

Acknowledgements

This work was supported by grants from Helmholtz Zentrum München, the University of Michigan Office of Research, the University of Michigan Energy Institute, the United States National Science Foundation (Grant #1724744) and the United States Department of Energy (Grant #DE-SC0018059).

References

Al Hasin, A., Gurman, S.J., Murphy, L.M., Perry, A., Smith, T.J., and Gardiner, P.H. (2010). Remediation of chromium(VI) by a methane-oxidizing bacterium. Environ. Sci. Technol. *44*, 400–405. https://doi.org/10.1021/es901723c

Amin, S.A., Moffett, J.W., Martens-Habbena, W., Jacquot, J.E., Han, Y., Devol, A., Ingalls, A.E., Stahl, D.A., and Armbrust, E.V. (2013). Copper requirements of the ammonia-oxidizing archaeon *Nitrosopumilus maritimus*

SCM1 and implications for nitrification in the marine environment. Limnol. Oceanogr. *58*, 2037-2045.

Anvar, S.Y., Frank, J., Pol, A., Schmitz, A., Kraaijeveld, K., den Dunnen, J.T., and Op den Camp, H.J. (2014). The genomic landscape of the verrucomicrobial methanotroph *Methylacidiphilum fumariolicum* SolV. BMC Genomics *15*, 914. https://doi.org/10.1186/1471-2164-15-914

Asenjo, J.A., and Suk, J.S. (1986). Microbial conversion of methane into poly-β-hydroxybutyrate (PHB): Growth and intracellular product accumulation in a type II methanotroph. J. Ferment. Technol. *64*, 271-278.

Baani, M., and Liesack, W. (2008). Two isozymes of particulate methane monooxygenase with different methane oxidation kinetics are found in *Methylocystis* sp. strain SC2. Proc. Natl. Acad. Sci. U.S.A. *105*, 10203–10208. https://doi.org/10.1073/pnas.0702643105

Balasubramanian, R., Levinson, B.T., and Rosenzweig, A.C. (2010). Secretion of flavins by three species of methanotrophic bacteria. Appl. Environ. Microbiol. *76*, 7356–7358. https://doi.org/10.1128/AEM.00935-10

Ballard, C., Gauthier, S., Cobett, A., Brayne, C., Aarsland, D., and Jones, E. (2011). Alzheimer's disease. The Lancet *377*, 1019 - 1031.

Bandow, N., Gilles, V.S., Freesmeier, B., Semrau, J.D., Krentz, B., Gallagher, W., McEllistrem, M.T., Hartsel, S.C., Choi, D.W., Hargrove, M.S., *et al.* (2012). Spectral and copper binding properties of methanobactin from the facultative methanotroph *Methylocystis* strain SB2. J. Inorg. Biochem. *110*, 72–82. https://doi.org/10.1016/j.jinorgbio.2012.02.002

Baral, B.S., Bandow, N.L., Vorobev, A., Freemeier, B.C., Bergman, B.H., Herdendorf, T.J., Fuentes, N., Ellias, L., Turpin, E., Semrau, J.D., *et al.* (2014). Mercury binding by methanobactin from *Methylocystis* strain SB2. J. Inorg. Biochem. *141*, 161–169.

Barkay, T., and Wagner-Döbler, I. (2005). Microbial transformations of mercury: potentials, challenges, and achievements in controlling mercury toxicity in the environment. Adv. Appl. Microbiol. *57*, 1–52.

Basu, P., Katterle, B., Kristoffer Andersson, K., and Dalton, H. (2003). The membrane-associated form of methane mono-oxygenase from *Methylococcus capsulatus* (Bath) is a copper/iron protein. Biochem. J. *369*, 417-427.

Behling, L.A., Hartsel, S.C., Lewis, D.E., DiSpirito, A.A., Choi, D.W., Masterson, L.R., Veglia, G., and Gallagher, W.H. (2008). NMR, mass spectrometry and chemical evidence reveal a different chemical structure for methanobactin that contains oxazolone rings. J. Am. Chem. Soc. *130*, 12604–12605. https://doi.org/10.1021/ja804747d

Bender, M., and Conrad, R. (1992). Kinetics of CH4 oxidation in oxic soils exposed to ambient air or high CH4 mixing ratios. FEMS Microbiol. Lett. *101*, 261-270.

Bodrossy, L., Holmes, E.M., Holmes, A.J., Kovács, K.L., and Murrell, J.C. (1997). Analysis of 16S rRNA and methane monooxygenase gene sequences reveals a novel group of thermotolerant and thermophilic methanotrophs, *Methylocaldum* gen. nov. Arch. Microbiol. *168*, 493–503.

Bodrossy, L., Kovács, K.L., McDonald, I.R., and Murrell, J.C. (1999). A novel thermophilic methane-oxidising γ-Proteobacterium. FEMS Microbiol. Lett. *170*, 335-341.

Bothe, H., Møller Jensen, K., Mergel, A., Larsen, J., Jørgensen, C., Bothe, H., and Jørgensen, L. (2002). Heterotrophic bacteria growing in association with *Methylococcus capsulatus* (Bath) in a single cell protein production process. Appl. Microbiol. Biotechnol. *59*, 33–39. https://doi.org/10.1007/s00253-002-0964-1

Bowman, J.P., Sly, L.I., and Stackebrandt, E. (1995). The phylogenetic position of the family *Methylococcaceae*. Int. J. Syst. Bacteriol. *45*, 182–185. https://doi.org/10.1099/00207713-45-1-182

Bowman, J.P., McCammon, S.A., and Skerratt, J.H. (1997). *Methylosphaera hansonii* gen. nov., sp. nov., a psychrophilic, group I methanotroph from Antarctic marine-salinity, meromictic lakes. Microbiology *143*, 1451–1459. https://doi.org/10.1099/00221287-143-4-1451

Brady, D.C., Crowe, M.S., Turski, M.L., Hobbs, G.A., Yao, X., Chaikuad, A., Knapp, S., Xiao, K., Campbell, S.L., Thiele, D.J., *et al.* (2014). Copper is required for oncogenic BRAF signalling and tumorigenesis. Nature *509*, 492–496. https://doi.org/10.1038/nature13180

Braun, V., Mahren, S., and Sauter, A. (2006). Gene regulation by transmembrane signaling. Biometals *19*, 103–113. https://doi.org/10.1007/s10534-005-8253-y

Brooks, B.E., and Buchanan, S.K. (2008). Signaling mechanisms for activation of extracytoplasmic function (ECF) sigma factors. Biochim. Biophys. Acta *1778*, 1930–1945.

Brown, K., Tegoni, M., Prudêncio, M., Pereira, A.S., Besson, S., Moura, J.J., Moura, I., and Cambillau, C. (2000). A novel type of catalytic copper cluster in nitrous oxide reductase. Nat. Struct. Biol. *7*, 191–195. https://doi.org/10.1038/73288

Brown, L.R., Strawinski, R.J., and Mccleskey, C.S. (1964). The isolation and characterization of *Methanomonas methanooxidans* Brown and Strawinski. Can. J. Microbiol. *10*, 791–799.

Bull, I.D., Parekh, N.R., Hall, G.H., Ineson, P., and Evershed, R.P. (2000). Detection and classification of atmospheric methane oxidizing bacteria in soil. Nature *405*, 175–178. https://doi.org/10.1038/35012061

Burrows, K.J., Cornish, A., Scott, D., and Higgins, I.J. (1984). Substrate specificities of the soluble and particulate methane mono-oxygenases of *Methylosinus trichosporium* OB3b. J. Gen. Microbiol. *130*, 3327–3333.

Cai, Y., Zheng, Y., Bodelier, P.L., Conrad, R., and Jia, Z. (2016). Conventional methanotrophs are responsible for atmospheric methane oxidation in paddy soils. Nat. Commun. *7*, 11728. https://doi.org/10.1038/ncomms11728

Cal, A.J., Sikkema, W.D., Ponce, M.I., Franqui-Villanueva, D., Riiff, T.J., Orts, W.J., Pieja, A.J., and Lee, C.C. (2016). Methanotrophic production of polyhydroxybutyrate-co-hydroxyvalerate with high hydroxyvalerate content. Int. J. Biol. Macromol. *87*, 302–307. https://doi.org/10.1016/j.ijbiomac.2016.02.056

Chang, J., Gu, W., Park, D., Semrau, J.D., DiSpirito, A.A., and Yoon, S. (2018). Methanobactin from *Methylosinus trichosporium* OB3b inhibits N2O reduction in denitrifiers. ISME J. *12*, 2086–2089. https://doi.org/10.1038/s41396-017-0022-8

Chidambarampadmavathy, K., Karthikeyan, O.P., and Heimann, K. (2015). Biopolymers made from methane in bioreactors. Eng. Life Sci. *15*, 689-699.

Chistoserdova, L., and Kalyuzhnaya, M.G. (2018). Current Trends in Methylotrophy. Trends Microbiol. *26*, 703–714.

Choi, D.W., Kunz, R.C., Boyd, E.S., Semrau, J.D., Antholine, W.E., Han, J.I., Zahn, J.A., Boyd, J.M., de la Mora, A.M., and DiSpirito, A.A. (2003). The membrane-associated methane monooxygenase (pMMO) and pMMO-NADH:quinone oxidoreductase complex from *Methylococcus capsulatus* Bath. J. Bacteriol. *185*, 5755–5764.

Choi, D.W., Antholine, W.E., Do, Y.S., Semrau, J.D., Kisting, C.J., Kunz, R.C., Campbell, D., Rao, V., Hartsel, S.C., and DiSpirito, A.A. (2005). Effect of methanobactin on the activity and electron paramagnetic resonance spectra of the membrane-associated methane monooxygenase in *Methylococcus capsulatus* Bath. Microbiology *151*, 3417-3426.

Choi, D.W., Do, Y.S., Zea, C.J., McEllistrem, M.T., Lee, S.W., Semrau, J.D., Pohl, N.L., Kisting, C.J., Scardino, L.L., Hartsel, S.C., *et al.* (2006a). Spectral and thermodynamic properties of Ag(I), Au(III), Cd(II), Co(II), Fe(III), Hg(II), Mn(II), Ni(II), Pb(II), U(IV), and Zn(II) binding by methanobactin from *Methylosinus trichosporium* OB3b. J. Inorg. Biochem. *100*, 2150–2161.

Choi, D.W., Zea, C.J., Do, Y.S., Semrau, J.D., Antholine, W.E., Hargrove, M.S., Pohl, N.L., Boyd, E.S., Geesey, G.G., Hartsel, S.C., *et al.* (2006b). Spectral, kinetic, and thermodynamic properties of Cu(I) and Cu(II) binding by methanobactin from *Methylosinus trichosporium* OB3b. Biochemistry *45*, 1442–1453. https://doi.org/10.1021/bi051815t

Chowdhury, T.R., and Dick, R.P. (2013). Ecology of aerobic methanotrophs in controlling methane fluxes from wetlands. Appl. Soil Ecol. *65*, 8–22.

Colby, J., Stirling, D.I., and Dalton, H. (1977). The soluble methane mono-oxygenase of *Methylococcus capsulatus* (Bath). Its ability to oxygenate n-alkanes, n-alkenes, ethers, and alicyclic, aromatic and heterocyclic compounds. Biochem. J. *165*, 395–402.

Crosa, J.H. (1997). Signal transduction and transcriptional and posttranscriptional control of iron-regulated genes in bacteria. Microbiol. Mol. Biol. Rev. *61*, 319–336.

Dalton, H., and Stirling, D.I. (1982). Co-metabolism. Philos. Trans. R. Soc. Lond. B Biol. Sci. *297*, 481–496.

Dalton, H., Prior, S.D., Leak, D.J., and Stanley, S.H. (1984). Regulation and Control of Methane Monooxygenase. In Microbial Growth in C1 Compounds, Crawford, R.L., and Hanson, R.S., eds. (American Society for Microbiology, Washington, D.C.).

Davies, K.M., Bohic, S., Carmona, A., Ortega, R., Cottam, V., Hare, D.J., Finberg, J.P., Reyes, S., Halliday, G.M., Mercer, J.F., *et al.* (2014). Copper pathology in vulnerable brain regions in Parkinson's disease. Neurobiol. Aging *35*, 858–866. https://doi.org/10.1016/j.neurobiolaging.2013.09.034

de Bie, P., Muller, P., Wijmenga, C., and Klomp, L.W. (2007). Molecular pathogenesis of Wilson and Menkes disease: correlation of mutations with molecular defects and disease phenotypes. J. Med. Genet. *44*, 673–688.

Dedysh, S.N., Liesack, W., Khmelenina, V.N., Suzina, N.E., Trotsenko, Y.A., Semrau, J.D., Bares, A.M., Panikov, N.S., and Tiedje, J.M. (2000). *Methylocella palustris* gen. nov., sp. nov., a new methane-oxidizing acidophilic bacterium from peat bogs, representing a novel subtype of serine-pathway methanotrophs. Int. J. Syst. Evol. Microbiol. *50*, 955–969. https://doi.org/10.1099/00207713-50-3-955

Dedysh, S.N., Khmelenina, V.N., Suzina, N.E., Trotsenko, Y.A., Semrau, J.D., Liesack, W., and Tiedje, J.M. (2002). *Methylocapsa acidiphila* gen. nov., sp. nov., a novel methane-oxidizing and dinitrogen-fixing acidophilic bacterium from Sphagnum bog. Int. J. Syst. Evol. Microbiol. *52*, 251–261.

Demidenko, A., Akberdin, I.R., Allemann, M., Allen, E.E., and Kalyuzhnaya, M.G. (2017). Fatty acid biosynthesis pathways in *Methylomicrobium buryatense* 5G(B1). Front. Microbiol. *7*, 2167.

Deutzmann, J.S., Hoppert, M., and Schink, B. (2014). Characterization and phylogeny of a novel methanotroph, *Methyloglobulus morosus* gen. nov., spec. nov. Syst. Appl. Microbiol. *37*, 165–169. https://doi.org/10.1016/j.syapm.2014.02.001

DiSpirito, A.A., Semrau, J.D., Murrell, J.C., Gallagher, W.H., Dennison, C., and Vuilleumier, S. (2016). Methanobactin and the Link between Copper and Bacterial Methane Oxidation. Microbiol. Mol. Biol. Rev. *80*, 387–409. https://doi.org/10.1128/MMBR.00058-15

El Ghazouani, A., Baslé, A., Firbank, S.J., Knapp, C.W., Gray, J., Graham, D.W., and Dennison, C. (2011). Copper-binding properties and structures of methanobactins from *Methylosinus trichosporium* OB3b. Inorg. Chem. *50*, 1378–1391. https://doi.org/10.1021/ic101965j

El Ghazouani, A., Baslé, A., Gray, J., Graham, D.W., Firbank, S.J., and Dennison, C. (2012). Variations in methanobactin structure influences copper utilization by methane-oxidizing bacteria. Proc. Natl. Acad. Sci. U.S.A. *109*, 8400–8404. https://doi.org/10.1073/pnas.1112921109

Ensign, S.A., Hyman, M.R., and Arp, D.J. (1993). In vitro activation of ammonia monooxygenase from *Nitrosomonas europaea* by copper. J. Bacteriol. *175*, 1971–1980.

Eswayah, A.S., Smith, T.J., Scheinost, A.C., Hondow, N., and Gardiner, P.H.E. (2017). Microbial transformations of selenite by methane-oxidizing bacteria. Appl. Microbiol. Biotechnol. *101*, 6713–6724. https://doi.org/10.1007/s00253-017-8380-8

Ettwig, K.F., Butler, M.K., Le Paslier, D., Pelletier, E., Mangenot, S., Kuypers, M.M., Schreiber, F., Dutilh, B.E., Zedelius, J., de Beer, D., *et al.* (2010). Nitrite-driven anaerobic methane oxidation by oxygenic bacteria. Nature *464*, 543–548. https://doi.org/10.1038/nature08883

Eyice, Ö., Myronova, N., Pol, A., Carrión, O., Todd, J.D., Smith, T.J., Gurman, S.J., Cuthbertson, A., Mazard, S., Mennink-Kersten, M.A., *et al.* (2018). Bacterial SBP56 identified as a Cu-dependent methanethiol oxidase widely distributed in the biosphere. ISME J. *12*, 145–160. https://doi.org/10.1038/ismej.2017.148

Fei, Q., Guarnieri, M.T., Tao, L., Laurens, L.M., Dowe, N., and Pienkos, P.T. (2014). Bioconversion of natural gas to liquid fuel: opportunities and challenges. Biotechnol. Adv. *32*, 596–614. https://doi.org/10.1016/j.biotechadv.2014.03.011

Fjellbirkeland, A., Kleivdal, H., Joergensen, C., Thestrup, H., and Jensen, H.B. (1997). Outer membrane proteins of *Methylococcus capsulatus* (Bath). Arch. Microbiol. *168*, 128–135.

Folkman, J., and Shing, Y. (1992). Angiogenesis. J. Biol. Chem. *267*, 10931–10934.

Foster, J.W., and Davis, R.H. (1966). A methane-dependent coccus, with notes on classification and nomenclature of obligate, methane-utilizing bacteria. J. Bacteriol. *91*, 1924–1931.

Fox, B.G., Froland, W.A., Dege, J.E., and Lipscomb, J.D. (1989). Methane monooxygenase from *Methylosinus trichosporium* Ob3b - purification and properties of a 3-component system with high specific activity from a type-II methanotroph. J. Biol. Chem. *264*, 10023–10033.

Fox, B.G., Borneman, J.G., Wackett, L.P., and Lipscomb, J.D. (1990). Haloalkene oxidation by the soluble methane monooxygenase from *Methylosinus trichosporium* OB3b: mechanistic and environmental implications. Biochemistry *29*, 6419–6427.

Gaggelli, E., Kozlowski, H., Valensin, D., and Valensin, G. (2006). Copper homeostasis and neurodegenerative disorders (Alzheimer's, Prion, and Parkinson's diseases and amyotrophic lateral sclerosisa). Chem. Rev. *106*, 1995–2044.

Gamez, P., and Caballero, A.B. (2015). Copper in Alzheimer's disease: implications in amyloid aggregation and neurotoxicity. AIP Advances 5, 092503.

Ganendra, G., Mercado-Garcia, D., Hernandez-Sanabria, E., Boeckx, P., Ho, A., and Boon, N. (2015). Methane biofiltration using autoclaved aerated concrete as the carrier material. Appl. Microbiol. Biotechnol. *99*, 7307–7320. https://doi.org/10.1007/s00253-015-6646-6

Garber, K. (2015). BIOMEDICINE. Targeting copper to treat breast cancer. Science *349*, 128–129. https://doi.org/10.1126/science.349.6244.128

Geymonat, E., Ferrando, L., and Tarlera, S.E. (2011). *Methylogaea oryzae* gen. nov., sp. nov., a mesophilic methanotroph isolated from a rice paddy field. Int. J. Syst. Evol. Microbiol. *61*, 2568–2572.

Ghashghavi, M., Jetten, M.S.M., and Lüke, C. (2017). Survey of methanotrophic diversity in various ecosystems by degenerate methane monooxygenase gene primers. AMB Express 7, 162. https://doi.org/10.1186/s13568-017-0466-2

Gitlin, J.D. (2003). Wilson disease. Gastroenterology *125*, 1868–1877.

Grosse, C., Friedrich, S., and Nies, D.H. (2007). Contribution of extracytoplasmic function sigma factors to transition metal homeostasis in *Cupriavidus metallidurans* strain CH34. J. Mol. Microbiol. Biotechnol. *12*, 227–240.

Gu, W. (2017). Metals and methanotrophs: 1. Genetic and biochemical characterization of the uptake and synthesis of methanobactin; 2. Bioinformatic analyses of the effect of rare earth elements on gene expression. PhD Thesis, of The University of Michigan.

Gu, W., Farhan Ul Haque, M., Baral, B.S., Turpin, E.A., Bandow, N.L., Kremmer, E., Flatley, A., Zischka, H., DiSpirito, A.A., and Semrau, J.D. (2016). A TonB-dependent transporter is responsible for methanobactin uptake by *Methylosinus trichosporium* OB3b. Appl. Environ. Microbiol. *82*, 1917–1923. https://doi.org/10.1128/AEM.03884-15

Gu, W., Baral, B.S., DiSpirito, A.A., and Semrau, J.D. (2017). An aminotransferase is responsible for the deamination of the N-terminal leucine and required for formation of oxazolone ring A in methanobactin of *Methylosinus trichosporium* OB3b. Appl. Environ. Microbiol. *83*, e02619-16.

Haft, D.H., Selengut, J.D., Richter, R.A., Harkins, D., Basu, M.K., and Beck, E. (2013). TIGRFAMs and genome properties in 2013. Nucleic Acids Res. *41*, D387–D395.

Han, J.I., and Semrau, J.D. (2004). Quantification of gene expression in methanotrophs by competitive reverse transcription-polymerase chain reaction. Environ. Microbiol. *6*, 388–399.

Hanson, R.S., and Hanson, T.E. (1996). Methanotrophic bacteria. Microbiol. Rev. *60*, 439–471.

Helland, R., Fjellbirkeland, A., Karlsen, O.A., Ve, T., Lillehaug, J.R., and Jensen, H.B. (2008). An oxidized tryptophan facilitates copper binding in *Methylococcus capsulatus*-secreted protein MopE. J. Biol. Chem. *283*, 13897–13904. https://doi.org/10.1074/jbc.M800340200

Henard, C.A., Smith, H., Dowe, N., Kalyuzhnaya, M.G., Pienkos, P.T., and Guarnieri, M.T. (2016). Bioconversion of methane to lactate by an obligate methanotrophic bacterium. Sci. Rep. *6*, 21585. https://doi.org/10.1038/srep21585

Heyer, J., Berger, U., Hardt, M., and Dunfield, P.F. (2005). *Methylohalobius crimeensis* gen. nov., sp. nov., a moderately halophilic, methanotrophic bacterium isolated from hypersaline lakes of Crimea. Int. J. Syst. Evol. Microbiol. *55*, 1817–1826.

Hirayama, H., Fuse, H., Abe, M., Miyazaki, M., Nakamura, T., Nunoura, T., Furushima, Y., Yamamoto, H., and Takai, K. (2013). *Methylomarinum vadi* gen. nov., sp. nov., a methanotroph isolated from two distinct marine environments. Int. J. Syst. Environ. Microbiol. *63*, 1073–1082.

Hirayama, H., Abe, M., Miyazaki, M., Nunoura, T., Furushima, Y., Yamamoto, H., and Takai, K. (2014). *Methylomarinovum caldicuralii* gen. nov., sp. nov., a moderately thermophilic methanotroph isolated from a shallow submarine hydrothermal system, and proposal of the family *Methylothermaceae* fam. nov. Int. J. Syst. Environ. Microbiol. *64*, 989–999.

Ho, A., de Roy, K., Thas, O., De Neve, J., Hoefman, S., Vandamme, P., Heylen, K., and Boon, N. (2014). The more, the merrier: heterotroph richness stimulates methanotrophic activity. ISME J. *8*, 1945–1948. https://doi.org/10.1038/ismej.2014.74

Ho, A., Reim, A., Kim, S.Y., Meima-Franke, M., Termorshuizen, A., de Boer, W., van der Putten, W.H., and Bodelier, P.L. (2015). Unexpected stimulation of soil methane uptake as emergent property of agricultural soils following bio-based residue application. Glob. Chang. Biol. *21*, 3864–3879. https://doi.org/10.1111/gcb.12974

Hoefman, S., van der Ha, D., Iguchi, H., Yurimoto, H., Sakai, Y., Boon, N., Vandamme, P., Heylen, K., and De Vos, P.

(2014). *Methyloparacoccus murrellii* gen. nov., sp. nov., a methanotroph isolated from pond water. Int. J. Syst. Evol. Microbiol. *64*, 2100–2107.

Hug, L.A., and Co, R. (2018). It takes a village: microbial communities thrive through interactions and metabolic handoffs. mSystems *3*, e00152–17. https://doi.org/10.1128/mSystems.00152-17

Iguchi, H., Yurimoto, H., and Sakai, Y. (2011a). *Methylovulum miyakonense* gen. nov., sp. nov., a type I methanotroph isolated from forest soil. Int. J. Syst. Evol. Microbiol. *61*, 810–815.

Iguchi, H., Yurimoto, H., and Sakai, Y. (2011b). Stimulation of methanotrophic growth in cocultures by cobalamin excreted by rhizobia. Appl. Environ. Microbiol. *77*, 8509–8515. https://doi.org/10.1128/AEM.05834-11

Jacquot, J.E., Horak, R.E.A., Amin, S.A., Devol, A.H., Ingalls, A.E., Armbrust, E.V., Stahl, D.A., and Moffett, J.W. (2014). Assessment of the potential for copper limitation of ammonia oxidation by Archaea in a dynamic estuary. Marine Chem. *162*, 37–49.

Jiang, H., Chen, Y., Jiang, P., Zhang, C., Smith, T.J., Murrell, J.C., and Xing, X.-H. (2010). Methanotrophs: Multifunctional bacteria with promising applications in environmental bioengineering. Biochem. Engin. J. *49*, 277–288.

Juliette, L.Y., Hyman, M.R., and Arp, D.J. (1995). Roles of bovine serum albumin and copper in the assay and stability of ammonia monooxygenase activity in vitro. J. Bacteriol. *177*, 4908–4913.

Kalyuzhnaya, M.G. (2016). Methane biocatalysis: selecting the right microbe A2 – Eckert, Carrie A. In Biotechnology for Biofuel Production and Optimization, Trinh, C.T., ed. (Elsevier, Amsterdam), pp. 353–383.

Kalyuzhnaya, M.G., Puri, A.W., and Lidstrom, M.E. (2015). Metabolic engineering in methanotrophic bacteria. Metab. Eng. *29*, 142–152.

Karlsen, O.A., Berven, F.S., Stafford, G.P., Larsen, Ø., Murrell, J.C., Jensen, H.B., and Fjellbirkeland, A. (2003). The surface-associated and secreted MopE protein of *Methylococcus capsulatus* (Bath) responds to changes in the concentration of copper in the growth medium. Appl. Environ. Microbiol. *69*, 2386–2388.

Kenney, G.E., and Rosenzweig, A.C. (2013). Genome mining for methanobactins. BMC Biol. *11*, 17. https://doi.org/10.1186/1741-7007-11-17

Kenney, G.E., Goering, A.W., Ross, M.O., DeHart, C.J., Thomas, P.M., Hoffman, B.M., Kelleher, N.L., and Rosenzweig, A.C. (2016). Characterization of Methanobactin from *Methylosinus* sp. LW4. J. Am. Chem. Soc. *138*, 11124–11127. https://doi.org/10.1021/jacs.6b06821

Khalifa, A., Lee, C.G., Ogiso, T., Ueno, C., Dianou, D., Demachi, T., Katayama, A., and Asakawa, S. (2015). Methylomagnum ishizawai gen. nov., sp. nov., a mesophilic type I methanotroph isolated from rice rhizosphere. Int. J. Syst. Evol. Microbiol. *65*, 3527–3534. https://doi.org/10.1099/ijsem.0.000451

Khmelenina, V.N., Sakharovskii, V.G., Reshetnikov, A.S., and Trotsenko, Y.A. (2000). Synthesis of osmoprotectants by halophilic and alkaliphilic methanotrophs. Microbiology *69*, 381–386.

Khmelenina, V.N., Rozova, O.N., But, S.Y., Mustakhimov, I.I., Reshetnikov, A.S., Beschastnyi, A.P., and Trotsenko, Y.A. (2015). Biosynthesis of secondary metabolites in methaotrophs: biochemical and genetic aspects (review). Appl. Biochem. Microbiol. *15*, 150–158.

Kim, H.J., Graham, D.W., DiSpirito, A.A., Alterman, M.A., Galeva, N., Larive, C.K., Asunskis, D., and Sherwood, P.M. (2004). Methanobactin, a copper-acquisition compound from methane-oxidizing bacteria. Science *305*, 1612–1615. https://doi.org/10.1126/science.1098322

Kits, K.D., Klotz, M.G., and Stein, L.Y. (2015). Methane oxidation coupled to nitrate reduction under hypoxia by the Gammaproteobacterium *Methylomonas denitrificans*, sp. nov. type strain FJG1. Environ. Microbiol. *17*, 3219–3232. https://doi.org/10.1111/1462-2920.12772

Knief, C. (2015). Diversity and habitat preferences of cultivated and uncultivated aerobic methanotrophic bacteria evaluated based on *pmoA* as molecular marker. Front. Microbiol. *6*, 1346. https://doi.org/10.3389/fmicb.2015.01346

Knief, C., and Dunfield, P.F. (2005). Response and adaptation of different methanotrophic bacteria to low methane mixing ratios. Environ. Microbiol. *7*, 1307–1317.

Knief, C., Lipski, A., and Dunfield, P.F. (2003). Diversity and activity of methanotrophic bacteria in different upland soils. Appl. Environ. Microbiol. *69*, 6703–6714.

Krentz, B.D., Mulheron, H.J., Semrau, J.D., Dispirito, A.A., Bandow, N.L., Haft, D.H., Vuilleumier, S., Murrell, J.C., McEllistrem, M.T., Hartsel, S.C., *et al.* (2010). A comparison of methanobactins from *Methylosinus trichosporium* OB3b and *Methylocystis* strain Sb2 predicts methanobactins are synthesized from diverse peptide precursors modified to create a common core for binding and reducing copper ions. Biochemistry *49*, 10117–10130. https://doi.org/10.1021/bi1014375

Lai, C.Y., Wen, L.L., Shi, L.D., Zhao, K.K., Wang, Y.Q., Yang, X., Rittmann, B.E., Zhou, C., Tang, Y., Zheng, P., *et al.* (2016). selenate and nitrate bioreductions using methane as the electron donor in a membrane biofilm reactor. Environ. Sci. Technol. *50*, 10179–10186. https://doi.org/10.1021/acs.est.6b02807

Lamont, I.L., Beare, P.A., Ochsner, U., Vasil, A.I., and Vasil, M.L. (2002). Siderophore-mediated signaling regulates virulence factor production in Pseudomonasaeruginosa. Proc. Natl. Acad. Sci. U.S.A. *99*, 7072–7077. https://doi.org/10.1073/pnas.092016999

Larsen, P.E., Field, D., and Gilbert, J.A. (2012). Predicting bacterial community assemblages using an artificial neural network approach. Nat. Methods *9*, 621–625. https://doi.org/10.1038/nmeth.1975

Lashof, D.A., and Ahuja, D.R. (1990). Relative contributions of greenhouse gas emissions to global warming. Nature *344*, 529–531.

Lee, S.W., Keeney, D.R., Lim, D.H., Dispirito, A.A., and Semrau, J.D. (2006). Mixed pollutant degradation by *Methylosinus trichosporium* OB3b expressing either soluble or particulate methane monooxygenase: can the tortoise beat the hare? Appl. Environ. Microbiol. *72*, 7503–7509.

Lichtmannegger, J., Leitinger, C., Winner, R., Schmitt, S., Schulz, S., Kabiri, Y., Eberhagen, C., Rieder, T., Janik, D., Neff, F., *et al.* (2016). Methanobactin: a new effective treatment strategy against acute liver failure in a Wilson disease rat model. J. Clin. Inves. *126*, 2721–2735.

Lieberman, R.L., and Rosenzweig, A.C. (2005). Crystal structure of a membrane-bound metalloenzyme that catalyses the biological oxidation of methane. Nature *434*, 177–182.

Lieberman, R.L., Shrestha, D.B., Doan, P.E., Hoffman, B.M., Stemmler, T.L., and Rosenzweig, A.C. (2003). Purified particulate methane monooxygenase from *Methylococcus capsulatus* (Bath) is a dimer with both mononuclear copper and a copper-containing cluster. Proc. Natl. Acad. Sci. U.S.A. *100*, 3820–3825. https://doi.org/10.1073/pnas.0536703100

Lipscomb, J.D. (1994). Biochemistry of the soluble methane monooxygenase. Annu. Rev. Microbiol. *48*, 371–399. https://doi.org/10.1146/annurev.mi.48.100194.002103

Lontoh, S., Zahn, J.A., DiSpirito, A.A., and Semrau, J.D. (2000). Identification of intermediates of in vivo trichloroethylene oxidation by the membrane-associated methane monooxygenase. FEMS Microbiol. Lett. *186*, 109–113.

Lu, X., Liu, Y., Johs, A., Zhao, L., Wang, T., Yang, Z., Lin, H., Elias, D.A., Pierce, E.M., Liang, L., *et al.* (2016). Anaerobic Mercury Methylation and Demethylation by *Geobacter bemidjiensis* Bem. Environ. Sci. Technol. *50*, 4366–4373. https://doi.org/10.1021/acs.est.6b00401

Lu, X., Gu, W., Zhao, L., Farhan Ul Haque, M., DiSpirito, A.A., Semrau, J.D., and Gu, B. (2017). Methylmercury uptake and degradation by methanotrophs. Sci. Adv. *3*, e1700041. https://doi.org/10.1126/sciadv.1700041

Macomber, L., and Imlay, J.A. (2009). The iron-sulfur clusters of dehydratases are primary intracellular targets of copper toxicity. Proc. Natl. Acad. Sci. U.S.A. *106*, 8344–8349. https://doi.org/10.1073/pnas.0812808106

Mahren, S., and Braun, V. (2003). The FecI extracytoplasmic-function sigma factor of *Escherichia coli* interacts with the beta' subunit of RNA polymerase. J. Bacteriol. *185*, 1796–1802.

Müller, H., Hellgren, L.I., Olsen, E., and Skrede, A. (2004). Lipids rich in phosphatidylethanolamine from natural gas-utilizing bacteria reduce plasma cholesterol and classes of phospholipids: a comparison with soybean oil. Lipids 39, 833–841.

Müller, J.C., Lichtmannegger, J., Zischka, H., Sperling, M., and Karst, U. (2018). High spatial resolution LA-ICP-MS demonstrates massive liver copper depletion in Wilson disease rats upon Methanobactin treatment. J. Trace Elem. Med. Biol. *49*, 119–127.

Murrell, J.C., McDonald, I.R., and Gilbert, B. (2000). Regulation of expression of methane monooxygenases by copper ions. Trends Microbiol. *8*, 221–225.

Myhre, G., Shindell, D., Bréon, F.-M., Collins, W., Fuglestvedt, J., Huang, J., Koch, D., Lamarque, J.-F., Lee, D., Mendoza, B., *et al.*, (2013). Anthropogenic and Natural Radiative Forcing. In Climate Change 2013: The Physical Science Basis. Contribution of Working Group I to the Fifth Assessment Report of the Intergovernmental Panel on Climate Change, Stocker, T.F., Qin, D., Plattner, G.-K., Tignor, M., Allen, S.K., Boschung, J., Nauels, A., Xia, Y., Bex, V., and Midgley, P.M., eds. (Cambridge University Press, Cambridge, UK; and New York, NY), pp. 659–740.

Myung, J., Galega, W.M., Van Nostrand, J.D., Yuan, T., Zhou, J., and Criddle, C.S. (2015). Long-term cultivation of a stable *Methylocystis*-dominated methanotrophic enrichment enabling tailored production of poly(3-hydroxybutyrate-co-3-hydroxyvalerate). Bioresour. Technol. *198*, 811–818. https://doi.org/10.1016/j.biortech.2015.09.094

Nzotungicimpaye, C.-M., and Zickfeld, K. (2017). The contribution from methane to the permafrost carbon feedback. Current Climate Change Reports 3, 58–68.

Op den Camp, H.J., Islam, T., Stott, M.B., Harhangi, H.R., Hynes, A., Schouten, S., Jetten, M.S., Birkeland, N.K., Pol, A., and Dunfield, P.F. (2009). Environmental, genomic and taxonomic perspectives on methanotrophic Verrucomicrobia. Environ. Microbiol. Rep. *1*, 293–306. https://doi.org/10.1111/j.1758-2229.2009.00022.x

Øverland, M., Schøyen, H.F., and Skrede, A. (2010a). Growth performance and carcase quality in broiler chickens fed on bacterial protein grown on natural gas. Br. Poult. Sci. *51*, 686–695. https://doi.org/10.1080/00071668.2010.522556

Øverland, M., Tauson, A.H., Shearer, K., and Skrede, A. (2010b). Evaluation of methane-utilising bacteria products as feed ingredients for monogastric animals. Arch. Anim. Nutr. *64*, 171–189. https://doi.org/10.1080/17450391003691534

Pan, Q., Kleer, C.G., van Golen, K.L., Irani, J., Bottema, K.M., Bias, C., De Carvalho, M., Mesri, E.A., Robins, D.M., Dick, R.D., *et al.* (2002). Copper deficiency induced by tetrathiomolybdate suppresses tumor growth and angiogenesis. Cancer Res. *62*, 4854–4859.

Pastor, J.M., Salvador, M., Argandoña, M., Bernal, V., Reina-Bueno, M., Csonka, L.N., Iborra, J.L., Vargas, C., Nieto, J.J., and Cánovas, M. (2010). Ectoines in cell stress protection: uses and biotechnological production. Biotechnol. Adv. *28*, 782–801. https://doi.org/10.1016/j.biotechadv.2010.06.005

Paszczynski, A.J., Paidisetti, R., Johnson, A.K., Crawford, R.L., Colwell, F.S., Green, T., Delwiche, M., Lee, H., Newby, D., Brodie, E.L., *et al.* (2011). Proteomic and targeted qPCR analyses of subsurface microbial communities for presence of methane monooxygenase. Biodegradation 22, 1045–1059. https://doi.org/10.1007/s10532-011-9462-4

Pham, A.N., Xing, G., Miller, C.J., and Waite, T.D. (2013). Fenton-like copper redox chemistry revisited: Hydrogen peroxide and superoxide mediation of copper-catalyzed oxidant production. J. Catal. *301*, 54–64.

Pieja, A.J., Rostkowski, K.H., and Criddle, C.S. (2011). Distribution and selection of poly-3-hydroxybutyrate production capacity in methanotrophic proteobacteria. Microb. Ecol. *62*, 564–573. https://doi.org/10.1007/s00248-011-9873-0

Pratscher, J., Vollmers, J., Wiegand, S., Dumont, M.G., and Kaster, A.K. (2018). Unravelling the identity, metabolic potential and global biogeography of the atmospheric methane-oxidizing upland soil cluster α. Environ. Microbiol. *20*, 1016–1029. https://doi.org/10.1111/1462-2920.14036

Prior, S.D., and Dalton, H. (1985a). Copper stress underlines the fundamental change in intracellular location of the membrane monooxygenase in methane oxidizing organisms: studies in batch and ontinuous culture. Biotechnol. Lett. *5*, 487–492.

Prior, S.D., and Dalton, H. (1985b). The effect of copper ions on membrane content and methane monooxygenase activity in methanol-grown cells of *Methylococcus capsulatus* (Bath). J. Gen. Microbiol. *131*, 155–163.

Rahalkar, M., Bussmann, I., and Schink, B. (2007). *Methylosoma difficile* gen. nov., sp. nov., a novel methanotroph enriched by gradient cultivation from littoral sediment of Lake Constance. Int. J. Syst. Evol. Microbiol. *57*, 1073–1080.

Reshetnikov, A.S., Khmelenina, V.N., and Trotsenko, Y.A. (2006). Characterization of the ectoine biosynthesis genes of haloalkalotolerant obligate methanotroph '*Methylomicrobium alcaliphilum* 20Z'. Arch. Microbiol. *184*, 286–297. https://doi.org/10.1007/s00203-005-0042-z

Reshetnikov, A.S., Khmelenina, V.N., Mustakhimov, I.I., Kalyuzhnaya, M., Lidstrom, M., and Trotsenko, Y.A. (2011a). Diversity and phylogeny of the ectoine biosynthesis genes in aerobic, moderately halophilic methylotrophic bacteria. Extremophiles *15*, 653–663. https://doi.org/10.1007/s00792-011-0396-x

Reshetnikov, A.S., Khmelenina, V.N., Mustakhimov, I.I., and Trotsenko, Y.A. (2011b). Genes and enzymes of ectoine biosynthesis in halotolerant methanotrophs. Meth. Enzymol. *495*, 15–30. https://doi.org/10.1016/B978-0-12-386905-0.00002-4

Roberts, E.A. (2011). Wilson's disease. Medicine *39*, 602–604.

Romarheim, O.H., Øverland, M., Mydland, L.T., Skrede, A., and Landsverk, T. (2011). Bacteria grown on natural gas prevent soybean meal-induced enteritis in Atlantic salmon. J. Nutr. *141*, 124–130. https://doi.org/10.3945/jn.110.128900

Rostkowski, K.H., Pfluger, A.R., and Criddle, C.S. (2013). Stoichiometry and kinetics of the PHB-producing Type II methanotrophs *Methylosinus trichosporium* OB3b and *Methylocystis parvus* OBBP. Bioresour. Technol. *132*, 71–77. https://doi.org/10.1016/j.biortech.2012.12.129

Saunois, M., Jackson, R.B., Bousquet, P., Poulter, B., and Canadell, J.G. (2016). The growing role of methane in anthropogenic climate change. Environ. Res. Lett. *11*, 120207.

Schilsky, M.L. (2014). Liver transplantation for Wilson's disease. Ann. N. Y. Acad. Sci. *1315*, 45–49. https://doi.org/10.1111/nyas.12454

Schuur, E.A.G., and Abbott, B. (2011). Climate change: high risk of permafrost thaw. Nature *480*, 32–33. https://doi.org/10.1038/480032a

Schwibbert, K., Marin-Sanguino, A., Bagyan, I., Heidrich, G., Lentzen, G., Seitz, H., Rampp, M., Schuster, S.C., Klenk, H.P., Pfeiffer, F., *et al.* (2011). A blueprint of ectoine metabolism from the genome of the industrial producer *Halomonas elongata* DSM 2581 T. Environ. Microbiol. *13*, 1973–1994. https://doi.org/10.1111/j.1462-2920.2010.02336.x

Semrau, J.D. (2011). Bioremediation via Methanotrophy: Overview of Recent Findings and Suggestions for Future Research. Front. Microbiol. *2*, 209. https://doi.org/10.3389/fmicb.2011.00209

Semrau, J.D., DiSpirito, A.A., and Yoon, S. (2010). Methanotrophs and copper. FEMS Microbiol. Rev. *34*, 496–531. https://doi.org/10.1111/j.1574-6976.2010.00212.x

Semrau, J.D., Jagadevan, S., DiSpirito, A.A., Khalifa, A., Scanlan, J., Bergman, B.H., Freemeier, B.C., Baral, B.S., Bandow, N.L., Vorobev, A., *et al.* (2013). Methanobactin and MmoD work in concert to act as the 'copper-switch' in methanotrophs. Environ. Microbiol. *15*, 3077–3086. https://doi.org/10.1111/1462-2920.12150

Semrau, J.D., DiSpirito, A.A., Gu, W., and Yoon, S. (2018). Metals and Methanotrophy. Appl. Environ. Microbiol. *84*, e02289–17.

Shukla, A.K., Vishwakarma, P., Upadhyay, S.N., Tripathi, A.K., Prasana, H.C., and Dubey, S.K. (2009). Biodegradation of trichloroethylene (TCE) by methanotrophic community. Bioresour. Technol. *100*, 2469–2474. https://doi.org/10.1016/j.biortech.2008.12.022

Singh, I., Sagare, A.P., Coma, M., Perlmutter, D., Gelein, R., Bell, R.D., Deane, R.J., Zhong, E., Parisi, M., Ciszewski, J., *et al.* (2013). Low levels of copper disrupt brain amyloid-β homeostasis by altering its production and clearance. Proc. Natl. Acad. Sci. U.S.A. *110*, 14771–14776. https://doi.org/10.1073/pnas.1302212110

Stafford, G., Scanlan, J., McDonald, I.R., and Murrell, J.C. (2003). *rpoN, mmoR* and *mmoG* genes involved in the expression of soluble methane monooxygenase in *Methylosinus trichosporium* OB3b. Microbiology *149*, 1771–1784.

Stainthorpe, A.C., Murrell, J.C., Salmond, G.P., Dalton, H., and Lees, V. (1989). Molecular analysis of methane monooxygenase from *Methylococcus capsulatus* (Bath). Arch. Microbiol. *152*, 154–159.

Stainthorpe, A.C., Lees, V., Salmond, G.P., Dalton, H., and Murrell, J.C. (1990a). The methane monooxygenase gene cluster of *Methylococcus capsulatus* (Bath). Gene *91*, 27–34.

Stainthorpe, A.C., Salmond, G.P.C., Dalton, H., and Murrell, J.C. (1990b). Screening of obligate methanotrophs for soluble methane monooxygenase genes. FEMS Microbiology Letters *70*, 211–215.

Stanley, S.H., Prior, S.D., Leak, D.J., and Dalton, H. (1983). Copper stress underlies the fundamental change in intracellular location of methane mono-oxygenase in methane-oxidizing oganisms - studies in batch and continuous cultures. Biotechnol. Lett. *5*, 487–492.

Stirling, D.I., Colby, J., and Dalton, H. (1979). Comparison of the substrate and electron-donor specificities of the methane mono-oxygenases from 3 strains of methane-oxidizing bacteria. Biochem. J. *177*, 361–364.

Stoecker, K., Bendinger, B., Schöning, B., Nielsen, P.H., Nielsen, J.L., Baranyi, C., Toenshoff, E.R., Daims, H., and Wagner, M. (2006). Cohn's Crenothrix is a filamentous methane oxidizer with an unusual methane monooxygenase. Proc. Natl. Acad. Sci. U.S.A. *103*, 2363–2367.

Strong, P.J., Xie, S., and Clarke, W.P. (2015). Methane as a resource: can the methanotrophs add value? Environ. Sci. Technol. *49*, 4001–4018. https://doi.org/10.1021/es504242n

Strong, P.J., Kalyuzhnaya, M., Silverman, J., and Clarke, W.P. (2016). A methanotroph-based biorefinery: Potential scenarios for generating multiple products from a single fermentation. Bioresour. Technol. *215*, 314–323.

Strong, P.J., Karthikeyan, O.P., Zhu, J., Clarke, W., and Wu, W. (2017). Methanotrophs: methane mitigation, denitrification and bioremediation. In Agro-Environmental Sustainability: Volume 2: Managing Environmental Pollution, J.S. Singh, and G. Seneviratne, eds. (Springer International Publishing, Cham), pp. 19-40.

Summer, K.H., Lichtmannegger, J., Bandow, N., Choi, D.W., DiSpirito, A.A., and Michalke, B. (2011). The biogenic methanobactin is an effective chelator for copper in a rat model for Wilson disease. J. Trace Elem. Med. Biol. *25*, 36–41.

Tavormina, P.L., Orphan, V.J., Kalyuzhnaya, M.G., Jetten, M.S., and Klotz, M.G. (2011). A novel family of functional operons encoding methane/ammonia monooxygenase-related proteins in gammaproteobacterial methanotrophs. Environ. Microbiol. Rep. *3*, 91–100. https://doi.org/10.1111/j.1758-2229.2010.00192.x

Tavormina, P.L., Hatzenpichler, R., McGlynn, S., Chadwick, G., Dawson, K.S., Connon, S.A., and Orphan, V.J. (2015). *Methyloprofundus sedimenti* gen. nov., sp. nov., an obligate methanotroph from ocean sediment belonging to the 'deep sea-1' clade of marine methanotrophs. Int. J. Syst. Environ. Microbiol. 65, 251–259.

Tonge, G.M., Harrison, D.E.F., and Higgins, I.L. (1977). Purification and properties fo the methane monooxygenase enzyme system from *Methylosinus trichosporium* OB3b. Biochem. J. *161*, 333–344.

Trotsenko, Y.A., and Murrell, J.C. (2008). Metabolic aspects of aerobic obligate methanotrophy. Adv. Appl. Microbiol. *63*, 183–229. https://doi.org/10.1016/S0065-2164(07)00005-6

Tsoi, R., Wu, F., Zhang, C., Bewick, S., Karig, D., and You, L. (2018). Metabolic division of labor in microbial systems. Proc. Natl. Acad. Sci. U.S.A. *115*, 2526–2531. https://doi.org/10.1073/pnas.1716888115

United States Energy Information Administration (2018). Natural gas prices. https://www.eia.gov/dnav/ng/NG_PRI_SUM_DCU_NUS_M.htm. Accessed July 6, 2018

van Teeseling, M.C., Pol, A., Harhangi, H.R., van der Zwart, S., Jetten, M.S., Op den Camp, H.J., and van Niftrik, L. (2014). Expanding the verrucomicrobial methanotrophic world: description of three novel species of *Methylacidimicrobium* gen. nov. Appl. Environ. Microbiol. *80*, 6782–6791. https://doi.org/10.1128/AEM.01838-14

Ve, T., Mathisen, K., Helland, R., Karlsen, O.A., Fjellbirkeland, A., Røhr, Å.K., Andersson, K.K., Pedersen, R.B., Lillehaug, J.R., and Jensen, H.B. (2012). The *Methylococcus capsulatus* (Bath) secreted protein, MopE*, binds both reduced and oxidized copper. PLOS ONE *7*, e43146. https://doi.org/10.1371/journal.pone.0043146

Vigliotta, G., Nutricati, E., Carata, E., Tredici, S.M., De Stefano, M., Pontieri, P., Massardo, D.R., Prati, M.V., De Bellis, L., and Alifano, P. (2007). *Clonothrix fusca* Roze 1896, a filamentous, sheathed, methanotrophic gamma-proteobacterium. Appl. Environ. Microbiol. 73, 3556–3565.

Visca, P., Leoni, L., Wilson, M.J., and Lamont, I.L. (2002). Iron transport and regulation, cell signalling and genomics: lessons from *Escherichia coli* and *Pseudomonas*. Mol. Microbiol. *45*, 1177–1190.

Vorobev, A.V., Baani, M., Doronina, N.V., Brady, A.L., Liesack, W., Dunfield, P.F., and Dedysh, S.N. (2011). *Methyloferula stellata* gen. nov., sp nov., an acidophilic, obligately methanotrophic bacterium that possesses only a soluble methane monooxygenase. Int. J. Syst. Evol. Microbiol. *61*, 2456–2463.

Vorobev, A., Jagadevan, S., Baral, B.S., Dispirito, A.A., Freemeier, B.C., Bergman, B.H., Bandow, N.L., and Semrau, J.D. (2013). Detoxification of mercury by methanobactin from *Methylosinus trichosporium* OB3b. Appl. Environ. Microbiol. *79*, 5918–5926. https://doi.org/10.1128/AEM.01673-13

Walker, C.B., de la Torre, J.R., Klotz, M.G., Urakawa, H., Pinel, N., Arp, D.J., Brochier-Armanet, C., Chain, P.S., Chan, P.P., Gollabgir, A., *et al.* (2010). *Nitrosopumilus maritimus* genome reveals unique mechanisms for nitrification and autotrophy in globally distributed marine crenarchaea. Proc. Natl. Acad. Sci. U.S.A. *107*, 8818–8823. https://doi.org/10.1073/pnas.0913533107

Wallar, B.J., and Lipscomb, J.D. (1996). Dioxygen Activation by Enzymes Containing Binuclear Non-Heme Iron Clusters. Chem. Rev. *96*, 2625–2658.

Walshe, J.M. (2007). Cause of death in Wilson disease. Mov. Disord. 22, 2216–2220. https://doi.org/10.1002/mds.21693

Wang, J.S., McElroy, M.B., Logan, J.A., Palmer, P.I., Chameides, W.L., Wang, Y., and Megretskaia, I.A. (2008). A quantitative assessment of uncertainties affecting estimates of global mean OH derived from methyl chloroform observations. J. Geophys. Res. Atmos. *113*, D12302.

Ward, N., Larsen, Ø., Sakwa, J., Bruseth, L., Khouri, H., Durkin, A.S., Dimitrov, G., Jiang, L., Scanlan, D., Kang, K.H., *et al.* (2004). Genomic insights into methanotrophy: the complete genome sequence of *Methylococcus capsulatus* (Bath). PLOS Biol. *2*, e303. https://doi.org/10.1371/journal.pbio.0020303

Weiss, K.H., and Stremmel, W. (2012). Evolving perspectives in Wilson disease: diagnosis, treatment and monitoring. Curr. Gastroenterol. Rep. *14*, 1–7. https://doi.org/10.1007/s11894-011-0227-3

Whittenbury, R., Phillips, K.C., and Wilkinson, J.F. (1970). Enrichment, isolation and some properties of methane-utilizing bacteria. J. Gen. Microbiol. *61*, 205–218. https://doi.org/10.1099/00221287-61-2-205

Wise, M.G., Vaun McArthur, J., and Shimkets, L.J. (2001). *Methylosarcina fibrata* gen. nov., sp. nov. and *Methylosarcina quisquiliarum* sp. nov., novel type I methanotrophs. Int. J. Syst. Evol. Microbiol. *51*, 611–621.

Wuebbles, D.J., and Hayhoe, K. (2002). Atmospheric methane and global change. Earth Sci. Rev. *57*, 177–210.

Xiao, Y., Angulo, M.T., Friedman, J., Waldor, M.K., Weiss, S.T., and Liu, Y.Y. (2017). Mapping the ecological networks of microbial communities. Nat. Commun. *8*, 2042. https://doi.org/10.1038/s41467-017-02090-2

Yan, X., Chu, F., Puri, A.W., Fu, Y., and Lidstrom, M.E. (2016). Electroporation-Based Genetic Manipulation in Type I Methanotrophs. Appl. Environ. Microbiol. *82*, 2062–2069. https://doi.org/10.1128/AEM.03724-15

Yarza, P., Yilmaz, P., Pruesse, E., Glöckner, F.O., Ludwig, W., Schleifer, K.H., Whitman, W.B., Euzéby, J., Amann, R., and Rosselló-Móra, R. (2014). Uniting the classification of cultured and uncultured bacteria and archaea using 16S rRNA gene sequences. Nat. Rev. Microbiol. *12*, 635–645. https://doi.org/10.1038/nrmicro3330

Ye, R.W., Yao, H., Stead, K., Wang, T., Tao, L., Cheng, Q., Sharpe, P.L., Suh, W., Nagel, E., Arcilla, D., *et al.* (2007). Construction of the astaxanthin biosynthetic pathway in a methanotrophic bacterium *Methylomonas* sp. strain 16a. J. Ind. Microbiol. Biotechnol. *34*, 289–299. https://doi.org/10.1007/s10295-006-0197-x

Yoon, S., Carey, J.N., and Semrau, J.D. (2009). Feasibility of atmospheric methane removal using methanotrophic biotrickling filters. Appl. Microbiol. Biotechnol. *83*, 949–956. https://doi.org/10.1007/s00253-009-1977-9

Zahn, J.A., and DiSpirito, A.A. (1996). Membrane-associated methane monooxygenase from *Methylococcus capsulatus* (Bath). J. Bacteriol. *178*, 1018–1029.

Zahniser, K.A. (2015). Why litigation-driven history matters. The Public Historian *37*, 46.

Zhang, T., Zhou, J., Wang, X., and Zhang, Y. (2017). Coupled effects of methane monooxygenase and nitrogen source on growth and poly-β-hydroxybutyrate (PHB) production of *Methylosinus trichosporium* OB3b. J. Environ. Sci. *52*, 49–57.

Zischka, H., Lichtmannegger, J., Schmitt, S., Jägemann, N., Schulz, S., Wartini, D., Jennen, L., Rust, C., Larochette, N., Galluzzi, L., *et al.* (2011). Liver mitochondrial membrane crosslinking and destruction in a rat model of Wilson disease. J. Clin. Invest. *121*, 1508–1518. https://doi.org/10.1172/JCI45401

Zlochower, I.A., and Green, G.M. (2009). The limiting oxygen concentration and flammability limits of gases and gas mixtures. J. Loss Prev. Process Ind. *22*, 499–505.

Diversity of Methane-cycling Microorganisms in Soils and Their Relation to Oxygen

2

Claudia Knief*

Institute of Crop Science and Resource Conservation – Molecular Biology of the Rhizosphere, University of Bonn, Bonn, Germany.

*Correspondence: knief@uni-bonn.de

https://doi.org/10.21775/9781912530045.02

Abstract

Microorganisms are important players in the global methane cycle. Anaerobic methanogenic archaea are largely responsible for methane production, while aerobic methanotrophic bacteria, as well as anaerobic methanotrophic bacteria and archaea, are involved in methane oxidation. In anoxic wetland soils, methanogens produce methane, while methanotrophs act as a filter and reduce methane emissions. In the predominantly oxic upland soils, aerobic methanotrophs oxidize atmospheric methane. This review gives an overview of the diversity of methanogenic and methanotrophic microorganisms, highlights recent discoveries and provides information concerning their occurrence in soils. Recent findings indicate that the methanogenic and methanotrophic lifestyles are more widespread in microorganisms than previously thought, and that the metabolic versatility of some methane-cycling organisms is broader than known from well-characterized cultivated organisms. It also turned out that the control of methanogenic and methanotrophic bacteria by oxygen is more complex than previously thought. The implications this finding may have for the life of these microorganisms in soils and on soil methane fluxes is discussed.

Introduction

Methane cycling microorganisms are of interest for microbiologists since more than a century. Research on these microorganisms was initially largely driven by the curiosity to understand their particular physiology that leads to the production or consumption of methane. While this interest is still a driver, the importance of methane as greenhouse gas has become another important factor, promoting further research on methanogenic and methanotrophic microorganisms. This leads to a continuously better understanding of their physiology and ecology, and it becomes evident that the processes of microbial methane production and consumption are mediated by more complex functional guilds than initially thought. The improved understanding is not only due to the constantly increasing diversity of methanogenic and methanotrophic microorganisms (e.g. Knief, 2015; Kallistova *et al.*, 2017); additionally, the metabolic versatility of these organisms appears to be much broader than previously thought. This became most evident during the last two decades, based on the study of enrichment cultures and isolates representing novel lineages of methanogens and methanotrophs, several of them with properties that have not been observed before in these organisms (Welte, 2018). The use of new high-throughput approaches for the analysis of organisms in culture or *in situ*, e.g. by deep metagenomic sequencing and the analysis of reconstructed genome information from individual organisms or near isogenic strains, allows the detection of methanogenic and methanotrophic potential in already known or new microbial taxa (Chistoserdova, 2015). This has resulted in the discovery of methanogenic and methanotrophic pathways in organisms that were not known before to represent methanogens or methanotrophs. Important for methane production or uptake in an ecosystem is not only the presence of methanogenic and methanotrophic organisms, but also their activity. Both presence and activity are largely controlled by diverse abiotic and biotic factors. Recent findings indicate that a strict categorization of the diverse organisms concerning their responses to specific environmental factors may not always be possible. In the present review, this will be exemplified focusing on oxygen dependence of methanogenic and methanotrophic microorganisms. Overall, the aims of this review are:

1. provide an update on the global methane budget and describe the role of soils in global methane cycling,
2. provide an update on the diversity of methanogenic archaea, aerobic methanotrophic bacteria and anaerobic methanotrophic archaea and bacteria,
3. present knowledge about the occurrence of these different groups of methane-cycling organisms in wetland and upland soils,
4. synthesize present knowledge about oxygen as a major environmental factor controlling the occurrence and activity of these groups of microorganisms.

The importance of soils as sources and sinks for atmospheric methane

Methane (CH_4) is the most abundant hydrocarbon in the atmosphere with a current mixing ratio of 1.85 ppmv (Dlugokencky, 2018). This exceeds the preindustrial levels of 0.7 ppmv by a factor of approximately 2.5 (Ciais *et al.*, 2013) and is higher than concentrations recorded in ice cores during the past 800,000 years (Loulergue *et al.*, 2008). From 2007 to 2017, the average yearly increase in atmospheric methane concentration was estimated to be 7 ppb, after emissions had transiently declined at the beginning of the 21st century (Dlugokencky, 2018). The reasons for this increase are under discussion, but a contribution of biogenic emissions, probably due to agricultural activities, appears likely (Saunois *et al.*, 2016b).

Increasing atmospheric methane concentrations are critical, because methane is the most important greenhouse gas after carbon dioxide (CO_2), contributing approximately 20% to global warming (Dlugokencky *et al.*, 2011; Kirschke *et al.*, 2013). This is related to its stronger global-warming potential, which is currently estimated to be 28 times stronger compared with CO_2 (Myhre *et al.*, 2013). Methane has a rather short lifetime of approximately 9 years in the atmosphere (Saunois *et al.*, 2016a), so that effective mitigation strategies could lead to near-term reductions in atmospheric concentrations and could complement CO_2 mitigation strategies (Saunois *et al.*, 2016b). Thus, methane is an interesting and important target to reduce global warming processes. However, in order to put mitigation strategies in action, knowledge about the sources and sinks of atmospheric methane and the underlying processes leading to methane production and consumption is needed.

Global budget calculations are performed based on different modelling approaches and with increasing accuracy. For this review, two recent calculations are considered (Kirschke *et al.*, 2013; Saunois *et al.*, 2016a). According to these studies, the total global methane emissions are around 560 Tg CH_4/year, while the total sink strength is 550 Tg CH_4/year, resulting in an atmospheric growth of approximately 10 Tg CH_4/year. This growth is with very high confidence linked to anthropogenic activities, which have been estimated to contribute about 60% to global emissions (Ciais *et al.*, 2013; Saunois *et al.*, 2016a).

Focusing on the sources, natural wetlands are the strongest individual source, contributing with 25–32% to global emissions (Fig. 2.1). Moreover, wetlands are assumed to be the main drivers of global inter-annual variability of methane emissions (Ciais *et al.*, 2013). Estimates for freshwaters (lakes, ponds, rivers, estuaries) show still a high uncertainty (Saunois *et al.*, 2016a). Further natural sources are of geological or oceanic origin or from animals (all ≤ 5%). Anthropogenic sources

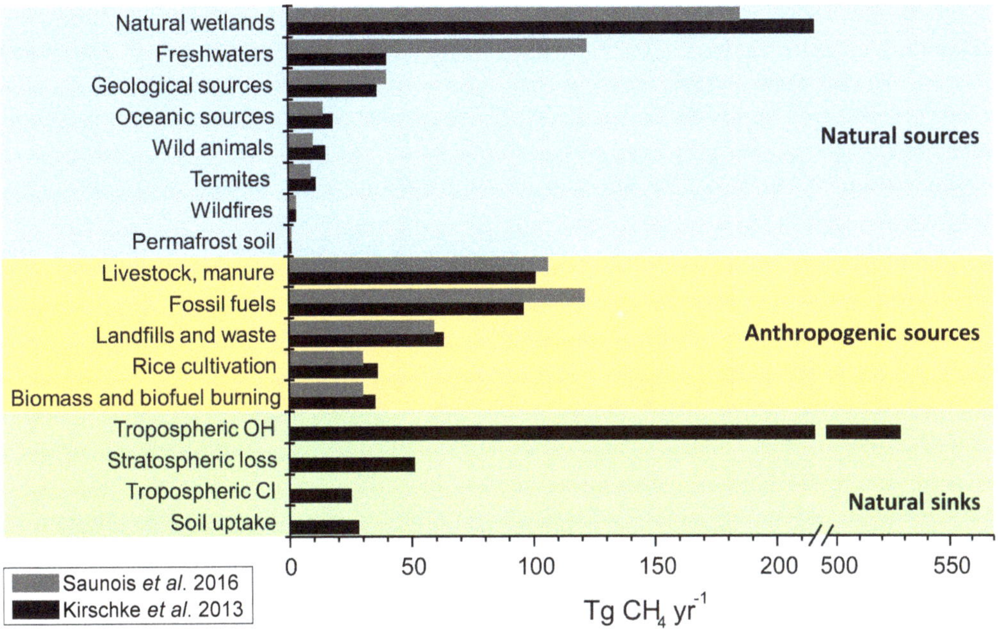

Figure 2.1 Sources and sinks of atmospheric methane. Data were taken from two recent publications, in which emissions were estimated from 2000 to 2009 (Kirschke *et al.*, 2013) and 2003 to 2012 (Saunois *et al.*, 2016a) based on different modelling approaches. Kirschke *et al.* (2013) presents data as provided in the IPCC report 2013. Different sinks were not resolved by Saunois *et al.* (2016a).

of atmospheric methane contribute between 50% and 60% to the total methane emissions and are predominantly from fossil fuel use and livestock farming (each approximately 15%), followed by landfills and waste treatment (9%), rice cultivation (5%) and biomass and biofuel burning (5%). Most of the atmospheric methane is eliminated by chemical reactions in the atmosphere (Fig. 2.1), whereby the chemical reaction with OH radicals in the troposphere is the predominant process (84%). Moreover, well-aerated soils serve as sink for atmospheric methane, contributing 4% to atmospheric methane oxidation (Kirschke *et al.*, 2013). These global budget calculations reveal that soils play an important role, especially as source of atmospheric methane, but also as sink. As sources, natural wetland soils are most relevant, followed by landfill soils and rice paddies. In contrast, well-aerated upland soils represent a relevant sink.

Microbial processes leading to methane production

Besides the classification of methane sources according to their natural or anthropogenic origin, they can be differentiated based on the underlying processes leading to methane formation. Thermogenic, pyrogenic and biogenic sources are differentiated and their source contribution can be estimated based on stable isotope analysis (Ciais *et al.*, 2013). Biogenic methane shows the strongest isotopic depletion and is the end product of organic matter degradation in the absence of oxygen or of other oxidants such as nitrate, sulfate or ferric iron. It is produced by methanogenic microorganisms (Conrad, 1996). This process is responsible for methane production in natural wetlands, freshwaters, organic waste deposits (landfills, waste, manure), rice paddies, ruminants, termites and wild animals, so that about 69% of the total atmospheric methane originates from the activity of methanogenic microorganisms (Conrad, 2009). The presence and activity of methanogens and therewith methane emissions are controlled by diverse environmental factors in soils, with substrate availability and concentration of oxygen being among the most relevant factors. These are to some extent linked to other factors such as the concentration and type of organic matter, soil redox potential, availability of electron acceptors, or water availability and water table. Moreover, temperature, soil pH, availability of nutrients and trace metals, salinity, vegetation, fertilizer and manure additions affect the development of methanogenic communities and their activity and therewith methane emissions (Dalal and Allen, 2008; Dalal *et al.*, 2008; Serrano-Silva *et al.*, 2014).

An update on the diversity of methanogenic Archaea

For decades, methane production was attributed to four specific classes of methanogenic *Euryarchaeota*, the *Methanomicrobia, Methanobacteria, Methanococci* and *Methanopyri*. The taxonomy, ecology and physiology of the members of these classes has been extensively reviewed (Garcia *et al.*, 2000; Conrad, 2007; Liu and Whitman, 2008; Thauer and Shima, 2008; Thauer *et al.*, 2008; Ferry, 2010; Nazaries *et al.*, 2013; Costa and Leigh, 2014; Serrano-Silva *et al.*, 2014). Many methanogens grow on hydrogen and carbon dioxide as substrates, while members of the genera *Methanosaeta* and *Methanosarcina* grow on acetate. The family *Methanosarcinaceae* is most versatile, i.e. most members are able to grow methylotrophically, for example on methanol or methylated compounds (Oren, 2014). Hydrogenotrophic and aceticlastic methanogenesis are most relevant for biogenic methane production (Conrad, 2005), and the relative contribution of each of these processes to methane production can vary substantially (Conrad, 1999). Methanogenic archaea that are typically found in anoxic soils include *Methanosaetaceae, Methanosarcinaceae* (especially *Methanosarcina*), *Methanomicrobiaceae* (especially *Methanoculleus, Methanomicrobium* and *Methanogenium*), *Methanoregulaceae, Methanospirillaceae, Methanocellaceae, Methanobacteriaceae* (especially *Methanobacterium* and *Methanobrevibacter*) and '*Candidatus* Methanoflorentaceae' (Garcia *et al.*, 2000; Conrad, 2007; Liu and Whitman, 2008). The candidate status of the family '*Ca.* Methanoflorentaceae' indicates that cultured representatives that underwent a formal description are currently lacking.

Recent discoveries indicate that the diversity and physiological versatility of methanogenic microorganisms is actually broader than previously thought, as highlighted by Welte (2018) and reviewed by Kallistova *et al.* (2017). Knowledge has either been gained due to cultivation of novel lineages or based on genome reconstructions from metagenomic data, which were obtained for enrichment cultures of uncultivated groups of methanogens that are merely known from the detection of specific marker genes in environmental samples. These markers are the 16S rRNA gene or the *mcrA* gene, which encodes a subunit of methyl coenzyme M reductase. It is the key enzyme of all methanogenic microorganisms and of some anaerobic methane-oxidizing Archaea. The currently known diversity of methanogenic archaea is illustrated in Fig. 2.2.

Recently discovered methanogenic taxa within the phylum *Euryarchaeota*

Within the *Euryarchaeota* different new groups of methanogens were discovered during the last years. The well-known rice cluster I (RC-I) organisms, which are important methane producers in rice paddies (Lu and Conrad, 2005; Conrad *et al.*, 2006), were brought into culture a decade ago and are in the meantime represented by three different species within the genus *Methanocella*, i.e. *Methanocella paludicola, Methanocella arvoryzae* and *Methanocella conradii*, all isolated from rice field soil (Lü and Lu, 2012).

Recently, rice cluster II (RC-II) (Großkopf *et al.*, 1998) has been characterized in more detail based on reconstructed genome data from a population inhabiting thawing permafrost soil. The family name '*Candidatus* Methanoflorentaceae' has been proposed for this group of organisms, belonging to the order *Methanocellales* (Fig. 2.2) (Mondav *et al.*, 2014). Similar as the other *Methanocellales*, '*Candidatus* Methanoflorens stordalenmirensis' is predicted to be hydrogenotrophic. It appears to be an important player in different ecosystems, but especially in cold wetlands (McCalley *et al.*, 2014; Mondav *et al.*, 2014; Kao-Kniffin *et al.*, 2015).

With the description of *Methanomassiliicoccus luminyensis* the first and currently only methanogenic isolate of the class *Thermoplasmata* was identified (Dridi *et al.*, 2012). It represents the new order *Methanomassiliicoccales* within this class (Iino *et al.*, 2013) and was formerly known as rice cluster III (RC-III) (Großkopf *et al.*, 1998). The class *Thermoplasmata* is the first one that harbours a methanogenic order as well as a non-methanogenic order. Besides the isolate *M. luminyensis*, several candidate genera representing *Methanomassiliicoccales*

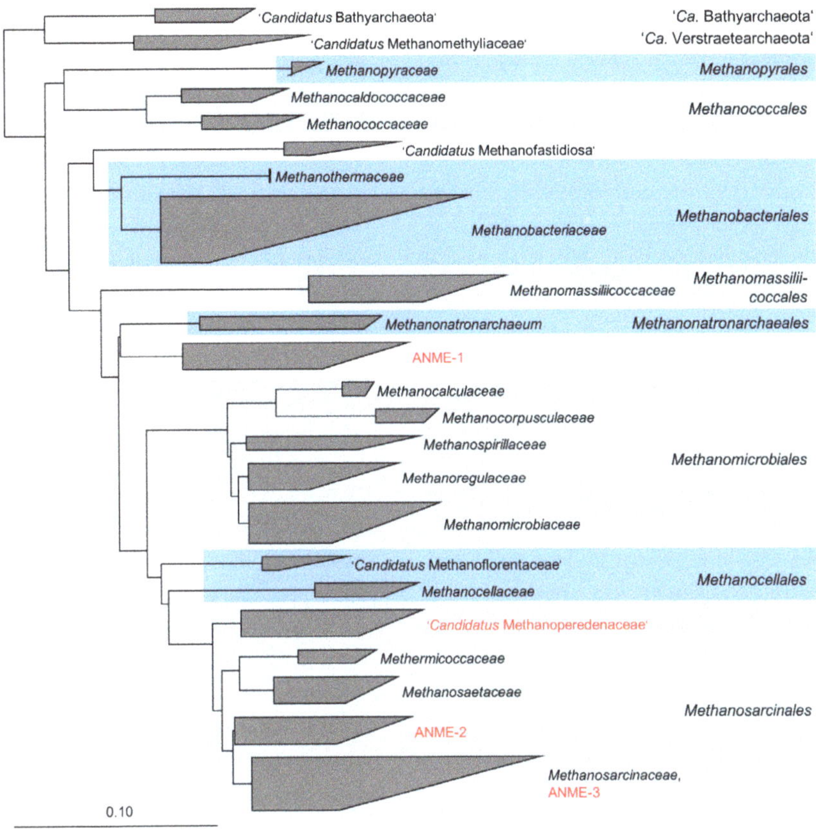

Figure 2.2 16S rRNA gene sequence based phylogenetic tree summarizing the diversity of methanogenic archaea (shown in black) and anaerobic methanotrophic archaea (marked in red). Sequences from type strains and a representative subset of sequences from uncultivated organisms are included in the tree as available in the SSURef_NR99_132_SILVA database. The tree was calculated in ARB using the neighbour joining algorithm with Jukes-Cantor correction and an archaeal filter (1450 nucleotide positions). Sequences were grouped at family level and the different orders of methanogenic *Euryarchaeota* are indicated as far as they have been classified at order level. For predicted methanogens outside of the phylum *Euryarchaeota* the candidate phylum name is given.

have been characterized based on metagenomic sequencing of highly enriched cultures, including '*Candidatus* Methanogranum caenicola', '*Candidatus* Methanomethylophilus alvus', '*Candidatus* Methanoplasma termitum' and '*Candidatus* Methanomassiliicoccus intestinalis' (Borrel *et al.*, 2013a,b; Iino *et al.*, 2013; Lang *et al.*, 2015). These methanogens were mostly obtained from the intestinal tract of humans or animals. 16S rRNA and *mcrA* gene sequence analyses including those from public databases indicate that methanogenic *Methanomassiliicoccales* form two distinct clades, the host-associated and the free-living clade (Paul *et al.*, 2012; Söllinger *et al.*, 2016; Borrel *et al.*, 2017). The free-living clade includes the isolate *Methanomassiliicoccus luminyensis* as well as sequences from diverse terrestrial habitats such as sediments, landfill leachates, wetland soils, hot springs, permafrost sediment, and rice paddies (Borrel *et al.*, 2013b; Iino *et al.*, 2013; Chojnacka *et al.*, 2015; Lang *et al.*, 2015; W. Li *et al.*, 2016; Merkel *et al.*, 2016; Söllinger *et al.*, 2016; Winkel *et al.*, 2018). In contrast to the other methanogenic *Euryarchaeota*, all (meta-)genome sequenced *Methanomassiliicoccales* lack the pathway for CO_2 reduction to methyl coenzyme M and gain energy by a hydrogen-dependent reduction of methanol or methylamines (Lang *et al.*, 2015; Y. Li *et al.*, 2016; Söllinger *et al.*, 2016).

The class *Methanonatronarchaeia* was only recently discovered, including the species *Methanonatronarchaeum thermophilum* and '*Candidatus* Methanohalarchaeum thermophilum' (Sorokin *et al.*, 2018). They represent the formerly uncultivated halophilic SA1 euryarchaeal group (Eder *et al.*, 2002). As implemented in the names, these organisms are adapted to hypersaline and moderately thermophilic conditions and were obtained from soda lakes. They are not monophyletic to other classes of methanogens, but are most closely related to the class *Halobacteria*. Likewise as the *Methanomassiliicoccales,* these organisms show a methylotrophic lifestyle, but in this case hydrogen or formate serve as electron donors and different C_1-compounds such as methanol or methylamines as acceptor (Sorokin *et al.*, 2017). These organisms lack some genes of the CO_2-reduction pathway, which is relevant for the formation of methane from CO_2.

The uncultivated methanogenic lineage WSA2 (or Arc I), which occurs in a wide range of natural and engineered environments but especially in wastewater treatment plants and marine sediments, is meanwhile represented by one candidate genus, proposed based on metagenome sequence analysis of four WSA2 populations, which were obtained from methanogenic bioreactors treating wastewater (Nobu *et al.*, 2016). It is referred to as '*Candidatus* Methanofastidiosum methylothiophilus' and represents a distinct class, '*Candidatus* Methanofastidiosa' (Nobu *et al* 2016). Likewise, as observed in several of the aforementioned new classes and orders, a complete pathway for CO_2 reduction to methane and for aceticlastic methanogenesis appears to be absent. Instead, these organisms seem to dependent on hydrogen as electron donor and methyl groups obtained from demethylation of methylated thiols, e.g. methylsulfide, as electron acceptors. Moreover, no carbon fixation pathway was identified, and a heterotrophic lifestyle with acetate, malonate or propionate as carbon sources was proposed for these organisms. Furthermore, they lack biosynthetic pathways for several amino acids. These peculiarities have likely contributed to the fact that these methanotrophs eluded cultivation so far, likewise as the host-associated *Methanomassiliicoccales*.

Novel potential methanogenic taxa beyond the phylum *Euryarchaeota*

Besides the isolation of new lineages of methanogens within the *Euryarchaeota*, metagenomic studies point to the existence of methanogens outside the *Euryarchaeaota*. Reconstructed genomes from metagenomic datasets revealed the presence of methanogenetic pathways in representatives from the archaeal candidate phyla '*Candidatus* Bathyarchaeota' and '*Candidatus* Verstraetearchaeota' (Evans *et al.*, 2015; Vanwonterghem *et al.*, 2016). These are representatives of a group of archaea with broad environmental distribution, harbouring species with diverse physiologies and ecological functions, also known as Miscellaneous Crenarchaeotal Group or Group 1.3 archaea (Lloyd, 2015).

Near-complete genome data from two distinct lineages of '*Ca.* Bathyarchaeota' were obtained in samples from formation water of deep coalbed methane wells in Australia (Evans *et al.*, 2015). The genome-sequenced representatives of '*Ca.* Verstraetearchaeota' were assigned to two new candidate genera, i.e. '*Candidatus* Methanomethylicus' and '*Candidatus* Methanosuratus' (Vanwonterghem *et al.*, 2016). They were detected in experimental anaerobic digesters set up with inocula from different natural and engineered anoxic environments with high methane flux (rumen, lake sediment, anaerobic digester and lagoon). From the reconstructed genomes the authors of both studies suggested that the organisms are methylotrophic methanogens. Remarkably for these organisms is the presence of metabolic pathways that appear to enable them to carry out fermentation processes using amino acids, fatty acids or sugars as substrates, a feature that has not been observed among archaeal methanogens. Moreover, Evans *et al.* (2015) stated that it might be possible that these organisms gain energy from anaerobic oxidation of methane. To further validate these predictions and prove the metabolic versatility, it will be necessary to study these organisms in more detail directly in the environment or after enrichment and, if possible, isolation of representative strains. A detailed analysis of the metabolic capabilities appears of particular relevance concerning the methane-cycling capabilities of '*Ca.* Bathyarchaeota', due to the finding that *mcrA* sequences of '*Candidatus* Syntrophoarchaeum', which are similar to those of '*Ca.* Bathyarchaeota', encode an MCR-like protein catalysing the formation of butyl-coenzyme M from butane (Laso-Pérez *et al.*, 2016). Similarly as the organisms sequenced by Evans *et al.* (2015), '*Ca.* Syntrophoarchaeum' possesses an almost-complete methanogenesis-related pathway and four complete *mcr* gene sets, but grows on butane, which is converted to butyl-coenzyme M, in analogy to the activation of methane to methyl-coenzyme M by anaerobic methanotrophic arachaea (ANME). However, formation of methyl-coenzyme M in the presence of methane was not observed for '*Ca.* Syntrophoarchaeum'. Thus, it can at the moment not be excluded that members of the '*Ca.* Bathyarchaeota' may actually be non-methane alkane oxidizers and the involvement in methane production or possibly oxidation needs to be carefully proven.

The occurrence of these two archaeal groups with methane-cycling potential in nature was further assessed by screening public databases for the presence of 16S rRNA genes in datasets from other studies. Sequences being similar to the sequenced members of both phyla, '*Ca.* Bathyarchaeota' and '*Ca.* Verstraetearchaeota' were indeed detected in different methane-rich habitats including freshwater wetland soils (Vanwonterghem *et al.*, 2016; Narrowe *et al.*, 2017). Vanwonterghem *et al.* (2016) concluded that anoxic conditions, high methane fluxes and a likelihood for increased concentrations of methylated compounds are common characteristics of the habitats in which members of '*Ca.* Verstraetearchaeota' are found. The phylum '*Ca.* Bathyarchaeota' includes diverse non-methanogenic members, which can even be present in methanogenic environments (He, Y. *et al.*, 2016; Lazar *et al.*, 2016; Maus *et al.*, 2018), so the mere detection of 16S rRNA gene sequences representing this phylum is not indicative for the presence of potential methane cycling microorganisms. Conclusions about the presence of these potential methane-cycling microorganisms should thus be drawn carefully and are most reliable if 16S rRNA gene sequences are found that are highly similar to those of the genome sequenced methane-cycling organisms.

Alternatively to the 16S rRNA gene, the *mcrA* gene is a useful target for the detection of methanogenic archaea. The *mcrA* gene was detected in all reconstructed genomes of '*Ca.* Bathyarchaeota' and '*Ca.* Verstraetearchaeota' (Evans *et al.*, 2015; Vanwonterghem *et al.*, 2016), with sequences of '*Ca.* Bathyarchaeota' being clearly distinct from those of the methanogenic *Euryarchaeota,* while those of '*Ca.* Verstraetearchaeota' are quite similar to those of other methanogenic *Euryarchaeota* (Vanwonterghem *et al.*, 2016). The detection of *mcrA* sequences highly similar to those of '*Ca.* Verstraetearchaeota' in metagenomic datasets from terrestrial mud volcanoes and palm oil mill effluent point to a broader distribution of these organisms (Vanwonterghem *et al.*, 2016). Moreover, *mcrA* genes of '*Ca.* Verstraetearchaeota' were found in geothermal spring sediments and at low abundance in a boreal lake sediment (McKay *et al.*, 2017; Rissanen *et al.*, 2017). Even *mcrA* gene expression was proven in these studies. Remarkably, no corresponding 16S rRNA gene sequences of '*Ca.* Verstraetearchaeota' were detected in the respective geothermal spring samples, indicating that the detected *mcrA* sequences are either present in another phylogenetic taxon besides '*Ca.* Vertraeterachaota', or that the 16S rRNA gene based primer was biased concerning the amplification of '*Ca.* Vertraetearchaota'. The *mcrA* genes highly similar to those of '*Ca.* Bathyarchaeota' were found in different high-methane flux environments, including other hydrocarbon seep samples, tar sand tailing ponds, petroleum reservoir sediments, several aquatic environments and geothermal spring sediments (Evans *et al.*, 2015; McKay *et al.*, 2017). Several of these habitats will provide other alkanes such as butane as carbon source, so that the function of this group of archaea in these habitats remains currently unclear.

Methane production in soils under oxic conditions

Recent findings indicate that methane production can also occur under oxic conditions, i.e. in upland soils or wetland soils that become temporarily or partially oxic. However, methanogens have long been considered to be strictly anaerobic and most of them are known to be sensitive to oxygen (Fetzer *et al.*, 1993; Whitman *et al.*, 2014). Despite this assumption, methanogenic archaea have been found in diverse upland soils, including soils from forests, meadows and grasslands, agricultural land, savannas, as well as cold and warm desert and sub-/alpine ecosystems (e.g. Peters and Conrad, 1995; Angel *et al.*, 2012; Aschenbach *et al.*, 2013; Praeg *et al.*, 2014; Hofmann *et al.*, 2016; Hernández *et al.*, 2017; Xie *et al.*, 2017). Moreover, methane production was observed upon incubation of these soils under anoxic conditions, indicating that the activity of the methanogenic archaea can be stimulated under appropriate conditions, and some studies reported methane production from oxic soils (Teh *et al.*, 2005; Kammann *et al.*, 2009). It is assumed that the activity of methanogens in these soils is temporally and spatially limited, occurring in anoxic microniches, which are formed by the soil structure or are provided by the soil fauna (Conrad, 1995; Kammann *et al.*, 2009, 2017). Their activity may support atmospheric methane-oxidizing bacteria, which rely otherwise on the very low concentrations of atmospheric methane.

Molecular analysis revealed that *Methanocella* and *Methanosarcina* were most commonly detected in upland soils. Besides, *Methanomassiliicoccus, Methanobacterium, Methanosaeta* and *Methanobrevibacter* were repeatedly detected (e.g. Angel *et al.*, 2011, 2012; Aschenbach *et al.*, 2013; Hu *et al.*, 2013; Praeg *et al.*, 2014; Hofmann *et al.*, 2016; Hernández *et al.*, 2017; Xie *et al.*, 2017). The activity of these genera in upland soils remains largely unexplored. Activity studies of methanogens have only been performed in different drained wetland soils. In oxic paddy soil, the same genera, i.e. *Methanosaeta, Methaosarcina, Methanobacterium* and *Methanocella,* were shown to be involved in organic matter degradation in a ^{13}C-labelling experiment (Lee *et al.*, 2012). Other studies with paddy soils revealed that oxygen exposure resulted in a strong decrease in *mcrA* gene expression (Yuan *et al.*, 2011; Liu *et al.*, 2018). After the aeration event, several of the above mentioned genera showed the best recovery, i.e. highest *mcrA* gene expression (Yuan *et al.*, 2011; Reim *et al.*, 2017; Liu *et al.*, 2018). In a freshwater wetland from a lakeshore, substantial methane production was reported from the oxygenated soil layer and attributed to the activity of '*Candidatus* Methanothrix paradoxum' (Angle *et al.*, 2017). The genus name *Methanothrix* is supposed to replace the genus name *Methanosaeta* (Garrity *et al.*, 2011), thus the finding is also in agreement with the observation concerning the presence of *Methanosaeta* in upland soils. The activity of this methanogen was assumed to be restricted to anoxic microsites within the overall oxic soil layer. This was concluded from the finding that gene expression analyses of the strain did not indicate the activation of genes involved in oxygen detoxification mechanisms, although this strain possesses a set of such genes.

The presence of genes involved in adaptation mechanisms to oxidative environments was recently analysed systematically in genome sequenced methanogenic archaea and revealed that the methanogens can be divided into two classes, those that have accumulated genes involved in oxygen resistance (*Methanocellales, Methanomicrobiales* and *Methanosarcinales*) and those that lack most of these genes (*Methanobacteriales, Methanococcales* and *Methanopyrales*) (Lyu and Lu, 2018). In agreement with this classification, nearly all taxa that were found in upland soils or in wetland soils under oxygen stress conditions belong to the group with the higher number of antioxidant features in the genome. Similarly, Lyu and Lu (2018) reported that the taxa with accumulation of antioxidant genes were more frequently detected in microaerophilic or oxic environments, including oceans, rice soils, subsurface soils, and diverse upland and wetland soils. Their observation was based on a meta-analysis, in which the occurrence of publicly available 16S rRNA genes was evaluated. Remarkably, the genera *Methanobrevibacter* and *Methanobacterium,* which were detected in some upland soils, do not belong to the group of methanogens harbouring a diverse set of antioxidant genes. Nevertheless, *Methanobrevibacter* has been shown to remain active in the presence of oxygen, being even able to reduce oxygen, as long as the concentration does not exceed the capacity for its removal (Tholen *et al.*, 2007). This indicates that the methanogenic archaea must have developed different strategies to survive oxic conditions in soils and may contribute to methane production even in environments that are mostly oxic. Methane production under apparently oxic conditions has not only been observed in soils, but also in lakes, where methane production was observed in the oxygenated water column, a phenomenon that is referred to as 'the methane paradox' (Grossart *et al.*, 2011; Bogard *et al.*, 2014; Tang *et al.*, 2014; Donis *et al.*, 2017).

Besides the activity of methanogenic archaea in oxic soils, fungi were recently reported to release methane from methionine as precursor and may contribute to methane production (Lenhart *et al.*, 2012). Moreover, different non-biogenic methane production processes are known to occur in soils (Wang *et al.*, 2013, 2017). These result from degradation processes of organic material, including photodegradation, thermal degradation, oxidation by reactive oxidation species, extracellular oxidative metabolism or inorganic chemical reactions. However, the contribution of these processes to methane emissions in upland soils or aerated wetland soils remains currently uncertain, but may be rather low (Lenhart *et al.*, 2012; Gu *et al.*, 2016; Wang *et al.*, 2017). The methane-oxidizing bacteria in the soils may be able to metabolize most of this methane before it reaches the atmosphere. Further work is needed to assess the importance of these different processes in the diverse continental ecosystems.

Microbial methane oxidation by aerobic methanotrophic bacteria

Biological aerobic methane oxidation is exclusively performed by bacteria. The methanotrophic bacteria, which gain carbon and energy from the oxidation of methane, inhabit diverse terrestrial, aquatic and marine habitats. Terrestrial ecosystems that are known to act as sources for atmospheric methane host diverse methane-oxidizing bacteria. These methanotrophic bacteria are found in high Arctic and tundra wetlands, peat bogs, rice paddies, landfill covers, sewage sludge and floodplains (Knief, 2015). In these ecosystems, aerobic methanotrophic bacteria usually inhabit the oxic/anoxic interfaces, where they oxidize the methane that is released by the methanogenic archaea. The filter capacity of the methanotrophic bacteria can lead to more than 80% reduction in methane emissions, especially in oceans, freshwaters and rice paddies, while the filter effect appears less efficient in wetlands and landfill soils (Conrad, 1996; Reeburgh, 2003).

In upland soils aerobic methanotrophic bacteria are responsible for atmospheric methane oxidation (Bender and Conrad, 1992; Dunfield, 2007; Kolb, 2009; Knief, 2015). They live on the expense of this atmospheric methane and, if available, endogenously produced methane in soil (see previous section). Moreover, some of them may profit from multi-carbon compounds (Pratscher *et al.*, 2011, 2018). The presence and activity of methanotrophic bacteria in soils is controlled by diverse environmental factors. As for the methanogenic archaea, substrate availability and concentration of oxygen are considered to be the most important factors, while water availability and water table, soil pH, availability of nutrients and trace metals (especially copper), temperature, salinity, vegetation, fertilizer and manure additions will further affect the abundance and activity of these microorganisms (Dalal and Allen, 2008; Dalal *et al.*, 2008; Semrau *et al.*, 2010; Aronson *et al.*, 2013; Serrano-Silva *et al.*, 2014). While the effect of these environmental factors on methane oxidation rates has been studied quite intensively, knowledge about the responses of the microbial groups involved in methane oxidation is less advanced.

An update on the diversity of cultivated aerobic methanotrophic bacteria

Aerobic methanotrophic bacteria are found within three bacterial classes, the *Alphaproteobacteria*, *Gammaproteobacteria* and *Methylacidiphilae*, the latter being members of the phylum *Verrucomicrobia*. An overview of the currently known diversity of cultivated methanotrophic bacteria is given in the phylogenetic tree (Fig. 2.3). Besides the classification of aerobic methanotrophic bacteria based on their phylogeny, grouping into type I (*Gammaproteobacteria*), type II (*Alphaproteobacteria*) and sometimes type III (*Methylacidiphilae*) methanotrophs is quite common, especially in cultivation-independent studies. This grouping into different types is not meant to encode specific phylogenetic information. The type I methanotrophs are further differentiated into type Ia to Id methanotrophs, whereby type Ia and Ib represent *Methylococcaceae*, type Ic *Methylothermaceae* and type Id an uncultivated lineage of methanotrophs, defined based on their *pmoA* sequences (Knief, 2015). Major characteristics of these methanotrophs were compiled in recent reviews (Knief, 2015; Dedysh and Knief, 2018). Thus, the focus in this section will be on recently obtained methanotrophic isolates that represent new genera of methanotrophic bacteria and their occurrence in soils.

Within the group of methanotrophic *Gammaproteobacteria*, a couple of different new isolates or enrichment cultures were recently obtained. *Methyloterricola oryzae* strain 73a^T was isolated from the lower part of stems from rice plants and is the first cultured representative of rice paddy cluster 1 (RPC1) (Frindte *et al.*, 2017). The isolate is a typical member of the *Methylococcaceae*, most closely related to the genera *Methylococcus* and '*Candidatus* Methylospira' (Fig. 2.2). RPC1 represents methanotrophic bacteria that are frequently detected

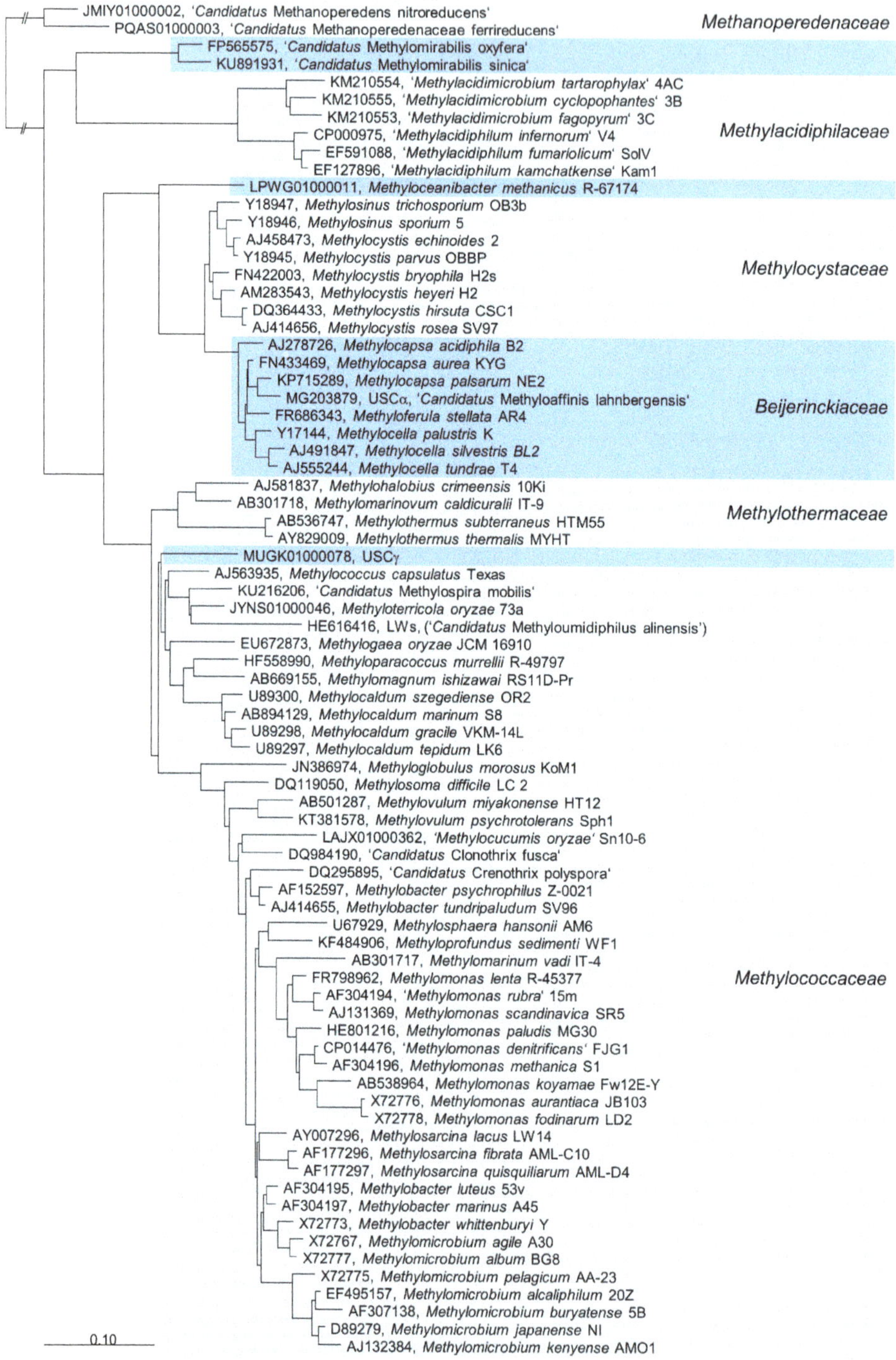

Figure 2.3 16S rRNA gene sequence based phylogenetic tree summarizing the diversity of aerobic methanotrophic bacteria and anaerobic methanotrophic bacteria and archaea. All type strains of validated aerobic methanotrophic species are included as well as methanotrophs that have been described as new species but have not yet been formally validated (names are hyphenated). Moreover, methanotrophs with candidate status are included, which are mostly available as enrichment culture, but not as pure culture isolates. This rules out their validation as new species based on current regulations. The tree was calculated in ARB using the neighbour joining algorithm with Jukes-Cantor correction and a bacterial filter (1565 nucleotide positions). Shorter sequences were added using the ARB parsimony quick-add tool. The assignment of the strains to different families is indicated as far as a classification at this taxonomic rank has been done.

in rice paddies, but also in aquatic ecosystems and wetlands (Knief, 2015). The cluster has been defined based on the detection of *pmoA* sequences. The *pmoA* gene encodes a subunit of the particulate methane monooxygenase and is the most widely used molecular marker for aerobic methanotrophic bacteria due to its presence in the fast majority of aerobic methanotrophs (Knief, 2015).

The methanotrophic isolate strain Sn10-6 is proposed to represent a new genus and species, '*Methylocucumis oryzae*' (Rahalkar *et al.*, 2016; Pandit *et al.*, 2018). It is distantly related to other type Ia methanotrophic bacteria and it was isolated from the rice rhizosphere. The preferred habitat of this genus remains currently largely unknown, because highly similar 16S rRNA or *pmoA* gene sequences are not present in the NCBI nucleotide collection database. A more specific search for similar sequences (>95% sequence identity) of these two marker genes in datasets obtained from rice ecosystems via high-throughput amplicon sequencing resulted in a few hits for both genes in two studies (Lee *et al.*, 2015; J. Liu *et al.*, 2017). In several other studies analysing paddy soil or the rice rhizosphere, it was not detected despite the higher sensitivity that is achieved when using next generation sequencing technologies (e.g. Lüke and Frenzel, 2011; Knief *et al.*, 2012; Ahn *et al.*, 2014; Vaksmaa *et al.*, 2017c; Shiau *et al.*, 2018). Such a rare detection of a newly isolated genus in cultivation-independent studies has previously been seen for a few other genera, mostly from marine environments (Knief, 2015). The relevance of these genera for global methane cycling remains thus largely unclear.

Another new candidate genus representing methanotrophic *Gammaproteobacteria*, '*Candidatus* Methylospira', could not yet be obtained in pure culture and has therefore the '*Candidatus*' status (Danilova *et al.*, 2016b). It was enriched from a *Sphagnum* dominated peat bog and is a representative of the *pmoA* OSC cluster, which is closely related to RPC I, as reflected by the relatedness of the 16S rRNA gene sequences of '*Candidatus* Methylospira mobilis' and *Methyloterricola oryzae* (Fig. 2.3). Sequences of this cluster were detected in different fen and bog ecosystems, but also in an organic-rich and a mineral soil. The study of Danilova *et al.* (2016b) reported its presence in different freshwater and lake sediments, thus it appears to colonize predominantly different aquatic and wetland ecosystems. Moreover, it was found as a dominant member within the methanotrophic community in a lichen-dominated patch of a boreal peatland ecosystem (Danilova *et al.*, 2016a). Characteristic for this genus are the spiral-shaped cells, which are so far unique among methanotrophs, and the preference to growth under micro-oxic conditions. The detection of this genus in different wetlands suggests that a microaerophilic lifestyle along with motility may be an important trait for this genus to establish a population in peatlands.

Recent insights obtained for major uncultivated groups of methanotrophic bacteria

Besides '*Candidatus* Methylospira mobilis', which exists as enrichment culture and has been described in detail, some further taxa of putative methanotrophs with *Candidatus* status have been proposed to exist. They were characterized based on information from genome reconstructions derived from metagenomic data. Considering the necessary requirements to refer to a taxon as *Candidatus*, these organisms should not be termed *Candidatus*, likewise as several of the above mentioned methanogens, because limited information is available concerning structural, metabolic or reproductive features. In some cases, a nearly complete 16S rRNA sequence is also lacking, although the availability of this sequence is currently another prerequisite for taxa with *Candidatus* status according to taxonomic rules (Murray and Stackebrandt, 1995). All groups of uncultivated methanotrophs that include a recently described strain or an uncultivated but genome sequenced representative are compiled in Table 2.1.

Upland soil cluster α (USCα) and MHP clade

The best-described group of uncultivated aerobic methanotrophic organisms is USCα, which is known as major player involved in atmospheric methane oxidation in upland soils (Dunfield, 2007; Kolb, 2009; Knief, 2015). For a genome analysis of this group of methanotrophs, cells were obtained upon artificial enrichment from forest soil samples known to harbour these organisms as dominant group of methanotrophs (Pratscher *et al.*, 2018). Phylogenetic placement of the 16S rRNA

Table 2.1 Recent insights obtained for major *pmoA* and *pmoA*-like sequence clusters

pmoA/*amoA* sequence cluster	Gained knowledge	Predominant habitat	Reference
RPC1	Isolate *Methyloterricola oryzae* strain 73a^T characterized, member of the *Methylococcaceae*	Rice paddies and aquatic habitats	Frindte *et al.* (2017)
OSC	Enrichment culture '*Candidatus* Methylospira mobilis' characterized, member of the *Methylococcaceae*	Peat bogs	Danilova *et al.* (2016b)
USCα	Genome reconstruction of a representative organism, member of the *Beijerinckiaceae*, proposition of the name '*Candidatus* Methyloaffinis lahnbergensis'	Upland soils, caves and lava tubes	Pratscher *et al.* (2018)
USCα, MHP clade	Genome reconstruction of two representative organisms, member of the *Beijerinckiaceae*	Peatlands	Singleton *et al.* (2018)
USCγ	Genome reconstruction of a representative organism, member of the *Gammaproteobacteria*	Upland soils, caves and lava tubes	Edwards *et al.* (2017)
LWs	Genome reconstruction of a representative organism, member of the *Methylococcaceae*, proposition of the name '*Candidatus* Methyloumidiphilus alinensis'	Aquatic habitats	Rissanen *et al.* (2018)
Crenothrix (*pmoA*/*amoA*)	Sequence cluster with 'unusual *pmoA*' of Crenothrix represents *amoA* sequences of *Nitrospira* species capable of comammox; Crenothrix harbour gammaproteobacterial *pmoA* gene sequences	Diverse terrestrial and aquatic habitats	Van Kessel *et al.* (2015), Daims *et al.* (2015), Oswald *et al.* (2017)

gene sequence and multi-locus sequence analyses indicated that this methanotroph represents a new genus within the family *Beijerinckiacea*, most closely related to *Methylocapsa* species, as already assumed based on *pmoA* gene sequence analysis (Fig. 2.4) and *pmo* operon analyses (Ricke *et al.*, 2005). Furthermore, genome analysis revealed the presence of a complete set of genes needed for C_1 metabolism and confirmed the possibility to grow on acetate as carbon source, as already proposed in an earlier study based on stable isotope incorporation with acetate as carbon substrate (Pratscher *et al.*, 2011). With the availability of a first known 16S rRNA gene sequence of USCα, the global distribution of this organism was reassessed, largely confirming *pmoA*-based findings, i.e. the recovery from diverse upland soils, in particular forest soils. Interestingly, highly similar 16S rRNA sequences were also recovered from subterranean environments, including caves and lava tubes, where they contributed up to 10% to the bacterial community composition, while this is ≤ 1% in upland soils (Kolb *et al.*, 2003; Pratscher *et al.*, 2018). The authors propose the name '*Candidatus* Methyloaffinis lahnbergensis' for this group of organisms.

Two further reconstructed genomes for USCα organisms were obtained from metagenomic data from permafrost soils (Singleton *et al.*, 2018). Here, the USCα methanotrophs appeared to be the predominant methanotrophs in palsa and were also abundant in the thawed bog samples. The palsa was mainly oxic and methane production was reported to be minimal (McCalley *et al.*, 2014), so that atmospheric methane oxidation by USCα is conceivable, while the detection of USCα in a bog sample is rather uncommon for USCα sequence types, with the exception of the MHP clade. The MHP clade is a subgroup within the large USCα cluster, which is typical for peatlands (Knief, 2015). Indeed, the most closely related *pmoA* sequences of the USCα contigs given in the study of Singleton *et al.* (2018) indicate that their USCα genomes represent this MHP clade. Thus, they represent a distinct subgroup of USCα methanotrophs compared to the organisms analysed by Pratscher *et al.* (2018). Analysis of the two reconstructed genomes

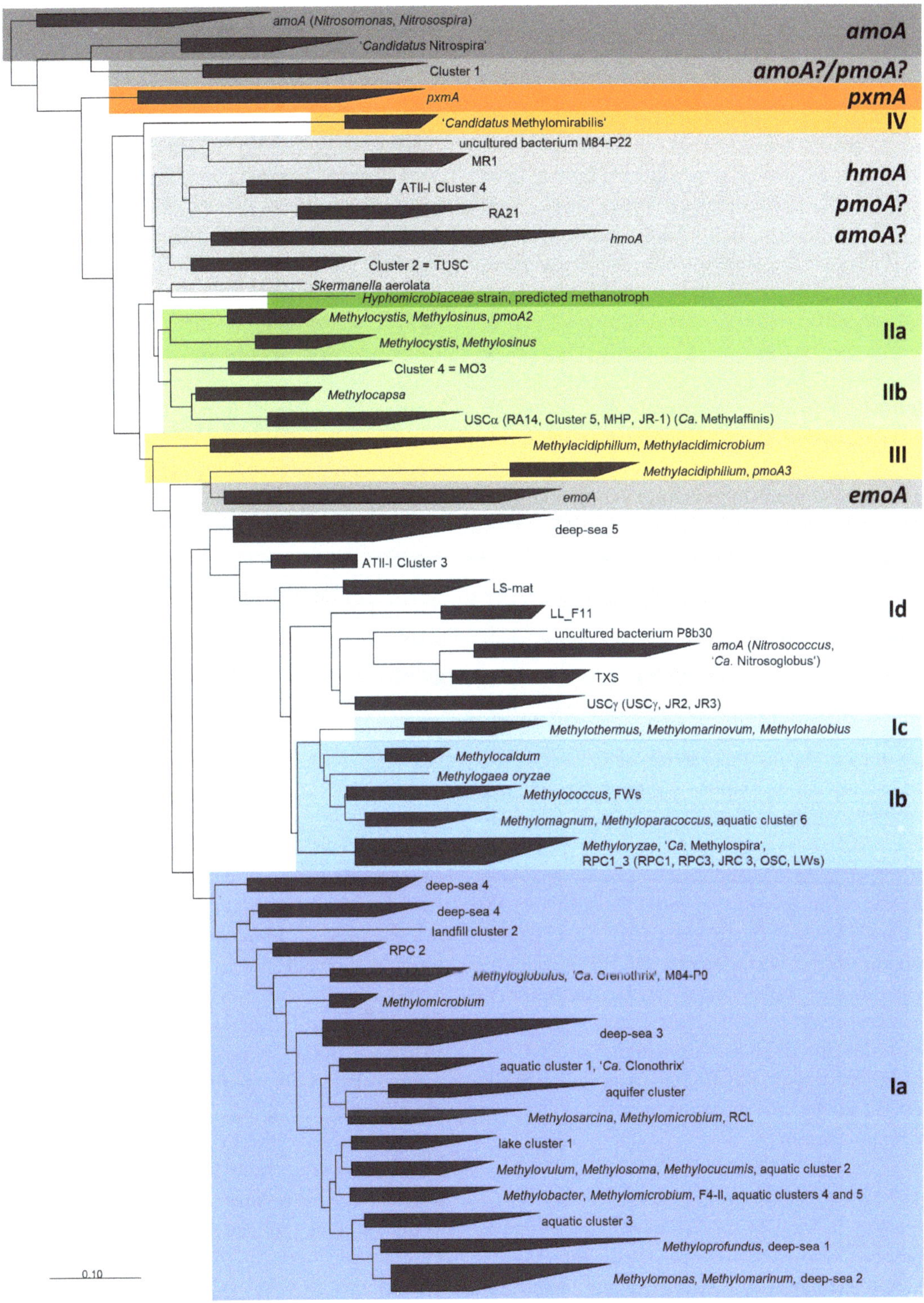

Figure 2.4 Phylogenetic tree showing the diversity of cultured methanotrophs and defined clusters of uncultivated aerobic methanotrophic bacteria based on *pmoA* gene sequence analysis. A backbone tree was used as presented earlier (Knief and Dedysh, 2018) and updated with sequences of recently published novel groups of methanotrophs using the ARB parsimony quick-add tool. The different groups of methanotrophic bacteria are labelled according to their grouping into type I to type IV.

confirmed the relationship of these USCα methanotrophs to *Methylocapsa* species. Furthermore, the presence of genes for pXMO homologue and a lanthanide-dependent XoxF-type methanol dehydrogenase as well as the absence of genes for a calcium-dependent MxaFI-type methanol dehydrogenase, which were reported by Pratscher *et al.* (2018), were confirmed in this study based on both genomes. Additionally, Singleton *et al.* (2018) report the presence of hydrogenase genes and genes involved in carbon monoxide oxidation, suggesting further metabolic versatility of these organisms besides a potential for growth on acetate. A hydrogenotrophic lifestyle was just recently reported for methanotrophic *Verrucomicrobia* (Carere *et al.*, 2017; Mohammadi *et al.*, 2017). The presence and activity of hydrogenases in methanotrophs is known since a long time (e.g. Chen and Yoch, 1987; Hanczár *et al.*, 2002). They were assumed to contribute to the generation of reductants, e.g. for methane monooxygenase, but not considered to allow growth under chemolithotrophic conditions, which demands a pathway for CO_2 fixation. Thus, it may only be an option for those type Ib methanotrophs that possess the capability to assimilate carbon via the Calvin cycle. While Pratscher *et al.* (2018) could not reconstruct a complete Calvin cycle for their organism, Singleton *et al.* (2018) report the presence of all relevant genes. As in other examples, the relevance of this predicted metabolic capability remains to be proven. Similarly, the involvement of the MHP clade in atmospheric methane oxidation has not yet been demonstrated. The study of Singleton *et al.*, 2018 included metatranscriptomic analyses, which did not reveal strong *pmoA* gene expression for these methanotrophs, so that the conditions under which they may be actively involved in methane oxidation in peatlands remain currently unclear.

Upland soil cluster γ

Likewise as for USCα, a reconstructed genome sequence was recently reported for USCγ (Edwards *et al.*, 2017). This *pmoA* sequence cluster is also known to be involved in atmospheric methane oxidation, but more frequently detected in pH neutral upland soils, while USCα is predominant in acidic soils (Knief *et al.*, 2003; Knief, 2015). A draft genome of a representative of this group of organisms was reconstructed from metagenomic data obtained from a mineral cryosoil sample. Only a partial 16S rRNA gene sequence is available, which indicates that USCγ methanotrophs are members of the class *Gammaproteobacteria* (Fig. 2.3). In the SILVA database used for tree reconstruction, the sequence clusters most closely to sequences of uncultured organisms, which form a branch separate from known families of *Gammaproteobacteria*. Interestingly, these closely related 16S rRNA sequences from environmental samples indicate its presence in caves and lava tubes, likewise as reported for USCα. Besides, some sequences were recovered from soils, especially from cold ecosystems. This is in agreement with the detection of the USCγ *pmoA* sequences, which were predominantly found in pH neutral and alkaline soils, and in soils from cold or dry ecosystems (Knief, 2015). Carbon assimilation of this group of organisms remains currently enigmatic, as neither a complete ribulose monophosphate pathway nor a complete serine cycle was reconstructed from the available genomic data. The presence of Calvin cycle genes, which could be another alternative for carbon assimilation, was not evaluated by Edwards *et al.* (2017).

Lake Washington cluster (LWs)

Another methanotrophic organism, proposed as '*Candidatus* Methyloumidiphilus alinensis', has recently been described based on genome reconstructions from a metagenomic dataset, which was obtained from a small oxygen-stratified humic lake (Rissanen *et al.*, 2018). A 16S rRNA gene sequence is not available from the reconstructed genome. Its phylogenetic placement was assessed based on a genomic comparison including a number of different genes, which indicated its relatedness to the genome sequenced *Methyloterricola oryzae* strain 73a^T. However, as other related organisms such as '*Candidatus* Methylospira mobilis' are not yet genome sequenced, its exact phylogenetic placement remains vague. Interestingly, its *pmoA* sequence indicates that the organism is a representative of the LWs cluster, which contains sequences from several different studies and is known to represent predominantly methanotrophs that inhabit freshwater lakes (Dumont *et al.*, 2014; Knief, 2015).

The Crenothrix cluster and a re-evaluation of the phylogenetic placement of Crenothrix *pmoA* sequences

A sequence cluster distantly related to *pmoA* sequences as well as to *amoA* sequences of nitrifying bacteria has been identified as Crenothrix cluster, based on the finding that *pmoA* sequences of the methane-oxidizing '*Candidatus* Crenothrix polyspora' fall into this cluster (Stoecker *et al.*, 2006). However, the identity and metabolic potential of the organisms represented by this sequence cluster has to be questioned based on two recent findings. First, different *Nitrospira* isolates were identified that possess an *amoA* gene with a sequence falling into this Crenothrix cluster (Daims *et al.*, 2015; van Kessel *et al.*, 2015). These *Nitrospira* species oxidize ammonia to nitrate, a process referred to as 'complete ammonium oxidation to nitrate', or comammox. This finding suggests that the sequence cluster is representing comammox bacteria of the genus *Nitrospira*. Second, a recent study reanalysed samples from the waterworks sand filter system, in which the 'unusual *pmoA*' sequences assigned to '*Ca.* Crenothrix polyspora' had initially been found. Genome reconstructions from metagenomic data revealed that the system harbours *Nitrospira* species with the comammox type of *amoA* sequence, and *Crenothrix* species with a gammaproteobacterial *pmoA* sequence (Oswald *et al.*, 2017). Such *pmoA* sequences were also detected in the initial study by Stoecker *et al.* (2006), but not linked to '*Ca.* Crenothrix polyspora' due to their low abundance. Taken together, these findings indicate that the 'Crenothrix cluster' represents *amoA*-like sequences of comammox organisms, while Crenothrix species harbour *pmoA* sequences similar to those of methanotrophic *Gammaproteobacteria*. They are actually similar to the *pmoA* sequences of *Methyloglobulus morosus* (Fig 2.4).

Besides the detection of a Crenothrix methanotroph with a *pmoA* sequence similar to that of *M. morosus*, Oswald *et al.* (2017) reported the existence of another '*Ca.* Crenothrix polyspora' organism from a stratified lake, which has a gammaproteobacterial *pmoA* sequence that is only distantly related to sequences of cultivated type Ia methanotrophs. Instead, its *pmoA* sequence clusters most closely to *pmoA* sequences of diverse uncultivated type Ia methanotrophs, several of them obtained from freshwater methane seeps. However, it does not consistently fall into a specific well-known cluster of uncultivated methanotrophs. Therefore, it is not highlighted in the phylogenetic tree in Fig. 2.4, where it is part of lake cluster 1. This inconsistent clustering is explained by a high number of sequences with equal similarities to each other among the type Ia methanotrophs. Thus, the clustering of sequences varies to some extent in dependence on the dataset and algorithm used for tree calculation. This also explains why the combination of sequences into larger clusters as presented in Fig. 2.4 is not in full agreement with trees shown in previous studies (Dedysh and Knief, 2018; Knief, 2015). Oswald *et al.* (2017) speculate that the freshwater lake Crenothrix has acquired the *pmoCAB* operon laterally from another methanotrophic *Gammaproteobacterium*, as the operon is flanked by transposase genes.

Evidence for methanotrophs in genera not yet known to include aerobic methanotrophic bacteria

The diversity of methanotrophic *Alphaproteobacteria* has recently been extended with the discovery of the methanotrophic strain '*Methyloceanibacter methanicus*' strain R-67174, a member of the order *Rhizobiales* (Vekeman *et al.*, 2016). Besides the *Methylocystaceae* and *Beijerinckiacae*, it represents a third group of methanotrophs within this order (the genus has not yet been classified at class level). The strain was isolated from a marine sediment sample. While the genus *Methyloceanibacter* is known to be methylotrophic, '*Methyloceanibacter methanicus*' strain R-67174 is currently the only known methanotrophic species and strain within this genus. This is the first example of a methanotroph within a non-methanotrophic though methylotrophic genus. The methane oxidation capacity of this strain is realized by the presence of a soluble methane monooxygenase, while genes encoding a membrane bound methane monooxygenase, which occurs almost consistently among methanotrophic bacteria, are absent. Besides *Methylocella* and *Methyloferula*, it is thus the third genus that possesses only the soluble form of the enzyme methane monooxygenase. The *mmoX* gene, which serves as molecular marker for methanotrophs harbouring a soluble methane monooxygenase, was related to those of *Methylocella* and *Methyloferula*, which are both representatives of the methanotrophic *Beijerinckiacea*. The authors speculate that *Methyloceanibacter* may have acquired the genes encoding the soluble methane monooxygenase by horizontal gene transfer. The occurrence of the genus *Methyloceanibacter* appears to be largely limited to marine environments, all isolates were obtained from marine environments (Takeuchi *et al.*, 2014; Vekeman *et al.*, 2016), and closely related 16S rRNA gene sequences were predominantly detected in marine environments. This is in agreement with the general finding that most marine methanotrophs are clearly distinct from methanotrophic taxa found in terrestrial or aquatic environments (Knief, 2015; Vekeman *et al.*, 2016). Thus, this methanotroph may not play a major role in soil. However, the observation that individual methylotrophs gain the capability to oxidize methane by acquiring genes encoding the soluble or particulate methane monooxygenase may apply to terrestrial microorganisms as well, especially if microorganisms reside in habitats where methanol as well as methane are available as carbon and energy sources, thus supporting the growth of methylotrophic and methanotrophic microorganisms.

Evidence for putative methanotrophic strains in another non-methanotrophic genus comes from the metagenomic study of Singleton *et al.* (2018), in which reconstructed genomes related to the photoheterotrophic *Rhodomicrobium* spp., members of the *Hyphomicrobiaceae*, were reported to harbour operons for both, the particulate and soluble methane monooxygenase. The *pmoA* sequences found in this group of organisms cluster basal to those of the *Methylocystaceae* and *Beijerinckiaceae* (Fig. 2.4) and reflect thus the 16S rRNA gene based phylogeny. Similarly, the MmoX sequences formed a novel cluster related to the *Beijerinckiaceae*. Likewise as some other alphaproteobacterial methylotrophs, this *Hyphomicrobiaceae* strain has genes for a thiol-dependent pathway for formaldehyde oxidation and the necessary equipment to perform carbon assimilation via the Calvin cycle. Furthermore, hydrogenase genes and a complete dissimilatory sulphate reduction pathway were identified, with *dsr* genes similar to those of *Rhodomicrobium*, indicating a broader metabolic versatility also for this group of organisms. In public sequence databases, its *pmoA* sequence type was available from a couple of metagenomes, indicating the presence in wetlands including peat and bog ecosystems. Moreover, 16S rRNA gene sequences related to *Rhodomicrobium* or *Hyphomicrobium* were detected in different ^{13}C-methane stable isotope labelling studies performed in peatlands (Morris *et al.*, 2002; Gupta *et al.*, 2012; Putkinen *et al.*, 2014; Deng *et al.*, 2016). Owing to the methylotrophic lifestyle realized by some members of the *Hyphomicrobiaceae*, labelling of these organisms was assumed to be the result of cross-feeding on partially oxidized C_1 compounds in such labelling experiments.

The results obtained for the *Mehyloceanibacter* isolate and the genome information of the *Hyphomicrobiaceae* strain suggest that methane oxidation must be more widespread among the *Alphaproteobacteria* than previously thought. As for USCα (family *Beijerinckiaceae*), these organisms appear to be metabolically more versatile than the *Methylocystaceae*, so that their impact on methane cycling in an ecosystem cannot yet be assessed, but it deserves more detailed studies to evaluate under which conditions these organisms may contribute to the methane oxidation process. Singleton *et al.* (2018) reported very weak expression of the genes for methane monooxygenases by the *Hyphomicrobiaceae* strain, based on metatranscriptomic data.

Oxygen dependence of aerobic methanotrophic bacteria

Methanotrophic bacteria are considered to be obligately aerobic due to their need of oxygen for respiration and for methane oxidation by the methane monooxygenase. However, studies accumulate that report the presence of aerobic methanotrophic *Gammaproteobacteria* in habitats with very low oxygen concentrations, especially in suboxic and anoxic layers or the sediments of stratified lakes (e.g. Biderre-Petit *et al.*, 2011; Blees *et al.*, 2014; Kojima *et al.*, 2014; Crevecoeur *et al.*, 2015; Hernandez *et al.*, 2015; Oswald *et al.*, 2016; Martinez-Cruz *et al.*, 2017; Naqvi *et al.*, 2018; Singleton *et al.*, 2018). In particular type Ia methanotrophs are consistently detected in these studies. More specifically, these are often reported to represent *Methylobacter* species. A specific assignment to the genus *Methylobacter* may in some cases be questionable, because the different species of *Methylobacter* are not forming monophyletic clusters in 16S rRNA and *pmoA* based trees, so that diverse sequence types exist in public databases that are termed 'uncultured *Methylobacter*'.

Such *pmoA* sequences cluster with the recently detected genera *Methylovulum* and *Methylosoma,* but are also similar to those of *Methylomicrobium* and *Methylosarcina,* or they fall into major groups of uncultivated organisms, representing the *pmoA* lake cluster or different aquatic clusters (Dumont *et al.*, 2014; Knief, 2015). Thus, the adaptation to low-oxygen conditions is most likely not strictly limited to the genus *Methylobacter.*

Focusing on soils, *Methylobacter* was reported to be the dominant active methanotroph with increasing depth and thus decreasing oxygen concentrations in an arctic peat soil (Tveit *et al.*, 2014). Moreover, *Methylobacter* was the dominant active methanotroph in the anoxic zone just below the oxic/anoxic interface in a rice paddy soil microcosm. Actually, a higher *pmoA* transcript to gene ratio was observed in the anoxic soil layer compared with the oxic top soil layer, and highest *pmoA* transcription was observed at the interface (Reim *et al.*, 2012). These findings indicate that *pmoA* gene expression and, linked to it, the activity of aerobic methanotrophic bacteria under microaerophilic or even anoxic conditions is not restricted to aquatic ecosystems, but occurs also in different wetland soils. Thus, aerobic methanotrophs in different ecosystems appear to be less dependent on (high) oxygen concentrations and oxygen availability may be a less strict regulatory factor for the occurrence and activity of at least some aerobic methanotrophs than initially thought.

In agreement with these observations is the enrichment and isolation of methanotrophs from lake sediment and a peat bog ecosystem that grow preferentially under microaerophilic conditions, i.e. *Methylosoma difficile, Methyloglobulus morosus* and '*Candidatus* Methylospira mobilis' (Rahalkar *et al.*, 2007; Deutzmann *et al.*, 2014; Danilova *et al.*, 2016b). While the adaptation mechanisms to microaerophilic conditions of these specific taxa are not yet known, different mechanisms are known from other aerobic methanotrophic bacteria. One strategy appears to be the use of alternative terminal oxidases with high affinity to oxygen. The presence of a cytochrome *bd* oxidase (K_m value for oxygen between 3 and 8 nM) has been reported for a strain of the family *Methylothermaceae,* related to *Methylohalobius crimeensis* and *Crenothrix* (Skennerton *et al.*, 2015; Oswald *et al.*, 2017). The genes for this oxidase were found in reconstructed genome information, assembled from metagenomic datasets. The activity and affinity of this oxidase in aerobic methanotrophic bacteria remains to be studied to validate its involvement in adaptation of methanotrophs to oxygen-limited conditions.

Furthermore, several aerobic methanotrophs have the genetic equipment to perform anaerobic respiration by denitrification. Methane-dependent denitrification activity using nitrate or nitrite as substrate was reported for '*Methylomonas denitrificans*' and *Methylomicrobium album* BG8, respectively (Kits *et al.*, 2015a,b) and relevant genes have been found in further gammaproteobacterial methanotrophs including *Methylobacter* or *Crenothrix* (Campbell *et al.*, 2011; Kalyuzhnaya *et al.*, 2015; Skennerton *et al.*, 2015; Oswald *et al.*, 2017). Moreover, the alphaproteobacterial *Methylocystis* sp. strain SC2 was shown to perform complete denitrification from nitrate to dinitrogen under anoxic conditions in the presence of methanol as growth substrate (Dam *et al.*, 2013). Since some *Methylocystis* strains are known to be facultative methanotrophs, even though with reduced growth capacities compared with growth on methane (Belova *et al.*, 2011; Im *et al.*, 2011), the combination of facultative methanotrophy with denitrification could be a way to overcome oxygen limitation for members of this genus. Besides anaerobic respiration, fermentation has recently been reported as an adaptation strategy for *Methylomicrobium alcaliphilum* strain 20Z (Kalyuzhnaya *et al.*, 2013). The authors observed the production of different organic acids in the presence of very low oxygen concentrations, which were still sufficient to perform methane oxidation. Finally, aerobic methanotrophic bacteria were reported to survive extended periods of anoxic conditions (Roslev and King, 1994).

All these observations demonstrate that aerobic methanotrophic bacteria have developed different strategies to live or at least to survive under conditions of severe oxygen limitation. However, the relevance of these different mechanisms under *in situ* conditions remains currently largely unclear. Likewise, mechanisms to overcome the need for oxygen for the initial methane oxidation step remain unknown (Chistoserdova, 2015). A putative oxygen scavenging protein was recently discussed in this context, due to the fact that gene expression of a cyanoglobin homologue was upregulated under hypoxic conditions and detected in '*Methylomonas denitrificans*' and the reconstructed genome of the *Methylothermaceae* strain (Kits *et al.*, 2015b; Skennerton *et al.*, 2015). Since bacterial globins can have different functions (Vinogradov *et al.*, 2013), this protein and its possible role in methanotrophs remains to be studied.

For freshwater ecosystems, further possibilities are under discussion to explain the activity of aerobic methanotrophs under anoxic conditions: (i) methanotrophic bacteria living in association with phototrophic microorganisms that produce oxygen, which is instantaneously consumed and thus not detectable (Milucka *et al.*, 2015; Oswald *et al.*, 2015), linked to it (ii) oxygen concentrations are below the detection limit of standard oxygen sensors and thus not detectable (Blees *et al.*, 2014), (iii) oxygen may episodically be transported into anoxic layers and therewith support aerobic methanotrophs (Blees *et al.*, 2014), (iv) the sedimentation of inactive cells from oxic layers (Schubert *et al.*, 2006), and (v) perhaps the use of alternative electron acceptors such as Mn(IV) or Fe(III) (Oswald *et al.*, 2016). However, these hypotheses remain to be proven and different explanations may be valid under different conditions.

Anaerobic methanotrophic archaea and bacteria

Besides the aerobic methanotrophic bacteria, anaerobic methane-oxidizing archaea and bacteria contribute to the reduction of methane emissions in ecosystems that act are sources for atmospheric methane. This is well known to be of particular relevance in marine ecosystems, where >90% of the produced methane is oxidized by anaerobic methanotrophs (Knittel and Boetius, 2009). Knowledge about the filter effect of anaerobic methanotrophs in terrestrial ecosystems is still limited, although evidence is accumulating concerning the relevance of this process. The process of anaerobic methane oxidation has been detected in diverse soils, especially in different natural wetlands (e.g. Smemo and Yavitt, 2007; Gupta *et al.*, 2012; Hu *et al.*, 2014; Gauthier *et al.*, 2015; Segarra *et al.*, 2015).

First global estimates about the relevance of anaerobic methane oxidation are available for natural wetlands, but still imprecise. According to two independent studies, between 4 and 200 Tg of methane is oxidized by these organisms per year (Hu *et al.*, 2014; Segarra *et al.*, 2015), which would correspond to a reduction in emissions between 2% and 50%, assuming a source strength of approximately 200 Tg of methane per year (Fig. 2.1). A global estimation for peatlands resulted also in a 50% reduction of global methane emissions due to anaerobic methane oxidation, corresponding to 41 Tg of methane per year that are anaerobically oxidized (Smemo and Yavitt, 2011). Another study reported an average of 24 Tg of methane consumption per year by anaerobic methanotrophs in peatlands, with very high variation between sites (Gupta *et al.*, 2012). Based on these first estimates, it appears that anaerobic methane oxidation contributes significantly to a reduction in methane emissions in wetlands and should therefore be considered as a relevant process.

In marine ecosystems, anaerobic methane oxidation is usually coupled with sulfate reduction and mediated by anaerobic methanotrophic archaea, which are often found in a consortium with sulfate-reducing bacteria (Knittel and Boetius, 2009). These anaerobic methane-oxidizing archaea are *Euryarchaeota* that are phylogenetically related to methanogenic archaea and referred to as ANME clusters (Fig. 2.2). In contrast, microorganisms performing anaerobic methane oxidation coupled with denitrification are more common in continental ecosystems. Two major groups of microorganisms have been characterized that are involved in this process, represented by '*Candidatus* Methylomirabilis oxyfera' and '*Candidatus* Methanoperedens nitroreducens' (Ettwig *et al.*, 2010; Haroon *et al.*, 2013). Moreover, a methanotrophic archaeon coupling anaerobic methane oxidation with iron reduction '*Candidatus* Methanoperedens ferrireducens' has been described (Cai *et al.*, 2018). All these different anaerobic methanotrophs can be successfully enriched in bioreactors despite their slow growth rates, but pure cultures are not available.

Characteristics of the different anaerobic methanotrophs have been summarized in diverse review articles (e.g. Chistoserdova, 2015; Cui *et al.*, 2015; Kallistova *et al.*, 2017), some of them with a focus on methanotrophs that oxidize methane coupled with sulfate reduction (Knittel and Boetius, 2009) and others with a focus on those that couple methane oxidation with denitrification (Shen *et al.*, 2015a; Welte *et al.*, 2016). Yet others discuss in detail the particular physiology of these organisms (Caldwell *et al.*, 2008; McGlynn, 2017; Timmers *et al.*, 2017). Within this review, relevant basic information about these organisms is given, but the focus will be on aspects related to their occurrence in soils.

Anaerobic methane oxidation coupled with sulfate reduction

Diversity and physiology of anaerobic methane oxidizers dependent on sulfate reduction

Three phylogenetic groups of anaerobic methane-oxidizing archaea are known that couple methane oxidation with sulfate reduction, ANME-1 to -3 (Knittel and Boetius, 2009; Timmers *et al.*, 2017). They were defined based on distinct clustering of their 16S rRNA and *mcrA* gene sequences in phylogenetic trees. The ANME-1 group has been reported to be distantly related to the *Methanosarcinales* and *Methanomicrobiales*. It now appears that they are also related to the recently described *Methanonatronarchaeales* (Fig. 2.2). The ANME-2 group is related to the *Methanosarcinales* and ANME-3 to the genus *Methanococcoides* within the order *Methanosarcinales*.

The ANME-1 group consists of subgroups a and b, ANME-2 of subgroups a, b and c. The groups ANME-2a and -2b form a coherent clade and are often grouped together as ANME-2a/b. Furthermore, an ANME-2d group has been proposed to exist, including '*Candidatus* Methanoperedens nitroreducens', which couples methane oxidation with denitrification (Haroon *et al.*, 2013). Earlier, this group was referred to as AOM associated archaea (AAA) group (Knittel and Boetius, 2009). In the literature, the name ANME-2d had been introduced once before for the related GOM Arc I sequence cluster, but the name was replaced by GOM Arc I because of missing evidence for anaerobic methane oxidation potential (Lloyd *et al.*, 2006). Today, the two clusters are sometimes combined again into a larger ANME–2d cluster, despite the missing evidence for methanotrophy in the GOM Arc I group, and this larger ANME-2d cluster is meanwhile divided into three subgroups based on 16S rRNA gene sequences (Welte *et al.*, 2016).

The ANME-1 and ANME-2 groups show a wide distribution in diverse marine environments, while ANME-3 has been predominantly detected in submarine mud volcanoes and marine methane seeps, indicating ecological niche separation (Knittel and Boetius, 2009; Cui *et al.*, 2015; Timmers *et al.*, 2017). The ANME organisms are frequently found in association with sulfate-reducing *Deltaproteobacteria*, ANME-1 and -2 with members of the *Desulfosarcina-Desulfococcus* group and ANME-3 preferentially with *Desulfobulbus* (Knittel and Boetius, 2009; Cui *et al.*, 2015). While the ANME organisms are performing methane oxidation to carbon dioxide, the sulfate reducers are responsible for sulfate reduction. Therefore, reducing equivalents are channelled between the two partners (McGlynn *et al.*, 2015; Wegener *et al.*, 2015). However, sometimes ANME groups have also been found in association with other bacterial taxa, and an association between sulfate reducers and ANME-1 is not consistently observed (Knittel and Boetius, 2009; Timmers *et al.*, 2017). This suggests that some ANME groups may perform the complete process alone, as proposed for ANME-2 organisms (Milucka *et al.*, 2012). Alternatively, they may couple methane oxidation to other not yet known reduction processes, either in association with a bacterial partner or possibly even alone. Moreover, it has been discussed that ANME-1 and ANME-2 organisms can perform methanogenesis instead of methane oxidation (House *et al.*, 2009; Bertram *et al.*, 2013). This metabolic versatility is the result of a genetic makeup that allows reverse methanogenesis for methane oxidation (Hallam *et al.*, 2004; Scheller *et al.*, 2010), but which can obviously still operate in the direction known from methanogenic archaea. Thus, the presence of ANME organisms can be linked to both, methane oxidation as well as methane production activity. The relative importance of a methanogenic activity needs to be studied in more detail in the future.

Evidence for anaerobic methane oxidation coupled to sulfate reduction in terrestrial ecosystems and soils

In continental ecosystems, anaerobic methane oxidation coupled with sulfate reduction is mostly considered to be of minor relevance, because sulfate concentrations are usually much lower than in marine ecosystems, rendering this process thermodynamically unfavourable (Smemo and Yavitt, 2011). However, internal redox-cycling of sulfur compounds, e.g. due to fluctuating water levels, might support this type of methane oxidation even in continental ecosystems at lower sulfate concentrations. Indirect evidence from mass balance approaches and measurements of methane oxidation rates in combination with sulfate reduction rates points to the existence of this process, e.g. in natural wetlands, rice paddies and groundwater at a landfill leachate plume (Murase and Kimura, 1994; Grossman *et al.*, 2002; Segarra *et al.*, 2015). However, final proof for a coupling of anaerobic methane oxidation with sulfate reduction is not given in most of these studies and oxidation rates were often considered to be quantitatively unimportant. To further validate the existence of this process in continental ecosystems, the presence of anaerobic methane-oxidizing microorganisms and their active involvement in the methane oxidation process need to be carefully proven (Timmers *et al.*, 2016).

The detection of anaerobic methanotrophic archaea of the clusters ANME-1, -2 and -3 was initially limited to anoxic, methane-rich, sulfate-containing marine sediments (Knittel *et al.*, 2005). In the meantime, 16S rRNA gene sequences of ANME-1 and -2, especially those of the ANME-1a and -2a sub-clusters, have repeatedly been detected in some specific continental environments such as freshwater subsurfaces and methane seeps, oilfield production waters and mud volcanoes (e.g. Knittel and Boetius, 2009; Niederberger *et al.*, 2010; Chang *et al.*, 2012). Moreover, their presence has been shown in the terrestrial subsurface, in soils from natural gas fields and a eutrophic freshwater lake (Eller *et al.*, 2005; Fry *et al.*, 2009; Miyashita *et al.*, 2009). As presence does not necessarily imply activity, anaerobic methane oxidation was demonstrated to be an active process in freshwater sediment samples based on 16S rRNA recovery and methane oxidation rate measurements, which were shown to be stimulated by sulfate amendments (Takeuchi *et al.*, 2011; Timmers *et al.*, 2016). The strongest evidence for a coupling of anaerobic methane oxidation with sulfate reduction was provided in the study by Timmers *et al.* (2016), who evaluated methane oxidation and sulfate reduction activity and detected ANME-2a/b sequences along with sequences of sulfate reducers in freshwater sediment samples. In contrast, no sulfate reducers were identified in the studies of Takeuchi *et al.* (2011), where ANME 1 sequences were found in the freshwater subsurface, and of Chang *et al.* (2012), who detected ANME-1a and -2a in a mud volcano and proposed a coupling to metal reduction rather than to sulfate. A detection of (active) ANME organisms in soils from natural wetlands or rice paddies remained usually unsuccessful (Miyashita *et al.*, 2009), indicating that the occurrence of ANME organisms in continental environments is largely limited to some freshwater systems and particular habitats such as mud volcanoes.

Anaerobic methane oxidation coupled with denitrification

The coupling of anaerobic methane oxidation with denitrification was first detected in an enrichment culture obtained from an anoxic freshwater sediment rich in nitrate (Raghoebarsing *et al.*, 2006). The microorganisms being responsible for this process were identified as bacteria of the candidate phylum NC10 and referred to as '*Ca.* Methylomirabilis oxyfera' (Ettwig *et al.*, 2010). These so-called 'NC10 bacteria' couple the oxidation of methane with nitrite reduction to dinitrogen.

A few years after the discovery of '*Ca.* Methylomirabilis oxyfera' it turned out that anoxic incubations with methane and nitrate (and nitrite or ammonium in addition) for the enrichment of denitrifying anaerobic methanotrophic bacteria support the establishment of another anaerobic methanotroph, the archaeal '*Ca.* Methanoperedens nitroreducens', representing the ANME-2d group (Haroon *et al.*, 2013). This organism is a member of the order *Methanosarcinales*, class '*Candidatus* Methanoperedenaceae'. It couples methane oxidation with nitrate reduction to nitrite. Depending on the nitrogen sources that are provided, '*Ca.* Methanoperedens nitroreducens' can form syntrophic associations either with '*Ca.* Methanoperedens nitroreducens' or with anaerobic ammonium-oxidizing (anammox) bacteria (Raghoebarsing *et al.*, 2006; Haroon *et al.*, 2013; Arshad *et al.*, 2015; Vaksmaa *et al.*, 2017a; Gambelli *et al.*, 2018). It is assumed that '*Ca.* Methylomirabilis oxyfera' or the anammox bacteria support the growth of '*Ca.* Methanoperedens nitroreducens' by eliminating nitrite, which is toxic at high concentrations (Welte *et al.*, 2016).

Physiology of '*Ca.* Methylomirabilis oxyfera' and '*Ca.* Methanoperedens nitroreducens'

Physiological properties of '*Ca.* Methanoperedens nitroreducens' were derived from reconstructed genome information, which is meanwhile available for a couple of enrichment cultures (Haroon *et al.*, 2013; Arshad *et al.*, 2015; Berger *et al.*, 2017; Vaksmaa *et al.*, 2017a). This indicates the coupling of nitrate reduction to nitrite with reverse methanogenesis. Nitrate reduction is performed by a membrane-bound nitrate reductase that appears to be of bacterial origin, possibly obtained via horizontal gene transfer. A membrane-bound nitrite reductase may be involved in the conversion of nitrite to ammonium, which may contribute to the elimination of nitrite.

Similarly as for '*Ca.* Methanoperedens nitroreducens', metagenomic sequencing helped to get insight into the metabolism of '*Ca.* Methylomirabilis oxyfera'. These organisms have established a completely different mechanism to couple methane oxidation with nitrite reduction to dinitrogen. A methane oxidation pathway as known from aerobic methanotrophic bacteria is present, while reverse methanogenesis does not play a role. The aerobic methane oxidation pathway includes methane oxidation via a particulate methane monooxygenase, which demands oxygen. It is assumed that this oxygen is derived from an intra-aerobic pathway, in which dinitrogen and oxygen are obtained by a dismutation reaction of two molecules of nitric oxide (Ettwig *et al.*, 2010, 2012). Oxygen that is not used for methane oxidation may be reduced by a terminal oxidase, allowing to gain additional energy by oxygen respiration (Wu *et al.*, 2011). Despite the need for oxygen, '*Ca.* Methylomirabilis oxyfera' is an anaerobic organism. In the presence of $\geq 2\%$ oxygen, methane oxidation and nitrite reduction activity decreased substantially and the cells encountered oxidative stress (Luesken *et al.*, 2012). However, it remains currently unclear whether they profit from external oxygen when available in trace amounts.

Diversity and environmental distribution of '*Ca.* Methylomirabilis oxyfera' and '*Ca.* Methanoperedens nitroreducens'

Besides '*Ca.* Methylomirabilis oxyfera', a second species of this candidate genus has been proposed in the meantime, '*Candidatus* Methylomirabilis sinica', enriched from paddy soil (He *et al.*, 2016). Several further reports describe the enrichment of bacteria coupling anaerobic methane oxidation with denitrification from diverse environmental samples, including paddy soils, peatland, river sediments, coastal sediments, wastewater and bioreactor sludges (Zhu *et al.*, 2012; He *et al.*, 2015; Bhattacharjee *et al.*, 2016; Chen *et al.*, 2016; Welte *et al.*, 2016; Vaksmaa *et al.*, 2017a). Furthermore, bacteria of the NC10 phylum were identified in several cultivation-independent studies (Shen *et al.*, 2015d; Chen *et al.*, 2016; Welte *et al.*, 2016). Such studies use the 16S rRNA or the *pmoA* gene as marker. Data interpretation of 16S rRNA gene marker based results has to be done with care, because the NC10 phylum includes four different 16S rRNA gene sequence clusters, but it is not yet clear whether all clusters represent anaerobic methanotrophic bacteria (Welte *et al.*, 2016). The use of *pmoA* as marker demands specific *pmoA* primers due to the distinct clustering of their *pmoA* sequences in phylogenetic trees (Fig. 2.4), and several different *pmoA* primer sets have been developed (Shen *et al.*, 2015d; Chen *et al.*, 2016). Analyses based on the 16S rRNA and *pmoA* marker genes indicate that methanotrophs of the NC10 phylum are present in diverse environments. The detection was successful in freshwater lake and river sediments, in different wetlands including peatlands and swamps, in paddy soils and a few times in other agricultural soils, in wastewater systems, in some marine and coastal sediments, and recently for the first time in the rumen fluid from goats (Chen *et al.*, 2016; Shen *et al.*, 2016b; Welte *et al.*, 2016; L. Liu *et al.*, 2017).

This broad occurrence is in line with predictions stating that anaerobic methane oxidation coupled to denitrification should be relevant in ecosystems with methane supply from anoxic compartments and availability of oxidized nitrogenous compounds (Thauer and Shima, 2008). Such conditions are found, for example, in wastewater treatment systems or near the oxic/anoxic interphase in paddy soils or freshwater ecosystems, especially when located in agricultural landscapes with high nitrogen input. Indeed, '*Ca.* Methylomirabilis oxyfera' can be detected at such oxic/anoxic interfaces in various wetlands (Raghoebarsing *et al.*, 2006; Zhu *et al.*, 2015). However, several studies report the predominant occurrence in deeper layers of wetlands and paddy soils (Zhu *et al.*, 2012; Hu *et al.*, 2014; Shen *et al.*, 2015b,c), where anoxic conditions are more stable. Moreover, NC10 methanotrophs were more abundant in freshwater samples such as reservoir and pond sediments with rather stable anoxic conditions than in wetland sediments or paddy soils (Shen *et al.*, 2016b). However, if '*Ca.* Methylomirabilis oxyfera' colonizes indeed preferably habitats that provide stable anoxic conditions, the detection in upland soils is surprising, as reported for tropical forest samples in 5–20 cm depth (Meng *et al.*, 2016) and agricultural soil samples in 50–60 cm depth (Hu and Ma, 2016; Shen *et al.*, 2016a). However, a detection in upland soils was not consistently observed (Zhu *et al.*, 2015). The relevance of '*Ca.* Methylomirabilis oxyfera' in such soils remains currently completely unclear.

Similarly as '*Ca.* Methylomirabilis oxyfera', '*Ca.* Methanoperedens nitroreducens', shows a broad occurrence. It has been detected in lake and river sediments, aquifers, paddy soils, peatlands, mud volcanoes, sewage treatment plants and, to limited extent, in the marine and brackish environment (Ding *et al.*, 2015; Welte *et al.*, 2016; Narrowe *et al.*, 2017). The cultivation-independent detection is done using group-specific 16S rRNA gene primers or *mcrA* gene primers (Ding *et al.*, 2015; Vaksmaa *et al.*, 2017b; Xu *et al.*, 2018). The obtained 16S rRNA gene sequences can be classified into three different subgroups, which are not all represented by enrichment cultures, so that the methane oxidation potential of some sequence clusters remains currently unclear (Welte *et al.*, 2016).

As presence does not necessarily imply activity, studies are needed to assess the activity of these methanotrophs in the different ecosystems. This has so far only been done in a few studies, either at the transcript (Padilla *et al.*, 2016) or protein level (Hanson and Madsen, 2015) in a marine and freshwater environment, respectively, demonstrating metabolic activity of '*Ca.* Methylomirabilis oxyfera'. Some other studies demonstrated by isotope experiments that an anoxic conversion of methane in the presence of nitrite or nitrate as electron acceptor is occurring in wetlands, lake sediment and a rice paddy soil. These analyses were combined with molecular approaches, confirming the presence of '*Ca.* Methylomirabilis oxyfera' (Deutzmann and Schink, 2011; Hu *et al.*, 2014; Shen *et al.*, 2014, 2015c) and '*Ca.* Methanoperedens nitroreducens' (Vaksmaa *et al.*, 2016) in the corresponding samples. In the paddy soil study the determined methane uptake rates were substantially higher than reported in an earlier study, indicating that anaerobic methane oxidation may play a significant role in these soils (Vaksmaa *et al.*, 2016). Differences concerning the activity of these methanotrophs were also reported for natural wetlands, i.e. an activity of anaerobic methanotrophs could not always be proven (Tveit *et al.*, 2013). In general, the potential contribution of anaerobic denitrifying methanotrophs in carbon and nitrogen cycling needs to be studied in more detail in the diverse environments to assess their relevance more precisely.

Alternative electron acceptors for anaerobic methane oxidation

Besides sulfate, nitrate and nitrite, further electron acceptors have been discussed to be of relevance in combination with anaerobic methane oxidation. A coupling with iron reduction has been demonstrated in '*Candidatus* Methanoperedens ferrireducens' (Cai *et al.*, 2018). This archaeon was enriched in a bioreactor fed with methane and ferrihydrite, which was set up with material from a freshwater reservoir. Metagenomic analysis predicts that methane oxidation occurs via reverse methanogenesis and iron reduction by multi-haem c-type cytochromes. The exact mechanism of the electron transfer to iron remains currently unclear. Earlier, a coupling of methane oxidation to iron reduction was already reported for an enrichment culture of '*Candidatus* Methanoperedens nitroreducens MPEBLZ', obtained from a freshwater sample and cultured in a reactor with methane and nitrate as substrates (Ettwig *et al.*, 2016). This culture was able to couple methane oxidation with nitrate or iron, but the rate of iron-based oxidation was 10-fold lower than in '*Ca.* Methanoperedens ferrireducens' (Cai *et al.*, 2018). Besides the use of nitrate and iron, '*Ca.* Methanoperedens nitroreducens MPEBLZ' was shown to reduce manganese. Obviously, '*Ca.* Methanoperedens' shows versatility concerning the use of electron acceptors, but may have substrate preferences. It remains currently unclear how commonly iron-reduction occurs within this candidate genus and how versatile the individual members are indeed.

The coupling of anaerobic methane oxidation with iron reduction has been demonstrated in incubation experiments with deep sea sediment material harbouring ANME-2c organisms (Scheller *et al.*, 2016) and by geochemical profiling or isotope tracer studies in environmental samples from lake water and sediment, marine sediments, paddy fields, a terrestrial mud volcano and a contaminated aquifer (Zandt *et al.*, 2018). The detection of the process and the presence of '*Ca.* Methanoperedens' species in such environments (Welte *et al.*, 2016; Narrowe *et al.*, 2017) suggests a high ecological relevance, especially in marine and freshwater ecosystems, where iron oxides are present in the sediments (Cai *et al.*, 2018), but more studies are needed that link the geochemical processes with microbiological data to demonstrate this relevance and identify the involved microorganisms.

In peatlands, methane oxidation coupled with iron reduction has also been discussed as an option because of low concentrations of available sulfate and nitrate, but could not yet be proven (Smemo and Yavitt, 2007, 2011; Gupta *et al.*, 2012). Alternatively, humic substances are considered to be of possible relevance, knowing that ANME organisms can transfer electrons to external acceptors (Scheller *et al.*, 2016) and that humic substances can act as electron acceptors (Scott *et al.*, 1998). Further evidence is provided by the observations that humic substances accumulate in peatlands and anaerobic methane oxidation was detected, but without strong evidence for a coupling with nitrate, sulfate or iron reduction (Smemo and Yavitt, 2007; Gupta *et al.*, 2012). Such a weak coupling with the known inorganic electron acceptors was also reported by Reed *et al.* (2017) for an eutrophic reservoir, along with the suggestion that organic acids, which are major constituents of organic matter, may serve as electron acceptors during anaerobic methane oxidation in eutrophic lakes and reservoirs.

These findings indicate that the geochemical characteristics of an ecosystem will have a major impact on the overall rate of anaerobic methane oxidation and the dominant type of methane oxidation process taking place. This is exemplified by the observation that anaerobic methane oxidation activity was quantitatively more important in nutrient-rich (minerotrophic) fens than in nutrient-poor (ombrotrophic) bogs (Smemo and Yavitt, 2007; Gupta *et al.*, 2012). Furthermore, Segarra *et al.* (2015) observed different dependencies on electron acceptors in different wetland systems. It can thus be concluded that anaerobic methane oxidation is of relevance in diverse terrestrial ecosystems. The coupling of anaerobic methane oxidation with specific reduction processes and the identity of the involved microorganisms appears to be ecosystem specific and requests a good understanding of the geochemical processes in combination with the physiology of the microorganisms inhabiting the respective ecosystems.

Conclusions and future perspectives

Microbiological processes leading to methane production and consumption are major drivers of the global methane cycle. For a long time, methane production was attributed to the activity of a few orders of methanogenic *Euryarchaeota*, while methane oxidation activity was ascribed to specific families of methanotrophic *Alpha-* and *Gammaproteobacteria*. Research of the last two decades has substantially extended the list of players in both groups. As described in this review, a couple of new orders of methanogens within the *Euryarchaeota* were discovered during the last years and metagenomic analyses suggest that methanogens may even exist beyond the phylum *Euryarchaeota*. Likewise, the methanotrophic lifestyle appears to be more widespread among microorganisms within the *Alpha-* and *Gammaproteobacteria* but also among other taxa, exemplified by the discovery of methanotrophy in the phylum *Verrucomicrobia*. Work of the last years also indicates that anaerobic methane oxidation plays an important role not only in marine ecosystems, but also in continental environments including natural as well as anthropogenic wetlands, which are known to represent major sources of atmospheric methane. Distinct groups of methanotrophs such as '*Ca.* Methylomirabilis' and '*Ca.* Methanoperedens' appear to be the major types responsible for this process in terrestrial ecosystems. In order to better understand the methane sink or source capacity of an ecosystem and the variation of emissions or uptake in space and over time, the activities of all groups involved in methane cycling need to be considered. Thus, there is a clear need for studies that integrate the activities of all players.

The spatial distribution, abundance and activity of the individual players is controlled by diverse environmental factors. Recent findings indicate that more controls need to be considered. Several of the recently discovered groups of methanogens and methanotrophs but also some of the well-known players show a broader metabolic versatility than previously thought. Thus their presence alone or their general metabolic activity does not necessarily allow direct conclusions about their involvement in methane cycling under given conditions. The methane oxidation activity of facultative or mixotrophic methanotrophs may for example be suppressed or modulated dependent on the availability of alternative carbon and energy sources such as acetate or dihydrogen. Similarly, the activity of anaerobic methanotrophs is largely dependent on the availability of suitable electron acceptors. This in turn can depend on the activities of other groups of organisms. Nitrite availability for '*Ca.* Methylomirabilis' will for example depend on the organisms that produce nitrite, e.g. aerobic nitrifiers in nearby oxic habitats or '*Ca.* Methanoperedens' in the same anoxic habitat, as well as on the presence and activity of potential competitors such as anammox bacteria. A good understanding of such dependencies helps to define the ecological niche of these bacteria in natural as well as in artificial ecosystems such as wastewater treatment systems, where they could be introduced to reduce methane emissions (van Kessel *et al.*, 2018).

This review also demonstrates that the control of methanotrophic activity by specific environmental factors has to be assessed at a finer scale, at least for some factors. Exemplarily, the dependency on oxygen is highlighted, due to the fact that recent studies revealed interesting new insights. On the one hand, several methanogenic archaea appear to be less sensitive to oxygen than previously thought and methane production may even occur under apparently oxic conditions, on the other hand, some aerobic methanotrophic bacteria seem to be less dependent on oxygen and may remain active under apparently anoxic conditions. Considering that anaerobic methane oxidation is a process that occurs in terrestrial ecosystems such as wetlands, the general and rather simple assumption that oxic conditions support methane oxidation activity and anoxic conditions methane production appears to be too simple. Such general assumptions are still used in many models (Gauthier *et al.*, 2015). Besides an improvement of current models, the effect of oxygen on the distribution and activity of aerobic methanotrophs needs to be studied in more detail in soils. It has so far mostly been analysed in well-stratified ecosystems, especially lakes, but may be of equal relevance for the control of methanotrophs and methanogens in wetland soils, which show a higher variation in oxygen availability in space and over time.

As evident from this review, work of the last years has substantially extended the known diversity of methanotrophic and methanogenic microorganisms as well as knowledge about their metabolic capabilities. The discovery of new putative methanogenic and methanotrophic organisms in taxonomic groups not yet known to include such organisms was mostly the result of genome sequencing efforts. Physiological capabilities of uncultured organisms can now be derived quite easily from such genome reconstructions. Thousands of genomes are meanwhile available (e.g. Whitman *et al.*, 2015; Anantharaman *et al.*, 2016; Parks *et al.*, 2017) and many more datasets can be expected to be generated in the near future, either by *de novo* sequencing of isolates, single-cell based approaches for the analysis of uncultivated bacteria or assembly of genomes from metagenomic data. The analysis of these data will provide further insight into the metabolic versatility of known methane cycling microorganisms and very likely lead to the identification of further groups of microorganisms harbouring methane cycling potential in phylogenetic groups that are not yet known to include methanogens or methanotrophs. It will be a major task to validate the methane production or oxidation capabilities of these organisms. Some of these microorganisms may have a broader metabolic versatility compared with the canonical methanotrophs and methanogens, so that their activities and therewith their contribution to methane cycling in an ecosystem needs to be assessed carefully, e.g. by applying multi-omics approaches including metatransriptomics, -proteomics and possibly -metabolomics. Such studies should be complemented by laboratory analyses of microcosms, enrichment cultures or isolates under controlled conditions to identify and to understand the regulatory mechanisms that determine methanogenic or methanotrophic activity rates of the different players. Genomic information available for uncultivated microorganisms may help to enrich and isolate these organisms, making them accessible for in-depth studies.

References

Ahn, J.H., Choi, M.Y., Kim, B.Y., Lee, J.S., Song, J., Kim, G.Y., and Weon, H.Y. (2014). Effects of water-saving irrigation on emissions of greenhouse gases and prokaryotic communities in rice paddy soil. Microb. Ecol. *68*, 271–283. https://doi.org/10.1007/s00248-014-0371-z

Anantharaman, K., Brown, C.T., Hug, L.A., Sharon, I., Castelle, C.J., Probst, A.J., Thomas, B.C., Singh, A., Wilkins, M.J., Karaoz, U., *et al.* (2016). Thousands of microbial genomes shed light on interconnected biogeochemical processes in an aquifer system. Nat. Commun. *7*, 13219. https://doi.org/10.1038/ncomms13219

Angel, R., Matthies, D., and Conrad, R. (2011). Activation of methanogenesis in arid biological soil crusts despite the presence of oxygen. PLOS ONE *6*, e20453. https://doi.org/10.1371/journal.pone.0020453

Angel, R., Claus, P., and Conrad, R. (2012). Methanogenic archaea are globally ubiquitous in aerated soils and become active under wet anoxic conditions. ISME J. *6*, 847–862. https://doi.org/10.1038/ismej.2011.141

Angle, J.C., Morin, T.H., Solden, L.M., Narrowe, A.B., Smith, G.J., Borton, M.A., Rey-Sanchez, C., Daly, R.A., Mirfenderesgi, G., Hoyt, D.W., *et al.* (2017). Methanogenesis in oxygenated soils is a substantial fraction of wetland methane emissions. Nat. Commun. *8*, 1567. https://doi.org/10.1038/s41467-017-01753-4

Aronson, E.L., Allison, S.D., and Helliker, B.R. (2013). Environmental impacts on the diversity of methane-cycling microbes and their resultant function. Front. Microbiol. *4*, 225. https://doi.org/10.3389/fmicb.2013.00225

Arshad, A., Speth, D.R., de Graaf, R.M., Op den Camp, H.J., Jetten, M.S., and Welte, C.U. (2015). A metagenomics-based metabolic model of nitrate-dependent anaerobic oxidation of methane by *Methanoperedens*-like archaea. Front. Microbiol. *6*, 1423. https://doi.org/10.3389/fmicb.2015.01423

Aschenbach, K., Conrad, R., Reháková, K., Doležal, J., Janatková, K., and Angel, R. (2013). Methanogens at the top of the world: occurrence and potential activity of methanogens in newly deglaciated soils in high-altitude cold deserts in the Western Himalayas. Front. Microbiol. *4*, 359. https://doi.org/10.3389/fmicb.2013.00359

Belova, S.E., Baani, M., Suzina, N.E., Bodelier, P.L., Liesack, W., and Dedysh, S.N. (2011). Acetate utilization as a survival strategy of peat-inhabiting *Methylocystis* spp. Environ. Microbiol. Rep. *3*, 36–46. https://doi.org/10.1111/j.1758-2229.2010.00180.x

Bender, M., and Conrad, R. (1992). Kinetics of CH_4 oxidation in oxic soils exposed to ambient air or high CH_4 mixing ratios. FEMS Microbiol. Ecol. *101*, 261–270. https://doi.org/10.1111/j.1574-6968.1992.tb05783.x

Berger, S., Frank, J., Dalcin Martins, P., Jetten, M.S.M., and Welte, C.U. (2017). High-quality draft genome sequence of '*Candidatus* Methanoperedens sp.' Strain BLZ2, a nitrate-reducing anaerobic methane-oxidizing archaeon enriched in an anoxic bioreactor. Genome Announc. *5*, e01159–17.

Bertram, S., Blumenberg, M., Michaelis, W., Siegert, M., Krüger, M., and Seifert, R. (2013). Methanogenic capabilities of ANME-archaea deduced from ^{13}C-labelling approaches. Environ. Microbiol. *15*, 2384–2393. https://doi.org/10.1111/1462-2920.12112

Bhattacharjee, A.S., Motlagh, A.M., Jetten, M.S.M., and Goel, R. (2016). Methane dependent denitrification-from ecosystem to laboratory-scale enrichment for engineering applications. Water Res. *99*, 244–252.

Biderre-Petit, C., Jézéquel, D., Dugat-Bony, E., Lopes, F., Kuever, J., Borrel, G., Viollier, E., Fonty, G., and Peyret, P. (2011). Identification of microbial communities involved in the methane cycle of a freshwater meromictic lake. FEMS Microbiol. Ecol. *77*, 533–545. https://doi.org/10.1111/j.1574-6941.2011.01134.x

Blees, J., Niemann, H., Wenk, C.B., Zopfi, J., Schubert, C.J., Kirf, M.K., Veronesi, M.L., Hitz, C., and Lehmann, M.F. (2014). Micro-aerobic bacterial methane oxidation in the chemocline and anoxic water column of deep south-Alpine Lake Lugano (Switzerland). Limnol. Oceanogr. *59*, 311–324. https://doi.org/10.4319/lo.2014.59.2.0311

Bogard, M.J., del Giorgio, P.A., Boutet, L., Chaves, M.C., Prairie, Y.T., Merante, A., and Derry, A.M. (2014). Oxic water column methanogenesis as a major component of aquatic CH_4 fluxes. Nat. Commun. *5*, 5350. https://doi.org/10.1038/ncomms6350

Borrel, G., Harriss, H.M.B., Parisot, N., Gaci, N., Tottey, W., Mihajlovski, A., Deane, J., Gribaldo, S., Bardot, O., Peyretaillade, E., *et al.* (2013a). Genome sequence of '*Candidatus* Methanomassiliicoccus intestinalis' Issoire-Mx1, a third *Thermoplasmatales*-related methanogenic archaeon from human feces. Genome Announc. *1*, e00453-00413. https://doi.org/10.1128/genomeA.00453-13

Borrel, G., O'Toole, P.W., Harris, H.M., Peyret, P., Brugère, J.F., and Gribaldo, S. (2013b). Phylogenomic data support a seventh order of methylotrophic methanogens and provide insights into the evolution of methanogenesis. Genome Biol. Evol. *5*, 1769–1780. https://doi.org/10.1093/gbe/evt128

Borrel, G., McCann, A., Deane, J., Neto, M.C., Lynch, D.B., Brugère, J.F., and O'Toole, P.W. (2017). Genomics and metagenomics of trimethylamine-utilizing *Archaea* in the human gut microbiome. ISME J. *11*, 2059–2074. https://doi.org/10.1038/ismej.2017.72

Cai, C., Leu, A.O., Xie, G.J., Guo, J., Feng, Y., Zhao, J.X., Tyson, G.W., Yuan, Z., and Hu, S. (2018). A methanotrophic archaeon couples anaerobic oxidation of methane to Fe(III) reduction. ISME J. *12*, 1929–1939. https://doi.org/10.1038/s41396-018-0109-x

Caldwell, S.L., Laidler, J.R., Brewer, E.A., Eberly, J.O., Sandborgh, S.C., and Colwell, F.S. (2008). Anaerobic oxidation of methane: mechanisms, bioenergetics, and the ecology of associated microorganisms. Environ. Sci. Technol. *42*, 6791–6799.

Campbell, M.A., Nyerges, G., Kozlowski, J.A., Poret-Peterson, A.T., Stein, L.Y., and Klotz, M.G. (2011). Model of the molecular basis for hydroxylamine oxidation and nitrous oxide production in methanotrophic bacteria. FEMS Microbiol. Lett. *322*, 82–89. https://doi.org/10.1111/j.1574-6968.2011.02340.x

Carere, C.R., Hards, K., Houghton, K.M., Power, J.F., McDonald, B., Collet, C., Gapes, D.J., Sparling, R., Boyd, E.S., Cook, G.M., *et al.* (2017). Mixotrophy drives niche expansion of verrucomicrobial methanotrophs. ISME J. *11*, 2599–2610. https://doi.org/10.1038/ismej.2017.112

Chang, Y.H., Cheng, T.W., Lai, W.J., Tsai, W.Y., Sun, C.H., Lin, L.H., and Wang, P.L. (2012). Microbial methane cycling in a terrestrial mud volcano in eastern Taiwan. Environ. Microbiol. *14*, 895–908. https://doi.org/10.1111/j.1462-2920.2011.02658.x

Chen, J., Dick, R., Lin, J.G., and Gu, J.D. (2016). Current advances in molecular methods for detection of nitrite-dependent anaerobic methane oxidizing bacteria in natural environments. Appl. Microbiol. Biotechnol. *100*, 9845–9860. https://doi.org/10.1007/s00253-016-7853-5

Chen, Y.P., and Yoch, D.C. (1987). Regulation of two nickel-requiring (inducible and constitutive) hydrogenases and their coupling to nitrogenase in *Methylosinus trichosporium* OB3b. J. Bacteriol. *169*, 4778–4783.

Chistoserdova, L. (2015). Methylotrophs in natural habitats: current insights through metagenomics. Appl. Microbiol. Biotechnol. *99*, 5763–5779. https://doi.org/10.1007/s00253-015-6713-z

Chojnacka, A., Szczęsny, P., Błaszczyk, M.K., Zielenkiewicz, U., Detman, A., Salamon, A., and Sikora, A. (2015). Noteworthy facts about a methane-producing microbial community processing acidic effluent from sugar beet molasses fermentation. PLOS ONE *10*, e0128008. https://doi.org/10.1371/journal.pone.0128008

Ciais, P., Sabine, C., Bala, G., Bopp, L., Brovkin, B., Canadell, J., Chhabra, A., DeFries, R., Galloway, J., Heimann, M., *et al.* (2013). Carbon and other biogeochemical cycles. In Climate Change 2013: The Physical Science Basis Contribution of Working Group I to the Fifth Assessment Report in the Intergovernmental Panel on Climate Change, T.F. Stocker, D. Qin, G.-K. Plattner, M. Tignor, S.K. Allen, J. Boschung, A. Nauels, Y. Xia, V. Bex, and P.M. Midgley, eds. (Cambridge University Press, Cambridge, UK, and New York, NY). https://doi.org/10.1017/CBO9781107415324.015

Conrad, R. (1995). Soil microbial processes involved in production and consumption of atmospheric trace gases. Adv. Microb. Ecol. *14*, 207–250. https://doi.org/10.1007/978-1-4684-7724-5_5

Conrad, R. (1996). Soil microorganisms as controllers of atmospheric trace gases (H_2, CO, CH_4, OCS, N_2O, and NO). Microbiol. Rev. *60*, 609–640.

Conrad, R. (1999). Contribution of hydrogen to methane production and control of hydrogen concentrations in methanogenic soils and sediments. FEMS Microbiol. Ecol. *28*, 193–202. https://doi.org/10.1111/j.1574-6941.1999.tb00575.x

Conrad, R. (2005). Quantification of methanogenic pathways using stable carbon isotopic signatures: a review and a proposal. Org. Geochem. *36*, 739–752. https://doi.org/10.1016/j.orggeochem.2004.09.006

Conrad, R. (2007). Microbial ecology of methanogens and methanotrophs. Adv. Agron. *96*, 1–63. https://doi.org/10.1016/S0065-2113(07)96005-8

Conrad, R. (2009). The global methane cycle: recent advances in understanding the microbial processes involved. Environ. Microbiol. Rep. *1*, 285–292. https://doi.org/10.1111/j.1758-2229.2009.00038.x

Conrad, R., Erkel, C., and Liesack, W. (2006). Rice Cluster I methanogens, an important group of *Archaea* producing greenhouse gas in soil. Curr. Opin. Biotechnol. *17*, 262–267.

Costa, K.C., and Leigh, J.A. (2014). Metabolic versatility in methanogens. Curr. Opin. Biotechnol. *29*, 70–75. https://doi.org/10.1016/j.copbio.2014.02.012

Crevecoeur, S., Vincent, W.F., Comte, J., and Lovejoy, C. (2015). Bacterial community structure across environmental gradients in permafrost thaw ponds: methanotroph-rich ecosystems. Front. Microbiol. *6*, 192. https://doi.org/10.3389/fmicb.2015.00192

Cui, M., Ma, A., Qi, H., Zhuang, X., and Zhuang, G. (2015). Anaerobic oxidation of methane: an 'active' microbial process. Microbiologyopen *4*, 1–11. https://doi.org/10.1002/mbo3.232

Daims, H., Lebedeva, E.V., Pjevac, P., Han, P., Herbold, C., Albertsen, M., Jehmlich, N., Palatinszky, M., Vierheilig, J., Bulaev, A., *et al.* (2015). Complete nitrification by *Nitrospira* bacteria. Nature *528*, 504–509. https://doi.org/10.1038/nature16461

Dalal, R.C., and Allen, D.E. (2008). Greenhouse gas fluxes from natural ecosystems. Austr. J. Bot. *56*, 369–407. https://doi.org/10.1071/bt07128

Dalal, R.C., Allen, D.E., Livesley, S.J., and Richards, G. (2008). Magnitude and biophysical regulators of methane emission and consumption in the Australian agricultural, forest, and submerged landscapes: a review. Plant Soil *309*, 43–76. https://doi.org/10.1007/s11104-007-9446-7

Dam, B., Dam, S., Blom, J., and Liesack, W. (2013). Genome analysis coupled with physiological studies reveals a diverse nitrogen metabolism in *Methylocystis* sp. strain SC2. PLOS ONE *8*, e74767. https://doi.org/10.1371/journal.pone.0074767

Danilova, O.V., Belova, S.E., Gagarinova, I.V., and Dedysh, S.N. (2016a). Microbial community composition and methanotroph diversity of a subarctic wetland in Russia. Mikrobiologiia *85*, 545–554.

Danilova, O.V., Suzina, N.E., Van De Kamp, J., Svenning, M.M., Bodrossy, L., and Dedysh, S.N. (2016b). A new cell morphotype among methane oxidizers: a spiral-shaped obligately microaerophilic methanotroph from northern low-oxygen environments. ISME J. *10*, 2734–2743. https://doi.org/10.1038/ismej.2016.48

Dedysh, S.N., and Knief, C. (2018). Diversity and phylogeny of described aerobic methanotrophs. In Methane Biocatalysis: Paving the Way to Sustainability, M.G. Kalyuzhanya, and X.-H. Xing, eds. (Springer, Cham, Switzerland), pp. 17–42. https://doi.org/10.1007/978-3-319-74866-5_2

Deng, Y., Cui, X., and Dumont, M.G. (2016). Identification of active aerobic methanotrophs in plateau wetlands using DNA stable isotope probing. FEMS Microbiol. Lett. *363*, fnw168. https://doi.org/10.1093/femsle/fnw168

Deutzmann, J.S., and Schink, B. (2011). Anaerobic oxidation of methane in sediments of Lake Constance, an oligotrophic freshwater lake. Appl. Environ. Microbiol. *77*, 4429–4436. https://doi.org/10.1128/AEM.00340-11

Deutzmann, J.S., Hoppert, M., and Schink, B. (2014). Characterization and phylogeny of a novel methanotroph, *Methyloglobulus morosus* gen. nov., spec. nov. Syst. Appl. Microbiol. *37*, 165–169. https://doi.org/10.1016/j.syapm.2014.02.001

Ding, J., Ding, Z.W., Fu, L., Lu, Y.Z., Cheng, S.H., and Zeng, R.J. (2015). New primers for detecting and quantifying denitrifying anaerobic methane oxidation archaea in different ecological niches. Appl. Microbiol. Biotechnol. *99*, 9805–9812. https://doi.org/10.1007/s00253-015-6893-6896.

Dlugokencky, E.J. (2018). NOAA/ESRL L (www.esrl.noaa.gov/gmd/ccgg/trends_ch4/).

Dlugokencky, E.J., Nisbet, E.G., Fisher, R., and Lowry, D. (2011). Global atmospheric methane: budget, changes and dangers. Philos. Trans. A Math. Phys. Eng. Sci. *369*, 2058–2072. https://doi.org/10.1098/rsta.2010.0341

Donis, D., Flury, S., Stöckli, A., Spangenberg, J.E., Vachon, D., and McGinnis, D.F. (2017). Full-scale evaluation of methane production under oxic conditions in a mesotrophic lake. Nat. Commun. *8*, 1661. https://doi.org/10.1038/s41467-017-01648-4

Dridi, B., Fardeau, M.L., Ollivier, B., Raoult, D., and Drancourt, M. (2012). *Methanomassiliicoccus luminyensis* gen. nov., sp nov., a methanogenic archaeon isolated from human faeces. Int. J. Syst. Evol. Microbiol. *62*, 1902–1907. https://doi.org/10.1099/ijs.0.033712-0

Dumont, M.G., Lüke, C., Deng, Y.C., and Frenzel, P. (2014). Classification of *pmoA* amplicon pyrosequences using BLAST and the lowest common ancestor method in MEGAN. Front. Microbiol. *5*, 34. https://doi.org/10.3389/fmicb.2014.00034

Dunfield, P.F. (2007). The soil methane sink. In Greenhouse Gas Sinks, D.S. Reay, C.N. Hewitt, K.A. Smith, and J. Grace, eds. (CABI Wallingford, UK), pp. 152–170. https://doi.org/10.1079/9781845931896.0152

Eder, W., Schmidt, M., Koch, M., Garbe-Schönberg, D., and Huber, R. (2002). Prokaryotic phylogenetic diversity and corresponding geochemical data of the brine-seawater interface of the Shaban Deep, Red Sea. Environ. Microbiol. *4*, 758–763.

Edwards, C.R., Onstott, T.C., Miller, J.M., Wiggins, J.B., Wang, W., Lee, C.K., Cary, S.C., Pointing, S.B., and Lau, M.C.Y. (2017). Draft genome sequence of uncultured upland soil cluster *Gammaproteobacteria* gives molecular insights into high-affinity methanotrophy. Genome Announc. *5*, e00047-17.

Eller, G., Känel, L., and Krüger, M. (2005). Cooccurrence of aerobic and anaerobic methane oxidation in the water column of Lake Plusssee. Appl. Environ. Microbiol. *71*, 8925–8928.

Ettwig, K.F., Butler, M.K., Le Paslier, D., Pelletier, E., Mangenot, S., Kuypers, M.M., Schreiber, F., Dutilh, B.E., Zedelius, J., de Beer, D., *et al.* (2010). Nitrite-driven anaerobic methane oxidation by oxygenic bacteria. Nature *464*, 543–548. https://doi.org/10.1038/nature08883

Ettwig, K.F., Speth, D.R., Reimann, J., Wu, M.L., Jetten, M.S., and Keltjens, J.T. (2012). Bacterial oxygen production in the dark. Front. Microbiol. *3*, 273. https://doi.org/10.3389/fmicb.2012.00273

Ettwig, K.F., Zhu, B., Speth, D., Keltjens, J.T., Jetten, M.S.M., and Kartal, B. (2016). Archaea catalyze iron-dependent anaerobic oxidation of methane. Proc. Natl. Acad. Sci. U.S.A. *113*, 12792–12796. https://doi.org/10.1073/pnas.1609534113

Evans, P.N., Parks, D.H., Chadwick, G.L., Robbins, S.J., Orphan, V.J., Golding, S.D., and Tyson, G.W. (2015). Methane metabolism in the archaeal phylum *Bathyarchaeota* revealed by genome-centric metagenomics. Science *350*, 434–438. https://doi.org/10.1126/science.aac7745

Ferry, J.G. (2010). The chemical biology of methanogenesis. Planet Space Sci. *58*, 1775–1783. https://doi.org/10.1016/j.pss.2010.08.014

Fetzer, S., Bak, F., and Conrad, R. (1993). Sensitivity of methanogenic bacteria from paddy soil to oxygen and desiccation. FEMS Microbiol. Ecol. *12*, 107–115.

Frindte, K., Maarastawi, S.A., Lipski, A., Hamacher, J., and Knief, C. (2017). Characterization of the first rice paddy cluster I isolate, *Methyloterricola oryzae* gen. nov., sp. nov. and amended description of *Methylomagnum ishizawai*. Int. J. Syst. Evol. Microbiol. *67*, 4507–4514. https://doi.org/10.1099/ijsem.0.002319

Fry, J.C., Horsfield, B., Skyes, R., Cragg, B.A., Heywood, C., Kim, G.T., Mangelsdorf, K., Mildenhall, D.C., Rinna, J., Vieth, A., *et al.* (2009). Prokaryotic populations and activities in an interbedded coal deposit, including a previously deeply buried section (1.6 - 2.3 km) above ~ 150 Ma basement rock. Geomicrobiol. J. *26*, 163–178. https://doi.org/10.1080/01490450902724832

Gambelli, L., Guerrero-Cruz, S., Mesman, R.J., Cremers, G., Jetten, M.S.M., Op den Camp, H.J.M., Kartal, B., Lueke, C., and van Niftrik, L. (2018). Community composition and ultrastructure of a nitrate-dependent anaerobic methane-oxidizing enrichment culture. Appl. Environ. Microbiol. *84*, e02186–17.

Garcia, J.L., Patel, B.K., and Ollivier, B. (2000). Taxonomic, phylogenetic, and ecological diversity of methanogenic Archaea. Anaerobe *6*, 205–226.

Garrity, G.M., Labeda, D.P., and Oren, A. (2011). XIIth international (IUMS) congress of bacteriology and applied microbiology. Int. J. Syst. Evol. Microbiol. *61*, 2775–2780. https://doi.org/10.1099/ijs.0.037366-0

Gauthier, M., Bradley, R.L., and Simek, M. (2015). More evidence that anaerobic oxidation of methane is prevalent in soils: Is it time to upgrade our biogeochemical models? Soil Biol. Biochem. *80*, 167–174. https://doi.org/10.1016/j.soilbio.2014.10.009

Grossart, H.P., Frindte, K., Dziallas, C., Eckert, W., and Tang, K.W. (2011). Microbial methane production in oxygenated water column of an oligotrophic lake. Proc. Natl. Acad. Sci. U.S.A. *108*, 19657–19661. https://doi.org/10.1073/pnas.1110716108

Großkopf, R., Stubner, S., and Liesack, W. (1998). Novel euryarchaeotal lineages detected on rice roots and in the anoxic bulk soil of flooded rice microcosms. Appl. Environ. Microbiol. *64*, 4983–4989.

Grossman, E.L., Cifuentes, L.A., and Cozzarelli, I.M. (2002). Anaerobic methane oxidation in a landfill-leachate plume. Environ. Sci. Technol. *36*, 2436–2442.

Gu, Q., Chang, S.X., Wang, Z.P., Feng, J.C., Chen, Q.S., and Han, X.G. (2016). Microbial versus non-microbial methane releases from fresh soils at different temperatures. Geoderma *284*, 178–184. https://doi.org/10.1016/j.geoderma.2016.08.027

Gupta, V., Smemo, K.A., Yavitt, J.B., and Basiliko, N. (2012). Active methanotrophs in two contrasting North American peatland ecosystems revealed using DNA-SIP. Microb. Ecol. *63*, 438–445. https://doi.org/10.1007/s00248-011-9902-z

Hallam, S.J., Putnam, N., Preston, C.M., Detter, J.C., Rokhsar, D., Richardson, P.M., and DeLong, E.F. (2004). Reverse methanogenesis: testing the hypothesis with environmental genomics. Science *305*, 1457–1462. https://doi.org/10.1126/science.1100025

Hanczár, T., Csáki, R., Bodrossy, L., Murrell, J.C., and Kovács, K.L. (2002). Detection and localization of two hydrogenases in *Methylococcus capsulatus* (Bath) and their potential role in methane metabolism. Arch. Microbiol. *177*, 167–172. https://doi.org/10.1007/s00203-001-0372-4

Hanson, B.T., and Madsen, E.L. (2015). In situ expression of nitrite-dependent anaerobic methane oxidation proteins by *Candidatus* Methylomirabilis oxyfera co-occurring with expressed anammox proteins in a contaminated aquifer. Environ. Microbiol. Rep. *7*, 252–264. https://doi.org/10.1111/1758-2229.12239

Haroon, M.F., Hu, S., Shi, Y., Imelfort, M., Keller, J., Hugenholtz, P., Yuan, Z., and Tyson, G.W. (2013). Anaerobic oxidation of methane coupled to nitrate reduction in a novel archaeal lineage. Nature *500*, 567–570. https://doi.org/10.1038/nature12375

He, Y., Li, M., Perumal, V., Feng, X., Fang, J., Xie, J., Sievert, S.M., and Wang, F. (2016). Genomic and enzymatic evidence for acetogenesis among multiple lineages of the archaeal phylum *Bathyarchaeota* widespread in marine sediments. Nat. Microbiol. *1*, 16035. https://doi.org/10.1038/nmicrobiol.2016.35

He, Z., Cai, C., Shen, L., Lou, L., Zheng, P., Xu, X., and Hu, B. (2015). Effect of inoculum sources on the enrichment of nitrite-dependent anaerobic methane-oxidizing bacteria. Appl. Microbiol. Biotechnol. *99*, 939–946. https://doi.org/10.1007/s00253-014-6033-8

He, Z., Cai, C., Wang, J., Xu, X., Zheng, P., Jetten, M.S., and Hu, B. (2016). A novel denitrifying methanotroph of the NC10 phylum and its microcolony. Sci. Rep. *6*, 32241. https://doi.org/10.1038/srep32241

Hernandez, M.E., Beck, D.A., Lidstrom, M.E., and Chistoserdova, L. (2015). Oxygen availability is a major factor in determining the composition of microbial communities involved in methane oxidation. PeerJ *3*, e801. https://doi.org/10.7717/peerj.801

Hernández, M., Conrad, R., Klose, M., Ma, K., and Lu, Y.H. (2017). Structure and function of methanogenic microbial communities in soils from flooded rice and upland soybean fields from Sanjiang plain, NE China. Soil Biol. Biochem. *105*, 81–91. https://doi.org/10.1016/j.soilbio.2016.11.010

Hofmann, K., Praeg, N., Mutschlechner, M., Wagner, A.O., and Illmer, P. (2016). Abundance and potential metabolic activity of methanogens in well-aerated forest and grassland soils of an alpine region. FEMS Microbiol. Ecol. *92*, fiv171. https://doi.org/10.1093/femsec/fiv171

House, C.H., Orphan, V.J., Turk, K.A., Thomas, B., Pernthaler, A., Vrentas, J.M., and Joye, S.B. (2009). Extensive carbon isotopic heterogeneity among methane seep microbiota. Environ. Microbiol. *11*, 2207–2215. https://doi.org/10.1111/j.1462-2920.2009.01934.x

Hu, B.L., Shen, L.D., Lian, X., Zhu, Q., Liu, S., Huang, Q., He, Z.F., Geng, S., Cheng, D.Q., Lou, L.P., *et al.* (2014). Evidence for nitrite-dependent anaerobic methane oxidation as a previously overlooked microbial methane sink in wetlands. Proc. Natl. Acad. Sci. U.S.A. *111*, 4495–4500. https://doi.org/10.1073/pnas.1318393111

Hu, H.W., Zhang, L.M., Yuan, C.L., and He, J.Z. (2013). Contrasting Euryarchaeota communities between upland and paddy soils exhibited similar pH-impacted biogeographic patterns. Soil Biol. Biochem. *64*, 18–27. https://doi.org/10.1016/j.soilbio.2013.04.003

Hu, Z., and Ma, R. (2016). Distribution and characteristic of nitrite-dependent anaerobic methane oxidation bacteria by comparative analysis of wastewater treatment plants and agriculture fields in northern China. PeerJ *4*, e2766. https://doi.org/10.7717/peerj.2766

Iino, T., Tamaki, H., Tamazawa, S., Ueno, Y., Ohkuma, M., Suzuki, K., Igarashi, Y., and Haruta, S. (2013). *Candidatus* Methanogranum caenicola: a novel methanogen from the anaerobic digested sludge, and proposal of *Methanomassiliicoccaceae* fam. nov. and *Methanomassiliicoccales* ord. nov., for a methanogenic lineage of the class *Thermoplasmata*. Microbes Environ. *28*, 244–250.

Im, J., Lee, S.W., Yoon, S., Dispirito, A.A., and Semrau, J.D. (2011). Characterization of a novel facultative *Methylocystis* species capable of growth on methane, acetate and ethanol. Environ. Microbiol. Rep. *3*, 174–181. https://doi.org/10.1111/j.1758-2229.2010.00204.x

in 't Zandt, M.H., de Jong, A.E., Slomp, C.P., and Jetten, M.S. (2018). The hunt for the most-wanted chemolithoautotrophic spookmicrobes. FEMS Microbiol. Ecol. *94*. https://doi.org/10.1093/femsec/fiy064

Kallistova, A.Y., Merkel, A.Y., Tarnovetskii, I.Y., and Pimenov, N.V. (2017). Methane formation and oxidation by prokaryotes. Microbiology *86*, 671–691. https://doi.org/10.1134/S0026261717060091

Kalyuzhnaya, M.G., Yang, S., Rozova, O.N., Smalley, N.E., Clubb, J., Lamb, A., Gowda, G.A., Raftery, D., Fu, Y., Bringel, F., *et al.* (2013). Highly efficient methane biocatalysis revealed in a methanotrophic bacterium. Nat. Commun. *4*, 2785. https://doi.org/10.1038/ncomms3785

Kalyuzhnaya, M.G., Lamb, A.E., McTaggart, T.L., Oshkin, I.Y., Shapiro, N., Woyke, T., and Chistoserdova, L. (2015). Draft genome sequences of gammaproteobacterial methanotrophs isolated from Lake Washington sediment. Genome Announc. *3*, e00103–15. https://doi.org/10.1128/genomeA.00103-15

Kammann, C., Hepp, S., Lenhart, K., and Mulller, C. (2009). Stimulation of methane consumption by endogenous CH_4 production in aerobic grassland soil. Soil Biol.

Biochem. *41*, 622–629. https://doi.org/10.1016/j.soilbio.2008.12.025

Kammann, C., Ratering, S., Görres, C.M., Guillet, C., and Muller, C. (2017). Stimulation of methane oxidation by CH_4-emitting rose chafer larvae in well-aerated grassland soil. Biol. Fertil. Soils 53, 491–499. https://doi.org/10.1007/s00374-017-1199-8

Kao-Kniffin, J., Woodcroft, B.J., Carver, S.M., Bockheim, J.G., Handelsman, J., Tyson, G.W., Hinkel, K.M., and Mueller, C.W. (2015). Archaeal and bacterial communities across a chronosequence of drained lake basins in Arctic Alaska. Sci. Rep. 5, 18165. https://doi.org/10.1038/srep18165

Kirschke, S., Bousquet, P., Ciais, P., Saunois, M., Canadell, J.G., Dlugokencky, E.J., Bergamaschi, P., Bergmann, D., Blake, D.R., Bruhwiler, L., *et al.* (2013). Three decades of global methane sources and sinks. Nat. Geosci. *6*, 813–823. https://doi.org/10.1038/Ngeo1955

Kits, K.D., Campbell, D.J., Rosana, A.R., and Stein, L.Y. (2015a). Diverse electron sources support denitrification under hypoxia in the obligate methanotroph *Methylomicrobium album* strain BG8. Front. Microbiol. 6, 1072. https://doi.org/10.3389/fmicb.2015.01072

Kits, K.D., Klotz, M.G., and Stein, L.Y. (2015b). Methane oxidation coupled to nitrate reduction under hypoxia by the Gammaproteobacterium *Methylomonas denitrificans*, sp. nov. type strain FJG1. Environ. Microbiol. *17*, 3219–3232. https://doi.org/10.1111/1462-2920.12772

Knief, C. (2015). Diversity and habitat preferences of cultivated and uncultivated aerobic methanotrophic bacteria evaluated based on *pmoA* as molecular marker. Front. Microbiol. *6*, 1346. https://doi.org/10.3389/fmicb.2015.01346

Knief, C., Lipski, A., and Dunfield, P.F. (2003). Diversity and activity of methanotrophic bacteria in different upland soils. Appl. Environ. Microbiol. 69, 6703–6714.

Knief, C., Delmotte, N., Chaffron, S., Stark, M., Innerebner, G., Wassmann, R., von Mering, C., and Vorholt, J.A. (2012). Metaproteogenomic analysis of microbial communities in the phyllosphere and rhizosphere of rice. ISME J. *6*, 1378–1390. https://doi.org/10.1038/ismej.2011.192

Knittel, K., and Boetius, A. (2009). Anaerobic oxidation of methane: progress with an unknown process. Annu. Rev. Microbiol. 63, 311–334. https://doi.org/10.1146/annurev.micro.61.080706.093130

Knittel, K., Lösekann, T., Boetius, A., Kort, R., and Amann, R. (2005). Diversity and distribution of methanotrophic archaea at cold seeps. Appl. Environ. Microbiol. *71*, 467–479.

Kojima, H., Tokizawa, R., Kogure, K., Kobayashi, Y., Itoh, M., Shiah, F.K., Okuda, N., and Fukui, M. (2014). Community structure of planktonic methane-oxidizing bacteria in a subtropical reservoir characterized by dominance of phylotype closely related to nitrite reducer. Sci. Rep. *4*, 5728. https://doi.org/10.1038/srep05728

Kolb, S. (2009). The quest for atmospheric methane oxidizers in forest soils. Environ. Microbiol. Rep. *1*, 336–346. https://doi.org/10.1111/j.1758-2229.2009.00047.x

Kolb, S., Knief, C., Stubner, S., and Conrad, R. (2003). Quantitative detection of methanotrophs in soil by novel *pmoA*-targeted real-time PCR assays. Appl. Environ. Microbiol. *69*, 2423–2429.

Lang, K., Schuldes, J., Klingl, A., Poehlein, A., Daniel, R., and Brunea, A. (2015). New mode of energy metabolism in the seventh order of methanogens as revealed by comparative genome analysis of '*Candidatus* methanoplasma termitum'. Appl. Environ. Microbiol. *81*, 1338–1352.

Laso-Pérez, R., Wegener, G., Knittel, K., Widdel, F., Harding, K.J., Krukenberg, V., Meier, D.V., Richter, M., Tegetmeyer, H.E., Riedel, D., *et al.* (2016). Thermophilic archaea activate butane via alkyl-coenzyme M formation. Nature 539, 396–401. https://doi.org/10.1038/nature20152

Lazar, C.S., Baker, B.J., Seitz, K., Hyde, A.S., Dick, G.J., Hinrichs, K.U., and Teske, A.P. (2016). Genomic evidence for distinct carbon substrate preferences and ecological niches of *Bathyarchaeota* in estuarine sediments. Environ. Microbiol. *18*, 1200–1211. https://doi.org/10.1111/1462-2920.13142

Lee, C.G., Watanabe, T., Murase, J., Asakawa, S., and Kimura, M. (2012). Growth of methanogens in an oxic soil microcosm: Elucidation by a DNA-SIP experiment using ^{13}C-labeled dried rice callus. Appl. Soil Ecol. *58*, 37–44. https://doi.org/10.1016/j.apsoil.2012.03.002

Lee, H.J., Jeong, S.E., Kim, P.J., Madsen, E.L., and Jeon, C.O. (2015). High resolution depth distribution of *Bacteria*, *Archaea*, methanotrophs, and methanogens in the bulk and rhizosphere soils of a flooded rice paddy. Front. Microbiol. *6*, 639. https://doi.org/10.3389/fmicb.2015.00639

Lenhart, K., Bunge, M., Ratering, S., Neu, T.R., Schüttmann, I., Greule, M., Kammann, C., Schnell, S., Müller, C., Zorn, H., *et al.* (2012). Evidence for methane production by saprotrophic fungi. Nat. Commun. 3, 1046. https://doi.org/10.1038/ncomms2049

Li, W., Guan, W., Chen, H., Liao, B.W., Hu, J., Peng, C.H., Rui, J.P., Tian, J.Q., Zhu, D., and He, Y.X. (2016). Archaeal communities in the sediments of different mangrove stands at Dongzhaigang, China. J. Soils Sediments *16*, 1995–2004. https://doi.org/10.1007/s11368-016-1427-0

Li, Y., Leahy, S.C., Jeyanathan, J., Henderson, G., Cox, F., Altermann, E., Kelly, W.J., Lambie, S.C., Janssen, P.H., Rakonjac, J., *et al.* (2016). The complete genome sequence of the methanogenic archaeon ISO4-H5 provides insights into the methylotrophic lifestyle of a ruminal representative of the *Methanomassiliicoccales*. Stand. Genomic Sci. *11*, 59. https://doi.org/10.1186/s40793-016-0183-5

Liu, D., Nishida, M., Takahashi, T., and Asakawa, S. (2018). Transcription of *mcrA* gene decreases upon prolonged non-flooding period in a methanogenic archaeal community of a paddy-upland rotational field soil. Microb. Ecol. 75, 751–760. https://doi.org/10.1007/s00248-017-1063-2

Liu, J., Xu, H., Jiang, Y., Zhang, K., Hu, Y., and Zeng, Z. (2017a). Methane emissions and microbial communities as influenced by dual cropping of *Azolla* along with early rice. Sci. Rep. 7, 40635. https://doi.org/10.1038/srep40635

Liu, L., Xu, X., Cao, Y., Cai, C., Cui, H., and Yao, J. (2017b). Nitrate decreases methane production also by increasing methane oxidation through stimulating NC10 population in ruminal culture. AMB Express *7*, 76. https://doi.org/10.1186/s13568-017-0377-2

Liu, Y., and Whitman, W.B. (2008). Metabolic, phylogenetic, and ecological diversity of the methanogenic archaea. Ann. N. Y. Acad. Sci. *1125*, 171–189. https://doi.org/10.1196/annals.1419.019

Lloyd, K. (2015). Microbiology. Beyond known methanogens. Science *350*, 384. https://doi.org/10.1126/science.aad4066

Lloyd, K.G., Lapham, L., and Teske, A. (2006). An anaerobic methane-oxidizing community of ANME-1b archaea in hypersaline Gulf of Mexico sediments. Appl. Environ. Microbiol. *72*, 7218–7230.

Loulergue, L., Schilt, A., Spahni, R., Masson-Delmotte, V., Blunier, T., Lemieux, B., Barnola, J.M., Raynaud, D., Stocker, T.F., and Chappellaz, J. (2008). Orbital and millennial-scale features of atmospheric CH4 over the past 800,000 years. Nature *453*, 383–386. https://doi.org/10.1038/nature06950

Lu, Y., and Conrad, R. (2005). In situ stable isotope probing of methanogenic archaea in the rice rhizosphere. Science *309*, 1088–1090.

Lü, Z., and Lu, Y. (2012). *Methanocella conradii* sp. nov., a thermophilic, obligate hydrogenotrophic methanogen, isolated from Chinese rice field soil. PLOS ONE *7*, e35279. https://doi.org/10.1371/journal.pone.0035279

Luesken, F.A., Wu, M.L., Op den Camp, H.J., Keltjens, J.T., Stunnenberg, H., Francoijs, K.J., Strous, M., and Jetten, M.S. (2012). Effect of oxygen on the anaerobic methanotroph '*Candidatus* Methylomirabilis oxyfera': kinetic and transcriptional analysis. Environ. Microbiol. *14*, 1024–1034. https://doi.org/10.1111/j.1462-2920.2011.02682.x

Lüke, C., and Frenzel, P. (2011). Potential of *pmoA* amplicon pyrosequencing for methanotroph diversity studies. Appl. Environ. Microbiol. *77*, 6305–6309. https://doi.org/10.1128/AEM.05355-11

Lyu, Z., and Lu, Y. (2018). Metabolic shift at the class level sheds light on adaptation of methanogens to oxidative environments. ISME J. *12*, 411–423. https://doi.org/10.1038/ismej.2017.173

Martinez-Cruz, K., Leewis, M.C., Herriott, I.C., Sepulveda-Jauregui, A., Anthony, K.W., Thalasso, F., and Leigh, M.B. (2017). Anaerobic oxidation of methane by aerobic methanotrophs in sub-Arctic lake sediments. Sci. Total Environ. 607-608, 23–31.

Maus, I., Rumming, M., Bergmann, I., Heeg, K., Pohl, M., Nettmann, E., Jaenicke, S., Blom, J., Pühler, A., Schlüter, A., *et al.* (2018). Characterization of *Bathyarchaeota* genomes assembled from metagenomes of biofilms residing in mesophilic and thermophilic biogas reactors. Biotechnol. Biofuels *11*, 167. https://doi.org/10.1186/s13068-018-1162-4

McCalley, C.K., Woodcroft, B.J., Hodgkins, S.B., Wehr, R.A., Kim, E.H., Mondav, R., Crill, P.M., Chanton, J.P., Rich, V.I., Tyson, G.W., *et al.* (2014). Methane dynamics regulated by microbial community response to permafrost thaw. Nature *514*, 478–481. https://doi.org/10.1038/nature13798

McGlynn, S.E. (2017). Energy metabolism during anaerobic methane oxidation in ANME archaea. Microbes Environ. *32*, 5–13. https://doi.org/10.1264/jsme2.ME16166

McGlynn, S.E., Chadwick, G.L., Kempes, C.P., and Orphan, V.J. (2015). Single cell activity reveals direct electron transfer in methanotrophic consortia. Nature *526*, 531–535. https://doi.org/10.1038/nature15512

McKay, L.J., Hatzenpichler, R., Inskeep, W.P., and Fields, M.W. (2017). Occurrence and expression of novel methyl-coenzyme M reductase gene (*mcrA*) variants in hot spring sediments. Sci. Rep. *7*, 7252. https://doi.org/10.1038/s41598-017-07354-x

Meng, H., Wang, Y.F., Chan, H.W., Wu, R.N., and Gu, J.D. (2016). Co-occurrence of nitrite-dependent anaerobic ammonium and methane oxidation processes in subtropical acidic forest soils. Appl. Microbiol. Biotechnol. *100*, 7727–7739. https://doi.org/10.1007/s00253-016-7585-6

Merkel, A.Y., Podosokorskaya, O.A., Sokolova, T.G., and Bonch-Osmolovskaya, E.A. (2016). Diversity of methanogenic archaea from the 2012 terrestrial hot spring (Valley of Geysers, Kamchatka). Microbiology *85*, 342–349. https://doi.org/10.1134/S0026261716030073

Milucka, J., Ferdelman, T.G., Polerecky, L., Franzke, D., Wegener, G., Schmid, M., Lieberwirth, I., Wagner, M., Widdel, F., and Kuypers, M.M. (2012). Zero-valent sulphur is a key intermediate in marine methane oxidation. Nature *491*, 541–546. https://doi.org/10.1038/nature11656

Milucka, J., Kirf, M., Lu, L., Krupke, A., Lam, P., Littmann, S., Kuypers, M.M., and Schubert, C.J. (2015). Methane oxidation coupled to oxygenic photosynthesis in anoxic waters. ISME J. *9*, 1991–2002. https://doi.org/10.1038/ismej.2015.12

Miyashita, A., Mochimaru, H., Kazama, H., Ohashi, A., Yamaguchi, T., Nunoura, T., Horikoshi, K., Takai, K., and Imachi, H. (2009). Development of 16S rRNA gene-targeted primers for detection of archaeal anaerobic methanotrophs (ANMEs). FEMS Microbiol. Lett. *297*, 31–37. https://doi.org/10.1111/j.1574-6968.2009.01648.x

Mohammadi, S., Pol, A., van Alen, T.A., Jetten, M.S., and Op den Camp, H.J. (2017). *Methylacidiphilum fumariolicum* SolV, a thermoacidophilic 'Knallgas' methanotroph with both an oxygen-sensitive and -insensitive hydrogenase. ISME J. *11*, 945–958. https://doi.org/10.1038/ismej.2016.171

Mondav, R., Woodcroft, B.J., Kim, E.H., McCalley, C.K., Hodgkins, S.B., Crill, P.M., Chanton, J., Hurst, G.B., VerBerkmoes, N.C., Saleska, S.R., *et al.* (2014). Discovery of a novel methanogen prevalent in thawing permafrost. Nat. Commun. *5*, 3212. https://doi.org/10.1038/ncomms4212

Morris, S.A., Radajewski, S., Willison, T.W., and Murrell, J.C. (2002). Identification of the functionally active methanotroph population in a peat soil microcosm by stable-isotope probing. Appl. Environ. Microbiol. *68*, 1446–1453.

Murase, J., and Kimura, M. (1994). Methane production and its fate in paddy fields. VII. Electron acceptors responsible for anaerobic methane oxidation. Soil Sci. Plant Nutr. *40*, 647–654.

Murray, R.G., and Stackebrandt, E. (1995). Taxonomic note: implementation of the provisional status *Candidatus* for incompletely described procaryotes. Int. J. Syst. Bacteriol. *45*, 186–187. https://doi.org/10.1099/00207713-45-1-186

Myhre, G., Shindell, D., Bréon, F.-M., Collins, W., Fuglestveldt, J., Huang, J., Koch, D., Lamarque, J.-F., Lee, D., Mendoza, B., *et al.* (2013). Anthropogenic and natural radiative forcing. In Climate Change 2013: The Physical Science Basis Contribution of Working Group I to the Fifth Assessment Report of the Intergovernmental Panel on Climate Change, T.F. Stocker, D. Qin, G.-K. Plattner, M. Tignor, S.K. Allen, J. Boschung, A. Nauels, Y. Xia, V. Bex, and P.M. Midgley, eds. (Cambridge University Press, Cambridge, UK and New York, NY). https://doi.org/10.1017/CBO9781107415324.018

Naqvi, S.W.A., Lam, P., Narvenkar, G., Sarkar, A., Naik, H., Pratihary, A., Shenoy, D.M., Gauns, M., Kurian, S., Damare, S., *et al.* (2018). Methane stimulates massive nitrogen loss from freshwater reservoirs in India. Nat. Commun. *9*, 1265. https://doi.org/10.1038/s41467-018-03607-z

Narrowe, A.B., Angle, J.C., Daly, R.A., Stefanik, K.C., Wrighton, K.C., and Miller, C.S. (2017). High-resolution sequencing reveals unexplored archaeal diversity in freshwater wetland soils. Environ. Microbiol. *19*, 2192–2209. https://doi.org/10.1111/1462-2920.13703

Nazaries, L., Murrell, J.C., Millard, P., Baggs, L., and Singh, B.K. (2013). Methane, microbes and models: fundamental understanding of the soil methane cycle for future predictions. Environ. Microbiol. *15*, 2395–2417. https://doi.org/10.1111/1462-2920.12149

Niederberger, T.D., Perreault, N.N., Tille, S., Lollar, B.S., Lacrampe-Couloume, G., Andersen, D., Greer, C.W., Pollard, W., and Whyte, L.G. (2010). Microbial characterization of a subzero, hypersaline methane seep in the Canadian High Arctic. ISME J. *4*, 1326–1339. https://doi.org/10.1038/ismej.2010.57

Nobu, M.K., Narihiro, T., Kuroda, K., Mei, R., and Liu, W.T. (2016). Chasing the elusive Euryarchaeota class WSA2: genomes reveal a uniquely fastidious methyl-reducing methanogen. ISME J. *10*, 2478–2487. https://doi.org/10.1038/ismej.2016.33

Oren, A. (2014). The family *Methanosarcinaceae*. In The Prokaryotes, E. Rosenberg, E.F. DeLong, S. Lory, E. Stackebrandt, and F. Thompson, eds. (Springer, Berlin, Heidelberg), pp. 259–281. https://doi.org/10.1007/978-3-642-38954-2_408

Oswald, K., Milucka, J., Brand, A., Littmann, S., Wehrli, B., Kuypers, M.M., and Schubert, C.J. (2015). Light-dependent aerobic methane oxidation reduces methane emissions from seasonally stratified lakes. PLOS ONE *10*, e0132574. https://doi.org/10.1371/journal.pone.0132574

Oswald, K., Milucka, J., Brand, A., Hach, P., Littmann, S., Wehrli, B., Kuypers, M.M.M., and Schubert, C.J. (2016). Aerobic gammaproteobacterial methanotrophs mitigate methane emissions from oxic and anoxic lake waters. Limnol. Oceanogr. *61*, S101–S118. https://doi.org/10.1002/lno.10312

Oswald, K., Graf, J.S., Littmann, S., Tienken, D., Brand, A., Wehrli, B., Albertsen, M., Daims, H., Wagner, M., Kuypers, M.M., *et al.* (2017). *Crenothrix* are major methane consumers in stratified lakes. ISME J. *11*, 2124–2140. https://doi.org/10.1038/ismej.2017.77

Padilla, C.C., Bristow, L.A., Sarode, N., Garcia-Robledo, E., Gómez Ramírez, E., Benson, C.R., Bourbonnais, A., Altabet, M.A., Girguis, P.R., Thamdrup, B., *et al.* (2016). NC10 bacteria in marine oxygen minimum zones. ISME J. *10*, 2067–2071. https://doi.org/10.1038/ismej.2015.262

Pandit, P.S., Hoppert, M., and Rahalkar, M.C. (2018). Description of *Methylocucumis oryzae* gen. nov., sp. nov., a type I methanotroph with large cells and pale pink colour, isolated from an Indian rice field. Anton. Leeuw. *111*, 2473–2484. https://doi.org/10.1007/s10482-018-1136-3

Parks, D.H., Rinke, C., Chuvochina, M., Chaumeil, P.A., Woodcroft, B.J., Evans, P.N., Hugenholtz, P., and Tyson, G.W. (2017). Recovery of nearly 8,000 metagenome-assembled genomes substantially expands the tree of life. Nat. Microbiol. *2*, 1533–1542. https://doi.org/10.1038/s41564-017-0012-7

Paul, K., Nonoh, J.O., Mikulski, L., and Brune, A. (2012). '*Methanoplasmatales*,' *Thermoplasmatales*-related archaea in termite guts and other environments, are the seventh order of methanogens. Appl. Environ. Microbiol. *78*, 8245–8253. https://doi.org/10.1128/AEM.02193-12

Peters, V., and Conrad, R. (1995). Methanogenic and other strictly anaerobic bacteria in desert soil and other oxic soils. Appl. Environ. Microbiol. *61*, 1673–1676.

Praeg, N., Wagner, A.O., and Illmer, P. (2014). Effects of fertilisation, temperature and water content on microbial properties and methane production and methane oxidation in subalpine soils. Eur. J. Soil Biol. *65*, 96–106. https://doi.org/10.1016/j.ejsobi.2014.10.002

Pratscher, J., Dumont, M.G., and Conrad, R. (2011). Assimilation of acetate by the putative atmospheric methane oxidizers belonging to the USC alpha clade. Environ. Microbiol. *13*, 2692–2701. https://doi.org/10.1111/j.1462-2920.2011.02537.x

Pratscher, J., Vollmers, J., Wiegand, S., Dumont, M.G., and Kaster, A.K. (2018). Unravelling the identity, metabolic potential and global biogeography of the atmospheric methane-oxidizing upland soil cluster α. Environ. Microbiol. *20*, 1016–1029. https://doi.org/10.1111/1462-2920.14036

Putkinen, A., Larmola, T., Tuomivirta, T., Siljanen, H.M., Bodrossy, L., Tuittila, E.S., and Fritze, H. (2014). Peatland succession induces a shift in the community composition of *Sphagnum*-associated active methanotrophs. FEMS Microbiol. Ecol. *88*, 596–611. https://doi.org/10.1111/1574-6941.12327

Raghoebarsing, A.A., Pol, A., van de Pas-Schoonen, K.T., Smolders, A.J., Ettwig, K.F., Rijpstra, W.I., Schouten, S., Damsté, J.S., Op den Camp, H.J., Jetten, M.S., *et al.* (2006). A microbial consortium couples anaerobic methane oxidation to denitrification. Nature *440*, 918–921.

Rahalkar, M., Bussmann, I., and Schink, B. (2007). *Methylosoma difficile* gen. nov., sp nov., a novel methanotroph enriched by gradient cultivation from littoral sediment of Lake Constance. Int. J. Syst. Evol. Microbiol. *57*, 1073–1080. https://doi.org/10.1099/ijs.0.64574-0

Rahalkar, M.C., Pandit, P.S., Dhakephalkar, P.K., Pore, S., Arora, P., and Kapse, N. (2016). Genome characteristics of a novel type I methanotroph (Sn10-6) isolated from a flooded Indian rice field. Microb. Ecol. *71*, 519–523. https://doi.org/10.1007/s00248-015-0699-z

Reeburgh, W.S. (2003). Global methane biogeochemistry. In Treatise on Geochemistry: The Atmosphere, R.F. Keeling, H.D. Holland, and K.K. Turekian, eds. (Elsevier-Pergamon, Oxford, UK), pp. 65–89. https://doi.org/10.1016/B0-08-043751-6/04036-6

Reed, D.C., Deemer, B.R., van Grinsven, S., and Harrison, J.A. (2017). Are elusive anaerobic pathways key methane sinks in eutrophic lakes and reservoirs? Biogeochem. *134*, 29–39. https://doi.org/10.1007/s10533-017-0356-3

Reim, A., Lüke, C., Krause, S., Pratscher, J., and Frenzel, P. (2012). One millimetre makes the difference: high-resolution analysis of methane-oxidizing bacteria and their specific activity at the oxic-anoxic interface in a flooded paddy soil. ISME J. *6*, 2128–2139. https://doi.org/10.1038/ismej.2012.57

Reim, A., Hernández, M., Klose, M., Chidthaisong, A., Yuttitham, M., and Conrad, R. (2017). Response of methanogenic microbial communities to desiccation stress in flooded and rain-fed paddy soil from Thailand. Front. Microbiol. *8*, 785. https://doi.org/10.3389/fmicb.2017.00785

Ricke, P., Kube, M., Nakagawa, S., Erkel, C., Reinhardt, R., and Liesack, W. (2005). First genome data from uncultured upland soil cluster alpha methanotrophs provide further evidence for a close phylogenetic relationship to *Methylocapsa acidiphila* B2 and for high-affinity methanotrophy involving particulate methane monooxygenase. Appl. Environ. Microbiol. *71*, 7472–7482.

Rissanen, A.J., Karvinen, A., Nykänen, H., Peura, S., Tiirola, M., Mäki, A., and Kankaala, P. (2017). Effects of alternative electron acceptors on the activity and community structure of methane-producing and consuming microbes in the sediments of two shallow boreal lakes. FEMS Microbiol. Ecol. *93*. https://doi.org/10.1093/femsec/fix078

Rissanen, A.J., Saarenheimo, J., Tiirola, M., Peura, S., Aalto, S.L., Karvinen, A., and Nykanen, H. (2018). Gammaproteobacterial methanotrophs dominate methanotrophy in aerobic and anaerobic layers of boreal lake waters. Aquat. Microb. Ecol. *81*, 257–276. https://doi.org/10.3354/ame01874

Roslev, P., and King, G.M. (1994). Survival and recovery of methanotrophic bacteria starved under oxic and anoxic conditions. Appl. Environ. Microbiol. *60*, 2602–2608.

Saunois, M., Bousquet, P., Poulter, B., Peregon, A., Ciais, P., Canadell, J.G., Dlugokencky, E.J., Etiope, G., Bastviken, D., Houweling, S., *et al.* (2016a). The global methane budget 2000–2012. Earth Syst. Sci. Data *8*, 697–751. https://doi.org/10.5194/essd-8-697-2016

Saunois, M., Jackson, R.B., Bousquet, P., Poulter, B., and Canadell, J.G. (2016b). The growing role of methane in anthropogenic climate change. Environ. Res. Lett. *11*, 120207. https://doi.org/10.1088/1748-9326/11/12/120207

Scheller, S., Goenrich, M., Boecher, R., Thauer, R.K., and Jaun, B. (2010). The key nickel enzyme of methanogenesis catalyses the anaerobic oxidation of methane. Nature *465*, 606–608. https://doi.org/10.1038/nature09015

Scheller, S., Yu, H., Chadwick, G.L., McGlynn, S.E., and Orphan, V.J. (2016). Artificial electron acceptors decouple archaeal methane oxidation from sulfate reduction. Science *351*, 703–707. https://doi.org/10.1126/science.aad7154

Schubert, C.J., Coolen, M.J., Neretin, L.N., Schippers, A., Abbas, B., Durisch-Kaiser, E., Wehrli, B., Hopmans, E.C., Damsté, J.S., Wakeham, S., *et al.* (2006). Aerobic and anaerobic methanotrophs in the Black Sea water column. Environ. Microbiol. *8*, 1844–1856.

Scott, D.T., McKnight, D.M., Blunt-Harris, E.L., Kolesar, S.E., and Lovley, D.R. (1998). Quinone moieties act as electron acceptors in the reduction of humic substances by humics-reducing microorganisms. Environ. Sci. Technol. *32*, 2984–2989. https://doi.org/10.1021/es980272q

Segarra, K.E., Schubotz, F., Samarkin, V., Yoshinaga, M.Y., Hinrichs, K.U., and Joye, S.B. (2015). High rates of anaerobic methane oxidation in freshwater wetlands reduce potential atmospheric methane emissions. Nat. Commun. *6*, 7477. https://doi.org/10.1038/ncomms8477

Semrau, J.D., DiSpirito, A.A., and Yoon, S. (2010). Methanotrophs and copper. FEMS Microbiol. Rev. *34*, 496–531. https://doi.org/10.1111/j.1574-6976.2010.00212.x

Serrano-Silva, N., Sarria-Guzmán, Y., Dendooven, L., and Luna-Guido, M. (2014). Methanogenesis and methanotrophy in soil: a review. Pedosphere *24*, 291–307. https://doi.org/10.1016/S1002-0160(14)60016-3

Shen, L.D., Liu, S., Huang, Q., Lian, X., He, Z.F., Geng, S., Jin, R.C., He, Y.F., Lou, L.P., Xu, X.Y., *et al.* (2014). Evidence for the cooccurrence of nitrite-dependent anaerobic ammonium and methane oxidation processes in a flooded paddy field. Appl. Environ. Microbiol. *80*, 7611–7619. https://doi.org/10.1128/AEM.02379-14

Shen, L.D., He, Z.F., Wu, H.S., and Gao, Z.Q. (2015a). Nitrite-dependent anaerobic methane-oxidising bacteria: unique microorganisms with special properties. Curr. Microbiol. *70*, 562–570. https://doi.org/10.1007/s00284-014-0762-x

Shen, L.D., Huang, Q., He, Z.F., Lian, X., Liu, S., He, Y.F., Lou, L.P., Xu, X.Y., Zheng, P., and Hu, B.L. (2015b). Vertical distribution of nitrite-dependent anaerobic methane-oxidising bacteria in natural freshwater wetland soils. Appl. Microbiol. Biotechnol. *99*, 349–357. https://doi.org/10.1007/s00253-014-6031-x

Shen, L.D., Liu, S., He, Z.F., Lian, X., Huang, Q., He, Y.F., Lou, L.P., Xu, X.Y., Zheng, P., and Hu, B.L. (2015c). Depth-specific distribution and importance of nitrite-dependent anaerobic ammonium and methane-oxidising bacteria in an urban wetland. Soil Biol. Biochem. *83*, 43–51. https://doi.org/10.1016/j.soilbio.2015.01.010

Shen, L.D., Wu, H.S., and Gao, Z.Q. (2015d). Distribution and environmental significance of nitrite-dependent anaerobic methane-oxidising bacteria in natural ecosystems. Appl. Microbiol. Biotechnol. *99*, 133–142. https://doi.org/10.1007/s00253-014-6200-y

Shen, L.D., Wu, H.S., Gao, Z.Q., Li, J., and Liu, X. (2016a). Presence of diverse *Candidatus* Methylomirabilis oxyfera-like bacteria of NC10 phylum in agricultural soils. J. Appl. Microbiol. *120*, 1552–1560. https://doi.org/10.1111/jam.13119

Shen, L.D., Wu, H.S., Gao, Z.Q., Liu, X., and Li, J. (2016b). Comparison of community structures of *Candidatus* Methylomirabilis oxyfera-like bacteria of NC10 phylum in different freshwater habitats. Sci. Rep. *6*, 25647. https://doi.org/10.1038/srep25647

Shiau, Y.-J., Cai, Y., Jia, Z., Chen, C.-L., and Chiu, C.-Y. (2018). Phylogenetically distinct methanotrophs modulate methane oxidation in rice paddies across Taiwan. Soil Biol. Biochem. *124*, 59–69. https://doi.org/10.1016/j.soilbio.2018.05.025

Singleton, C.M., McCalley, C.K., Woodcroft, B.J., Boyd, J.A., Evans, P.N., Hodgkins, S.B., Chanton, J.P., Frolking, S., Crill, P.M., Saleska, S.R., *et al.* (2018). Methanotrophy across a natural permafrost thaw environment. ISME J. *12*, 2544–2558. https://doi.org/10.1038/s41396-018-0065-5

Skennerton, C.T., Ward, L.M., Michel, A., Metcalfe, K., Valiente, C., Mullin, S., Chan, K.Y., Gradinaru, V., and Orphan, V.J. (2015). Genomic reconstruction of an uncultured hydrothermal vent gammaproteobacterial methanotroph (family *Methylothermaceae*) indicates multiple adaptations to oxygen limitation. Front. Microbiol. *6*, 1425. https://doi.org/10.3389/fmicb.2015.01425

Smemo, K.A., and Yavitt, J.B. (2007). Evidence for anaerobic CH4 oxidation in freshwater peatlands. Geomicrobiology *24*, 583–597. https://doi.org/10.1080/01490450701672083

Smemo, K.A., and Yavitt, J.B. (2011). Anaerobic oxidation of methane: an underappreciated aspect of methane cycling in peatland ecosystems? Biogeoscience *8*, 779–793. https://doi.org/10.5194/bg-8-779-2011

Söllinger, A., Schwab, C., Weinmaier, T., Loy, A., Tveit, A.T., Schleper, C., and Urich, T. (2016). Phylogenetic and genomic analysis of *Methanomassiliicoccales* in wetlands and animal intestinal tracts reveals clade-specific habitat preferences. FEMS Microbiol. Ecol. 92, fiv149. https://doi.org/10.1093/femsec/fiv149

Sorokin, D.Y., Makarova, K.S., Abbas, B., Ferrer, M., Golyshin, P.N., Galinski, E.A., Ciordia, S., Mena, M.C., Merkel, A.Y., Wolf, Y.I., *et al.* (2017). Discovery of extremely halophilic, methyl-reducing euryarchaea provides insights into the evolutionary origin of methanogenesis. Nat. Microbiol. 2, 17081. https://doi.org/10.1038/nmicrobiol.2017.81

Sorokin, D.Y., Merkel, A.Y., Abbas, B., Makarova, K.S., Rijpstra, W.I.C., Koenen, M., Sinninghe Damsté, J.S., Galinski, E.A., Koonin, E.V., and van Loosdrecht, M.C.M. (2018). *Methanonatronarchaeum thermophilum* gen. nov., sp. nov. and *'Candidatus* Methanohalarchaeum thermophilum', extremely halo(natrono)philic methyl-reducing methanogens from hypersaline lakes comprising a new euryarchaeal class *Methanonatronarchaeia* classis nov. Int. J. Syst. Evol. Microbiol. *68*, 2199–2208. https://doi.org/10.1099/ijsem.0.002810

Stoecker, K., Bendinger, B., Schöning, B., Nielsen, P.H., Nielsen, J.L., Baranyi, C., Toenshoff, E.R., Daims, H., and Wagner, M. (2006). Cohn's Crenothrix is a filamentous methane oxidizer with an unusual methane monooxygenase. Proc. Natl. Acad. Sci. U.S.A. *103*, 2363–2367. https://doi.org/10.1073/pnas.0506361103

Takeuchi, M., Yoshioka, H., Seo, Y., Tanabe, S., Tamaki, H., Kamagata, Y., Takahashi, H.A., Igari, S., Mayumi, D., and Sakata, S. (2011). A distinct freshwater-adapted subgroup of ANME-1 dominates active archaeal communities in terrestrial subsurfaces in Japan. Environ. Microbiol. *13*, 3206–3218. https://doi.org/10.1111/j.1462-2920.2011.02517.x

Takeuchi, M., Katayama, T., Yamagishi, T., Hanada, S., Tamaki, H., Kamagata, Y., Oshima, K., Hattori, M., Marumo, K., Nedachi, M., *et al.* (2014). *Methyloceanibacter caenitepidi* gen. nov., sp nov., a facultatively methylotrophic bacterium isolated from marine sediments near a hydrothermal vent. Int. J. Syst. Evol. Microbiol. *64*, 462–468. https://doi.org/10.1099/ijs.0.053397-0

Tang, K.W., McGinnis, D.F., Frindte, K., Bruchert, V., and Grossart, H.P. (2014). Paradox reconsidered: Methane oversaturation in well-oxygenated lake waters. Limnol. Oceanogr. *59*, 275–284. https://doi.org/10.4319/lo.2014.59.1.0275

Teh, Y.A., Silver, W.L., and Conrad, M.E. (2005). Oxygen effects on methane production and oxidation in humid tropical forest soils. Global Change Biol. *11*, 1283–1297. https://doi.org/10.1111/j.1365-2486.2005.00983.x

Thauer, R.K., and Shima, S. (2008). Methane as fuel for anaerobic microorganisms. Ann. N. Y. Acad. Sci. *1125*, 158–170.

Thauer, R.K., Kaster, A.K., Seedorf, H., Buckel, W., and Hedderich, R. (2008). Methanogenic archaea: ecologically relevant differences in energy conservation. Nat. Rev. Microbiol. *6*, 579–591. https://doi.org/10.1038/nrmicro1931

Tholen, A., Pester, M., and Brune, A. (2007). Simultaneous methanogenesis and oxygen reduction by *Methanobrevibacter cuticularis* at low oxygen fluxes. FEMS Microbiol. Ecol. *62*, 303–312.

Timmers, P.H., Suarez-Zuluaga, D.A., van Rossem, M., Diender, M., Stams, A.J., and Plugge, C.M. (2016). Anaerobic oxidation of methane associated with sulfate reduction in a natural freshwater gas source. ISME J. *10*, 1400–1412. https://doi.org/10.1038/ismej.2015.213

Timmers, P.H., Welte, C.U., Koehorst, J.J., Plugge, C.M., Jetten, M.S., and Stams, A.J. (2017). Reverse methanogenesis and respiration in methanotrophic Archaea. Archaea *2017*, 1654237. https://doi.org/10.1155/2017/1654237

Tveit, A., Schwacke, R., Svenning, M.M., and Urich, T. (2013). Organic carbon transformations in high-Arctic peat soils: key functions and microorganisms. ISME J. 7, 299–311. https://doi.org/10.1038/ismej.2012.99

Tveit, A.T., Urich, T., and Svenning, M.M. (2014). Metatranscriptomic analysis of arctic peat soil microbiota. Appl. Environ. Microbiol. *80*, 5761–5772. https://doi.org/10.1128/AEM.01030-14

Vaksmaa, A., Lüke, C., van Alen, T., Valè, G., Lupotto, E., Jetten, M.S., and Ettwig, K.F. (2016). Distribution and activity of the anaerobic methanotrophic community in a nitrogen-fertilized Italian paddy soil. FEMS Microbiol. Ecol. *92*, fiw181.

Vaksmaa, A., Guerrero-Cruz, S., van Alen, T.A., Cremers, G., Ettwig, K.F., Lüke, C., and Jetten, M.S.M. (2017a). Enrichment of anaerobic nitrate-dependent methanotrophic '*Candidatus* Methanoperedens nitroreducens' archaea from an Italian paddy field soil. Appl. Microbiol. Biotechnol. *101*, 7075–7084. https://doi.org/10.1007/s00253-017-8416-0

Vaksmaa, A., Jetten, M.S., Ettwig, K.F., and Lüke, C. (2017b). *McrA* primers for the detection and quantification of the anaerobic archaeal methanotroph '*Candidatus* Methanoperedens nitroreducens'. Appl. Microbiol. Biotechnol. *101*, 1631–1641. https://doi.org/10.1007/s00253-016-8065-8

Vaksmaa, A., van Alen, T.A., Ettwig, K.F., Lupotto, E., Valè, G., Jetten, M.S.M., and Lüke, C. (2017c). Stratification of diversity and activity of methanogenic and methanotrophic microorganisms in a nitrogen-fertilized Italian paddy soil. Front. Microbiology *8*, 2127. https://doi.org/10.3389/fmicb.2017.02127

van Kessel, M.A., Speth, D.R., Albertsen, M., Nielsen, P.H., Op den Camp, H.J., Kartal, B., Jetten, M.S., and Lücker, S. (2015). Complete nitrification by a single microorganism. Nature *528*, 555–559. https://doi.org/10.1038/nature16459

van Kessel, M.A., Stultiens, K., Slegers, M.F., Guerrero Cruz, S., Jetten, M.S., Kartal, B., and Op den Camp, H.J. (2018). Current perspectives on the application of N-damo and anammox in wastewater treatment. Curr. Opin. Biotechnol. *50*, 222–227.

Vanwonterghem, I., Evans, P.N., Parks, D.H., Jensen, P.D., Woodcroft, B.J., Hugenholtz, P., and Tyson, G.W. (2016). Methylotrophic methanogenesis discovered in the archaeal phylum *Verstraetearchaeota*. Nat. Microbiol. *1*, 16170. https://doi.org/10.1038/nmicrobiol.2016.170

Vekeman, B., Kerckhof, F.M., Cremers, G., de Vos, P., Vandamme, P., Boon, N., Op den Camp, H.J., and Heylen, K. (2016). New *Methyloceanibacter* diversity from North Sea sediments includes methanotroph containing solely the soluble methane monooxygenase. Environ. Microbiol. *18*, 4523–4536. https://doi.org/10.1111/1462-2920.13485

Vinogradov, S.N., Tinajero-Trejo, M., Poole, R.K., and Hoogewijs, D. (2013). Bacterial and archaeal globins - a revised perspective. Biochim. Biophys. Acta *1834*, 1789–1800. https://doi.org/10.1016/j.bbapap.2013.03.021

Wang, B., Lerdau, M., and He, Y. (2017). Widespread production of nonmicrobial greenhouse gases in soils. Glob. Chang. Biol. *23*, 4472–4482. https://doi.org/10.1111/gcb.13753

Wang, Z.P., Chang, S.X., Chen, H., and Han, X.G. (2013). Widespread non-microbial methane production by organic compounds and the impact of environmental stresses. Earth Sci. Rev. *127*, 193–202. https://doi.org/10.1016/j.earscirev.2013.10.001

Wegener, G., Krukenberg, V., Riedel, D., Tegetmeyer, H.E., and Boetius, A. (2015). Intercellular wiring enables electron transfer between methanotrophic archaea and bacteria. Nature *526*, 587–590. https://doi.org/10.1038/nature15733

Welte, C.U. (2018). Revival of archaeal methane microbiology. mSystems *3*, e00181–17. https://doi.org/10.1128/mSystems.00181-17

Welte, C.U., Rasigraf, O., Vaksmaa, A., Versantvoort, W., Arshad, A., Op den Camp, H.J., Jetten, M.S., Lüke, C., and Reimann, J. (2016). Nitrate- and nitrite-dependent anaerobic oxidation of methane. Environ. Microbiol. Rep. *8*, 941–955. https://doi.org/10.1111/1758-2229.12487

Whitman, W.B., Bowden, T.L., and Boone, D.R. (2014). The methanogenic bacteria. In The Prokaryotes, E. Rosenberg, E.F. DeLong, S. Lory, E. Stackebrandt, and F. Thompson, eds. (Springer, Berlin, Heidelberg), pp. 123–163. https://doi.org/10.1007/0-387-30743-5_9

Whitman, W.B., Woyke, T., Klenk, H.P., Zhou, Y., Lilburn, T.G., Beck, B.J., De Vos, P., Vandamme, P., Eisen, J.A., Garrity, G., *et al.* (2015). Genomic encyclopedia of bacterial and archaeal type strains, phase III: the genomes of soil and plant-associated and newly described type strains. Stand. Genomic Sci. *10*, 26. https://doi.org/10.1186/s40793-015-0017-x

Winkel, M., Mitzscherling, J., Overduin, P.P., Horn, F., Winterfeld, M., Rijkers, R., Grigoriev, M.N., Knoblauch, C., Mangelsdorf, K., Wagner, D., *et al.* (2018). Anaerobic methanotrophic communities thrive in deep submarine permafrost. Sci. Rep. *8*, 1291. https://doi.org/10.1038/s41598-018-19505-9

Wu, M.L., de Vries, S., van Alen, T.A., Butler, M.K., Op den Camp, H.J., Keltjens, J.T., Jetten, M.S., and Strous, M. (2011). Physiological role of the respiratory quinol oxidase in the anaerobic nitrite-reducing methanotroph '*Candidatus* Methylomirabilis oxyfera'. Microbiology *157*, 890–898. https://doi.org/10.1099/mic.0.045187-0

Xie, W., Jiao, N., Ma, C., Fang, S., Phelps, T.J., Zhu, R., and Zhang, C. (2017). The response of archaeal species to seasonal variables in a subtropical aerated soil: insight into the low abundant methanogens. Appl. Microbiol. Biotechnol. *101*, 6505–6515. https://doi.org/10.1007/s00253-017-8349-7

Xu, S., Cai, C., Guo, J., Lu, W., Yuan, Z., and Hu, S. (2018). Different clusters of *Candidatus* 'Methanoperedens nitroreducens'-like archaea as revealed by high-throughput sequencing with new primers. Sci. Rep. *8*, 7695. https://doi.org/10.1038/s41598-018-24974-z

Yuan, Y., Conrad, R., and Lu, Y. (2011). Transcriptional response of methanogen *mcrA* genes to oxygen exposure of rice field soil. Environ. Microbiol. Rep. *3*, 320–328. https://doi.org/10.1111/j.1758-2229.2010.00228.x

Zhu, B., van Dijk, G., Fritz, C., Smolders, A.J., Pol, A., Jetten, M.S., and Ettwig, K.F. (2012). Anaerobic oxidization of methane in a minerotrophic peatland: enrichment of nitrite-dependent methane-oxidizing bacteria. Appl. Environ. Microbiol. *78*, 8657–8665. https://doi.org/10.1128/AEM.02102-12

Zhu, G., Zhou, L., Wang, Y., Wang, S., Guo, J., Long, X.E., Sun, X., Jiang, B., Hou, Q., Jetten, M.S., *et al.* (2015). Biogeographical distribution of denitrifying anaerobic methane oxidizing bacteria in Chinese wetland ecosystems. Environ. Microbiol. Rep. *7*, 128–138. https://doi.org/10.1111/1758-2229.1221

Metagenomic Approaches Unearth Methanotroph Phylogenetic and Metabolic Diversity

3

Garrett J. Smith[1]* and Kelly C. Wrighton[1,2]*

[1]Department of Microbiology, The Ohio State University, Columbus, OH, USA.
[2]Department of Soil and Crop Science, Colorado State University, Fort Collins, CO, USA.

*Correspondence: smith.10284@osu.edu and kelly.wrighton@colostate.edu

https://doi.org/10.21775/9781912530045.03

Abstract

Methanotrophic microorganisms utilize methane as an electron donor and a carbon source. To date, the capacity to oxidize methane is restricted to microorganisms from three bacterial and one archaeal phyla. Most of our knowledge of methanotrophic metabolism has been obtained using highly enriched or pure cultures grown in the laboratory. However, many methanotrophs currently evade cultivation, thus metagenomics provides a complementary approach for gaining insight into currently unisolated microorganisms. Here we synthesize the studies using metagenomics to glean information about methanotrophs. We complement this summary with an analysis of methanotroph marker genes from 235 publicly available metagenomic datasets. We analyse the phylogenetic and environmental distribution of methanotrophs sampled by metagenomics. We also highlight metabolic insights that methanotroph genomes assembled from metagenomes are illuminating. In summary, metagenomics has increased methanotrophic foliage within the tree of life, as well as provided new insights into methanotroph metabolism, which collectively can guide new cultivation efforts. Lastly, given the importance of methanotrophs for biotechnological applications and their capacity to filter greenhouse gases from a variety of ecosystems, metagenomics will continue to be an important component in the arsenal of tools needed for understanding methanotroph diversity and metabolism in both engineered and natural systems.

Introduction

Ferdinand Cohn described a filamentous microorganism that quickly earned infamy for flourishing in European waterworks and obstructing the flow of water in the late 1800s (Cohn, 1870). More than 100 years after the initial discovery of *Crenothrix polyspora*, this organism still evades cultivation. In the past 40 years, ultrastructure evidence (Völker *et al.*, 1977) and the presence of key functional genes (Stoecker *et al.*, 2006) hinted that these pervasive filamentous cells grew using methane as a carbon and energy source. In 2017, metagenomic sequencing, or the untargeted sequencing of DNA directly from a microbial community, led to the first insight into *Crenothrix* genomic contents (Oswald *et al.*, 2017). In combination with other chemical and imaging technologies, here metagenomics confirmed that these mysterious, filamentous cells were methanotrophs. This is only one example of how metagenomics, especially when combined with other methodologies, has demystified the physiology of uncultivated microorganisms (Brown *et al.*, 2015; Solden *et al.*, 2016).

Methanotrophic microorganisms play immense roles in mediating biogeochemical cycles. These organisms can filter 50–97% of methane from ecosystems (Segarra *et al.*, 2015), and therefore have a large impact on the release of this potent greenhouse gas to the atmosphere. In addition to the carbon cycle, it has also been shown that methanotrophs can aid in the removal of 90–99% of the nitrogen during wastewater treatment (Jewell *et al.*, 1992; He *et al.*, 2015). More recently, methanotrophic strains have been harnessed to degrade organic pollutants, produce a variety of industrial precursors, and scavenge metals such as copper (Sullivan *et al.*, 1998; Semrau *et al.*, 2010; Puri *et al.*, 2015; Strong *et al.*, 2015, 2016). Given their important biogeochemical and biotechnological applications, metagenomics is a critical research tool for improved understanding of methanotroph diversity and physiology in natural and manmade settings.

Today, with the power of recovering near-complete to closed genomes directly from microbial communities, axenic cultivation is no longer a requirement for investigating a microorganism's metabolic capabilities (Solden *et al.*, 2016). Metagenomics, which provides an inventory of the metabolic potential of a community, is especially powerful when utilized in parallel with metatranscriptomics (community RNA sequencing) or metaproteomics (community protein analyses). These latter technologies provide information on gene and protein expression in a microbial community, indicating metabolic processes and organisms that are responding to specific environmental conditions. Here our objective is to describe how metagenomics and enabled 'omic' technologies (multi-omics) have altered perspectives on methanotroph environmental distribution, phylogenetic diversity, and metabolism. Leveraging recent metagenomic studies, we (i) summarize key findings from a variety of ecosystems, (ii) describe the methanotroph genomic foliage added to the tree of life, and (iii) discuss how metagenomic analyses of methanotrophs has impacted other scientific disciplines.

Overview of methanotroph phylogenetic and metabolic diversity

Methanotrophs are currently assigned to three bacterial and one archaeal phyla (Fig. 3.1). Today all known Bacterial methanotrophs encode methane monooxygenase (MMO), the enzyme complex that oxidizes a C–H bond in methane using oxygen. There are two forms of this enzyme, particulate (pMMO) and soluble (sMMO), and at least one of these is required for aerobic oxidation of methane (Dalton, 1983). Functional marker genes for the particulate (*pmoA*) and soluble (*mmoX*) forms are commonly used to probe for bacterial methanotrophs (McDonald *et al.*, 2008; Knief, 2015).

The most well studied bacterial methanotrophs belong to the phylum Proteobacteria specifically within the Gammaproteobacteria and Alphaproteobacteria classes. The Gammaproteobacteria methanotrophs belong to the order Methylococcales, and are often referred to as 'Type I' and 'Type X' methanotrophs. Within the Alphaproteobacteria, methanotrophs are associated with the order Rhizobiales, referred to as 'Type II' methanotrophs (Dalton, 1983). These two Proteobacteria methanotroph classes are distinguished not only by phylogenetic assignment, but also have differences in carbon assimilation pathways and intra-cytoplasmic membrane ultrastructure (Dalton, 1983; Knief, 2015).

The second bacterial phylum containing methanotrophs is the Verrucomicrobia. Here members of the family Methylacidiphilales, have core methane oxidation pathways resembling Proteobacteria methanotrophs. However, unlike the Proteobacteria methanotrophs, these taxa may also utilize hydrogen as an alternative electron donor and require lanthanide metals for growth (Pol *et al.*, 2007, 2014; Mohammadi *et al.*, 2017). The third bacterial phylum of methanotrophs is the Methylomirabilota, which contains the genus '*Candidatus* Methylomirabilis' and were previously classified as members of the candidate NC10 phylum (Ettwig *et al.*, 2010; Glöckner *et al.*, 2017; Parks *et al.*, 2018). These methanotrophs are proposed to generate molecular oxygen for methane oxidation by reducing nitrite, followed by dismutation of nitric oxide into dinitrogen and dioxygen (Ettwig *et al.*, 2008, 2010; Wu *et al.*, 2015). Thus, these bacteria are proposed to couple aerobic methane oxidation with

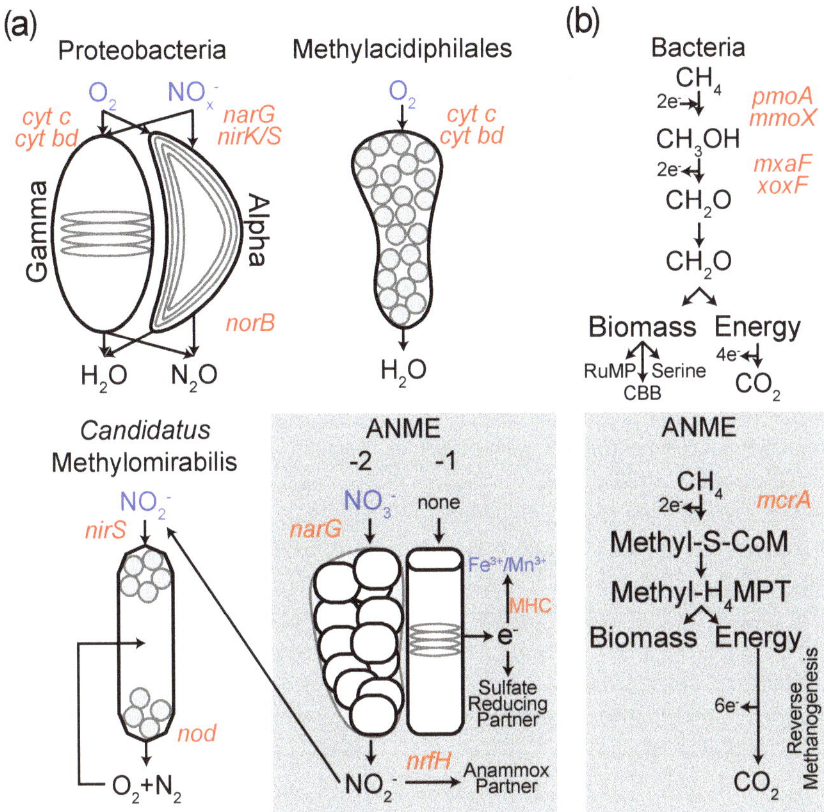

Figure 3.1 Overview of energy generation and methane oxidation by known methanotrophic clades. (a) Energy generation pathway(s) and (b) methane metabolism. Typical morphologies for a model in each clade are shown. Electron acceptors for respiration (blue) and marker genes (red) are labelled for each pathway. Anaerobic metabolisms are identified by the grey background. Acronyms are alphabetically as follows: CBB, Calvin-Benson-Bassham cycle; *cyt c*, cytochrome *c* oxidase; *cyt bd*, cytochrome *bd* ubiquinol:oxidreductase; *mcrA*, methyl coenzyme reductase subunit A; *mmoX*, soluble methane monooxygenase component A; *mxaF*, methanol dehydrogenase alpha subunit; *narG*, respiratory nitrate reductase subunit alpha; *nirK/S*, copper-containing (K) or cytochrome (S) nitrite reductase; *norB*; nitric oxide reductase; *nod*, nitric oxide dismutase; *nrfH*, tetraheme NapC/NirT-type cytochrome c; *pmoA*, particulate methane monooxygenase subunit A; RuMP, Ribulose MonoPhoshate cycle; *xoxF*, PQQ-dependent ethanol/methanol dehydrogenase.

denitrification through an intra-aerobic pathway. Across all the bacterial methanotrophs sampled to date, while much of the enzymatic machinery for aerobic methane oxidation is strongly conserved (e.g. MMO), energy generation and other pathways differ between the phylum (Fig. 3.1) or even strains (Hoefman *et al.*, 2014; Heylen *et al.*, 2016).

Unlike the bacterial methanotrophs, the archaeal methanotrophs do not employ MMO. Archaeal ANaerobic MEthanotrophs (ANME) instead reverse the pathway used by the methanogens to produce methane, hereby oxidizing methane in the absence of molecular oxygen. Archaeal methane oxidation can be coupled to numerous electron acceptors including sulfate, nitrate, nitrite, humic acids, iron or manganese. These organisms thrive in anoxic environments, and, like methanogens, are subjects to oxygen toxicity (Haroon *et al.*, 2013; Scheller *et al.*, 2016; Cai *et al.*, 2018). The variation in electron acceptors, methane inputs, or biomass differences may partially explain the highly variable rates of anaerobic methane oxidation for these taxa across ecosystems (Knittel and Boetius, 2009).

The genes for the enzyme responsible for anaerobic oxidation of methane by ANME, methyl co-enzyme reductase (MCR), can be used in conjunction with 16S rRNA genes to phylogenetically identify these methanotrophs, and can differentiate them from related methanogens (Hallam *et al.*, 2003). Using these marker genes, all currently known ANME lineages are members within the Methanomicrobia class of the Euryarchaeota phylum. Within the Methanomicrobia, the ANME-1 appear to be a distinct order, several lineages of ANME-2 form a monophyletic clade within the Methanosarcinales order, while the ANME-3 likely represent a specific genus within the Methanosarcinaceae family (Knittel and Boetius, 2009; Glöckner *et al.*, 2017; Timmers *et al.*, 2017).

Single marker gene sequencing of the 16S rRNA genes along with bacterial MMO (*pmoA*, *mmoX*) and archaeal MCR (*mcrA*) genes have sampled methanotrophs from a wide range of environments. From these studies it is inferred that a minor fraction of methanotrophic phylogenetic diversity is represented by axenic cultures. In fact, entire clades of methanotrophs still lack a single isolated representative, including ANME and *Ca.* Methylomirabilis. By comparing small subunit ribosomal protein S3 (rpS3) genes recovered in metagenomes with known reference isolate genomic sequences, we highlight that metagenomics continues to recover potentially novel species, i.e. 251 out of 641 recovered genes shared < 95% amino acid identity to methanotroph isolate genomes (see Web resources). Furthermore, this includes the identification of numerous potentially novel species, using the same criteria, even among the well-sampled phylogenetic groups such as the Gammaproteobacteria, e.g. 170 out of 539 rpS3 genes recovered from metagenomes (see Web resources). Recent metaomic studies continue to uncover and define new lineages of methanotrophs, as well as provide necessary metabolic and ecological context, thereby playing a critical role in advancing our understanding of methanotrophy from single genes, to microbial communities, to entire ecosystems.

Metagenomic approaches used to study methanotrophs *in situ* and *in vitro*

A 'metagenome' is the theoretical collection of genomes or genomic information from members of a given microbial community. While metagenomic methods and sequencing technologies have changed over time, the core principle remains: the untargeted sequencing of DNA from a microbial community. The earliest metagenomic methanotroph studies relied on sequencing libraries containing large inserts of DNA cloned into plasmids (Hallam *et al.*, 2004; Meyerdierks *et al.*, 2005; Ricke *et al.*, 2005; Dumont *et al.*, 2006; Chen and Murrell, 2010). Contemporary metagenomics leverage next-generation sequencing advancements, typically resulting in fragments (reads) with increased sequencing depth per sample (Escobar-Zepeda *et al.*, 2015). However, as sequencing technology continues to improve, metagenomics show increasing promise in complete genome recovery from complex environmental samples.

Metagenomic data can be analysed using two main approaches: gene-centric or genome-centric (Fig. 3.2). In one gene-centric approach, sequencing reads are aligned (mapped) to a database of microbial genomes or reference genes to assign reads to closely related genes, pathways or genomes contained in a database. While both time- and cost-effective, this approach may miss novel genomic information, owing to insufficient representation in databases. In an alternative gene-centric approach, sequencing reads can be *de novo* assembled into larger genome fragments, which are then compared to nearest neighbour genomes. Here the advantage is that novel functional and phylogenetic markers can be reconstructed from an environmental sample. The limitation of both gene-centric approaches is the lack of a holistic genome content and organization, hindering the precise understanding of the metabolic pathways or phylogenetic signals essential in description of novel taxa and pathways.

In contrast, in the genome-centric approach, the assembled genomic fragments (contiguous sequences, or contigs) are clustered together to form distinct, population-level genome bins. In the past 15 years, these Metagenome-Assembled Genome bins (MAGs) have yielded near-complete or essentially closed genomes from a myriad of environmental samples (Tyson *et al.*, 2004; Sharon and Banfield, 2013; Parks *et al.*, 2017). This approach allows for genome wide comparisons, which provide expanded phylogenetic resolution through the use of conserved housekeeping genes, residing at various locations in the genome (Rinke *et al.*, 2013; Hug *et al.*, 2016; Yeoh *et al.*, 2016; Parks *et al.*, 2017), as well as for the extended analysis of the metabolic pathway content (Anantharaman *et al.*, 2016), virus–host linkages via CRISPR-Cas loci, etc. (Edwards *et al.*, 2016). A combination

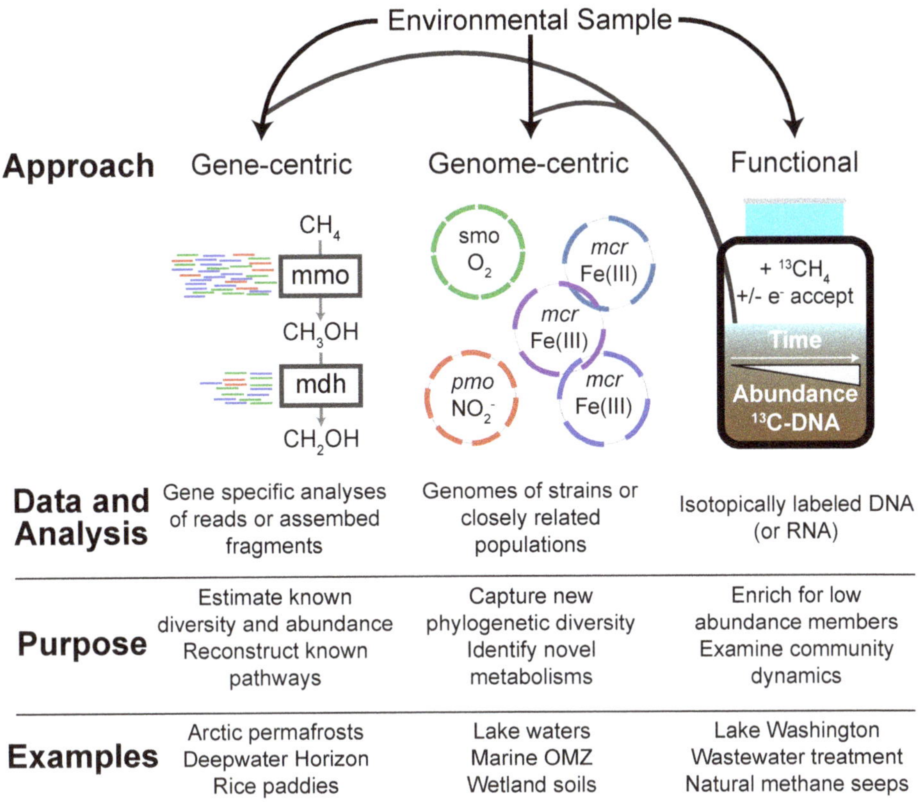

Figure 3.2 Schematic overview of metagenomics approaches. There are two primary methods of metagenomic analysis: gene-centric that maps reads, genes, or contigs (represented by coloured dashes) to a reference database (boxed genes), and genome-centric that reconstructs genomes (broken circles) independent of reference databases. Functional metagenomics precedes these other methods using an initial enrichment, with or without a label, to more directly target methanotrophs, e.g. ^{13}C-methane. The utility of these methods and example ecosystems are outlined for each approach. Acronyms are alphabetically as follows: mmo, methane monooxygenase genes; mdh, methanol dehydrogenase genes; *mcr*, methyl coenzyme M reductase genes; *pmo*, particulate methane monooxygenase genes; smo, soluble methane monooxygenase genes.

of the established bioinformatics platforms and the affordable sequencing technologies now allow for prevalent use of the genome-centric approach. Given sufficient depth of sequencing, this method can recover genomic bins of the quality comparable to the quality of draft genomes from isolate cultures (Bowers *et al.*, 2017), and thus this method represents an important avenue for discovering novel methanotrophs. Here we catalogue the methanotroph MAGs from multiple databases, to provide facile access to these genomic data, and summarize their diversity and ecosystem and high-level genomic information (see Web resources).

Given that all methanotrophs, by definition, are able to consume methane, metagenomic methods have been successfully combined with stable isotope probing (SIP), or incubation of an environmental sample with ^{13}C-methane, followed by the analysis of ^{13}C-DNA (Friedrich, 2006). Sequencing of the labelled DNA, referred to as functional metagenomics, followed by gene- or genome-centric analysis, directly links methane consumption to specific taxa (McDonald *et al.*, 2005; Kalyuzhnaya *et al.*, 2008; Chen and Murrell, 2010). Functional metagenomics, with or without SIP, have been proven to be exceptionally useful for examining slow-growing or consortia-dependent methanotroph taxa, providing a majority of genomic information to date for lineages such as ANME and *Ca.* Methylomirabilis that lack pure culture representatives (discussed below). The difficulties facing both methanotroph isolation and metagenomic sequencing of complex samples can be simplified or overcome by the use of enrichment-based metagenomics.

Metagenomic sampling of methanotroph phylogeny and physiology across diverse habitats

We queried 741 metagenomes and metatranscriptomes associated with methane cycling from the Integrated Microbial Genomes database (IMG, 4th June 2018), provided by the Joint Genome Institute. We used ribosomal protein S3 (rpS3) genes from know methanotroph genomes to mine a total of 566 methanotroph rpS3 proteins, from 235 metagenomic samples, spanning a variety of taxa and ecosystems (see Web resources). In addition to this gene-specific search, we also recovered the rpS3 genes from 75 methanotroph MAGs (37 inferred methanotroph MAGs lacked this gene) (see Web resources). These two metagenome-derived rpS3 data sets, along with reference rpS3 proteins from the known methanotroph isolate lineages, were used to construct a phylogenetic tree visualizing the methanotroph phylogenetic diversity captured by both isolation and metagenomic sampling (Fig. 3.3). Auxiliary rings on the phylogenetic tree provide additional information for class level assignment, the type of sequencing performed, and the ecosystem from which the sequence was recovered. The dissimilarity between reference isolate genomes and MAGs' rpS3 genes shows that many MAGs are not well represented by isolates – in only 22 out of 112 MAGs did the rpS3 gene peptides share > 95% amino acid identity with reference genomes. MAGs continue to play an important role in expanding the methanotroph foliage on the tree of life. Using our meta-analysis as a guide, below we highlight some of the prominent metagenomic studies performed across a range of different habitats, and indicate the insights gleaned about methanotrophs from these studies.

Marine and terrestrial hydrocarbon and methane seeps

Gene-centric metagenomics have provided genomic insights into the anaerobic methane-oxidizing taxa from natural hydrocarbon seeps at Coal Oil Point (California) (Håvelsrud *et al.*, 2011), Troll petroleum reservoir (Norway) (Håvelsrud *et al.*, 2012), Lei-Gong-Hou mud volcano (Taiwan) (Tu *et al.*, 2017), and the Mississippi Canyon (Gulf of Mexico) (Vigneron *et al.*, 2013). In these ecosystems, reaching up to 1100 metres in depth, the ANME types were the most abundant methanotroph taxa. Here, it was inferred that the electron acceptor for methane oxidation was ferric hydroxides or sulfate, the latter reduced through a partnership with sulfate-reducing bacteria (SRB). These results are consistent with much of our understanding of extremely diverse ANME metabolism derived primarily from naturally and artificially enriched vent samples (Knittel and Boetius, 2009), discussed in more detail in sections below. Beyond anaerobic methane oxidation, some additional evidence was provided that aerobic methanotrophs may also be important in marine methane seep systems, as reads mapping to aerobic methanotroph genomes from the Proteobacteria (Methylococcales and Rhizobiales), Verrucomicrobia (Methylacidiphilales), and Methylomirabilota (*Ca.* Methylomirabilis) were also present in many of the metagenomes (Håvelsrud *et al.*, 2011, 2012; Tu *et al.*, 2017). Additionally, gene-centric metagenomics performed on a hydrocarbon seep in the Red Sea reconstructed functional genes for methane oxidation not detected in clone libraries, and was one of the first studies showing that sMMO genes were more abundant than pMMO genes in a natural ecosystem (Abdallah *et al.*, 2014).

In addition to the naturally occurring methane seeps, anthropogenic sources of hydrocarbons also occur in marine and terrestrial systems, and represent another focus for the methanotrophy research. Following the Deepwater Horizon spill in 2010, several metagenomic studies revealed that aerobic Gammaproteobacteria methanotrophs of the order Methylococcales must have consumed nearly all the methane in the water column, mitigating methane release to the atmosphere (Mason *et al.*, 2014; King *et al.*, 2015; Yergeau *et al.*, 2015). The methane consumption activity also caused an oxygen anomaly as well as depleted certain metals in the water column (Kessler *et al.*, 2011; King *et al.*, 2015; Shiller *et al.*, 2017). Taken together, these studies demonstrate both the positive and potentially the negative impacts methanotroph activity may have on an ecosystem scale.

An additional hydrocarbon source is oil sand tailing ponds, which are small natural freshwater bodies contaminated with hydrocarbon waste after industrial petroleum extraction from bituminous sands. SIP-enabled gene-centric metagenomics of Canadian tailings pond sediments recovered

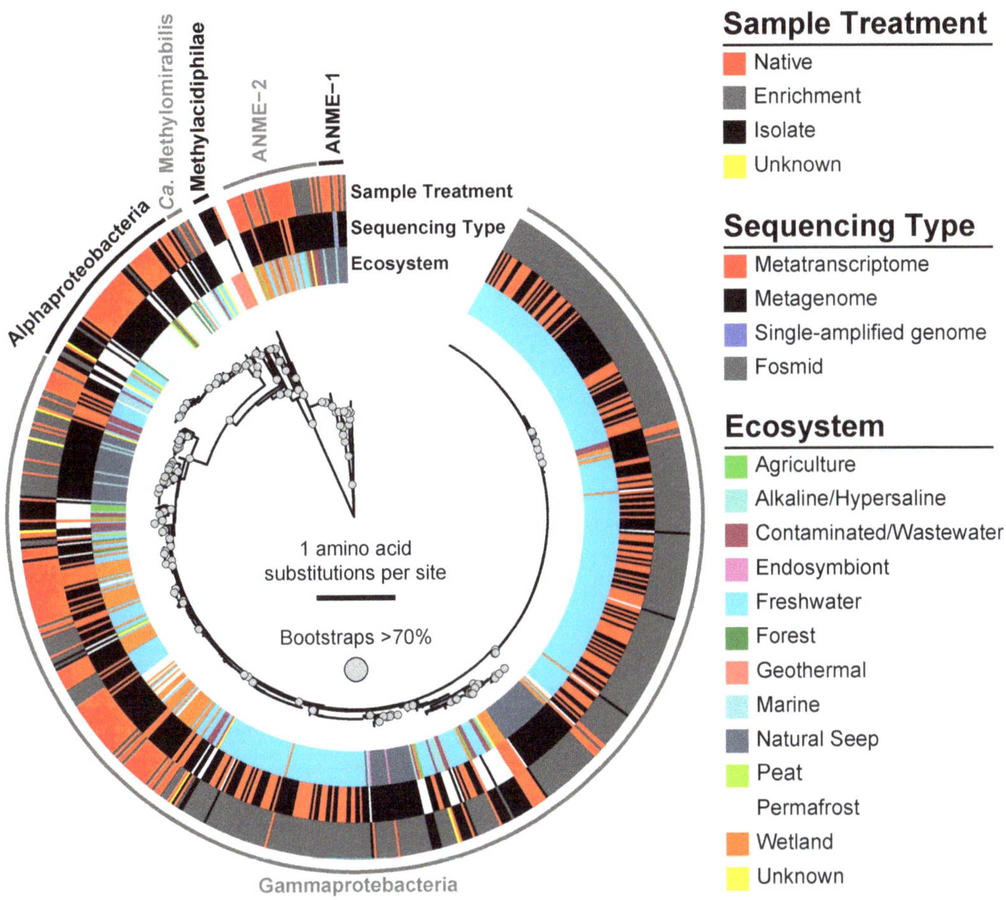

Figure 3.3 Phylogeny of methanotroph ribosomal protein S3 (rpS3) genes identified in metagenomic datasets and isolates. A total 566 recovered and 152 reference – 77 isolates and 75 Metagenome-Assembled Genomes (MAGs) – rpS3 genes were used for analysis if the rpS3 gene was greater than 180 amino acids, approximately 50% of the maximum reference sequence length. Protein sequences were aligned using MUSCLE and modified to remove predominantly gapped, clade-specific positions resulting in 183 residues. The maximum-likelihood phylogenetic tree was generated using RAxML with the GAMMA model of rate heterogeneity and WAG substitution matrix, with 100 bootstrap replicates. Auxiliary layers around the tree are colour-coded for metadata accompanying the genomic information and indicate cultivation status (Sample Treatment), the type of sequencing performed (Sequencing Type), and the environmental source (Ecosystem) of the samples and genomic information. No information is provided for non-methanotrophic out-group sequences.

pMMO genes most similar to the Gammaproteobacteria Methylococcales member of the genus *Methylocaldum,* inferred to cross-feed with non-methanotrophic methylotrophic taxa via methanol (Saidi-Mehrabad *et al.,* 2013). Methanol-mediated communal metabolism of methane has been demonstrated in other ecosystems (Beck *et al.,* 2013; Krause *et al.,* 2017), suggesting that, while methanotroph and methylotroph genera may differ between ecosystems, the metabolic exchanges may be conserved. Although generally in low quantities in the most marine and terrestrial environments, naturally and artificially hydrocarbon- or methane-enriched environments support diverse communities of methanotrophs that mitigate methane release into the atmosphere.

Atmospheric methane sinks in dry soils

Forest soils have long been of interest for the study of methanotrophs, due to their ability to act as net methane sinks by consuming trace (<2 parts per million) quantities of methane present in the atmosphere. Several studies have inferred that this phenomenon, named high-affinity methane oxidation (HAMO), likely occurred through the activity of uncultivated Alphaproteobacteria methanotrophs (Dunfield *et al.*, 1999; Holmes *et al.*, 1999; Henckel *et al.*, 2000). One of the early SIP-enabled metagenomics projects generated sequences of Bacterial Artificial Chromosomes (BACs) from forest soil, recovering fragments of genomes most closely related to members of the Alphaproteobacteria genus *Methylocystis* (Dumont *et al.*, 2006). This demonstrated the application of this technique for recovering the genomic contents of methanotrophs that are generally present at low abundance in complex soil communities. In a separate study, sequencing BACs from forest soils provided the first genomic information for the Alphaproteobacterial Upland Soil Cluster (USCα) methanotrophs, a prevalent lineage in forest soils implicated in oxidizing methane at atmospheric pressure levels (Ricke *et al.*, 2005). The gene-centric approach revealed that, based on the pMMO and neighbouring genes, the proposed USCα methanotroph was most similar to the known Alphaproteobacteria methanotrophs (Ricke *et al.*, 2005).

Recently, Pratscher *et al.* (2018) provided the first morphological information for the USCα, and recovered the first draft genome of USCα methanotroph, by combining targeted cell sorting and metagenomic sequencing. The resulting MAG lacked sMMO genes, but contained several pMMO gene variants (Pratscher *et al.*, 2011, 2018). It was proposed that these different pMMO variants may function in a manner similar to variants previously identified in *Methylocystis* isolates, showing different substrate affinities towards methane (Baani and Liesack, 2008; Dam *et al.*, 2013). Additionally, the USCα genome encoded pathways for acetate utilization, signifying the significance of the capability to utilize alternative carbon substrates as a means of adaptation for HAMO (Pratscher *et al.*, 2011, 2018). Mining of 16S rRNA gene-databases posited that USCα species are not restricted to forest soils, but are also present in lava cave biofilms and arctic permafrost (Pratscher *et al.*, 2018). Similar to these findings, Singleton *et al.* (2018) recovered a partial USCα methanotroph genome from partially thawed permafrost soils in Sweden. The much wider distribution of USCα methanotrophs, including non-arid environments such as permafrosts, suggests the importance of HAMO on a global scale.

In addition to the USCα methanotrophs, the first nearly complete genome of the divergent Gammaproteobacterial Upland Soil Cluster (USCγ) methanotroph has recently been recovered from cold, arid soils. This genome was shown to be significantly divergent from the genomes of known Methylococcales (Edwards *et al.*, 2017) (Fig. 3.3; see Web resources), while encoding typical methylotrophy metabolic pathways. Since their initial discovery in forest soils, the recovery of USCγ from Taylor Dry Valley in Antarctica demonstrates their distribution beyond forested environments (Edwards *et al.*, 2017). Importantly, for both USCα and USCγ lineages, these MAGs alone cannot explain the specific metabolic requirements or the enzymatic differences that enable HAMO. Given the broad environmental distribution of these organisms (northern latitude forests, Antarctic deserts, and permafrost ecosystems), there is a need for cultivation and physiological studies targeting these lineages.

Freshwater lake sediments

One of the best-characterized ecosystems for microbial C_1 cycling is freshwater lake sediments. These systems have anoxic conditions in the deeper sediments, resulting in methanogenesis, with methane diffusing to oxic sediment–water interface, where aerobic methanotrophy occurs. One of the earliest uses of SIP in combination with metagenomics investigated the microbial taxa involved in methylotrophy in Lake Washington sediments (Kalyuzhnaya *et al.*, 2008). Here sediments incubated with ^{13}C-methane largely recaptured the data from the cultivation efforts, showing that *Methylobacter,* of the Gammaproteobacteria Methylococcales clade, were the dominant methane-oxidizing taxa (Auman *et al.*, 2000). Additionally, these methane-fed communities were enriched for Betaproteobacteria species *Methylotenera,* which were likely labelled through consuming methanol, a by-product of methane oxidation. This metabolic cross-feeding spurred a series of experiments showing that at lower oxygen:methane mixing ratios, *Methylobacter,* consistently outcompeted other methanotrophs, and also produced methanol to cross-feed *Methylotenera* (Beck *et al.*, 2011; Oshkin *et al.*, 2015; Krause *et al.*, 2017).

Alternatively, under high oxygen:methane mixing ratios, members of the same family but different genus (*Methylosarcina*) cross-fed different Betaproteobacteria microorganisms (*Methylophilus*) (Hernandez *et al.*, 2015). These findings highlight the role of cooperative metabolic exchanges fuelled by methanotrophy occurring across ecosystems, and the value of both laboratory and field based studies.

Thawing arctic permafrost

There is growing interest in the metabolic activity of methanotrophs in arctic peat and permafrost soils. These environments store approximately 1300–1600 petagrams of carbon that is susceptible to release under predicted climate warming scenarios (Dean *et al.*, 2015). Warming temperature trends threaten to thaw permafrost, making the previously inaccessible carbon newly available. The newly accessible organic carbon can be degraded to yield methane, which may be consumed by the methanotrophs or be emitted into the atmosphere, the latter with future climatic feedbacks (MacKelprang *et al.*, 2011).

Several studies by Tveit *et al.* (2013, 2014) clearly articulated the importance of aerobic Gammaproteobacteria Methylococcales in warming permafrost. In the first two studies, metatranscriptomic data were collected from permafrost soils in Svalbard, Norway. Here 14 nearly full-length assembled 16S rRNA genes (> 1330 nucleotides) in the metatranscriptome were assigned to the genus *Methylobacter*. Metatranscriptomic reads were mapped to pMMO, methanol dehydrogenase, and formaldehyde oxidation pathways from reference genomes, further demonstrating that members of *Methylobacter* were primarily responsible for methane oxidation in these permafrost soils. Notably, genes from other methanotrophs were absent, or present at negligible counts (Tveit *et al.*, 2013, 2014), demonstrating the metabolic prevalence of this one genus in these soils. Laboratory mesocosm incubations designed to examine the effects of thaw gradient (temperatures ranging from 1 to 30°C) on these soils also demonstrated that *Methylobacter* remained the dominant methanotroph (Tveit *et al.*, 2015). Taking these studies together, it is possible that in these permafrost soils, active methanotrophy today and under future warming scenarios may be mediated by a single genus of methanotrophic bacteria (Tveit *et al.*, 2015).

Other studies have implicated Alphaproteobacteria as key methanotrophs to respond to permafrost thaw. Mackelprang *et al.* (2011) used changes in the relative abundance of 16S rRNA and pMMO genes during *in vitro* thaw simulation experiment, to infer members of the genus *Methylocystis* as responsive species. A large-scale study at Stordalen Mire in Sweden found both classes of Proteobacteria methanotrophs to be present, but these were differentially distributed across the thaw gradient (Singleton *et al.*, 2018). Metagenomes from partially thawed and intact permafrosts contained mostly reads mapping to Alphaproteobacteria methanotrophs, particularly *Methylocystis* and USC*α* MAGs, while fully thawed permafrost metagenomes harboured reads that mapped mostly to Gammaproteobacteria Methylococcales MAGs. This trend was matched by the distribution and diversity of pMMO and sMMO genes identified in the metagenomes, as well as metatranscriptomic reads mapping to these genes (Singleton *et al.*, 2018).

Two other studies in the Canadian High Arctic also used metagenomics technologies to investigate methanotroph diversity and activity in permafrost soils. One gene-centric metagenomics study concluded that Gammaproteobacterial Methylococcales were the dominant methanotrophs in the active layer and up to 2 m depth into the arctic permafrost (Yergeau *et al.*, 2010). In contrast, a paired metagenomic, metatranscriptomic, and metaproteomic investigation of another permafrost site noted that Alphaproteobacterial USC*α* methanotrophs recruited most reads and peptide fragments in all datasets (Lau *et al.*, 2015). It was inferred that the relative aridity and atmospheric methane consumption in this site, compared to other permafrost field locations, may have contributed to the prevalence of USC*α*. Similarly, Siberian permafrosts were also dominated by Alphaproteobacteria members of the genus *Methylocystis,* but Gammaproteobacteria methanotrophs could also be detected (Rivkina *et al.*, 2016).

In summary, while the active methane-oxidizing communities across several different permafrost habitats were dominated by Proteobacteria methanotrophs, it appeared that the thaw stage was a major controlling factor for the distribution and activity of different clades of methanotrophs in these permafrost ecosystems. Therefore, understanding the physiological constraints, e.g. temperature, hydrology, methane oxidation rates, etc. of Proteobacteria methanotrophs will be critically important for improving predictions of methane release from Northern thawing permafrost ecosystems. Understanding these dynamics is important for predicting the efficacy of methanotrophs in reducing methane release from ecosystems with large amounts of stored carbon.

Temperate hydric soils and sediments (peats or wetlands)

Temperate wetlands can be defined as flooded or seasonally flooded soils located in regions with relatively mild climates, and these include freshwater, coastal, and peatland soils. Despite their small land coverage, on a global scale, these systems account for a disproportionate fraction of the total methane budget (Nazaries *et al.*, 2013; Dean *et al.*, 2018). Below we summarize some of the recent metagenomic investigations of these habitats.

In a boreal peat bog in Minnesota, SIP-enabled metagenomes revealed enrichment for Alphaproteobacteria *Methylocystis* and Gammaproteobacterial *Methylomonas* pMMO genes, at a ratio of about 3:1 (Esson *et al.*, 2016). At increasing depths, the *Methylomonas* signal was eventually below detection, consistent with qPCR and microarray data from the same study site (Esson *et al.*, 2016). Other studies also found that pMMO genes affiliated with *Methylocystis* and other Alphaproteobacteria methanotrophs tended to be more abundant, possibly owing to the acidic pH or to copper limitation (Chen *et al.*, 2008; Kolb and Horn, 2012; Lin *et al.*, 2014; Graham *et al.*, 2017). Corroborating decades of interest in peatland systems, gene-centric analyses have reaffirmed the importance of Alphaproteobacteria methanotrophs in many of these acidic environments as natural filters of methane release.

In a restored peat ecosystem in California, metagenomes from samples collected up to 25 cm into the soil column revealed methanotroph relative abundances ranging from ≈ 1% to 3% of total microbial community (S. He *et al.*, 2015). Both the number of pMMO genes and the reads mapped to components of methane oxidation pathways were enriched in plant-associated samples, dominated primarily by Alphaproteobacteria *Methylocystis* and *Methylosinus*. Data collected from grasslands, forests, mangroves, and systems replete in plants corroborated the finding that *Methylocystis* types were the most prevalent methanotrophs in soils within the rhizosphere (Xu *et al.*, 2014). Therefore, further understanding of the physiology and metabolism of *Methylocystis* is important for improving models for methane cycling in vegetated environments.

The Great Lakes region on the border between the United States and Canada stores large amounts of carbon (Nahlik and Fennessy, 2016), and wetlands adjacent to Lake Erie are known to be net methane emitters, with seasonal variations in methane output (Angle *et al.*, 2017; Rey-Sanchez *et al.*, 2018). Gene-centric approaches, 16S rRNA gene abundances and pMMO gene transcript inventories, have shown that *Methylobacter* species dominated the methanotroph community (Smith *et al.*, 2018). Furthermore, pMMO transcription was season-dependent, and found to be increased in autumn relative to summer. In contrast, methanogen gene expression was relatively stable over the sampling period, suggesting that the greater methane emissions observed in the summer months might be due to the reduced methanotroph activity and not to increased methanogenesis activity (Smith *et al.*, 2018). A biogeographical survey conducted by Smith *et al.* (Smith *et al.*, 2018) detected pMMO genes similar to the those of *Methylobacter* lineages found in many methane-emitting Northern latitude environments that include Lake Washington (Chistoserdova, 2011a), the restored Twitchell Island peatlands (He, S. *et al.*, 2015), and the Prairie Potholes region of North Dakota (Dalcin Martins *et al.*, 2017). Despite major differences in the latitude, the prevalence of *Methylobacter* genotypes and their genomic conservation across geographically distinct wetland and permafrost habitats suggests that further studies of the representatives of this genus are relevant for global climate change research (Tveit *et al.*, 2013).

Rice paddies and agricultural soils

Rice paddies represent inundated agricultural soils that contribute 4–20% to the atmospheric methane budget (Nazaries *et al.*, 2013; Dean *et al.*, 2018), despite most of the methane generated (up to 90%) being consumed by the methanotrophs (Bao *et al.*,

2014; Lee *et al.*, 2015). Given the large error in current predictions for methane emissions from these habitats, there is a desire to better understand how agricultural land management impacts the activity of methane-cycling microorganisms. In particular, rice paddies are characterized by drastic oxyclines that occur along the water depth profile and along the distance from the plant roots, conducive of both aerobic and anaerobic methanotrophy (Reim *et al.*, 2012; Lee *et al.*, 2015). A study on Indian rice paddies with different fertilizers by Bhattacharyya *et al.* taxonomically classified reads, finding that Gammaproteobacteria Methylococcales, particularly *Methylobacter* and *Methylococcus*, were the most abundant methanotroph taxa, regardless of fertilizer treatment (Philippot *et al.*, 2010; Bhattacharyya *et al.*, 2017). Alphaproteobacteria and Verrucomicrobia methanotrophs were also detected, but were less abundant. The reconstruction of the formaldehyde assimilation pathways showed predominance of the serine cycle genes, which are present in both Alphaproteobacteria and Gammaproteobacteria methanotrophs (Ward *et al.*, 2004; Takeuchi *et al.*, 2014).

Profiling active methane metabolism along rice paddy soil compartments, including water, soil, phyllo-, rhizo-, and endo-spheres, is a current research area. One study that applied metagenomics in combination with metaproteomics to both the phyllosphere and the rhizosphere has found that proteins for both methane oxidation and methanogenesis were significantly enriched in the rhizospheres. Both Alphaproteobacteria and Gammaproteobacteria methanotrophs were detected, and both sMMO and pMMO were expressed at approximately equal peptide spectra (Knief *et al.*, 2012). In contrast, Bao *et al.* (2014) failed to recover any peptides mapping to Methylococcales and found that rice root metaproteomes mainly represented methanotrophic Alphaproteobacteria of the Methylocystaceae family, which carry out both methanotrophy and dinitrogen fixation in the vascular bundles and epidermal cells of rice roots. A separate metatranscriptomic analysis of rice paddy laboratory incubations found Methylococcales to be the predominant methane oxidizer, but was restricted to the very narrow 2 mm thick oxic surface layers, and no transcripts affiliated with methanotrophs were detected in the anoxic zone at 6–8 cm (Kim and Liesack, 2015).

Seasonal dynamics likely play a role in constraining methanotroph abundance and activity in rice paddies. During the drought season, rice paddies typically serve as net sinks of methane (Harriss *et al.*, 1982; Kelley *et al.*, 1995; Melling *et al.*, 2007; Kolb and Horn, 2012). Incubation of rice paddy soils at high versus near-atmospheric methane mixing rations found a strong enrichment for Gammaproteobacteria genus *Methylosarcina* and mild enrichment of Alphaproteobacteria *Methylocystis* (Cai *et al.*, 2016). From a combination of metatranscriptomics and SIP experiments, it was thought that *Methylosarcina* responded to the initial spikes of methane (≈9000 ppm methane), consuming methane to produce metabolites that are necessary to support HAMO (≈2 ppm methane) by the Alphaproteobacteria methanotrophs. Interestingly, this study noted that neither of the USC clades were detected in the metagenomes or metatranscriptomes, therefore HAMO observed in these samples was dependent on the known methanotrophs and not on the expected USC-type methanotrophs.

Subsurface ecosystems sustained by methane

Microbial life in caves and subsurface ecosystems, which typically lack organic carbon input by plant primary productivity, is dependent on chemolithotrophic metabolisms. Methane is a relatively rich source of energy and carbon in a variety of geographically unrelated subsurface ecosystems (Hutchens *et al.*, 2004; Brankovits *et al.*, 2017; Karwautz *et al.*, 2018). A model cave system in Romania, Movile Cave, undergoes seasonal extremes of methane concentrations due to consumption by up to four distinct bacterial clades of methanotrophs – USCα, USCγ, and two potentially novel lineages (Waring *et al.*, 2017). Metagenomic analysis revealed that aerobic methanotrophic marker genes were present in both the oxic mat and the anoxic sediment, and also showed that the predominant MMO types varied between niches (Kumaresan *et al.*, 2018). In another subsurface environment, an artificial mine in a granitic system in Japan, methane oxidation appeared to entirely rely on the ANME type Archaea, and a nearly complete genome of an ANME-2d organism has been reconstructed that differed from the previously reconstructed ANME-2d genomes by lacking any genes for denitrification or metal reduction (Ino *et al.*, 2018). Instead, the genome encoded functions for a putative respiratory H_2 oxidation as well as a possible pathway for utilizing zero-valent sulfur to exchange electrons with syntrophic partners (Milucka *et al.*, 2012; Ino *et al.*, 2018). Overall, caves and other subsurface ecosystems are relatively understudied environments, but the few examples in the literature suggest that these ecosystems are home to interesting methanotrophs, and that methane oxidation occurs differentially along ecological niches.

Metagenome-derived genomes reveal methanotroph metabolic variability

In this review, we compiled the genomic information for all 80 isolates and 112 MAGs available via the IMG or the National Center for Biotechnology Information (NCBI), assigned to methanotroph taxa or taxa described as a putative methanotrophs in publications (Fig. 3.4; see Web resources). Reconstruction of MAGs provides genomic information for major clades entirely lacking a single pure culture representative, e.g. ANME, USCα, USCγ or *Ca.* Methylomirabilis. These MAGs provide essential information about these microorganisms that are phylogenetically and physiologically divergent from the known, isolated methanotrophs. Genome-centric metagenomics approaches have more than doubled methanotroph genomic database sampling in as little time as a decade. Below, we summarize the major discoveries that analyses of MAGs have added to the phylogenetic and metabolic understanding of methanotrophs in natural as well as artificial settings.

MAGs represent the only genomic information from some uncultivated methanotroph lineages

To date there is not a single ANME representative in axenic culture. Genomic information for ANME-1 was first obtained in 2005, linking methanogenesis and 16S rRNA genes on the same contig (Meyerdierks *et al.*, 2005). This finding was followed by the first MAG described for ANME-1 lineage, recovered in 2010 by Meyerdierks *et al.* (2010) by fosmid sequencing of microbial mats in the Black Sea. From this MAG, the authors reconstructed the first near-complete pathway for Archaeal methane oxidation, inferring the methanogenesis pathway operating in reverse (Hallam *et al.*, 2003; Meyerdierks *et al.*, 2010). Additionally, they recovered the first [Fe–Fe] hydrogenases in Archaea, as well as the multiple multi-haem *c* type Cytochromes (MHC) that were proposed to act as the electron shuttling intermediates involved in interactions with the sulfate-reducing bacteria (SRB) or in metal reduction (Meyerdierks *et al.*, 2010). These genomes were critical, providing the first evidence for the types of anaerobic energy generation mechanisms encoded by the archaeal methanotrophs. Shortly after, combined metagenomic and metaproteomic/metatranscriptomic studies corroborated the genomic potential postulated from the first ANME-1 genome (Stokke *et al.*, 2012; Krukenberg *et al.*, 2018).

Shortly after the recovery of the first ANME-1 genome, a MAG representative of the ANME-2d lineage was reconstructed from a methane consuming and denitrifying bioreactor inoculated with a mixture of waste sludge and freshwater lake sediments (Haroon *et al.*, 2013). This near-complete genome is referred to as *Candidatus* Methanoperedens nitroreducens, named after the inferred metabolic capacity to couple methane oxidation, via reverse methanogenesis, to dissimilatory nitrate reduction (Haroon *et al.*, 2013). In contrast to ANME-1, the *Ca.* Methanoperedens genome contained a full reverse methanogenesis pathway. An additional unique feature of this genome was the presence of only the first gene for denitrification, *nar,* which would only allow *Ca.* Methanoperedens nitroreducens to reduce nitrate to nitrite (Haroon *et al.*, 2013). Metatranscriptomic coupled to ^{15}N-nitrate SIP analysis applied to the bioreactor microbial community suggested that the nitrite generated was likely consumed by the nitrite-utilizing microorganisms, including *Ca.* Methylomirabilis, or by anaerobic ammonia-oxidizing microorganisms (Raghoebarsing *et al.*, 2006; Haroon *et al.*, 2013). Subsequent ANME-2 MAGs from enriched freshwater canal sediments in the Netherlands found that other *Ca.* Methanoperedens nitroreducens MAGs encoded the capacity to perform dissimilatory nitrate reduction to Ammonia (DNRA, *nrfH*) (Arshad *et al.*, 2015). Taken together, these studies are important for highlighting the coupling of anaerobic methane oxidation to nitrate reduction.

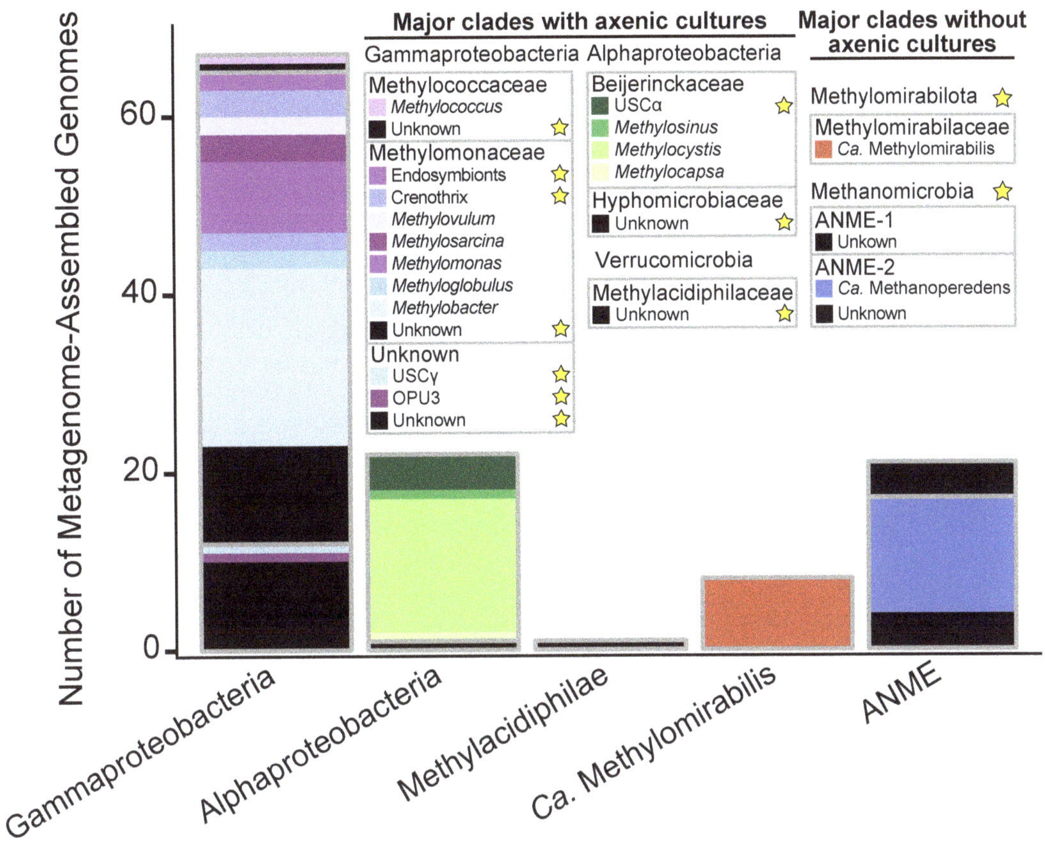

Figure 3.4 Phylogenetic composition of rpS3 genes recovered in putative methanotrophic Metagenome-Assembled Genomes (MAGs). Distribution of MAGs at the genus level, if known, among the major clades of methanotrophs and grouped by family according to SILVA (SSU refNR 132) and the Genome Taxonomy DataBase (GTDB). Yellow stars indicate the lack of axenic, i.e. pure, cultures of specific methanotrophic genera (Proteobacteria and Verrucomicrobia) and classes (Methylomirabilota and Methanomicrobia).

Since their initial discovery, approximately 20 genomes have been reconstructed for the ANME lineages, including 4 ANME-1 and 15 ANME-2 representatives (Fig. 3.4, see Web resources). MAGs of both ANME-1 and ANME-2 lineages contain a surprising number of MHC genes, suggesting that both groups may be capable of coupling methane oxidation to the reduction of metals (Meyerdierks *et al.*, 2010; Wang *et al.*, 2014; Cai *et al.*, 2018), or Direct Interspecies Electron Transfer (DIET) to bacterial consortia (Wegener *et al.*, 2015). Studies investigating anaerobic methane metabolism have shown that ANME-1 increase extracellular cytochrome production and form cellular structures connecting methane oxidizers to their partnering SRBs (Wegener *et al.*, 2015). Others have shown that ANME-2, possessing MHC with 41 haem-binding motifs, coupled methane oxidation to iron reduction (Ettwig *et al.*, 2016). Furthermore, these analyses also suggested that MHC was the prevailing mode of interaction between the ANME and the SRB among both the thermophilic and the psychrotolerant communities of the, respectively, Guaymas Basin and the Crimean Peninsula, and the mid-Norwegian margin (Meyerdierks *et al.*, 2010; Stokke *et al.*, 2012; Krukenberg *et al.*, 2018), suggesting that this function is important across temperature and geographic ranges. Genomic inferences suggested that ANME-2 have the potential for nitrogen fixation, a finding confirmed by the cellular incorporation of labelled dinitrogen gas (Pernthaler *et al.*, 2008), and in contrast to the other ANME-2 MAGs reconstructed from subsurface ecosystems (Ino *et al.*, 2018). Taken together, metagenomics, often in combination with other techniques, have provided a wealth of information on the metabolic diversity maintained among ANME lineages.

These reconstructed ANME MAGs have also enabled improved detection of the ANME-type methanotrophs from a wide range of ecosystems. For instance, sequences have been recruited to ANME genomes from deep marine ecosystems and wastewater treatment reactors (Raghoebarsing *et al.*, 2006; Håvelsrud *et al.*, 2011, 2012; Lee *et al.*, 2016; Welte *et al.*, 2016; Cai *et al.*, 2018), but also from rice paddies, wetlands, and peatlands soils, as well as from subsurface locations (Takeuchi *et al.*, 2011; Vaksmaa *et al.*, 2017; Ino *et al.*, 2018). The impact of ANME in ecosystems exposed to dynamic oxygen fluctuations is poorly understood relative to marine and wastewater treatment environments, but more genomic representatives in databases may result in promising new areas for future research.

In addition to ANME, methanotrophic representatives of the Methylomirabilota, formerly of the candidate NC10 phylum, the *Ca.* Methylomirabilis genus also lack any pure culture isolates. However, genomic insights into the metabolism of this group are enabled by at least eight MAGs. These genomes were reconstructed from denitrifying wastewater treatment reactors (Raghoebarsing *et al.*, 2006). These anoxic reactor fluids, when supplemented with nitrate and methane, became enriched with two anaerobic methanotroph populations: ANME-2, described above, and *Ca.* Methylomirabilis (Ettwig *et al.*, 2010). The exact mechanism of nitrogen transformation in this system was largely unknown until the recovery of MAGs from the bioreactor. Analysis of the *Ca.* Methylomirabilis MAGs identified a paradoxical lack of the nitrous oxide reductase (*nos*) genes that were expected to be responsible for the stoichiometric conversions of nitrite to dinitrogen gas, observed in previous studies (Ettwig *et al.*, 2010, 2012). To explain this phenomenon, Ettwig *et al.* (2010) hypothesized that an unknown enzyme, a nitric oxide dismutase (NOD), was responsible for disproportionation of two nitric oxide molecules into one molecular nitrogen and one molecular oxygen. While the biochemistry of this hypothesized enzyme remains to be fully elucidated (Reimann *et al.*, 2015), intracellularly generated oxygen could explain the production dioxygen in cultures with no other exogenous sources, and enable the traditional aerobic methane oxidation using oxygen-dependent pMMO in anoxic conditions (Ettwig *et al.*, 2010).

MAGs demonstrate the breadth of phylogenetic diversity within the Proteobacteria methanotrophs

Beyond the lineages comprised exclusively of uncultivated methanotrophs, MAGs have also sampled lineages represented by or closely related to the cultivated methanotrophs (Fig. 3.4). Multiple MAGs have been recovered related to the members of *Methylobacter*, *Methylomonas*, and *Methylocystis* (Fig. 3.4; see Web resources). Additionally, MAGs reconstructed from marine oxygen minimum zone and marine endosymbiont samples have sampled novel lineages within the Gammaproteobacteria. In this section, we highlight the metabolic information derived MAG representatives of uncultivated Proteobacteria methanotroph lineages.

In marine systems, endosymbiotic methanotrophs have been identified, forming thus far a monophyletic lineage within Gammaproteobacteria Methylococcales (Petersen and Dubilier, 2009). One of the most well-known marine animals to host the methanotrophs in its gill tissues is *Bathymodiolus*, a mussel found at the bottom of the ocean (Fisher *et al.*, 1987; Nix *et al.*, 1995). Because endosymbiotic bacteria are difficult to isolate, knowledge on their metabolism relies on these culture-independent methods. Phylogenetic analyses using 16S rRNA and pMMO genes and ultrastructure affiliated the endosymbiotic methanotrophs with the Methylococcales genus *Methyloprofundus* (Spiridonova *et al.*, 2006; Tavormina *et al.*, 2015; Sun *et al.*, 2017).

Two recent publications independently reconstructed MAGs from distinct species of *Bathymodiolus* (Ponnudurai *et al.*, 2017; Takishita *et al.*, 2017). The rpS3 genes recovered from these genomes were most closely related to each other and to *Methyloprofundus* (Fig. 3.3; see Web resources). These MAGs were not only enlightening for resolving the identity of the endosymbionts, but also for their function(s) within the host. Ponnundurai *et al.* (2017) concluded that the methanotroph supplied carbon dioxide for assimilation by other non-methanotrophic endosymbionts, as well as synthesizing amino acids and necessary cofactors for the host.

Takishita *et al.* (2017) found that the methanotrophic endosymbiont synthesized unique cholesterol intermediates that the mussel could not produce on its own. Further investigations into these relationships using metagenomics will likely uncover additional critical functions these endosymbiotic methanotrophs perform for the host, as well as further insights into the evolution of methanotroph-animal endosymbioses.

Recovery of MAGs from low-oxygen marine ecosystems have enabled the identification of multiple uncultivated, novel Gammaproteobacteria clades. From deep-sea marine vents at the Lau Basin, Methylothermaceae B42 is the only genomic representative of a likely novel genus in the Gammaproteobacteria Methylococcales (Skennerton *et al.*, 2015). This genome shares only 94% 16S rRNA sequence identity with its closest isolated relative in the genus *Methylohalobius*. Similarly, the first genome from the uncultivated Operational *PmoA* Unit 3 (OPU3) lineage was recovered from Oxygen Minimum Zone (OMZ) of Gulfo Dulce, sharing <93% 16S rRNA sequence identity with any Methylococcales isolates (Hayashi *et al.*, 2007; Padilla *et al.*, 2017). OPU3 appear to be cosmopolitan in marine systems and are enriched in some OMZ ecosystems and during the Deepwater Horizon oil spill (Tavormina *et al.*, 2010; Lesniewski *et al.*, 2012; Li *et al.*, 2014). Both Methylothermaceae *B42* and *OPU3* MAGs were sampled from oxygen-limited environments, and other 16S rRNA gene surveys indicate that the OPU3 are found in deep waters (100–2,000 m depth), suggesting tolerance to low dissolved oxygen concentrations (Tavormina *et al.*, 2010), as discussed below. These two studies have both recovered genomic information for uncultivated lineages in marine systems, which likely signify the beginning of future discoveries of novel methanotrophic Gammaproteobacteria taxa in marine environments through metagenomics.

Although most information regarding the distribution and ecology of USC methanotrophs comes from forest soils, MAGs representing both USCα and USCγ have been reconstructed from polar latitudes (Edwards *et al.*, 2017; Pratscher *et al.*, 2018; Singleton *et al.*, 2018). There is increasingly strong evidence that the USCα are a divergent lineage closely related to Alphaproteobacteria genus *Methylocapsa* (Ricke *et al.*, 2005; Pratscher *et al.*, 2018; Singleton *et al.*, 2018), but evolutionary history of the USCγ is less clear (Edwards *et al.*, 2017). While the pMMO marker genes are similar to Methylococcales genus *Methylocaldum*, the 16S rRNA gene recovered in the MAG was most similar to microorganisms within the Gammaproteobacteria class Chromatiales (Edwards *et al.*, 2017), similar to its placement upon the fringe of the Methylococcales clade in the rpS3 gene phylogeny (Fig. 3.3). To our knowledge, there are no other close isolated relatives to the USCγ MAG, therefore more cultivation-based sampling is needed to make inferences about metabolic or phylogenetic relationships for this genome (Edwards *et al.*, 2017).

Interestingly, the recovery of USCα MAGs in thawing permafrost by Singleton *et al.* (2018) uncovered more phylogenetic novelty within the Alphaproteobacteria. This includes the presence of pMMO and sMMO genes in a novel genus of the non-methanotrophic but methylotrophic Alphaproteobacteria family Hyphomicrobiaceae (Singleton *et al.*, 2018). This is not the first occurrence of MMO detected in non-methanotrophic methylotrophic clade; a species of non-methanotrophic isolate genus *Methyloceanibacter* was isolated that encoded pMMO genes and could utilize methane (Vekeman *et al.*, 2016a). These genome-centric studies in polar regions have not only improved the phylogenetic resolution of the USC methanotrophs, but have additionally posited a novel taxa capable of oxidizing methane.

Despite being studied for decades, novel lineages of methanotroph within the Proteobacteria are constantly being discovered. This is true of ecosystems ranging from the bottom and deep waters of the ocean to terrestrial ecosystems across the globe. Interestingly, much of the new diversity is found within the existing methanotroph families, e.g. Beijerinckaceae, Methylocystaceae, and Methylococcaceae. Further analyses of these genomes may provide information on the strain-level variations enabling adaptation to different environmental conditions across the globe. It is apparent that natural ecosystems host many novel methanotrophs, and the employment of metagenomics will continue to resolve the vast diversity of methanotrophs by reconstructing MAG representatives of uncultivated lineages.

Methylococcales MAGs widely encode the denitrification potential, potentially enabling methane oxidation under low oxygen conditions

Using pure cultures, Stein and colleagues provided evidence that Gammaproteobacteria methanotrophs may conserve oxygen for methane oxidation and use nitrate, rather than oxygen, to generate a proton motive force for energy generation (Kits *et al.*, 2015). This metabolic potential has been inferred from multiple, diverse MAGs, located across a range of ecosystems. It is thought that this capacity may enable methane oxidation in environments with low or below detectable oxygen concentrations.

In marine oxygen minimum zones, both the Methylothermaceae B42 and OPU3 MAGs encode genes for denitrification. This includes the capacity to reduce nitrate to nitrous oxide (*nar, nap, nir, nor*). Both Methylothermaceae B42 and OPU3 MAGs contained *nar* genes that are divergent from the genes in other Methylococcales methanotrophs, suggesting a recent horizontal transfer from non-methanotrophs (Skennerton *et al.*, 2015; Padilla *et al.*, 2017). It was theorized that denitrification can support methane oxidation in marine waters where dissolved oxygen is commonly below detection limits. Interestingly, transcripts for one of the *nar* genes, encoded by the OPU3 MAG, were detected in anoxic marine waters (Padilla *et al.*, 2017). This is so far the only environmental evidence that these denitrification pathways, encoded by aerobic Gammaproteobacteria methanotrophs, are transcribed or expressed *in situ* (Padilla *et al.*, 2017).

Beyond marine systems, the genomic potential for denitrification has been widely reported in methanotroph MAGs from terrestrial environments. For example, Dumont *et al.* (2011) observed expression of *Methylobacter*-like denitrification genes in oxic incubations of lake sediments. Additionally, the first genome representative of the mysterious *Crenothrix* clade encoded the metabolic capacity for denitrification, possibly explaining the methane consumption activity in anoxic incubations of freshwater lake waters (Oswald *et al.*, 2017). Lastly, *Methylobacter* MAGs from freshwater wetland soils also possessed the potential for denitrification, but the transcripts for this pathway was absent in metatranscriptomes from these soils (Smith *et al.*, 2018).

The denitrification metabolic potential of Methylococcales isolates and MAGs has been recently inventoried several times (Padilla *et al.*, 2017; Smith *et al.*, 2018). The most recent analysis by Smith *et al.* (2018) recovered at least partial denitrification (*nar, nap,* or *nir*) encoded in over 45 out of 57 combined MAG and isolate genomes. Moreover, 77% of the 18 reconstructed MAGs contained the capacity for reducing nitrate (Smith *et al.*, 2018). Notably, both analyses failed to identify genes for the final step in denitrification, the reduction of nitrous oxide to dinitrogen gas by nitrous oxide reductase (*nos*), in any Methylococcales genomes (Padilla *et al.*, 2017; Smith *et al.*, 2018). Therefore, it remains unclear whether the greenhouse gas mitigation benefits from methane consumption activity across wider oxygen gradients or whether it might be offset by the production of a more potent greenhouse gas nitrous oxide. The prevalence and conservation of this metabolism among Methylococcales suggests it is likely a valuable metabolic strategy, however more environmental context and expression data are required to truly understand the impacts of methanotroph-mediated denitrification.

Beyond denitrification, methanotrophs may also be able to tolerate much lower oxygen concentrations than previously thought. Genome analysis of MAGs and isolate genomes has identified the presence of high-affinity cytochromes, e.g. cytochrome *bd* ubiquinol–oxidoreductase, in addition to the traditional low-affinity cytochrome *c* oxidase. The use of these high-affinity oxidases would enable methane oxidation coupled to aerobic respiration in nanomolar range of oxygen (Skennerton *et al.*, 2015; Oswald *et al.*, 2017; Smith *et al.*, 2018). Thus, it is possible that methane oxidation by aerobic Proteobacteria may be occurring in environments with lower dissolved oxygen than currently assumed in global models (Riley *et al.*, 2011; Xu *et al.*, 2016).

Another possible adaptation for maintaining methane consumption in low oxygen conditions is 'micro-aerobic fermentation', which has been reported in pure cultures of *Methylomicrobium buryatense* (Kalyuzhnaya *et al.*, 2013; Gilman *et al.*, 2015). Oswald *et al.* (2017) and Smith *et al.* (2018) both reconstructed pathways for fermentation in, respectively, *Crenothrix* and *Methylobacter* MAGs. These fermentation pathways would enable the re-oxidation of NAD(P), and would ultimately lead to the secretion of mixed fermentation products including acetate, lactate, succinate, ethanol, and/or hydrogen. However, discerning the likelihood of this fermentative metabolism from metagenomic or metatranscriptomic data alone is difficult, and therefore future physiological investigations in these and other lineages are warranted to better understand micro-aerobic fermentation by methanotrophs.

A final possible mechanism for consuming methane when oxygen is relatively depleted is by scavenging oxygen via a bacteriohemerythrin. Hemerythrins are able to bind molecular oxygen, and they have been shown to interact with the pMMO enzyme, suggesting a role in directly providing oxygen to the methane oxidation machinery (Karlsen *et al.*, 2005; Kao *et al.*, 2008; Chen *et al.*, 2012). Bacteriohemerythrin genes have been identified in 80% of the methanotroph genomes, including at least four Methylococcales MAGs, 18 Methylococcales, six Alphaproteobacteria, and one Verrucomicrobia methanotroph genomes (Rahalkar and Bahulikar, 2018). Their wide distribution strongly suggests that they provide an important and conserved mechanism for sustaining methane consumption during oxygen limitation across the aerobic methanotrophic phyla. In fact, up-regulated transcription of this gene has been reported under oxygen-limited conditions in *Methylomicrobium* (Gilman *et al.*, 2017). From these studies, it is clear that MAGs play an important role in elucidating the repertoire of mechanisms supporting methane oxidation under a variety of redox conditions, across a range of environments.

Methanotroph MAGs and isolate genomes indicate the importance of novel metal biochemistry

Methanol DeHydrogenases (MDHs) catalyse the oxidation of methanol to formaldehyde, an essential step in aerobic methane oxidation. There are two known MDH enzymes, the canonical and long-studied MxaF type and the novel type, XoxF (Chistoserdova *et al.*, 2009). Despite similar functions, these two different MDH require different metal cofactors: the MxaFI complex requiring calcium, while the XoxF enzyme requiring a light lanthanide series metal (known as Rare Earth Elements, REEs) (Keltjens *et al.*, 2014; Pol *et al.*, 2014; Chistoserdova, 2015, 2016). Low concentrations of REEs have been reported to strongly repress expression of the *mxa* genes (Chu and Lidstrom, 2016; Chu *et al.*, 2016), suggesting that XoxF may be the 'preferred' MDH. Microorganisms using XoxF-type enzymes are now of current interest not only for the ability to transform multi-carbon compounds (Good *et al.*, 2016; Wehrmann *et al.*, 2017), or alter trophic interactions (Krause *et al.*, 2017), but also of for their potential in extracting REEs from electronics (Martinez-Gomez *et al.*, 2016),.

Verrucomicrobia methanotrophs rely on the XoxF MDH as all genomes thus far lack MxaF genes and isolate grow poorly in the absence of lanthanide metals. So far, Verrucomicrobia methanotrophs appear to be nearly absent in metagenomes assembled from mesophilic grassland and freshwater environments (Kielak *et al.*, 2010; Cabello-Yeves *et al.*, 2017; He *et al.*, 2017) and are primarily detected in thermo- and acidophilic geothermal ecosystems (Op den Camp *et al.*, 2009; Sharp *et al.*, 2014; Knief, 2015). One MAG putatively assigned to a Methylacidiphilaceae (Methylacidiphilaceae UBA1321) was reconstructed by Parks *et al.* (2017) from the Twitchell Island restored wetland. However, both the MAG and the metagenome lack the MMO genes (S. He *et al.*, 2015), suggesting this might not be a bona fide Verrucomicrobia methanotroph. This clade is not well understood given the few number of isolates and limited metagenome-derived genomic information, however their reliance on XoxF type MDH, hydrogen metabolism (Carere *et al.*, 2017; Mohammadi *et al.*, 2017), and heat and acid tolerances make them physiologically unique and interesting.

Early investigations of *xoxF* distribution found the gene present in all the Proteobacteria methylotrophs, often in addition to the *mxaFI*-encoded MDH (Chistoserdova, 2011b). Environmental surveys suggested that *xoxF* might be more prevalent and widespread among uncultivated lineages in natural ecosystems than *mxaF* (Ramachandran and Walsh, 2015; Taubert *et al.*, 2015). Metagenomic analysis of enrichment cultures of two closely related Methylococcales and native samples from marine ecosystems found MAGs containing only *xoxF* MDH (Vekeman *et al.*, 2016b; Padilla *et al.*, 2017). However, as is the case when making inferences from non-closed genomes, caution must be used when inferring absence of genes from MAGs. Interestingly, metagenomes from freshwater and terrestrial systems have failed to detect *mxaF* in the metagenomes using both gene- and genome-centric approaches (Oswald *et al.*, 2017; Pratscher *et al.*, 2018; Singleton *et al.*, 2018; Smith *et al.*, 2018). This evidence, composed of both isolate and metagenomic information, suggests that the *xoxF* is an environmentally relevant MDH, and therefore cultivation strategies for novel methanotrophs should be modified to include lanthanides in enrichment medium.

Conclusions/discussion

Metagenomics has greatly advanced the field of methane oxidation by providing an unbiased sequencing approach to successfully capture gene- and genomic-level information for novel lineages, improving our understanding of both the phylogenetic and metabolic diversity encompassed by methanotrophs. In the past decade metagenomics has more than doubled methanotroph genomic databases by reconstructing MAGs, which now includes representatives of clades that contain few or no isolated representatives. Specifically, these enigmatic lineages include members of the Proteobacteria (e.g. USCα, USCγ, and *Crenothrix*), members of the anaerobic methanotrophs (ANME-1, ANME-2), and of the genus *Ca.* Methylomirabilis formerly of the candidate NC10 phylum. Additionally, MAGs have been recovered that are closely related to the well characterized methanotrophs (e.g. *Methylobacter*), which open up the possibility of investigating finely resolved differences in genotypes that allow for specific environmental adaption.

Most importantly, metagenomic surveys have been critical to furthering our understanding of the methanotroph distribution across ecosystems. These studies have highlighted roles for methane oxidation from the depths of the ocean, to the tissues of endosymbionts, to agricultural soils. From MAGs assembled across these ecosystems, it is now possible to extend physiological insights from cultivation studies to the field scale. For instance, recent appreciation for the versatility of methanotroph energy generation pathways, has highlighted the capacity of these organisms to withstand fluctuating redox conditions. Moreover, metagenomic studies from marine, freshwater, and polar systems have found surprisingly low numbers of *mxaF* genes, compared with the number of *xoxF*, potentially further signifying the importance of this enzyme and informing cultivation efforts. It is clear that metagenomics will continue to play an important role in furthering understanding of methanotroph physiology and phylogenetic diversity.

Future trends

While we acknowledge that our analysis was not comprehensive of all current metagenomic sequencing projects (741 available in IMG before 1 June 2016 queried), geographic bias was obvious in our meta-analysis (Fig. 3.5). For example, far more methanotroph genomes have been recovered from metagenomes from the northern hemisphere, mostly in the United States and Europe, than anywhere else. The limitation in knowledge about methanotrophs in more equatorial latitudes may have important global ramifications, as tropical regions account for ≈50–60% of the emissions from wetland systems (Dean *et al.*, 2018). Substantially increased metagenomic sequencing of more equatorial regions is likely to uncover novel methanotrophs involved in mediating the balance between methane source and sinks.

We also note that metagenomic sampling of methanotrophs in Eastern and Southern regions were similarly sparse. These include major rice cultivation areas such as China and India. Rice paddies are important ecosystems that provide food for a large global population, but also represent major sources of methane to the atmosphere. It is imperative that we continue to study the distribution and activity of methanotrophs in these and other agricultural systems, to better understand methane oxidation with different agricultural practices and climatic changes. As metagenomic sequencing becomes more cost efficient and commonplace, it is anticipated that methanotroph genomes will be better sampled across the globe in the near future.

Technological developments that continue to improve ease and efficiency of metagenomic sequencing and analysis will enable better gene- and genome-centric approaches. Sequencing technologies that generate long reads, for example

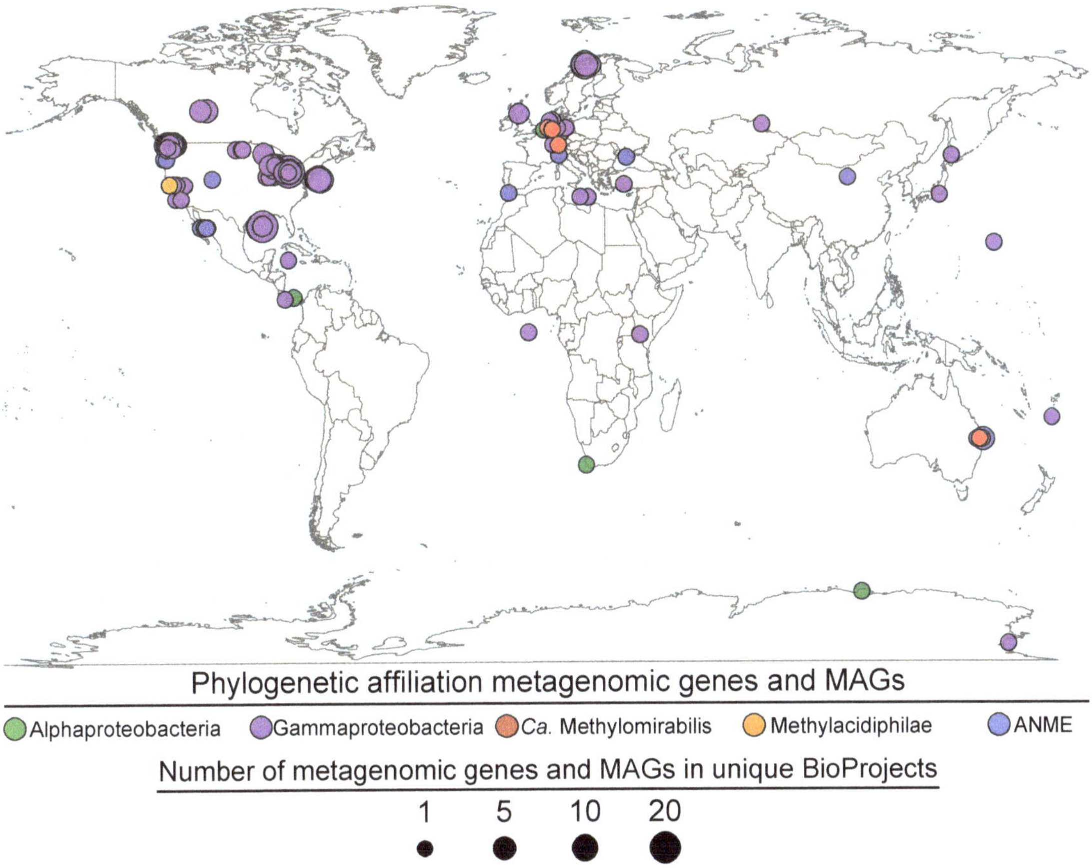

Figure 3.5 Geographic distribution of methanotrophs detected in metagenomic and metatranscriptomic datasets across the globe. Each point represents one of 327 unique BioProjects associated with a metagenome containing a methanotroph rpS3 gene or a putative methanotrophic Metagenome-Assembled Genome (MAG). Phylogenetic affiliation of rpS3 genes not in MAGs were assigned to the class of the most similar reference sequence determined by BLASTp and confirmed by topological coherence with reference genes. The colours indicate the phylogenetic affiliation of the rpS3 gene, and the size of the point represents the number of recovered sequences per unique BioProject. Sixteen MAGs lacked identifiable location information.

Nanopore (Feng *et al.*, 2015) and PacBio (Rhoads and Au, 2015), are already showing great promise for improved metagenomic assemblies, yielding greater recovery of genes, pathways, and MAGs among complex, diverse samples. In addition, we are hopeful that pairing metagenomics with other tools (e.g. metatranscriptomics, metaproteomics, SIP, Nano-Sims, fluorescent *in situ* microscopy, or exometabolomics) will further elucidate the physical and metabolic interactions between methanotrophs and other community members. Coupling metagenomics with measurements of gene or protein expression have already proven invaluable for firmly confirming methane oxidation by the *Crenothrix* filaments (Oswald *et al.*, 2017), nuances of trace metal metabolism in ANME (Glass *et al.*, 2014), and will likely continue to prove to be exceptionally powerful for exploring methanotroph ecology and function.

According to metabolic reconstructions of ANME and *Ca.* Methylomirabilis MAGs to date, members of both lineages should be capable of coupling methane oxidation to respiration independently of consortia, although no isolated representatives currently exist. Increased sampling of these lineages both in the laboratory and the field may offer new physiological variants that better withstand axenic growth. For instance, MAGs from ANME and *Ca.* Methylomirabilis have been sampled from wastewater treatment and natural methane seep ecosystems (Welte *et al.*, 2016); however, these same taxa are also detected and often prevalent in wetlands, rice paddies, and freshwater lakes (Figs. 3.3 and 3.5) (Haroon *et al.*, 2013; Welte *et al.*, 2016; Narrowe *et al.*, 2017; in 't Zandt *et al.*, 2018; Graf *et al.*, 2018). These latter habitats represent an untapped potential for examination of the metabolic versatility of these uncultivated methanotroph lineages. Additionally, further genomic investigations of stable consortia may also offer insights into factors that hinder cultivation of these strains. Isolation of these microorganisms may be inhibited by their extremely slow growth rates, but this is unlikely to be the sole cause. Additional physiological-genomic studies may elucidate dependencies on other microorganisms, for example amino acid or vitamin auxotrophies, or the detoxification of toxic metabolites.

MAGs have highlighted the metabolic versatility that may enable methane oxidation along broader redox gradients, however much of this flexibility is currently only theoretical or poorly understood. An example of a currently only hypothetical metabolism is the reverse methanogenesis by Bathyarchaeota, encoding a divergent MCR gene and using pathways similar to the ones in ANME (Evans *et al.*, 2015; Zhou *et al.*, 2018). While there is no direct evidence for this metabolism, it is expected to yield acetate that could be exchanged with SRB. Further understanding of these exchanges may help explain isotopic anomalies already observed in some habitats (Zhou *et al.*, 2018).

The role of dissimilatory nitrate reduction in the aerobic Methylococcales requires further investigation. Leveraging the large number of cultivated representatives within this clade to better understand oxygen responses in the Methylococcales seems warranted, given the broad distribution and activity of these taxa across the globe. For instance, some of these genomes encode up to four different, but possibly overlapping energy generation mechanisms: low-affinity cytochromes for oxic aerobic respiration, high-affinity cytochromes for suboxic aerobic respiration, respiratory nitrate reduction, and micro-aerobic fermentation (Kits *et al.*, 2015; Skennerton *et al.*, 2015; Padilla *et al.*, 2017; Smith *et al.*, 2018). It is currently almost completely unknown how these metabolic pathways are regulated, or what effects they have on methane oxidation rates. New discoveries could drastically alter our perception of the redox gradients across which Methylococcales can consume methane, as well as the metabolites secreted by these methanotrophs.

Additionally, there is increasing evidence that the metabolic versatility of methanotrophs is greater than was historically assumed. For example, the utilization of hydrogen has been observed in Verrucomicrobia, and may be possible in some Proteobacteria methanotrophs (Carere *et al.*, 2017; Mohammadi *et al.*, 2017; Singleton *et al.*, 2018; Smith *et al.*, 2018). While the role of hydrogen metabolisms, evident in Verrucomicrobia methanotrophs, is still unclear, hydrogen production or consumption by methanotrophs may expand the niches occupied by methanotrophs as well as their function in microbial communities. These proposed metabolisms require verification using axenic cultures, however metagenomic sequencing coupled with activity quantification of specific genes or pathways may continue to elucidate the distribution and impact of these genes or pathways under *in situ* conditions.

Web resources

We would like to acknowledge of the Organization of Methanotroph Genome Analysis (OMeGA) for their invaluable contributions to the genomic databases of methane oxidizing microorganisms. Methanotroph genomes sequenced and analysed by OMeGA are available online at IMG/MER (https://img.jgi.doe.gov/cgi-bin/mer/main.cgi) or at the National Center for Biotechnology (https://www.ncbi.nlm.nih.gov/). We would like to thank IMG/MER for hosting the repository of metagenomic information that can be readily mined, and Simon Roux for assistance and guidance for the data mining.

Online supplemental sequences, metadata, tree generation commands, and R scripts used for the analyses presented here are available at the Wrighton laboratory's GitHub website: https://github.com/TheWrightonLab/Methanotroph_rpS3Analyses_SmithWrighton2018. This repository includes minimally modified rpS3 amino acids sequences for the methanotroph reference and metagenomic sequences, the alignment used for phylogenetic tree generation, the phylogenetic tree displayed in Fig. 3.3, and the metadata tables used as inputs for information displayed in Figs 3.3, 3.4, and 3.5, and finally the R script for recreation of the analyses and figures.

References

Abdallah, R.Z., Adel, M., Ouf, A., Sayed, A., Ghazy, M.A., Alam, I., Essack, M., Lafi, F.F., Bajic, V.B., El-Dorry, H., *et al.* (2014). Aerobic methanotrophic communities at the Red Sea brine-seawater interface. Front. Microbiol. *5*, 487. https://doi.org/10.3389/fmicb.2014.00487

Anantharaman, K., Brown, C.T., Hug, L.A., Sharon, I., Castelle, C.J., Probst, A.J., Thomas, B.C., Singh, A., Wilkins, M.J., Karaoz, U., *et al.* (2016). Thousands of microbial genomes shed light on interconnected biogeochemical processes in an aquifer system. Nat. Commun. *7*, 13219. https://doi.org/10.1038/ncomms13219

Angle, J.C., Morin, T.H., Solden, L.M., Narrowe, A.B., Smith, G.J., Borton, M.A., Rey-Sanchez, C., Daly, R.A., Mirfenderesgi, G., Hoyt, D.W., *et al.* (2017). Methanogenesis in oxygenated soils is a substantial fraction of wetland methane emissions. Nat. Commun. *8*, 1567. https://doi.org/10.1038/s41467-017-01753-4

Arshad, A., Speth, D.R., de Graaf, R.M., Op den Camp, H.J., Jetten, M.S., and Welte, C.U. (2015). A metagenomics-based metabolic model of nitrate-dependent anaerobic oxidation of methane by Methanoperedens-like Archaea. Front. Microbiol. *6*, 1423. https://doi.org/10.3389/fmicb.2015.01423

Auman, A.J., Stolyar, S., Costello, A.M., and Lidstrom, M.E. (2000). Molecular characterization of methanotrophic isolates from freshwater lake sediment. Appl. Environ. Microbiol. *66*, 5259–5266.

Baani, M., and Liesack, W. (2008). Two isozymes of particulate methane monooxygenase with different methane oxidation kinetics are found in Methylocystis sp. strain SC2. Proc. Natl. Acad. Sci. U.S.A. *105*, 10203–10208. https://doi.org/10.1073/pnas.0702643105

Bao, Z., Okubo, T., Kubota, K., Kasahara, Y., Tsurumaru, H., Anda, M., Ikeda, S., and Minamisawa, K. (2014). Metaproteomic identification of diazotrophic methanotrophs and their localization in root tissues of field-grown rice plants. Appl. Environ. Microbiol. *80*, 5043–5052. https://doi.org/10.1128/AEM.00969-14

Beck, D.A., Hendrickson, E.L., Vorobev, A., Wang, T., Lim, S., Kalyuzhnaya, M.G., Lidstrom, M.E., Hackett, M., and Chistoserdova, L. (2011). An integrated proteomics/transcriptomics approach points to oxygen as the main electron sink for methanol metabolism in *Methylotenera mobilis*. J. Bacteriol. *193*, 4758–4765. https://doi.org/10.1128/JB.05375-11

Beck, D.A., Kalyuzhnaya, M.G., Malfatti, S., Tringe, S.G., Glavina Del Rio, T., Ivanova, N., Lidstrom, M.E., and Chistoserdova, L. (2013). A metagenomic insight into freshwater methane-utilizing communities and evidence for cooperation between the *Methylococcaceae* and the *Methylophilaceae*. PeerJ *1*, e23. https://doi.org/10.7717/peerj.23

Bhattacharyya, P., Roy, K.S., Nayak, A.K., Shahid, M., Lal, B., Gautam, P., and Mohapatra, T. (2017). Metagenomic assessment of methane production-oxidation and nitrogen metabolism of long term manured systems in lowland rice paddy. Sci. Total Environ. *586*, 1245–1253.

Bowers, R.M., Kyrpides, N.C., Stepanauskas, R., Harmon-Smith, M., Doud, D., Reddy, T.B.K., Schulz, F., Jarett, J., Rivers, A.R., Eloe-Fadrosh, E.A., *et al.* (2017). Minimum information about a single amplified genome (MISAG) and a metagenome-assembled genome (MIMAG) of bacteria and archaea. Nat. Biotechnol. *35*, 725–731. https://doi.org/10.1038/nbt.3893

Brankovits, D., Pohlman, J.W., Niemann, H., Leigh, M.B., Leewis, M.C., Becker, K.W., Iliffe, T.M., Alvarez, F., Lehmann, M.F., and Phillips, B. (2017). Methane- and dissolved organic carbon-fueled microbial loop supports a tropical subterranean estuary ecosystem. Nat. Commun. *8*, 1835. https://doi.org/10.1038/s41467-017-01776-x

Brown, C.T., Hug, L.A., Thomas, B.C., Sharon, I., Castelle, C.J., Singh, A., Wilkins, M.J., Wrighton, K.C., Williams, K.H., and Banfield, J.F. (2015). Unusual biology across a group comprising more than 15% of domain Bacteria. Nature *523*, 208–211. https://doi.org/10.1038/nature14486

Cabello-Yeves, P.J., Ghai, R., Mehrshad, M., Picazo, A., Camacho, A., and Rodriguez-Valera, F. (2017). Reconstruction of diverse verrucomicrobial genomes from metagenome datasets of freshwater reservoirs. Front. Microbiol. *8*, 2131. https://doi.org/10.3389/fmicb.2017.02131

Cai, C., Leu, A.O., Xie, G.J., Guo, J., Feng, Y., Zhao, J.X., Tyson, G.W., Yuan, Z., and Hu, S. (2018). A methanotrophic archaeon couples anaerobic oxidation of methane to Fe(III) reduction. ISME J. *12*, 1929–1939. https://doi.org/10.1038/s41396-018-0109-x

Cai, Y., Zheng, Y., Bodelier, P.L., Conrad, R., and Jia, Z. (2016). Conventional methanotrophs are responsible for atmospheric methane oxidation in paddy soils. Nat. Commun. *7*, 11728. https://doi.org/10.1038/ncomms11728

Carere, C.R., Hards, K., Houghton, K.M., Power, J.F., McDonald, B., Collet, C., Gapes, D.J., Sparling, R., Boyd, E.S., Cook, G.M., *et al.* (2017). Mixotrophy drives niche expansion of verrucomicrobial methanotrophs. ISME J. *11*, 2599–2610. https://doi.org/10.1038/ismej.2017.112

Chen, K.H., Wu, H.H., Ke, S.F., Rao, Y.T., Tu, C.M., Chen, Y.P., Kuei, K.H., Chen, Y.S., Wang, V.C., Kao, W.C., *et al.* (2012). Bacteriohemerythrin bolsters the activity of the particulate methane monooxygenase (pMMO) in Methylococcus capsulatus (Bath). J. Inorg. Biochem. *111*, 10–17. https://doi.org/10.1016/j.jinorgbio.2012.02.019

Chen, Y., and Murrell, J.C. (2010). When metagenomics meets stable-isotope probing: progress and perspectives. Trends Microbiol. *18*, 157–163. https://doi.org/10.1016/j.tim.2010.02.002

Chen, Y., Dumont, M.G., Neufeld, J.D., Bodrossy, L., Stralis-Pavese, N., McNamara, N.P., Ostle, N., Briones, M.J., and Murrell, J.C. (2008). Revealing the uncultivated majority: combining DNA stable-isotope probing, multiple displacement amplification and metagenomic analyses of uncultivated Methylocystis in acidic peatlands. Environ. Microbiol. *10*, 2609–2622. https://doi.org/10.1111/j.1462-2920.2008.01683.x

Chistoserdova, L. (2011a). Methylotrophy in a lake: from metagenomics to single-organism physiology. Appl. Environ. Microbiol. *77*, 4705–4711. https://doi.org/10.1128/AEM.00314-11

Chistoserdova, L. (2011b). Modularity of methylotrophy, revisited. Environ. Microbiol. *13*, 2603–2622. https://doi.org/10.1111/j.1462-2920.2011.02464.x

Chistoserdova, L. (2015). Methylotrophs in natural habitats: current insights through metagenomics. Appl. Microbiol. Biotechnol. *99*, 5763–5779. https://doi.org/10.1007/s00253-015-6713-z

Chistoserdova, L. (2016). Lanthanides: New life metals? World J. Microbiol. Biotechnol. *32*, 138. https://doi.org/10.1007/s11274-016-2088-2

Chistoserdova, L., Kalyuzhnaya, M.G., and Lidstrom, M.E. (2009). The expanding world of methylotrophic metabolism. Annu. Rev. Microbiol. *63*, 477–499. https://doi.org/10.1146/annurev.micro.091208.073600

Chu, F., and Lidstrom, M.E. (2016). XoxF acts as the predominant methanol dehydrogenase in the Type I methanotroph *Methylomicrobium buryatense*. J. Bacteriol. *198*, 1317–1325. https://doi.org/10.1128/JB.00959-15

Chu, F., Beck, D.A., and Lidstrom, M.E. (2016). MxaY regulates the lanthanide-mediated methanol dehydrogenase switch in *Methylomicrobium buryatense*. PeerJ *4*, e2435. https://doi.org/10.7717/peerj.2435

Cohn, F. (1870). Über den Brunnenfaden (Crenothrix polyspora) mit Bemerkungen über die mikroskopische Analyse des Brunnenwassers (J.U. Kern's Verlag, Wrocław).

Dalcin Martins, P., Hoyt, D.W., Bansal, S., Mills, C.T., Tfaily, M., Tangen, B.A., Finocchiaro, R.G., Johnston, M.D., McAdams, B.C., Solensky, M.J., *et al.* (2017). Abundant carbon substrates drive extremely high sulfate reduction rates and methane fluxes in Prairie Pothole Wetlands. Glob. Chang. Biol. *23*, 3107–3120. https://doi.org/10.1111/gcb.13633

Dalton, H. (1983). The Biochemistry of Methylotrophs (Academic Press, London).

Dam, B., Dam, S., Blom, J., and Liesack, W. (2013). Genome analysis coupled with physiological studies reveals a diverse nitrogen metabolism in *Methylocystis* sp. strain SC2. PLOS ONE *8*, e74767. https://doi.org/10.1371/journal.pone.0074767

Dean, J.F., Middelburg, J.J., Röckmann, T., Aerts, R., Blauw, L.G., Egger, M., Jetten, M.S.M., de Jong, A.E.E., Meisel, O.H., Rasigraf, O., *et al.* (2018). Methane feedbacks to the global climate system in a warmer world. Rev. Geophys. *56*, 207–250.

Dean, S.L., Farrer, E.C., Porras-Alfaro, A., Suding, K.N., and Sinsabaugh, R.L. (2015). Assembly of root-associated bacteria communities: interactions between abiotic and biotic factors. Environ. Microbiol. Rep. *7*, 102–110.

Dumont, M.G., Radajewski, S.M., Miguez, C.B., McDonald, I.R., and Murrell, J.C. (2006). Identification of a complete methane monooxygenase operon from soil by combining stable isotope probing and metagenomic analysis. Environ. Microbiol. *8*, 1240–1250.

Dumont, M.G., Pommerenke, B., Casper, P., and Conrad, R. (2011). DNA-, rRNA- and mRNA-based stable isotope probing of aerobic methanotrophs in lake sediment. Environ. Microbiol. *13*, 1153–1167. https://doi.org/10.1111/j.1462-2920.2010.02415.x

Dunfield, P.F., Liesack, W., Henckel, T., Knowles, R., and Conrad, R. (1999). High-affinity methane oxidation by a soil enrichment culture containing a type II methanotroph. Appl. Environ. Microbiol. *65*, 1009–1014.

Edwards, C.R., Onstott, T.C., Miller, J.M., Wiggins, J.B., Wang, W., Lee, C.K., Cary, S.C., Pointing, S.B., and Lau, M.C.Y. (2017). draft genome sequence of uncultured upland soil cluster *gammaproteobacteria* gives molecular insights into high-affinity methanotrophy. Genome Announc. *5*, e00047-17.

Edwards, R.A., McNair, K., Faust, K., Raes, J., and Dutilh, B.E. (2016). Computational approaches to predict bacteriophage-host relationships. FEMS Microbiol. Rev. *40*, 258–272. https://doi.org/10.1093/femsre/fuv048

Escobar-Zepeda, A., Vera-Ponce de León, A., and Sanchez-Flores, A. (2015). the road to metagenomics: from microbiology to DNA sequencing technologies and bioinformatics. Front. Genet. *6*, 348. https://doi.org/10.3389/fgene.2015.00348

Esson, K.C., Lin, X., Kumaresan, D., Chanton, J.P., Murrell, J.C., and Kostka, J.E. (2016). Alpha- and gammaproteobacterial methanotrophs codominate the active methane-oxidizing communities in an acidic boreal peat bog. Appl. Environ. Microbiol. *82*, 2363–2371. https://doi.org/10.1128/AEM.03640-15

Ettwig, K.F., Shima, S., van de Pas-Schoonen, K.T., Kahnt, J., Medema, M.H., Op den Camp, H.J., Jetten, M.S., and Strous, M. (2008). Denitrifying bacteria anaerobically oxidize methane in the absence of Archaea. Environ. Microbiol. *10*, 3164–3173. https://doi.org/10.1111/j.1462-2920.2008.01724.x

Ettwig, K.F., Butler, M.K., Le Paslier, D., Pelletier, E., Mangenot, S., Kuypers, M.M., Schreiber, F., Dutilh, B.E., Zedelius, J., de Beer, D., *et al.* (2010). Nitrite-driven anaerobic methane oxidation by oxygenic bacteria. Nature *464*, 543–548. https://doi.org/10.1038/nature08883

Ettwig, K.F., Speth, D.R., Reimann, J., Wu, M.L., Jetten, M.S., and Keltjens, J.T. (2012). Bacterial oxygen production in the dark. Front. Microbiol. *3*, 273. https://doi.org/10.3389/fmicb.2012.00273

Ettwig, K.F., Zhu, B., Speth, D., Keltjens, J.T., Jetten, M.S.M., and Kartal, B. (2016). Archaea catalyze iron-dependent anaerobic oxidation of methane. Proc. Natl. Acad. Sci. U.S.A. *113*, 12792–12796. https://doi.org/10.1073/pnas.1609534113

Evans, P.N., Parks, D.H., Chadwick, G.L., Robbins, S.J., Orphan, V.J., Golding, S.D., and Tyson, G.W. (2015). Methane metabolism in the archaeal phylum Bathyarchaeota revealed by genome-centric metagenomics. Science *350*, 434–438. https://doi.org/10.1126/science.aac7745

Feng, Y., Zhang, Y., Ying, C., Wang, D., and Du, C. (2015). Nanopore-based fourth-generation DNA sequencing technology. Genomics Proteomics Bioinformatics *13*, 4–16. https://doi.org/10.1016/j.gpb.2015.01.009

Fisher, C.R., Childress, J.J., Oremland, R.S., and Bidigare, R.R. (1987). The importance of methane and thiosulfate in the metabolism of the bacterial symbionts of two deep-sea mussels. Mar. Biol. *96*, 59–71.

Friedrich, M.W. (2006). Stable-isotope probing of DNA: insights into the function of uncultivated microorganisms from isotopically labeled metagenomes. Curr. Opin. Biotechnol. *17*, 59–66.

Gilman, A., Laurens, L.M., Puri, A.W., Chu, F., Pienkos, P.T., and Lidstrom, M.E. (2015). Bioreactor performance parameters for an industrially-promising methanotroph *Methylomicrobium buryatense* 5GB1. Microb. Cell Fact. *14*, 182. https://doi.org/10.1186/s12934-015-0372-8

Gilman, A., Fu, Y., Hendershott, M., Chu, F., Puri, A.W., Smith, A.L., Pesesky, M., Lieberman, R., Beck, D.A.C., and Lidstrom, M.E. (2017). Oxygen-limited metabolism in the methanotroph *Methylomicrobium buryatense* 5GB1C. PeerJ *5*, e3945. https://doi.org/10.7717/peerj.3945

Glass, J.B., Yu, H., Steele, J.A., Dawson, K.S., Sun, S., Chourey, K., Pan, C., Hettich, R.L., and Orphan, V.J. (2014). Geochemical, metagenomic and metaproteomic insights into trace metal utilization by methane-oxidizing microbial consortia in sulphidic marine sediments. Environ. Microbiol. *16*, 1592–1611. https://doi.org/10.1111/1462-2920.12314

Glöckner, F.O., Yilmaz, P., Quast, C., Gerken, J., Beccati, A., Ciuprina, A., Bruns, G., Yarza, P., Peplies, J., Westram, R., *et al.* (2017). 25 years of serving the community with ribosomal RNA gene reference databases and tools. J. Biotechnol. *261*, 169–176. https://doi.org/10.1016/jbiotec.2017.06.1198

Good, N.M., Vu, H.N., Suriano, C.J., Subuyuj, G.A., Skovran, E., and Martinez-Gomez, N.C. (2016). Pyrroloquinoline quinone ethanol dehydrogenase in *Methylobacterium extorquens* AM1 extends lanthanide-dependent metabolism to multicarbon substrates. J. Bacteriol. *198*, 3109–3118.

Graf, J.S., Mayr, M.J., Marchant, H.K., Tienken, D., Hach, P.F., Brand, A., Schubert, C.J., Kuypers, M.M.M., and Milucka, J. (2018). Bloom of a denitrifying methanotroph, '*Candidatus* Methylomirabilis limnetica', in a deep stratified lake. Environ. Microbiol. *20*, 2598–2614. https://doi.org/10.1111/1462-2920.14285

Graham, L.E., Graham, J.M., Knack, J.J., Trest, M.T., Piotrowski, M.J., and Arancibia-Avila, P. (2017). A sub-Antarctic peat moss metagenome indicates microbiome resilience to stress and biogeochemical functions of early paleozoic terrestrial ecosystems. Int. J. Plant Sci. *178*, 618–628.

Hallam, S.J., Girguis, P.R., Preston, C.M., Richardson, P.M., and DeLong, E.F. (2003). Identification of methyl coenzyme M reductase A (mcrA) genes associated with methane-oxidizing archaea. Appl. Environ. Microbiol. *69*, 5483–5491.

Hallam, S.J., Putnam, N., Preston, C.M., Detter, J.C., Rokhsar, D., Richardson, P.M., and DeLong, E.F. (2004). Reverse methanogenesis: testing the hypothesis with environmental genomics. Science *305*, 1457–1462. https://doi.org/10.1126/science.1100025

Haroon, M.F., Hu, S., Shi, Y., Imelfort, M., Keller, J., Hugenholtz, P., Yuan, Z., and Tyson, G.W. (2013). Anaerobic oxidation of methane coupled to nitrate reduction in a novel archaeal lineage. Nature *500*, 567–570. https://doi.org/10.1038/nature12375

Harriss, R.C., Sebacher, D.I., and Day, F.P. (1982). Methane flux in the Great Dismal Swamp. Nature *297*, 673–674.

Håvelsrud, O.E., Haverkamp, T.H., Kristensen, T., Jakobsen, K.S., and Rike, A.G. (2011). A metagenomic study of methanotrophic microorganisms in Coal Oil Point seep sediments. BMC Microbiol. *11*, 221. https://doi.org/10.1186/1471-2180-11-221

Håvelsrud, O.E., Haverkamp, T.H., Kristensen, T., Jakobsen, K.S., and Rike, A.G. (2012). Metagenomic and geochemical characterization of pockmarked sediments overlaying the Troll petroleum reservoir in the North Sea. BMC Microbiol. *12*, 203. https://doi.org/10.1186/1471-2180-12-203

Hayashi, H., Obata, T. Gamo, Y. Sano, T.N. (2007). Distribution and phylogenetic characteristics ofthe genes encoding enzymes relevant to methane oxidation in oxygen minimum zones of the Eastern Pacific Ocean. Res. J. Environ. Sci. *1*, 275–284.

He, S., Malfatti, S.A., McFarland, J.W., Anderson, F.E., Pati, A., Huntemann, M., Tremblay, J., Glavina del Rio, T., Waldrop, M.P., Windham-Myers, L., *et al.* (2015). Patterns in wetland microbial community composition and functional gene repertoire associated with methane emissions. MBio *6*, e00066–15. https://doi.org/10.1128/mBio.00066-15

He, S., Stevens, S.L.R., Chan, L.K., Bertilsson, S., Glavina Del Rio, T., Tringe, S.G., Malmstrom, R.R., and McMahon, K.D. (2017). Ecophysiology of Freshwater Verrucomicrobia Inferred from Metagenome-Assembled Genomes. mSphere *2*, e00277–17. https://doi.org/10.1128/mSphere.00277-17

He, Z., Wang, J., Zhang, X., Cai, C., Geng, S., Zheng, P., Xu, X., and Hu, B. (2015). Nitrogen removal from wastewater by anaerobic methane-driven denitrification in a lab-scale reactor: heterotrophic denitrifiers associated with denitrifying methanotrophs. Appl. Microbiol. Biotechnol. *99*, 10853–10860. https://doi.org/10.1007/s00253-015-6939-9

Henckel, T., Jäckel, U., Schnell, S., and Conrad, R. (2000). Molecular analyses of novel methanotrophic communities in forest soil that oxidize atmospheric methane. Appl. Environ. Microbiol. *66*, 1801–1808.

Hernandez, M.E., Beck, D.A., Lidstrom, M.E., and Chistoserdova, L. (2015). Oxygen availability is a major factor in determining the composition of microbial communities involved in methane oxidation. PeerJ *3*, e801. https://doi.org/10.7717/peerj.801

Heylen, K., De Vos, P., and Vekeman, B. (2016). Draft genome sequences of eight obligate methane oxidizers occupying distinct niches based on their nitrogen metabolism. Genome Announc. *4*, e00421–16. https://doi.org/10.1128/genomeA.00421-16

Hoefman, S., van der Ha, D., Boon, N., Vandamme, P., De Vos, P., and Heylen, K. (2014). Niche differentiation in nitrogen metabolism among methanotrophs within an operational taxonomic unit. BMC Microbiol. *14*, 83. https://doi.org/10.1186/1471-2180-14-83

Holmes, A.J., Roslev, P., McDonald, I.R., Iversen, N., Henriksen, K., and Murrell, J.C. (1999). Characterization of methanotrophic bacterial populations in soils showing atmospheric methane uptake. Appl. Environ. Microbiol. *65*, 3312–3318.

Hug, L.A., Baker, B.J., Anantharaman, K., Brown, C.T., Probst, A.J., Castelle, C.J., Butterfield, C.N., Hernsdorf, A.W., Amano, Y., Ise, K., *et al.* (2016). A new view of the tree of life. Nat. Microbiol. *1*, 16048. https://doi.org/10.1038/nmicrobiol.2016.48

Hutchens, E., Radajewski, S., Dumont, M.G., McDonald, I.R., and Murrell, J.C. (2004). Analysis of methanotrophic bacteria in Movile Cave by stable isotope probing. Environ. Microbiol. *6*, 111–120.

in 't Zandt, M.H., de Jong, A.E.E., Slomp, C.P., and Jetten, M.S.M. (2018). The hunt for the most-wanted chemolithoautotrophic spookmicrobes. FEMS Microbiol. Ecol. *94*, 1–17.

Ino, K., Hernsdorf, A.W., Konno, U., Kouduka, M., Yanagawa, K., Kato, S., Sunamura, M., Hirota, A., Togo, Y.S., Ito, K., *et al.* (2018). Ecological and genomic profiling of anaerobic methane-oxidizing archaea in a deep granitic environment. ISME J. *12*, 31–47. https://doi.org/10.1038/ismej.2017.140

Jewell, W.J., Nelson, Y.M., and Wilson, M.S. (1992). Methanotrophic bacteria for nutrient removal from waste-water: attached film system. Water Environ. Res. *64*, 756–765.

Kalyuzhnaya, M.G., Lapidus, A., Ivanova, N., Copeland, A.C., McHardy, A.C., Szeto, E., Salamov, A., Grigoriev, I.V., Suciu, D., Levine, S.R., *et al.* (2008). High-resolution metagenomics targets specific functional types in complex microbial communities. Nat. Biotechnol. *26*, 1029–1034. https://doi.org/10.1038/nbt.1488

Kalyuzhnaya, M.G., Yang, S., Rozova, O.N., Smalley, N.E., Clubb, J., Lamb, A., Gowda, G.A., Raftery, D., Fu, Y., Bringel, F., *et al.* (2013). Highly efficient methane biocatalysis revealed in a methanotrophic bacterium. Nat. Commun. *4*, 2785. https://doi.org/10.1038/ncomms3785

Kao, W.C., Wang, V.C., Huang, Y.C., Yu, S.S., Chang, T.C., and Chan, S.I. (2008). Isolation, purification and characterization of hemerythrin from *Methylococcus capsulatus* (Bath). J. Inorg. Biochem. *102*, 1607–1614. https://doi.org/10.1016/j.jinorgbio.2008.02.008

Karlsen, O.A., Ramsevik, L., Bruseth, L.J., Larsen, Ø., Brenner, A., Berven, F.S., Jensen, H.B., and Lillehaug, J.R. (2005). Characterization of a prokaryotic haemerythrin from the methanotrophic bacterium *Methylococcus capsulatus* (Bath). FEBS J. 272, 2428–2440.

Karwautz, C., Kus, G., Stöckl, M., Neu, T.R., and Lueders, T. (2018). Microbial megacities fueled by methane oxidation in a mineral spring cave. ISME J. *12*, 87–100. https://doi.org/10.1038/ismej.2017.146

Kelley, C.A., Martens, C.S., and Ussler, W. (1995). Methane dynamics across a tidally flooded riverbank margin. Limnol. Oceanogr. *40*, 1112–1129.

Keltjens, J.T., Pol, A., Reimann, J., and Op den Camp, H.J. (2014). PQQ-dependent methanol dehydrogenases: rare-earth elements make a difference. Appl. Microbiol. Biotechnol. *98*, 6163–6183. https://doi.org/10.1007/s00253-014-5766-8

Kessler, J.D., Valentine, D.L., Redmond, M.C., Du, M., Chan, E.W., Mendes, S.D., Quiroz, E.W., Villanueva, C.J., Shusta, S.S., Werra, L.M., *et al.* (2011). A persistent oxygen anomaly reveals the fate of spilled methane in the deep Gulf of Mexico. Science *331*, 312–315. https://doi.org/10.1126/science.1199697

Kielak, A., Rodrigues, J.L., Kuramae, E.E., Chain, P.S., van Veen, J.A., and Kowalchuk, G.A. (2010). Phylogenetic and metagenomic analysis of Verrucomicrobia in former agricultural grassland soil. FEMS Microbiol. Ecol. *71*, 23–33. https://doi.org/10.1111/j.1574-6941.2009.00785.x

Kim, Y., and Liesack, W. (2015). Differential assemblage of functional units in paddy soil microbiomes. PLOS ONE *10*, e0122221. https://doi.org/10.1371/journal.pone.0122221

King, G.M., Kostka, J.E., Hazen, T.C., and Sobecky, P.A. (2015). Microbial responses to the *Deepwater Horizon* oil spill: from coastal wetlands to the deep sea. Ann. Rev. Mar. Sci. *7*, 377–401. https://doi.org/10.1146/annurev-marine-010814-015543

Kits, K.D., Klotz, M.G., and Stein, L.Y. (2015). Methane oxidation coupled to nitrate reduction under hypoxia by the Gammaproteobacterium *Methylomonas denitrificans*, sp. nov. type strain FJG1. Environ. Microbiol. *17*, 3219–3232. https://doi.org/10.1111/1462-2920.12772

Knief, C. (2015). Diversity and habitat preferences of cultivated and uncultivated aerobic methanotrophic bacteria evaluated based on *pmoA* as molecular marker. Front. Microbiol. *6*, 1346. https://doi.org/10.3389/fmicb.2015.01346

Knief, C., Delmotte, N., Chaffron, S., Stark, M., Innerebner, G., Wassmann, R., von Mering, C., and Vorholt, J.A. (2012). Metaproteogenomic analysis of microbial communities in the phyllosphere and rhizosphere of rice. ISME J. *6*, 1378–1390. https://doi.org/10.1038/ismej.2011.192

Knittel, K., and Boetius, A. (2009). Anaerobic oxidation of methane: progress with an unknown process. Annu. Rev. Microbiol. *63*, 311–334. https://doi.org/10.1146/annurev.micro.61.080706.093130

Kolb, S., and Horn, M.A. (2012). Microbial CH(4) and N(2)O Consumption in Acidic Wetlands. Front. Microbiol. *3*, 78. https://doi.org/10.3389/fmicb.2012.00078

Krause, S.M., Johnson, T., Samadhi Karunaratne, Y., Fu, Y., Beck, D.A., Chistoserdova, L., and Lidstrom, M.E. (2017). Lanthanide-dependent cross-feeding of methane-derived carbon is linked by microbial community interactions. Proc. Natl. Acad. Sci. U.S.A. *114*, 358–363. https://doi.org/10.1073/pnas.1619871114

Krukenberg, V., Riedel, D., Gruber-Vodicka, H.R., Buttigieg, P.L., Tegetmeyer, H.E., Boetius, A., and Wegener, G. (2018). Gene expression and ultrastructure of meso- and thermophilic methanotrophic consortia. Environ. Microbiol. *20*, 1651–1666. https://doi.org/10.1111/1462-2920.14077

Kumaresan, D., Stephenson, J., Doxey, A.C., Bandukwala, H., Brooks, E., Hillebrand-Voiculescu, A., Whiteley, A.S., and Murrell, J.C. (2018). Aerobic proteobacterial methylotrophs in Movile Cave: genomic and metagenomic analyses. Microbiome *6*, 1. https://doi.org/10.1186/s40168-017-0383-2

Lau, M.C., Stackhouse, B.T., Layton, A.C., Chauhan, A., Vishnivetskaya, T.A., Chourey, K., Ronholm, J., Mykytczuk, N.C., Bennett, P.C., Lamarche-Gagnon, G., *et al.* (2015). An active atmospheric methane sink in high Arctic mineral cryosols. ISME J. *9*, 1880–1891. https://doi.org/10.1038/ismej.2015.13

Lee, H.J., Jeong, S.E., Kim, P.J., Madsen, E.L., and Jeon, C.O. (2015). High resolution depth distribution of Bacteria, Archaea, methanotrophs, and methanogens in the bulk and rhizosphere soils of a flooded rice paddy.

Front. Microbiol. *6*, 639. https://doi.org/10.3389/fmicb.2015.00639

Lee, J.W., Kwon, K.K., Bahk, J.J., Lee, D.H., Lee, H.S., Kang, S.G., and Lee, J.H. (2016). Metagenomic analysis reveals the contribution of anaerobic methanotroph-1b in the oxidation of methane at the Ulleung Basin, East Sea of Korea. J. Microbiol. *54*, 814–822. https://doi.org/10.1007/s12275-016-6379-y

Lesniewski, R.A., Jain, S., Anantharaman, K., Schloss, P.D., and Dick, G.J. (2012). The metatranscriptome of a deep-sea hydrothermal plume is dominated by water column methanotrophs and lithotrophs. ISME J. *6*, 2257–2268. https://doi.org/10.1038/ismej.2012.63

Li, M., Jain, S., Baker, B.J., Taylor, C., and Dick, G.J. (2014). Novel hydrocarbon monooxygenase genes in the metatranscriptome of a natural deep-sea hydrocarbon plume. Environ. Microbiol. *16*, 60–71. https://doi.org/10.1111/1462-2920.12182

Lin, X., Tfaily, M.M., Steinweg, J.M., Chanton, P., Esson, K., Yang, Z.K., Chanton, J.P., Cooper, W., Schadt, C.W., and Kostka, J.E. (2014). Microbial community stratification linked to utilization of carbohydrates and phosphorus limitation in a boreal peatland at Marcell Experimental Forest, Minnesota, USA. Appl. Environ. Microbiol. *80*, 3518–3530. https://doi.org/10.1128/AEM.00205-14

Mackelprang, R., Waldrop, M.P., DeAngelis, K.M., David, M.M., Chavarria, K.L., Blazewicz, S.J., Rubin, E.M., and Jansson, J.K. (2011). Metagenomic analysis of a permafrost microbial community reveals a rapid response to thaw. Nature *480*, 368–371. https://doi.org/10.1038/nature10576

Martinez-Gomez, N.C., Vu, H.N., and Skovran, E. (2016). Lanthanide chemistry: from coordination in chemical complexes shaping our technology to coordination in enzymes shaping bacterial metabolism. Inorg. Chem. *55*, 10083–10089. https://doi.org/10.1021/acs.inorgchem.6b00919

Mason, O.U., Scott, N.M., Gonzalez, A., Robbins-Pianka, A., Bælum, J., Kimbrel, J., Bouskill, N.J., Prestat, E., Borglin, S., Joyner, D.C., *et al.* (2014). Metagenomics reveals sediment microbial community response to Deepwater Horizon oil spill. ISME J. *8*, 1464–1475. https://doi.org/10.1038/ismej.2013.254

McDonald, I.R., Radajewski, S., and Murrell, J.C. (2005). Stable isotope probing of nucleic acids in methanotrophs and methylotrophs: a review. Org. Geochem. *36*, 779–787.

McDonald, I.R., Bodrossy, L., Chen, Y., and Murrell, J.C. (2008). Molecular ecology techniques for the study of aerobic methanotrophs. Appl. Environ. Microbiol. *74*, 1305–1315. https://doi.org/10.1128/AEM.02233-07

Melling, L., Hatano, R., and Goh, K.J. (2007). Nitrous oxide emissions from three ecosystems in tropical peatland of Sarawak, Malaysia. Soil Sci. Plant Nutr. 53, 792–805.

Meyerdierks, A., Kube, M., Lombardot, T., Knittel, K., Bauer, M., Glöckner, F.O., Reinhardt, R., and Amann, R. (2005). Insights into the genomes of archaea mediating the anaerobic oxidation of methane. Environ. Microbiol. 7, 1937–1951.

Meyerdierks, A., Kube, M., Kostadinov, I., Teeling, H., Glöckner, F.O., Reinhardt, R., and Amann, R. (2010). Metagenome and mRNA expression analyses of anaerobic methanotrophic archaea of the ANME-1 group. Environ. Microbiol. *12*, 422–439. https://doi.org/10.1111/j.1462-2920.2009.02083.x

Milucka, J., Ferdelman, T.G., Polerecky, L., Franzke, D., Wegener, G., Schmid, M., Lieberwirth, I., Wagner, M., Widdel, F., and Kuypers, M.M. (2012). Zero-valent sulphur is a key intermediate in marine methane oxidation. Nature *491*, 541–546. https://doi.org/10.1038/nature11656

Mohammadi, S., Pol, A., van Alen, T.A., Jetten, M.S., and Op den Camp, H.J. (2017). Methylacidiphilum fumariolicum SolV, a thermoacidophilic 'Knallgas' methanotroph with both an oxygen-sensitive and -insensitive hydrogenase. ISME J. *11*, 945–958. https://doi.org/10.1038/ismej.2016.171

Nahlik, A.M., and Fennessy, M.S. (2016). Carbon storage in US wetlands. Nat. Commun. *7*, 13835. https://doi.org/10.1038/ncomms13835

Narrowe, A.B., Angle, J.C., Daly, R.A., Stefanik, K.C., Wrighton, K.C., and Miller, C.S. (2017). High-resolution sequencing reveals unexplored archaeal diversity in freshwater wetland soils. Environ. Microbiol. *19*, 2192–2209. https://doi.org/10.1111/1462-2920.13703

Nazaries, L., Murrell, J.C., Millard, P., Baggs, L., and Singh, B.K. (2013). Methane, microbes and models: fundamental understanding of the soil methane cycle for future predictions. Environ. Microbiol. *15*, 2395–2417. https://doi.org/10.1111/1462-2920.12149

Nix, E.R., Fisher, C.R., Vodenichar, J., and Scott, K.M. (1995). Physiological ecology of a mussel with methanotrophic endosymbionts at three hydrocarbon seep sites in the Gulf of Mexico. Mar. Biol. *122*, 605–617.

Op den Camp, H.J., Islam, T., Stott, M.B., Harhangi, H.R., Hynes, A., Schouten, S., Jetten, M.S., Birkeland, N.K., Pol, A., and Dunfield, P.F. (2009). Environmental, genomic and taxonomic perspectives on methanotrophic Verrucomicrobia. Environ. Microbiol. Rep. *1*, 293–306. https://doi.org/10.1111/j.1758-2229.2009.00022.x

Oshkin, I.Y., Beck, D.A., Lamb, A.E., Tchesnokova, V., Benuska, G., McTaggart, T.L., Kalyuzhnaya, M.G., Dedysh, S.N., Lidstrom, M.E., and Chistoserdova, L. (2015). Methane-fed microbial microcosms show differential community dynamics and pinpoint taxa involved in communal response. ISME J. *9*, 1119–1129. https://doi.org/10.1038/ismej.2014.203

Oswald, K., Graf, J.S., Littmann, S., Tienken, D., Brand, A., Wehrli, B., Albertsen, M., Daims, H., Wagner, M., Kuypers, M.M., *et al.* (2017). Crenothrix are major methane consumers in stratified lakes. ISME J. *11*, 2124–2140. https://doi.org/10.1038/ismej.2017.77

Padilla, C.C., Bertagnolli, A.D., Bristow, L.A., Sarode, N., Glass, J.B., Thamdrup, B., and Stewart, F.J. (2017). Metagenomic binning recovers a transcriptionally active Gammaproteobacterium linking methanotrophy to partial denitrification in an anoxic oxygen minimum zone. Front. Mar. Sci. *4*, 23.

Parks, D.H., Rinke, C., Chuvochina, M., Chaumeil, P.A., Woodcroft, B.J., Evans, P.N., Hugenholtz, P., and Tyson, G.W. (2017). Recovery of nearly 8,000 metagenome-assembled genomes substantially expands the tree of life. Nat. Microbiol. *2*, 1533–1542. https://doi.org/10.1038/s41564-017-0012-7

Parks, D.H., Chuvochina, M., Waite, D.W., Rinke, C., Skarshewski, A., Chaumeil, P.-A., and Hugenholtz,

P. (2018). A standardized bacterial taxonomy based on genome phylogeny substantially revises the tree of life. Nat. Biotechnol. *36*, 996–1004. https://doi.org/10.1038/nbt.4229

Pernthaler, A., Dekas, A.E., Brown, C.T., Goffredi, S.K., Embaye, T., and Orphan, V.J. (2008). Diverse syntrophic partnerships from deep-sea methane vents revealed by direct cell capture and metagenomics. Proc. Natl. Acad. Sci. U.S.A. *105*, 7052–7057. https://doi.org/10.1073/pnas.0711303105

Petersen, J.M., and Dubilier, N. (2009). Methanotrophic symbioses in marine invertebrates. Environ. Microbiol. Rep. *1*, 319–335. https://doi.org/10.1111/j.1758-2229.2009.00081.x

Philippot, L., Andersson, S.G., Battin, T.J., Prosser, J.I., Schimel, J.P., Whitman, W.B., and Hallin, S. (2010). The ecological coherence of high bacterial taxonomic ranks. Nat. Rev. Microbiol. *8*, 523–529. https://doi.org/10.1038/nrmicro2367

Pol, A., Heijmans, K., Harhangi, H.R., Tedesco, D., Jetten, M.S., and Op den Camp, H.J. (2007). Methanotrophy below pH 1 by a new Verrucomicrobia species. Nature *450*, 874–878.

Pol, A., Barends, T.R., Dietl, A., Khadem, A.F., Eygensteyn, J., Jetten, M.S., and Op den Camp, H.J. (2014). Rare earth metals are essential for methanotrophic life in volcanic mudpots. Environ. Microbiol. *16*, 255–264. https://doi.org/10.1111/1462-2920.12249

Ponnudurai, R., Kleiner, M., Sayavedra, L., Petersen, J.M., Moche, M., Otto, A., Becher, D., Takeuchi, T., Satoh, N., Dubilier, N., *et al.* (2017). Metabolic and physiological interdependencies in the *Bathymodiolus azoricus* symbiosis. ISME J. *11*, 463–477. https://doi.org/10.1038/ismej.2016.124

Pratscher, J., Dumont, M.G., and Conrad, R. (2011). Assimilation of acetate by the putative atmospheric methane oxidizers belonging to the USCα clade. Environ. Microbiol. *13*, 2692–2701. https://doi.org/10.1111/j.1462-2920.2011.02537.x

Pratscher, J., Vollmers, J., Wiegand, S., Dumont, M.G., and Kaster, A.K. (2018). Unravelling the identity, metabolic potential and global biogeography of the atmospheric methane-oxidizing upland soil cluster α. Environ. Microbiol. *20*, 1016–1029. https://doi.org/10.1111/1462-2920.14036

Puri, A.W., Owen, S., Chu, F., Chavkin, T., Beck, D.A., Kalyuzhnaya, M.G., and Lidstrom, M.E. (2015). Genetic tools for the industrially promising methanotroph *Methylomicrobium buryatense*. Appl. Environ. Microbiol. *81*, 1775–1781. https://doi.org/10.1128/AEM.03795-14

Raghoebarsing, A.A., Pol, A., van de Pas-Schoonen, K.T., Smolders, A.J., Ettwig, K.F., Rijpstra, W.I., Schouten, S., Damsté, J.S., Op den Camp, H.J., Jetten, M.S., *et al.* (2006). A microbial consortium couples anaerobic methane oxidation to denitrification. Nature *440*, 918–921.

Rahalkar, M.C., and Bahulikar, R.A. (2018). Hemerythrins are widespread and conserved for methanotrophic guilds. Gene Rep. *11*, 250–254.

Ramachandran, A., and Walsh, D.A. (2015). Investigation of XoxF methanol dehydrogenases reveals new methylotrophic bacteria in pelagic marine and freshwater ecosystems. FEMS Microbiol. Ecol. *91*, fiv105. https://doi.org/10.1093/femsec/fiv105

Reim, A., Lüke, C., Krause, S., Pratscher, J., and Frenzel, P. (2012). One millimetre makes the difference: high-resolution analysis of methane-oxidizing bacteria and their specific activity at the oxic-anoxic interface in a flooded paddy soil. ISME J. *6*, 2128–2139. https://doi.org/10.1038/ismej.2012.57

Reimann, J., Jetten, M.S., and Keltjens, J.T. (2015). Metal enzymes in 'impossible' microorganisms catalyzing the anaerobic oxidation of ammonium and methane. Met. Ions Life Sci. *15*, 257–313. https://doi.org/10.1007/978-3-319-12415-5_7

Rey-Sanchez, A.C., Morin, T.H., Stefanik, K.C., Wrighton, K., and Bohrer, G. (2018). Determining total emissions and environmental drivers of methane flux in a Lake Erie estuarine marsh. Ecol. Eng. *114*, 7–15.

Rhoads, A., and Au, K.F. (2015). PacBio sequencing and its applications. Genomics Proteomics Bioinformatics *13*, 278–289. https://doi.org/10.1016/j.gpb.2015.08.002

Ricke, P., Kube, M., Nakagawa, S., Erkel, C., Reinhardt, R., and Liesack, W. (2005). First genome data from uncultured upland soil cluster alpha methanotrophs provide further evidence for a close phylogenetic relationship to *Methylocapsa acidiphila* B2 and for high-affinity methanotrophy involving particulate methane monooxygenase. Appl. Environ. Microbiol. *71*, 7472–7482.

Riley, W.J., Subin, Z.M., Lawrence, D.M., Swenson, S.C., Torn, M.S., Meng, L., Mahowald, N.M., and Hess, P. (2011). Barriers to predicting changes in global terrestrial methane fluxes: Analyses using CLM4Me, a methane biogeochemistry model integrated in CESM. Biogeosciences *8*, 1925–1953.

Rinke, C., Schwientek, P., Sczyrba, A., Ivanova, N.N., Anderson, I.J., Cheng, J.F., Darling, A., Malfatti, S., Swan, B.K., Gies, E.A., *et al.* (2013). Insights into the phylogeny and coding potential of microbial dark matter. Nature *499*, 431–437. https://doi.org/10.1038/nature12352

Rivkina, E., Petrovskaya, L., Vishnivetskaya, T., Krivushin, K., Shmakova, L., Tutukina, M., Meyers, A., and Kondrashov, F. (2016). Metagenomic analyses of the late Pleistocene permafrost – Additional tools for reconstruction of environmental conditions. Biogeosciences *13*, 2207–2219.

Saidi-Mehrabad, A., He, Z., Tamas, I., Sharp, C.E., Brady, A.L., Rochman, F.F., Bodrossy, L., Abell, G.C., Penner, T., Dong, X., *et al.* (2013). Methanotrophic bacteria in oilsands tailings ponds of northern Alberta. ISME J. *7*, 908–921. https://doi.org/10.1038/ismej.2012.163

Scheller, S., Yu, H., Chadwick, G.L., McGlynn, S.E., and Orphan, V.J. (2016). Artificial electron acceptors decouple archaeal methane oxidation from sulfate reduction. Science *351*, 703–707. https://doi.org/10.1126/science.aad7154

Segarra, K.E., Schubotz, F., Samarkin, V., Yoshinaga, M.Y., Hinrichs, K.U., and Joye, S.B. (2015). High rates of anaerobic methane oxidation in freshwater wetlands reduce potential atmospheric methane emissions. Nat. Commun. *6*, 7477. https://doi.org/10.1038/ncomms8477

Semrau, J.D., DiSpirito, A.A., and Yoon, S. (2010). Methanotrophs and copper. FEMS Microbiol. Rev.

34, 496–531. https://doi.org/10.1111/j.1574-6976.2010.00212.x

Sharon, I., and Banfield, J.F. (2013). Microbiology. Genomes from metagenomics. Science *342*, 1057–1058. https://doi.org/10.1126/science.1247023

Sharp, C.E., Smirnova, A.V., Graham, J.M., Stott, M.B., Khadka, R., Moore, T.R., Grasby, S.E., Strack, M., and Dunfield, P.F. (2014). Distribution and diversity of Verrucomicrobia methanotrophs in geothermal and acidic environments. Environ. Microbiol. *16*, 1867–1878. https://doi.org/10.1111/1462-2920.12454

Shiller, A.M., Chan, E.W., Joung, D.J., Redmond, M.C., and Kessler, J.D. (2017). Light rare earth element depletion during Deepwater Horizon blowout methanotrophy. Sci. Rep. *7*, 10389. https://doi.org/10.1038/s41598-017-11060-z

Singleton, C.M., McCalley, C.K., Woodcroft, B.J., Boyd, J.A., Evans, P.N., Hodgkins, S.B., Chanton, J.P., Frolking, S., Crill, P.M., Saleska, S.R., *et al.* (2018). Methanotrophy across a natural permafrost thaw environment. ISME J. *12*, 2544–2558. https://doi.org/10.1038/s41396-018-0065-5

Skennerton, C.T., Ward, L.M., Michel, A., Metcalfe, K., Valiente, C., Mullin, S., Chan, K.Y., Gradinaru, V., and Orphan, V.J. (2015). Genomic reconstruction of an uncultured hydrothermal vent gammaproteobacterial methanotroph (family Methylothermaceae) indicates multiple adaptations to oxygen limitation. Front. Microbiol. *6*, 1425. https://doi.org/10.3389/fmicb.2015.01425

Smith, G.J., Angle, J.C., Solden, L.M., Morin, T.H., Borton, M.A., Daly, R.A., Johnston, M.D., Stefanik, K.C., Wolfe, R., Bohrer, G., *et al.* (2018). Members of the Methylobacter are inferred to account for the majority of aerobic methane oxidation in oxic soils from a freshwater wetland. MBio *9*, 1–17.

Solden, L., Lloyd, K., and Wrighton, K. (2016). The bright side of microbial dark matter: lessons learned from the uncultivated majority. Curr. Opin. Microbiol. *31*, 217–226.

Spiridonova, E.M., Kuznetsov, B.B., Pimenov, N.V., and Tourova, T.P. (2006). Phylogenetic characterization of endosymbionts of the hydrothermal vent mussel Bathymodiolus azoricus by analysis of the 16S rRNA, cbbL, and pmoA genes. Microbiology *75*, 694–701.

Stoecker, K., Bendinger, B., Schöning, B., Nielsen, P.H., Nielsen, J.L., Baranyi, C., Toenshoff, E.R., Daims, H., and Wagner, M. (2006). Cohn's Crenothrix is a filamentous methane oxidizer with an unusual methane monooxygenase. Proc. Natl. Acad. Sci. U.S.A. *103*, 2363–2367.

Stokke, R., Roalkvam, I., Lanzen, A., Haflidason, H., and Steen, I.H. (2012). Integrated metagenomic and metaproteomic analyses of an ANME-1-dominated community in marine cold seep sediments. Environ. Microbiol. *14*, 1333–1346. https://doi.org/10.1111/j.1462-2920.2012.02716.x

Strong, P.J., Xie, S., and Clarke, W.P. (2015). Methane as a resource: can the methanotrophs add value? Environ. Sci. Technol. *49*, 4001–4018. https://doi.org/10.1021/es504242n

Strong, P.J., Kalyuzhnaya, M., Silverman, J., and Clarke, W.P. (2016). A methanotroph-based biorefinery: Potential scenarios for generating multiple products from a single fermentation. Bioresour. Technol. *215*, 314–323.

Sullivan, J.P., Dickinson, D., and Chase, H.A. (1998). Methanotrophs, *Methylosinus trichosporium* OB3b, sMMO, and their application to bioremediation. Crit. Rev. Microbiol. *24*, 335–373. https://doi.org/10.1080/10408419891294217

Sun, Y., Wang, M., Li, L., Zhou, L., Wang, X., Zheng, P., Yu, H., Li, C., and Sun, S. (2017). Molecular identification of methane monooxygenase and quantitative analysis of methanotrophic endosymbionts under laboratory maintenance in *Bathymodiolus platifrons* from the South China Sea. PeerJ *5*, e3565. https://doi.org/10.7717/peerj.3565

Takeuchi, M., Yoshioka, H., Seo, Y., Tanabe, S., Tamaki, H., Kamagata, Y., Takahashi, H.A., Igari, S., Mayumi, D., and Sakata, S. (2011). A distinct freshwater-adapted subgroup of ANME-1 dominates active archaeal communities in terrestrial subsurfaces in Japan. Environ. Microbiol. *13*, 3206–3218. https://doi.org/10.1111/j.1462-2920.2011.02517.x

Takeuchi, M., Kamagata, Y., Oshima, K., Hanada, S., Tamaki, H., Marumo, K., Maeda, H., Nedachi, M., Hattori, M., Iwasaki, W., *et al.* (2014). *Methylocaldum marinum* sp. nov., A thermotolerant, methane-oxidizing bacterium isolated from marine sediments, And emended description of the genus methylocaldum. Int. J. Syst. Evol. Microbiol. *64*, 3240–3246.

Takishita, K., Takaki, Y., Chikaraishi, Y., Ikuta, T., Ozawa, G., Yoshida, T., Ohkouchi, N., and Fujikura, K. (2017). Genomic evidence that methanotrophic endosymbionts likely provide deep-sea bathymodiolus mussels with a sterol intermediate in cholesterol biosynthesis. Genome Biol. Evol. *9*, 1148–1160. https://doi.org/10.1093/gbe/evx082

Taubert, M., Grob, C., Howat, A.M., Burns, O.J., Dixon, J.L., Chen, Y., and Murrell, J.C. (2015). XoxF encoding an alternative methanol dehydrogenase is widespread in coastal marine environments. Environ. Microbiol. *17*, 3937–3948. https://doi.org/10.1111/1462-2920.12896

Tavormina, P.L., Ussler, W., Joye, S.B., Harrison, B.K., and Orphan, V.J. (2010). Distributions of putative aerobic methanotrophs in diverse pelagic marine environments. ISME J. *4*, 700–710. https://doi.org/10.1038/ismej.2009.155

Tavormina, P.L., Hatzenpichler, R., McGlynn, S., Chadwick, G., Dawson, K.S., Connon, S.A., and Orphan, V.J. (2015). *Methyloprofundus sedimenti* gen. nov., sp. nov., an obligate methanotroph from ocean sediment belonging to the 'deep sea-1' clade of marine methanotrophs. Int. J. Syst. Evol. Microbiol. *65*, 251–259.

Timmers, P.H., Welte, C.U., Koehorst, J.J., Plugge, C.M., Jetten, M.S., and Stams, A.J. (2017). Reverse methanogenesis and respiration in methanotrophic Archaea. Archaea *2017*, 1654237. https://doi.org/10.1155/2017/1654237

Tu, T.H., Wu, L.W., Lin, Y.S., Imachi, H., Lin, L.H., and Wang, P.L. (2017). Microbial community composition and functional capacity in a terrestrial ferruginous, sulfate-depleted mud volcano. Front. Microbiol. *8*, 2137. https://doi.org/10.3389/fmicb.2017.02137

Tveit, A., Schwacke, R., Svenning, M.M., and Urich, T. (2013). Organic carbon transformations in high-Arctic peat soils: key functions and microorganisms. ISME J. 7, 299–311. https://doi.org/10.1038/ismej.2012.99

Tveit, A.T., Urich, T., and Svenning, M.M. (2014). Metatranscriptomic analysis of arctic peat soil microbiota. Appl. Environ. Microbiol. *80*, 5761–5772. https://doi.org/10.1128/AEM.01030-14

Tveit, A.T., Urich, T., Frenzel, P., and Svenning, M.M. (2015). Metabolic and trophic interactions modulate methane production by Arctic peat microbiota in response to warming. Proc. Natl. Acad. Sci. U.S.A. *112*, E2507–16. https://doi.org/10.1073/pnas.1420797112

Tyson, G.W., Chapman, J., Hugenholtz, P., Allen, E.E., Ram, R.J., Richardson, P.M., Solovyev, V.V., Rubin, E.M., Rokhsar, D.S., and Banfield, J.F. (2004). Community structure and metabolism through reconstruction of microbial genomes from the environment. Nature *428*, 37–43. https://doi.org/10.1038/nature02340

Vaksmaa, A., Guerrero-Cruz, S., van Alen, T.A., Cremers, G., Ettwig, K.F., Lüke, C., and Jetten, M.S.M. (2017). Enrichment of anaerobic nitrate-dependent methanotrophic 'Candidatus Methanoperedens nitroreducens' archaea from an Italian paddy field soil. Appl. Microbiol. Biotechnol. *101*, 7075–7084. https://doi.org/10.1007/s00253-017-8416-0

Vekeman, B., Kerckhof, F.M., Cremers, G., de Vos, P., Vandamme, P., Boon, N., Op den Camp, H.J., and Heylen, K. (2016a). New Methyloceanibacter diversity from North Sea sediments includes methanotroph containing solely the soluble methane monooxygenase. Environ. Microbiol. *18*, 4523–4536. https://doi.org/10.1111/1462-2920.13485

Vekeman, B., Speth, D., Wille, J., Cremers, G., De Vos, P., Op den Camp, H.J., and Heylen, K. (2016b). Genome characteristics of two novel Type I methanotrophs enriched from North Sea sediments containing exclusively a lanthanide-dependent XoxF5-type methanol dehydrogenase. Microb. Ecol. *72*, 503–509. https://doi.org/10.1007/s00248-016-0808-7

Vigneron, A., Cruaud, P., Pignet, P., Caprais, J.C., Cambon-Bonavita, M.A., Godfroy, A., and Toffin, L. (2013). Archaeal and anaerobic methane oxidizer communities in the Sonora Margin cold seeps, Guaymas Basin (Gulf of California). ISME J. *7*, 1595–1608. https://doi.org/10.1038/ismej.2013.18

Völker, H., Schweisfurth, R., and Hirsch, P. (1977). Morphology and ultrastructure of Crenothrix polyspora Cohn. J. Bacteriol. *131*, 306–313.

Wang, F.P., Zhang, Y., Chen, Y., He, Y., Qi, J., Hinrichs, K.U., Zhang, X.X., Xiao, X., and Boon, N. (2014). Methanotrophic archaea possessing diverging methane-oxidizing and electron-transporting pathways. ISME J. *8*, 1069–1078. https://doi.org/10.1038/ismej.2013.212

Ward, N., Larsen, Ø., Sakwa, J., Bruseth, L., Khouri, H., Durkin, A.S., Dimitrov, G., Jiang, L., Scanlan, D., Kang, K.H., *et al.* (2004). Genomic insights into methanotrophy: the complete genome sequence of *Methylococcus capsulatus* (Bath). PLOS Biol. *2*, e303. https://doi.org/10.1371/journal.pbio.0020303

Waring, C.L., Hankin, S.I., Griffith, D.W.T., Kertesz, M.A., Kobylski, V., Wilson, N.L., Coleman, N.V., Kettlewell, G., Zlot, R., Bosse, M., *et al.* (2017). Seasonal total methane depletion in limestone caves. Sci. Rep. 7, 8314. https://doi.org/10.1038/s41598-017-07769-6

Wegener, G., Krukenberg, V., Riedel, D., Tegetmeyer, H.E., and Boetius, A. (2015). Intercellular wiring enables electron transfer between methanotrophic archaea and bacteria. Nature *526*, 587–590. https://doi.org/10.1038/nature15733

Wehrmann, M., Billard, P., Martin-Meriadec, A., Zegeye, A., and Klebensberger, J. (2017). Functional role of lanthanides in enzymatic activity and transcriptional regulation of pyrroloquinoline quinone-dependent alcohol dehydrogenases in *Pseudomonas putida* KT2440. MBio *8*, e00570–17.

Welte, C.U., Rasigraf, O., Vaksmaa, A., Versantvoort, W., Arshad, A., Op den Camp, H.J., Jetten, M.S., Lüke, C., and Reimann, J. (2016). Nitrate- and nitrite-dependent anaerobic oxidation of methane. Environ. Microbiol. Rep. *8*, 941–955. https://doi.org/10.1111/1758-2229.12487

Wu, M.L., Wessels, J.C., Pol, A., Op den Camp, H.J., Jetten, M.S., and van Niftrik, L. (2015). XoxF-type methanol dehydrogenase from the anaerobic methanotroph '*Candidatus* Methylomirabilis oxyfera'. Appl. Environ. Microbiol. *81*, 1442–1451.

Xu, X., Yuan, F., Hanson, P.J., Wullschleger, S.D., Thornton, P.E., Riley, W.J., Song, X., Graham, D.E., Song, C., and Tian, H. (2016). Reviews and syntheses: Four decades of modeling methane cycling in terrestrial ecosystems. Biogeosciences *13*, 3735–3755.

Xu, Z., Hansen, M.A., Hansen, L.H., Jacquiod, S., and Sørensen, S.J. (2014). Bioinformatic approaches reveal metagenomic characterization of soil microbial community. PLOS ONE *9*, e93445. https://doi.org/10.1371/journal.pone.0093445

Yeoh, Y.K., Sekiguchi, Y., Parks, D.H., and Hugenholtz, P. (2016). Comparative genomics of candidate phylum TM6 suggests that parasitism is widespread and ancestral in this lineage. Mol. Biol. Evol. 33, 915–927. https://doi.org/10.1093/molbev/msv281

Yergeau, E., Hogues, H., Whyte, L.G., and Greer, C.W. (2010). The functional potential of high Arctic permafrost revealed by metagenomic sequencing, qPCR and microarray analyses. ISME J. *4*, 1206–1214. https://doi.org/10.1038/ismej.2010.41

Yergeau, E., Maynard, C., Sanschagrin, S., Champagne, J., Juck, D., Lee, K., and Greer, C.W. (2015). Microbial community composition, functions, and activities in the Gulf of Mexico 1 year after the *Deepwater Horizon* accident. Appl. Environ. Microbiol. *81*, 5855–5866. https://doi.org/10.1128/AEM.01470-15

Zhou, Z., Pan, J., Wang, F., Gu, J.D., and Li, M. (2018). Bathyarchaeota: globally distributed metabolic generalists in anoxic environments. FEMS Microbiol. Rev. *42*, 639–655. https://doi.org/10.1093/femsre/fuy023

Metabolic Features of Aerobic Methanotrophs: News and Views

4

Valentina N. Khmelenina*, Sergey Y. But, Olga N. Rozova and Yuri A. Trotsenko

G.K. Skryabin Institute of Biochemistry and Physiology of Microorganisms, Russian Academy of Sciences, Pushchino, Russia.

*Correspondence: khmelenina@ibpm.pushchino.ru

https://doi.org/10.21775/9781912530045.04

Abstract

This review is focused on recent studies of carbon metabolism in aerobic methanotrophs that specifically addressed the properties, distribution and phylogeny of some of the key enzymes involved in assimilation of carbon from methane. These include enzymes involved in sugar synthesis and cleavage, conversion of intermediates of the tricarboxylic acid cycle, as well as in osmoadaptation in halotolerant methanotrophs.

Introduction

Methanotrophs inhabit a variety of ecosystems including soils, fresh and marine waters and sediments, saline and alkaline lakes, hot springs, rice paddies, peatlands, and tissues of higher organisms. They play a major role in both global carbon and global nitrogen cycles (Hanson and Hanson, 1996; McDonald *et al.*, 2008, Ettwig *et al.*, 2010; Khadem *et al.*, 2010). The metabolic potential of methanotrophs as industrial platforms for bioconversion of methane into value added compounds is currently under intense investigation, aimed at organisms that are best suited for such applications. The recent genomic and biochemical studies demonstrate high metabolic flexibility of methanotrophs, and high specialization of the enzymes of their central metabolic pathways towards methane utilization and also towards extreme condition adaptations. From the multitude of the characterized methanotrophs, halotolerant representatives are especially promising for industrial applications, as they are genetically tractable due to the availability of a variety of genetic manipulation tools (Kalyuzhnaya *et al.*, 2015; Mustakhimov *et al.*, 2015; Fu and Lidstrom, 2017; Garg *et al.*, 2018). The adaptation of halo- and thermotolerant methanotrophs to extreme environmental conditions includes the acquisition of the specific mechanisms for synthesis and reutilization of the compatible solutes. This review focuses on the recent studies of carbon metabolism in three model species of the aerobic methanotrophs *Methylomicrobium alcaliphilum* 20Z, *Methylosinus trichosporium* OB3b, and *Methylococcus capsulatus* Bath.

The carbon assimilation pathways

Methanotrophs obtain energy for growth predominantly via oxidation of methane to CO_2 and assimilate carbon at the level of formaldehyde, formate and/or CO_2 via three biochemical pathways that have been originally deciphered by Professor J. R. Quayle and colleagues in the middle of the 20th century. For more details, we refer the reader to the recent reviews (Bowmann, 2006; Kelly *et al.*, 2014; Webb *et al.*, 2014). The known aerobic methanotrophs belong to the phyla Proteobacteria, Verrucomicrobia, and the yet unnamed NC10 phylum. The proteobacterial methanotrophs are classified as type I or type II, according to whether they belong to the Alpha- or Gammaproteobacteria class. The consensus is that gammaproteobacterial methanotrophs, represented by the families *Methylococcaceae, Methylothermaceae* and *Candidatus* family *Crenotrichaceae,* assimilate formaldehyde via the ribulose monophosphate (RuMP) pathway, where hexosephosphates are the early products formed by the condensation of formaldehyde and ribulose-5-phosphate. The alphaproteobacterial methanotrophs of the families *Methylocystaceae* and *Beijerinckiaceae* assimilate carbon from methane via the serine pathway, where the amino acid serine is firstly formed. The methanotrophic representatives of the Verrucomicrobia (family *Methylacidiphilaceae*) and the NC10 phyla use methane as the energy source and fix carbon at the CO_2 level, via the Calvin–Benson–Bassham (CBB) cycle (Dunfield *et al.*, 2007; Islam *et al.*, 2008; Khadem *et al.*, 2011; van Teeseling *et al.*, 2014; Rasigraf *et al.*, 2014). In this chapter we will concentrate on the heterotrophic methanotrophs that use reasonably efficient RuMP or serine cycles.

The serine cycle

Genomic analysis revealed that the genes encoding the enzymes of the serine cycle occur in both type I and type II methanotrophs, though the functionality and the role of the serine cycle in type I methanotrophs remain not fully understood (But *et al.*, 2017). In both alpha- and gammaproteobacterial methanotrophs, genes encoding hydroxypyruvate reductase (*hpr*), serine-glyoxylate aminotransferase (*sga*), glycerate-2-kinase (*gck2*), malate thiokinase (*mtkAB*), malyl-CoA-lyase (*mcl*), malate dehydrogenase (*mdh*), serine-transhydroxymethylase (*glyA*), as well as methylene tetrahydromethanopterin/methylene tetrahydrofolate (H_4F) dehydrogenase (*mtdA*) and formate-tetrahydrofolate lygase (*ftfL*) are adjacently located in a single, two or several clusters (Fig. 4.1), thus corroborating the 'module' organization of the serine cycle genes in methylotrophs (Chistoserdova, 2011). In all type II methanotrophs, the cluster additionally contains genes encoding phosphoenolpyruvate carboxylase (*ppc*) and methylene H_4F cyclohydrolase (*fch*), whereas among type I methanotroph genomes, only the *Methylomonas* species harbour the *ppc* genes. It is possible that other representatives of type I methanotrophs possess non-homologous enzymes which can perform PEP carboxylation, one possibility being a PPi-dependent PEP carboxykinase (Chiba *et al.*, 2015; Khmelenina *et al.*, 2018). Among the methanotrophs, the glyoxylate regeneration cycle associated with the poly-β-hydroxybutyrate (PHB) pathway is present only in members of the *Methylocystaceae* family, whereas in the *Beiyerinckiaceae* family, this function can be fulfilled by the glyoxylate shunt (Chen *et al.*, 2010; Matsen *et al.*, 2013). However, gammaproteobacterial methanotrophs presumably lack this pathway.

Two key serine cycle enzymes, hydroxypyruvate reductase (Hpr) and serine-glyoxylate aminotransferase (Sga), have been purified from three methanotrophic bacteria: *Mm. alcaliphilum* 20Z, *Mc. capsulatus* Bath and *Ms. trichosporium* OB3b (But *et al.*, 2017, 2018a). The biochemical properties of methanotrophic Hprs differed from those of other organisms. Unlike plants or non-methanotrophic bacteria (Chistoserdova *et al.*, 1991; Ho *et al.*, 1999; Ali *et al.*, 2003), the three methanotrophic Hprs catalysed the irreversible NAD(P)H-dependent reduction of hydroxypyruvate and glyoxylate, but were unable to oxidize of glycerate and glycolate (But *et al.*, 2017). Such peculiarity, along with high activity and affinity to hydroxypyruvate, suggest that these enzymes must be of primary importance for C1 assimilation. Like other Hprs characterized to date, Hprs from *Mm. alcaliphilum* 20Z and *Mc. capsulatus* Bath are homodimeric, whereas the *Ms. trichosporium* Hpr is tetrameric (But *et al.*, 2017). The three enzymes displayed similar pH optima *in vitro,* while these strains are either neutrophilic (strains OB3b and Bath) or alkaliphilic (strain 20Z). Hpr from *Ms. trichosporium* OB3b displayed higher activity and affinity than the enzymes from *Mm. alcaliphilum* 20Z and *Mc. capsulatus* Bath. Hpr from *Mm. alcaliphilum* 20Z was shown to be a subject to allosteric regulation, its activity increasing in the presence of the intermediates of glycolysis and the TCA cycle, serine, ADP, ATP and inorganic pyrophosphate, but decreasing in the presence of Pi, pyruvate, acetyl-CoA, phosphoenolpyruvate and ribulose-5-phosphate. The capacity of the enzyme to reduce glyoxylate into glycolate, in combination with FAD-dependent glycolate oxidase (*ADVE02_v2_14253-14256; MCA1499-150* and *MALCv4_1676-1678*) (Ward *et al.*, 2004; Stain *et al.*, 2010; Vuilleumier *et al.*, 2012), suggests that

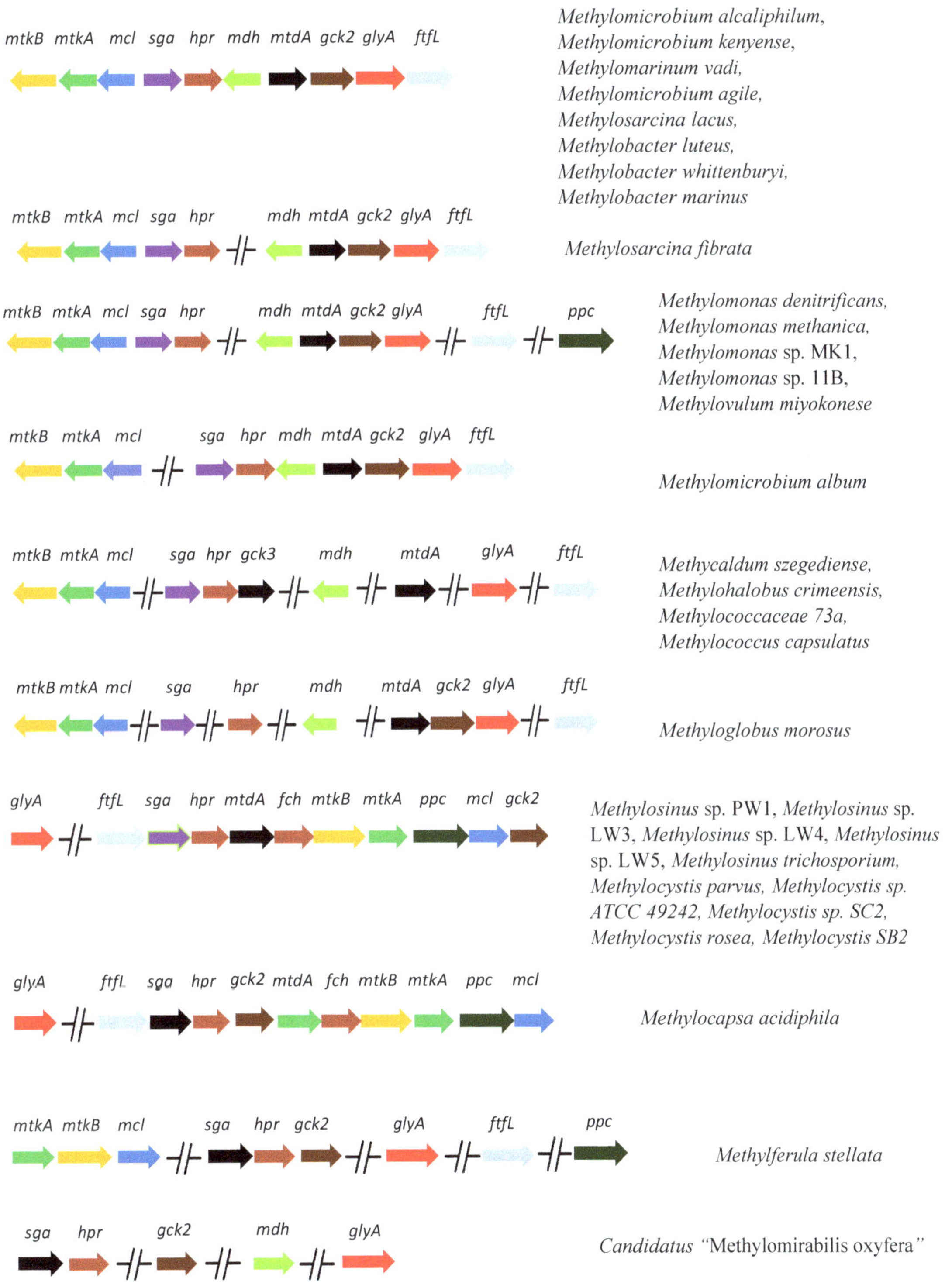

Figure 4.1 Organization of gene clusters encoding enzymes of the serine cycle in alpha- and gammaproteobacterial methanotrophs. Hydroxypyruvate reductase (*hpr*), serine-glyoxylate aminotransferase (*sga*), glycerate-2-kinase (*gck2*), glycerate-3-kinase (*gck3*), malate thiokinase (*mtkAB*), malyl-CoA-lyase (*mcl*), malate dehydrogenase (*mdh*), serine-transhydroxymethylase (*glyA*), tetrahydromethanopterine/methylentetrahydrofolate dehydrogenase (*mtdA*), formate-tetrahydrofolate ligase (*ftfL*), phosphoenolpyruvate carboxylase (*ppc*), methylene H_4F cyclohydrolase (*fch*).

Hpr could be involved in the regulation of the reduced equivalents content (But *et al.*, 2017). Phylogenetically, methanotrophic Hpr's can be divided into two divergent groups, and these share low identities with enzymes from non-methanotrophic heterotrophs (37–43%), algae (32–37%), plants (35–65%) or animals (26–46%). The phylogenetic tree of the methanotrophic Hprs implies a vertical evolution of the enzyme in the methanotrophs, with none or few instances of horizontal gene transfers (But *et al.*, 2017).

Another key enzyme of the serine cycle is serine-glyoxylate aminotransferase (Sga, EC 2.6.1.45), and these were also obtained from *Mm. alcaliphilum* 20Z, *Ms. trichosporium* OB3b and *Mc. capsulatus* Bath (But *et al.*, 2018a). The three enzymes showed similar biochemical properties, and those were also similar to the properties of Sga enzymes from other microbes and from plants (Kendziorek *et al.*, 2008; Kameya *et al.*, 2010). The methanotrophic Sgas can also transfer the amino group from serine to pyruvate, but display the highest catalytic efficiencies in the reaction between serine and glyoxylate, which would be consistent with a role in C_1 assimilation. The enzymes from strains 20Z and Bath also transferred the amino group from serine to α-ketoglutarate and from alanine to glyoxylate. Disruption of *sga* or simultaneously *sga* and *hpr* genes caused retardation of growth of *Mm. alcaliphilum* and an increase in lag-phase duration after the passage from methane to methanol. The mutant strain was shown to accumulate formaldehyde in the growth medium, suggesting that the serine cycle plays a role in type I methanotrophs, serving as one of the formaldehyde detoxification pathways (But *et al.*, 2018a).

The RuMP cycle

In the chromosomes of gammaproteobacteral methanotrophs, there are up to three copies of *hps/phi* operons coding for hexulosephosphate synthase (HPS) and phosphohexulose isomerase (PHI), key enzymes of the RuMP pathway. About half of the known type I methanotroph genomes additionally contain the fused gene *hps-phi* coding for bi-domain HPS-PHI protein (Rozova *et al.*, 2017). HPS and HPS-PHI purified from *Mm. alcaliphilum* 20Z are homodimeric enzymes (2 × 20 kDa and 2 × 40 kDa, respectively) catalysing condensation of formaldehyde with ribulose-5-phosphate with different activities (172 and 22 U/mg) and affinities to formaldehyde (K_m 0.98 mM versus 0.64 mM). AMP and ADP were shown to be powerful inhibitors for both enzymes. According to the k_{cat}/K_m ratio, HPS had a higher catalytic efficiency compared with HPS-PHI. Unexpectedly, the HPS-PHI fused enzyme lacked the isomerase activity. HPS but not the fused enzyme is prerequisite for carbon assimilation, since the insertion of a kanamycin cassette into the *hps-phi* fused gene in *Mm. alcaliphilum* 20Z did not alter growth rate of the mutant strain, whereas attempts to obtain a viable strain missing either *hps* or *phi* were unsuccessful (Rozova *et al.*, 2017). Although HPS-PHI was inactive as the isomerase, when expressed separately, both domains of the fused enzyme possessed the respective synthase and isomerase activities. The absence of the isomerase activity in the dimeric HPS-PHI could be due to the inactive enzyme configuration, since only tetrameric PHI have been characterized so far as active enzymes, whose catalytic centres were composed of four monomers (Orita *et al.*, 2005).

The unique membrane-bound hexulosephosphate synthase purified from *Mc. capsulatus* Bath by traditional chromatographic methods displayed very high activity and affinity to formaldehyde (Ferenci *et al.*, 1974). Its high molecular mass (310 kDa) implied that the bi-domain, hexameric enzyme was characterized. Importantly, besides the *hps-phi* fused gene, two complete *hps/phi* operons are present in the genome of *Mc. capsulatus* Bath (Ward *et al.*, 2004).

Surprisingly, none of the methanotroph genomes sequenced so far harbour the *hps-phi* fused gene alone. In contrast, none of the methylotrophic bacteria unable to use methane as a growth substrate encode the HPS-PHI fused enzyme. Interestingly, the genome of the model methanotroph *Methylomicrobium buryatense* possesses the *hps-phi* fused gene along with three copies of *hps/phi* operons. It remains to be demonstrated whether there is any effect of HPS multiplication on the methanotrophic growth (Rozova *et al.*, 2017). High homology between HPS and the synthase domain of the fused enzyme in the same methanotroph (73% identity in *Mm. alcaliphilum* 20Z) imply relatively recent division of HPS-PHI into separate enzymes (Rozova *et al.*, 2017). The sequences of the methanotrophic HPS comprise a tight phylogenetic cluster, distant from those of archaea (sharing 38–40% identities), in which this enzyme functions in synthesis of pentose phosphates from C_6-phosphosugars (Yurimoto *et al.*, 2002; Mitsui *et al.*, 2003; Orita *et al.*, 2005, 2006). The bi-domain organization of the enzymes in thermophilic archaea is considered to provide enhanced stability to the enzyme against thermal inactivation (Orita *et al.*, 2005). However, such a property would not be very relevant in mesophilic bacteria.

PPi-dependent glycolysis

At least three biochemical routes for sugar cleavage can operate in gammaproteobacterial methanotrophs: the Entner–Doudoroff pathway, the glycolysis and the phosphoketolase pathway (Fig. 4.2). The cause and effect of excessive metabolic flexibility in methanotrophs is not clear. Methanotrophs obtain energy for growth predominantly from oxidation of C_1 substrates to CO_2 and can acquire ATP from oxidative phosphorylation, based on several experimental studies devoted to deciphering the respiratory mechanisms in methanotrophs (Tonge *et al.*, 1977; Chetina *et al.*, 1986). In addition, glycolysis and fermentation have been proven to be essential sources of energy in at least gammaproteobacterial methanotrophs. These methanotrophs operate a modified version of the Embden–Meyerhoff–Parnas pathway, where PPi instead of ATP serves as the phosphate donor in an energy consuming reaction of fructose-6-phosphate phosphorylation. Along the PP_i-dependent 6-phosphofructokinase (PP_i-PFK, EC 2.7.1.90), they possess another glycolytic enzyme, pyruvate phosphate dikinase (PPDK), catalysing the PPi-dependent reversible interconversion between phosphoenolpyruvate (PEP) and pyruvate, accompanied by interconversion between ATP and AMP. These substitutions make the glycolytic pathway in these organisms more efficient with respect to ATP production, compared with the classical version of the pathway. Owing to the PPi-dependent reactions, the glycolysis in methanotrophs is fully reversible. It therefore can operate as the

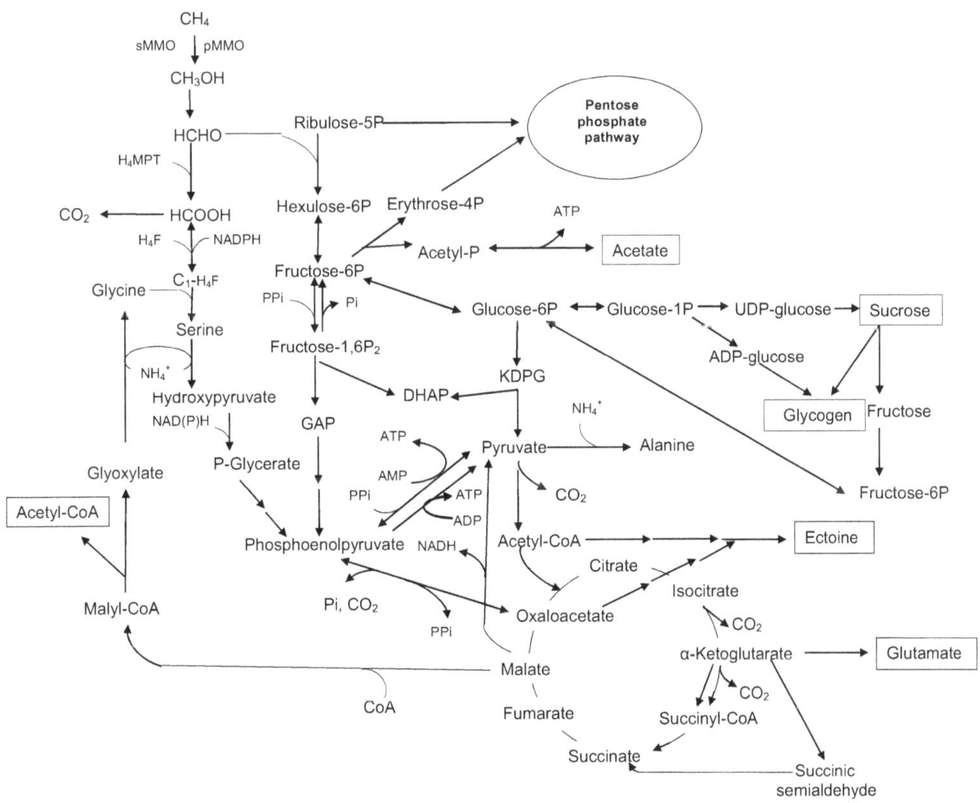

Figure 4.2 Pathways of carbon metabolism in gammaproteobacterial (Type I) methanotrophs. Modified from Rozova *et al.* (2015). H_4MPT, tetrahydromethanopterin; THF, tetrahydrofolate.

energy-inert gluconeogenic route where the energy contained in ATP can be saved in PPi molecules. In *Mm. alcaliphilum* 20Z, due to the presence pyruvate kinase, the glycolytic direction is prevailing (Kalyuzhnaya *et al.*, 2013). Considering the presence, in addition, of gluconeogenic enzymes, fructose bisphosphatase and PEP synthase, carbohydrate metabolism in type I methanotrophs is exceedingly flexible (Fig. 4.2).

PPi-PFKs of methanotrophs catalysing the reversible reaction of phosphorylation of fructose-6-phosphate (Fr6P) somewhat vary in their biochemical properties. PPi-PFK from *Methylomonas methanica* (2x45 kDa) and *Mm. alcaliphilum* 20Z (4x45 kDa) are very active enzymes that are not regulated allosterically (Reshetnikov *et al.*, 2008; Rozova *et al.*, 2010). PPi-PFK from *Mc. capsulatus* Bath is highly active as a sedoheptulose-7-phosphate (S7P) kinase, phosphorylating S7P with much higher activity than fructose-6-phosphate as a substrate. It can also phosphorylate Ru5P. In *Mc. capsulatus* Bath, which, in addition to the RuMP pathway and the serine cycle, encodes reactions of the Calvin cycle, PPi-PFK is involved in synthesis of RuBP, a primary acceptor for CO_2. Owing to modulations of activity of the ribulosebisphosphate carboxylase/oxygenase (RuBisCO) in response to temperature fluctuations (Eshinimaev *et al.*, 2004), an intracellular content of RuBP can be occasionally increased. In the case of temperature decrease or CO_2 deficiency, RuBP can be returned to the basic metabolism, retaining metabolic energy.

In at least several methanotrophs, the *pfp* gene is co-transcribed with the gene coding for a V-type H^+-pyrophosphatase (H^+-PPi-ase, EC 3.6.1.1). Therefore, direct involvement of PPi in membrane energization and ATP synthesis has been proposed (Reshetnikov *et al.*, 2008; Khmelenina *et al.*, 2018).

In alphaproteobacterial methanotrophs, the gluconeogenic function of PPi-PFK is obvious. The hexameric PPi-PFK from *Ms. trichosporium* OB3b (6 × 45 kDa) performs dephosphorylation of FBP with higher efficiency in comparison to phosphorylation of Fr6P (Rozova *et al.*, 2012). Unexpectedly, *Ms. trichosporium* PPi-PFK also phosphorylates S7P and Ru5P, although with much lower activities and affinities that are displayed with Fr6P. This bacterium also encodes ATP-dependent phosphoribulokinase (ID 2507407232), while it lacks the RuBisCO encoding genes. This metabolic feature implies an evolutionary link between methanotrophy and autotrophy and corroborates the recently uncovered essential regulatory functions of phosphoribulokinase and RuBP in methylotrophy (Ochsner *et al.*, 2017).

Among the methanotrophs, the representatives of the *Beijerenkiaceae* family and the phylum *Verrucomicrobia* lack PP_i-PFK (Khmelenina *et al.* 2018). In contrast, PPi-PFK is essential for methanotrophs possessing the intracytoplasmic membranes (ICM), suggesting a special function for PPi in their bioenergetics. The two oxygen-consuming processes, methane oxidation by particulate methane monooxygenase (pMMO) and oxidative phosphorylation, are both associated with the ICMs, suggesting complex regulation. At the same time, the reactive oxygen species generated by the electron transfer to pMMO (Murrell *et al.*, 2000; Kamachi and Okura, 2018) and the oxygen radicals generated from the respiratory chain may disturb membrane integrity and therefore hamper ATP synthesis or methane oxidation. The use of PPi instead of ATP as a phosphoryl donor in some cytoplasmic reactions can mitigate this bio-energetic problem. Analogous problem is faced by chemoautotrophs oxidizing NH_4^+ via ammonia monooxygenase, which is structurally and functionally similar to pMMO. This corroborates the supposition that the PP_i-PFK is specific to organisms that are energetically constrained (Mertens, 1991). The exceptions from this rule are acidophilic methanotrophs having pMMO but not possessing PPi-PFK-encoding genes, the *Methylacidiphilum* species that do not appear to form ICMs and only possess ATP-PFK (Fig. 4.3), and *Methylocapsa acidiphila* that forms ICMs but does not harbour a phosphofructokinase. The anaerobic methanotroph *Methylomirabilis oxyfera* possesses both pMMO and a very divergent PPi-PFK (Fig. 4.3), while apparently lacking ICMs (Wu *et al.*, 2012a). In this polygon-shaped bacterium able to couple anaerobic methane oxidation to denitrification, pMMO appears to be located in the cytoplasmic membrane (Wu *et al.*, 2012a,b). Further studies are needed to unravel the evolutionary history and the physiological roles of PPi and PPi-dependent enzymes in the methanotrophs.

Phylogenetically, methanotrophic PPi-PFKs belong to several divergent groups, and these share identities with the enzymes from non-methanotrophic heterotrophs and the protozoans (Fig. 4.3).

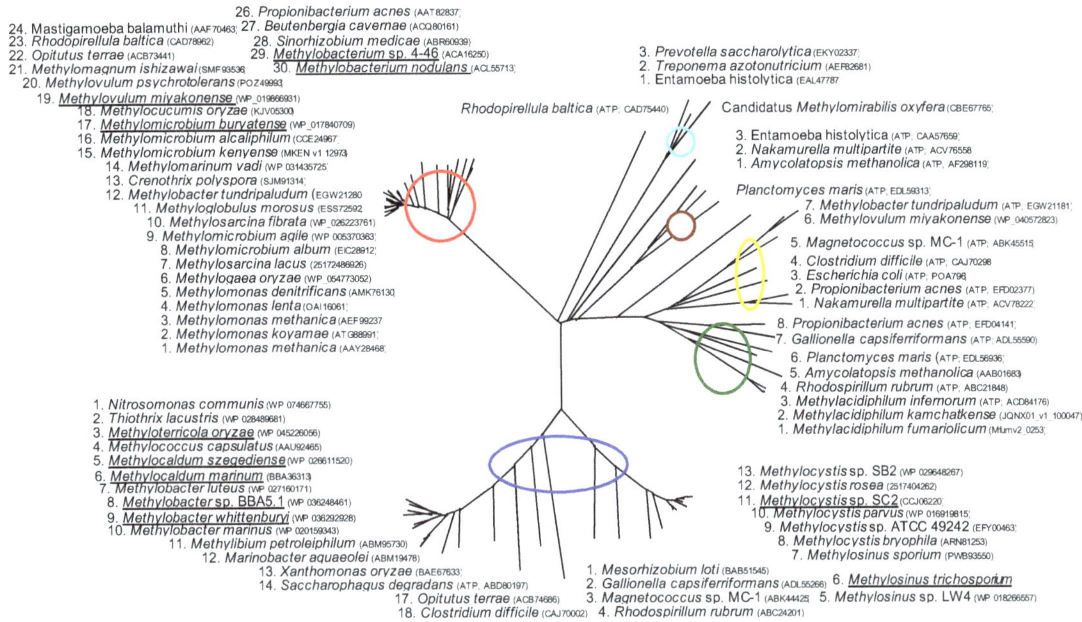

Figure 4.3 The phylogenetic tree of PPi- and ATP-dependent 6-phosphofructokinases. The phosphate donor (ATP) is indicated for ATP-PFKs in the brackets. Monophyletic groups are named according to the classification (Müller *et al.*, 2001; Bapteste *et al.* 2003) as 'B2' (dark-blue), 'BI' (orange), 'III' (green), 'X' (brown), 'LONG' (blue), and 'P' (red). Amino acid sequences were obtained from the NCBI database (http://www.ncbi.nlm.nih.gov), the Microbial Genome Annotation and Analysis Platform (www.genoscope.cns.fr), and from the Integrated Microbial Genomes and Microbiomes system (img.jgi.doe.gov) by BLAST. The tree was constructed using the MEGA 4 program (computed from 100 independent trials).

Such distribution on the phylogenetic tree corroborates hypothesis of 'random' lateral transfer of genes coding PPi-PFKs in microorganisms (Bapteste *et al.*, 2003). Nevertheless, some correlations between the phylogenetic position and the type of carbon metabolism could also be observed.

The phosphoketolase pathway

Phosphoketolase (PKT, EC 4.1.2.9/EC 4.1.2.22) catalyses the cleavage of xylulose-5-phosphate and/or fructose-6-phosphate into acetyl-phosphate and glyceraldehyde-3-phosphate or erythrose-4-phosphate (Sánchez *et al.*, 2010). PKT-encoding genes occurs in all gammaproteobacterial methanotrophs, and at least two PKT isoforms are present in the genomes of halotolerant methanotrophs (Rozova *et al.*, 2015). The PKT pathway produces acetyl-CoA, which is the precursor of the compatible solute ectoine that is accumulated by halotolerant methanotrophs in response to increased medium salinity. The methanotroph *M. buryatense* expresses all the PKT pathway components (Henard *et al.*, 2017). In this pathway, the C_2 compounds synthesized bypass glycolysis and pyruvate decarboxylation where the C–C bond would be cleaved. The PKT pathway can produce ATP, as acetate kinase is co-expressed with PKT (Rozova *et al.*, 2015a). Acetate kinase catalyses the reversible reaction of dephosphorylation of acetyl-P into acetate with higher catalytic efficiency in the direction of ATP formation, as compared to acetate phosphorylation. Moreover, PKT in *Mm. buryatense* has proven to be involved in the global regulatory alterations of the methanotrophic metabolism (Henard *et al.*, 2017). PKT pathway engineering has been successfully used to increase yields of an array of acetyl-CoA-derived products, in diverse microbial biocatalysts. Owing to overexpression of the PKT isoform, *Mm. buryatense* has an increased biomass yields and the lipid content (Herand and Guarnieri, 2018).

In the methanotrophs, the PKT-shunted glycolytic pathway operates along with the PPi-dependent glycolysis, another energy efficient pathway, thus providing the metabolic flexibility and allowing the gammaproteobacterial methanotrophs to survive in micro-aerobic ecosystems. Numerous environmental studies indicate that methanotrophs thrive at oxic–anoxic interfaces. The discovery of the ATP-producing assimilatory routes raised the possibility of fermentation in methanotrophs, as a new mode of methane utilization at low O_2 tensions, where formate, acetate, succinate, lactate, hydroxybutyrate and hydrogen can be the end products (Kalyuzhnaya *et al.*, 2013). This discovery challenges our understanding of methanotrophy as microbial metabolism linked solely to respiration, and it has major implications for the indispensable role of methanotrophic bacteria in removing the greenhouse gas methane in O_2-limited environments, including landfields. In such specific environments, methanotrophs drive the conversion of methane to excreted organic products and hydrogen, which are then used and transformed by non-methanotrophs.

The TCA cycle

In methanotrophs generating metabolic energy by the direct oxidation of reduced C_1 compounds or by formaldehyde fermentation, the predominantly anabolic role of the tricarboxylic acid (TCA) cycle is evident. While no or negligible activity of 2-oxoglutarate dehydrogenase has been found in methanotrophs where tested, the complete set of the genes coding for this enzyme have been identified in the genomes, and the functionality of the oxidative TCA cycle has been further proven in some methanotrophs (Matsen *et al.*, 2013; Fu *et al.*, 2017). Based on the genomic evidence, in type I methanotrophs, additional routes may bypass the need for the classic 2-oxoglutarate dehydrogenase complex, such as 2-oxoglutarate ferredoxin oxidoreductase, catalysing the reversible oxidative decarboxylation of 2-oxoglutarate to succinyl-CoA, as well as 2-oxoglutarate oxidase + succinate semialdehyde dehydrogenase (Fu *et al.*, 2017) (Fig. 4.2). In *Mm. buryatense* 5GB1, genes for both of these enzymes are expressed at the levels similar to the ones of other TCA cycle genes.

Malate dehydrogenases (MDH, L-malate: NAD oxidoreductase, EC 1.1.1.37) purified from *Ms. trichosporium* OB3b and *Mm. alcaliphilum* 20Z displayed properties reflecting their roles in respective primary metabolisms (Rozova *et al.*, 2015b). The high catalytic efficiency of MDH from *Ms. trichosporium* OB3b in the direction of malate synthesis would be reflective of the central position of malate in the serine pathway, where malate activation to malyl-CoA and its further cleavage into acetyl-CoA and glyoxylate are key reactions in glycine regeneration. Acetyl-CoA enters the ethylmalonyl-CoA (EMC) cycle, where it is transformed to glyoxylate (Chistoserdova *et al.*, 2009) or is directed to the syntheses of fatty acids and the storage compound poly-β-hydroxybutyrate (PHB). The main source of OAA for the MDH reaction is carboxylation of PEP by PEP carboxylase, which is a highly active enzyme in type II methanotrophs. Some methanotrophs, such as *Ms. trichosporium* OB3b, possess two PEP carboxylases, sharing 33% identity at the amino acid level (Matsen *et al.*, 2013).

Alternatively, MDH from *Mm. alcaliphilum* 20Z shows preference for malate oxidation (OAA formation) versus malate synthesis, as follows from respective k_{cat}/K_m ratios (Rozova *et al.*, 2015b). Such enzyme features are in agreement with the high demand for aspartate in this organism, as a precursor of the osmoprotectant ectoine (Reshetnikov *et al.*, 2011).

The malic enzyme (MaE) is another enzyme that is thought to be involved in anaplerotic CO_2 fixation and malate biosynthesis. However, MaE from *Ms. trichosporium* OB3b (EC 1.1.1.40) predominantly catalyses $NADP^+$-dependent reaction of malate decarboxylation, while it has much lower activity and affinity to pyruvate (Rozova *et al.*, 2019). $NADP^+$-dependency and high efficiency in the direction of decarboxylation suggest that MaE is the ‘lipogenic enzyme’ important for the production of $NADPH_2$ that is needed for biosynthesis of long-chain fatty acids and/or steroids (Fig. 4.4). In contrast, *Mm. alcaliphilum* MaE (EC 1.1.1.38) catalyses non-reversible NAD^+-dependent reaction of malate decarboxylation into pyruvate but is unable to fix CO_2. Evidently, the source of malate for this reaction could be the reactions of the TCA cycle, which, in turn, can be replenished by CO_2 fixation. Although type I methanotrophs lack the PEP carboxylase (with an exception of the *Methylomonas* species), those may possess unusual PPi-dependent PEP-carboxykinases catalysing the reversible reaction of CO_2 fixation. However, the functionality and the roles of such enzymes remain to be proven. As seen in Fig. 4.5, reactions catalysed by PEP carboxykinase, MDH, MaE and PPDK can produce two PPi molecules, to trigger the modified the Emden-Meyerhof-Parnas glycolysis.

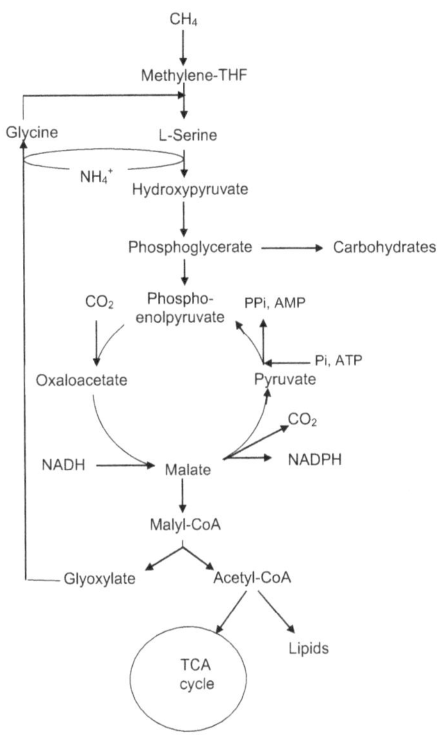

Figure 4.4 Pathways for interconversion of phosphoenolpyruvate and pyruvate in Type II methanotrophs and the role of the malic enzyme in NADPH production. TCA, tricarboxylic acid (modified from Rozova *et al.*, 2019).

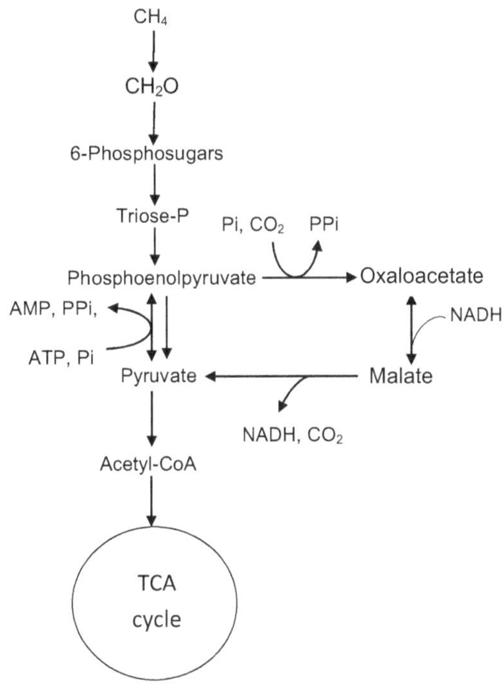

Figure 4.5 Pathways for conversion of phosphoenolpyruvate to pyruvate in Type I methanotrophs and the role of the malic enzyme in the synthesis of PPi. TCA, tricarboxylic acid.

Sucrose metabolism

The ability to synthesize sucrose is inherent in phototrophic bacteria and plants, and it has recently been revealed in non-photosynthetic bacteria, including aerobic methylotrophs that assimilate carbon via the RuMP pathway (But *et al.*, 2013, 2015). In the methylotrophs, the biochemical pathway for sucrose synthesis from fructose-6-phosphate and UDP-glucose is rather similar to the one in other pro- and eukaryotic organisms, being composed of two enzymes, sucrose phosphate synthase (Sps, EC 2.4.1.14) and sucrose phosphate phosphatase (Spp, EC 3.1.3.24), with sucrose-6-phosphate as an intermediate. These enzymes are encoded by the operon *sps–spp–fruK–ams*, which also encodes, in addition to Sps and Spp, enzymes for sucrose metabolism amylosucrase (Ams, EC 2.4.1.4) and fructokinase (FruK, EC 2.7.1.4). Amylosucrase transfers the glycosyl residue from sucrose onto glycogen primer, and it also catalyses cleavage of sucrose to fructose and glucose. Glucose and fructose need to be phosphorylated by the respective ATP-dependent kinases, to enter into central metabolic pathways (But *et al.*, 2013, 2015). The glucokinase from *Mm. alcaliphilum* 20Z (EC 2.7.1.1) also phosphorylates glucosamine (Mustakhimov *et al.*, 2017), whereas FruK (EC 2.7.1.4) is highly specific to fructose and ATP (But *et al.*, 2012). Two molecules of nucleotide triphosphate (NTP) are needed for attaching the glucosyl residue to the glycogen primer by Ams, and only one molecule of NTP is required for elongation of glycogen via glucose pyrophosphorylase (GlgC) and glycogen synthase (GlgA). Therefore, the sucrose cycle, where glycogen synthesis involves sucrose as an intermediate, could be envisaged as a dynamic mechanism that balances the intracellular levels of ATP.

Sucrose synthesis can provide some advantages to the methylotrophs. Being a carbon storage compound, sucrose also serves as a compatible solute, contributing to cell turgor maintenance at elevated salinity. However, sucrose accumulation ensures only moderate salt tolerance by methylotrophic bacteria (But *et al.*, 2013). Of note, the genomes of most gammaproteobacterial methanotrophs isolated from soils possess the sucrose biosynthesis genes, whereas methanotrophs obtained from freshwater or sewage systems lack these genes (But *et al.*, 2015). The halotolerant methylotrophs are known to tolerate extraordinarily high concentrations of methanol, being able to grow at 7% methanol (v/v; Eshinimaev *et al.*, 2002). Since sucrose synthesis starts with early products of the RuMP pathway, high rates of formaldehyde outflow may be responsible for the methanol tolerance.

In the thermophilic *Md. szegediense* O12, sucrose synthase (Sus) is involved in metabolism of sucrose in place of Ams, cleaving disaccharide into fructose and ADP-/or UDP-glucose (But *et al.*, 2018b). The clear preference of Sus for ADP implies a connection between sucrose and glycogen metabolism, since ADP-glucose is a substrate for glycogen synthesis. So far, no genes for sucrose metabolism have been identified in methanotrophs assimilating methane via the serine pathway (phylum Alphaproteobacteria) or through the Calvin cycle (Verrucomicrobia and NC10 phyla) (But *et al.*, 2015).

Adaptation of methanotrophs to saline environments

Methanotrophs isolated from soda and hypersaline lakes, marine waters and soils are often halotolerant or halophilic, and these species belong to the genera *Methylomicrobium*, *Methylobacter* and *Methylohalobius.* All halotolerant/halophilic methanotrophs accumulate in their cells ectoine, sucrose and glutamate, which all participate in balancing osmotic pressure between cytoplasm and the cell surroundings. The cyclic imino acid ectoine (1,4,5,6-tetrahydro-2-methyl-4-pyrimidine carboxylic acid) is the major compatible solute accumulating at concentration of up to ≈1 mol per 1 l of intracellular water content. The halophilic/tolerant methanotrophs are the promising producers of ectoine, which is a multifunctional bioprotectant useful in pharmaceutical applications (Graf *et al.*, 2008). The biochemical pathway for ectoine biosynthesis in methanotrophs is almost identical to that in other halophilic prokaryotes. Three specific enzymes, diaminobutyric acid (DABA) aminotransferase (EctB), DABA acetyltransferase (EctA, and ectoine synthase (EctC) compose this pathway encoded by the *ectABC* operon, which, in *Mm. alcaliphilum* 20Z, also includes a gene for negative transcriptional regulator EctR, of the MarR family, and a gene for an additional aspartokinase. DABA aminotransferase (EctB) from *Mm. alcaliphilum* 20Z is a hexameric pyridoxal-5′-phosphate-dependent enzyme (300 kDa) synthesizing diaminobutyric acid (DABA) from D,L-aspartyl semialdehyde using L-glutamate as a donor of an amino group. It requires K^+ for activity and stability, being more active in the presence of 0.01–0.5 M KCl (Reshetnikov *et al.*, 2011).

DABA acetyltransferase (EctA) catalyses the acetylation of DABA, yielding γ-N-acetyl-α,γ-diaminobutyric acid. The properties of EctA from *Mm. alcaliphilum* 20Z and non-methanotrophic methylotrophs *Methylophaga alcalica* ATCC 35842 and *Methylophaga thalassica* ATCC 33146 represent striking examples of bacterial eco-physiological adaptation at the enzyme level (Mustakhimov *et al.*, 2008). The EctA from the neutrophilic *M. thalassica* has optimal pH between 8.2 or 9.0, whereas the enzymes from the alkaliphilic *Mm. alcaliphilum* 20Z and *M. alcalica* have pH optima at ≥ 9.5. Activities of EctA from *M. alcalica* and *Mm. alcaliphilum* 20Z, which have been isolated from soda lakes, were increased in the presence of carbonate ions, whereas EctA from a marine isolate *M. thalassica* was considerably inhibited by carbonates. Copper ions completely inhibited the enzyme activity from *M. alcalica* but had no effect on the enzyme from *Mm. alcaliphilum* 20Z. Ectoine synthase (35 kDa) catalyses the cyclization of N-acetyl-DABA into ectoine. The methylotrophs synthesizing ectoine also encode a pathway for ectoine degradation, first proposed for a non-methylotrophic bacterium *Halomonas elongata* (Schwibbert *et al.*, 2011).

Methylohalobius crimeensis 10Ki displays the highest halotolerance recognized among the halophilic methanotrophs, and it is able to grow at 12% NaCl (Sharp *et al.*, 2015). It possesses an array of mechanisms for osmoadaptation. In addition to the *ectABCD* gene cluster, it encodes ectoine hydroxylase (EctD), an enzyme responsible for biosynthesis of hydroxyectoine. It also possesses genes for a high-affinity importer of choline/glycine betaine, driven by a sodium-motive force. It also harbours three copies of a gene for choline dehydrogenase and a gene encoding betaine aldehyde dehydrogenase, indicating possible glycine betaine synthesis from choline. In addition, it encodes a pathway for sucrose synthesis and degradation/reutilization, which includes Sps, Spp, FruK and sucrose synthase instead of amylosucrase characteristic of other halotolerant methanotrophs. The use of a sodium-motive force and Na^+ export have been suggested by the presence of the genes encoding a putative Na^+/H^+ antiporter localized within the gene cluster encoding an ATP synthase, and a complete *nqr* gene cluster encoding Na^+-pumping NADH:quinone oxidoreductase (Sharp *et al.*, 2015).

Future trends

Despite the confirmation of the fundamental biochemical pathways for methanotrophy, which have been originally described a while ago, many important details of the methanotrophic metabolism still remain unclear. Future studies are needed to decipher the energy generation mechanisms, the basis for the obligate dependency on C1 compounds in many species, as well as the potential significance of the unusual secondary metabolites (Puri *et al.*, 2018) for central metabolism or organism's physiology. A better understanding of methanotrophy, including in the halotolerant methanotrophs that are known for robust growth, through closing the extant gaps in our knowledge, will allow for employing these organisms as novel platforms for synthetic biology applications, forming a base for microbial factories that will utilize methane, the major component of natural gas, as a feedstock for the synthesis of high-value commodity chemicals.

Acknowledgement

This material is based upon work supported by the Russian Scientific Foundation (18-04-00326).

References

Ali, V., Shigeta, Y., and Nozaki, T. (2003). Molecular and structural characterization of NADPH-dependent D-glycerate dehydrogenase from the enteric parasitic protist *Entamoeba histolytica*. Biochem. J. *375*, 729–736.

Bapteste, E., Moreira, D., and Philippe, H. (2003). Rampant horizontal gene transfer and phospho-donor change in the evolution of the phosphofructokinase. Gene *318*, 185–191.

Baxter, N.J., Hirt, R.P., Bodrossy, L., Kovacs, K.L., Embley, T.M., Prosser, J.I., and Murrell, J.C. (2002). The ribulose-1,5-bisphosphate carboxylase/oxygenase gene cluster of *Methylococcus capsulatus* (Bath). Arch. Microbiol. *177*, 279–289. https://doi.org/10.1007/s00203-001-0387-x

Bowman, J. (2006). The methanotrophs: the families *Methylococcaceae* and *Methylocystaceae*. In The Prokaryotes, Dworkin, M., ed. (Springer, New York, NY), pp. 266–289.

But, S.Y., Rozova, O.N., Khmelenina, V.N., Reshetnikov, A.S., and Trotsenko, Y.A. (2012). Properties of recombinant ATP-dependent fructokinase from halotolerant methanotroph *Methylomicrobium alcaliphilum* 20Z. Biochem. Mosc. 77, 372–377. https://doi.org/10.1134/S0006297912040086

But, S.Y., Khmelenina, V.N., Reshetnikov, A.S., and Trotsenko, Y.A. (2013). Bifunctional sucrose phosphate synthase/phosphatase is involved in the sucrose biosynthesis by *Methylobacillus flagellatus* KT. FEMS Microbiol. Lett. *347*, 43–51. https://doi.org/10.1111/1574-6968.12219

But, S.Y., Khmelenina, V.N., Reshetnikov, A.S., Mustakhimov, I.I., Kalyuzhnaya, M.G., and Trotsenko, Y.A. (2015). Sucrose metabolism in halotolerant methanotroph *Methylomicrobium alcaliphilum* 20Z. Arch. Microbiol. *197*, 471–480. https://doi.org/10.1007/s00203-015-1080-9

But, S.Y., Egorova, S.V., Khmelenina, V.N., and Trotsenko, Y.A. (2017). Biochemical properties and phylogeny of hydroxypyruvate reductases from methanotrophic bacteria with different C1-assimilation pathways. Biochem. Mosc. *82*, 1295–1303. https://doi.org/10.1134/S0006297917110074

But, S.Y., Egorova, S.V., Khmelenina, V.N., and Trotsenko, Y.A. (2018a). Serine-glyoxylate aminotransferases from methanotrophs realizing different C1-assimilation pathways. Antonie Van Leeuwenhoek [Epub ahead of print]. https://doi.org/10.1007/s10482-018-1208-14

But, S.Y., Solntseva, N.P., Egorova, S.V., Mustakhimov, I.I., Khmelenina, V.N., Reshetnikov, A., and Trotsenko, Y.A. (2018b). The genes and enzymes of sucrose metabolism in moderately thermophilic methanotroph *Methylocaldum szegediense* O12. Extremophiles *22*, 433–445. https://doi.org/10.1007/s00792-018-1006-y

Chen, Y., Crombie, A., Rahman, M.T., Dedysh, S.N., Liesack, W., Stott, M.B., Alam, M., Theisen, A.R., Murrell, J.C., and Dunfield, P.F. (2010). Complete genome sequence of the aerobic facultative methanotroph *Methylocella silvestris* BL2. J. Bacteriol. *192*, 3840–3841. https://doi.org/10.1128/JB.00506-10

Chetina, E.V., and Trotsenko, Y.A. (1986). The activity of oxidative phosphorylation in the membranes of methylotrophic bacteria. Microbiology (Russian) *4*, 539–542.

Chiba, Y., Kamikawa, R., Nakada-Tsukui, K., Saito-Nakano, Y., and Nozaki, T. (2015). Discovery of PPi-type Phosphoenolpyruvate Carboxykinase Genes in Eukaryotes and Bacteria. J. Biol. Chem. *290*, 23960–23970. https://doi.org/10.1074/jbc.M115.672907

Chistoserdova, L. (2011). Modularity of methylotrophy, revisited. Environ. Microbiol. *13*, 2603–2622. https://doi.org/10.1111/j.1462-2920.2011.02464.x

Chistoserdova, L. (2016). Lanthanides: New life metals? World J. Microbiol. Biotechnol. *32*, 138. https://doi.org/10.1007/s11274-016-2088-2

Chistoserdova, L.V., and Lidstrom, M.E. (1991). Purification and characterization of hydroxypyruvate reductase from the facultative methylotroph *Methylobacterium extorquens* AM1. J. Bacteriol. *173*, 7228–7232.

Chistoserdova, L., Kalyuzhnaya, M.G., and Lidstrom, M.E. (2009). The expanding world of methylotrophic metabolism. Annu. Rev. Microbiol. *63*, 477–499. https://doi.org/10.1146/annurev.micro.091208.073600

Crowther, G.J., Kosály, G., and Lidstrom, M.E. (2008). Formate as the main branch point for methylotrophic metabolism in *Methylobacterium extorquens* AM1. J. Bacteriol. *190*, 5057–5062. https://doi.org/10.1128/JB.00228-08

de la Torre, A., Metivier, A., Chu, F., Laurens, L.M., Beck, D.A., Pienkos, P.T., Lidstrom, M.E., and Kalyuzhnaya, M.G. (2015). Genome-scale metabolic reconstructions and theoretical investigation of methane conversion in *Methylomicrobium buryatense* strain 5G(B1). Microb. Cell Fact. *14*, 188. https://doi.org/10.1186/s12934-015-0377-3

Dunfield, P.F., Yuryev, A., Senin, P., Smirnova, A.V., Stott, M.B., Hou, S., Ly, B., Saw, J.H., Zhou, Z., Ren, Y., *et al.* (2007). Methane oxidation by an extremely acidophilic bacterium of the phylum Verrucomicrobia. Nature *450*, 879–882.

Eshinimaev, B.T.s., Khmelenina, V.N., Sakharovskiĭ, V.G., Suzina, N.E., and Trotsenko, I.u.A. (2002). [Physiological, biochemical, and cytological characteristics of a halotolerant and alkalitolerant methanotroph grown on methanol.] Mikrobiologiia *71*, 596–603.

Eshinimaev, B.T.s., Medvedkova, K.A., Khmelenina, V.N., Suzina, N.E., Osipov, G.A., Lysenko, A.M., and Trotsenko, I.u.A. (2004). [New thermophilic methanotrophs of the genus *Methylocaldum*.] Mikrobiologiia *73*, 530–539.

Ettwig, K.F., Butler, M.K., Le Paslier, D., Pelletier, E., Mangenot, S., Kuypers, M.M., Schreiber, F., Dutilh, B.E., Zedelius, J., de Beer, D., *et al.* (2010). Nitrite-driven anaerobic methane oxidation by oxygenic bacteria. Nature *464*, 543–548. https://doi.org/10.1038/nature08883

Ferenci, T., Strom, T., and Quayle, J.R. (1974). Purification and properties of 3-hexulose phosphate synthase and phospho-3-hexuloisomerase from *Methylococcus capsulatus*. Biochem. J. *144*, 477–486.

Fu, Y., Li, Y., and Lidstrom, M. (2017). The oxidative TCA cycle operates during methanotrophic growth of the Type I methanotroph *Methylomicrobium buryatense* 5GB1. Metab. Eng. *42*, 43–51.

Garg, S., Wu, H., Clomburg, J.M., and Bennett, G.N. (2018). Bioconversion of Methane to C-4 Carboxylic Acids Using Carbon Flux Through Acetyl-CoA in Engineered *Methylomicrobium buryatense* 5GB1C, Metab. Eng. *48*, 175–183.

Gilman, A., Laurens, L.M., Puri, A.W., Chu, F., Pienkos, P.T., and Lidstrom, M.E. (2015). Bioreactor performance parameters for an industrially-promising methanotroph *Methylomicrobium buryatense* 5GB1. Microb. Cell Fact. *14*, 182. https://doi.org/10.1186/s12934-015-0372-8

Gilman, A., Fu, Y., Hendershott, M., Chu, F., Puri, A.W., Smith, A.L., Pesesky, M., Lieberman, R., Beck, D.A.C., and Lidstrom, M.E. (2017). Oxygen-limited metabolism in the methanotroph *Methylomicrobium buryatense* 5GB1C. PeerJ *5*, e3945. https://doi.org/10.7717/peerj.3945

Graf, R., Anzali, S., Buenger, J., Pfluecker, F., and Driller, H. (2008). The multifunctional role of ectoine as a natural cell protectant. Clin. Dermatol. *26*, 326–333. https://doi.org/10.1016/j.clindermatol.2008.01.002

Hanson, R.S., and Hanson, T.E. (1996). Methanotrophic Bacteria. Microbiol. Rev. *60*, 439–471.

Henard, C.A., and Guarnieri, M.T. (2018). Metabolic engineering of methanotrophic bacteria for industrial biomanufacturing. In Methane Biocatalysis: Paving the Way to Sustainability, Kalyuzhnaya, M.G., Xing, X.H., eds. (Springer, International Publishing AG, Switzerland), pp. 117–132. https://doi.org/10.1007/978-3-319-74866-5.

Henard, C.A., Smith, H., Dowe, N., Kalyuzhnaya, M.G., Pienkos, P.T., and Guarnieri, M.T. (2016). Bioconversion of methane to lactate by an obligate methanotrophic bacterium. Sci. Rep. *6*, 21585. https://doi.org/10.1038/srep21585

Henard, C.A., Smith, H.K., and Guarnieri, M.T. (2017). Phosphoketolase overexpression increases biomass and lipid yield from methane in an obligate methanotrophic biocatalyst. Metab. Eng. *41*, 152–158.

Ho, C.L., Noji, M., Saito, M., and Saito, K. (1999). Regulation of serine biosynthesis in *Arabidopsis*. Crucial role of plastidic 3-phosphoglycerate dehydrogenase in non-photosynthetic tissues. J. Biol. Chem. *274*, 397–402.

Hou, S., Makarova, K.S., Saw, J.H., Senin, P., Ly, B.V., Zhou, Z., Ren, Y., Wang, J., Galperin, M.Y., Omelchenko, M.V., *et al.* (2008). Complete genome sequence of the extremely acidophilic methanotroph isolate V4, *Methylacidiphilum infernorum*, a representative of the bacterial phylum *Verrucomicrobia*. Biol. Direct *3*, 26. https://doi.org/10.1186/1745-6150-3-26

Islam, T., Jensen, S., Reigstad, L.J., Larsen, O., and Birkeland, N.K. (2008). Methane oxidation at 55 degrees C and pH 2 by a thermoacidophilic bacterium belonging to the *Verrucomicrobia* phylum. Proc. Natl. Acad. Sci. U.S.A. *105*, 300–304. https://doi.org/10.1073/pnas.0704162105

Izumi, Y., Yoshida, T., Kanzaki, H., Toki, S., Miyazaki, S.S., and Yamada, H. (1990). Purification and characterization of hydroxypyruvate reductase from a serine-producing methylotroph, *Hyphomicrobium methylovorum* GM2. Eur. J. Biochem. *190*, 279–284.

Kalyuzhnaya, M.G. (2016). Methane biocatalysis: selecting the right microbe. In Biotechnology for Biofuel Production and Optimization, Eckert, C. and Trinh, C.T. eds. (Elsevier, Amsterdam), pp. 353–383.

Kalyuzhnaya, M.G., Yang, S., Rozova, O.N., Smalley, N.E., Clubb, J., Lamb, A., Gowda, G.A., Raftery, D., Fu, Y., Bringel, F., *et al.* (2013). Highly efficient methane biocatalysis revealed in a methanotrophic bacterium. Nat. Commun. *4*, 2785. https://doi.org/10.1038/ncomms3785

Kalyuzhnaya, M.G., Puri, A.W., and Lidstrom, M.E. (2015). Metabolic engineering in methanotrophic bacteria. Metab. Eng. *29*, 142–152.

Kamachi, T., and Okura, I. Methanol biosynthesis using methanotrophs. In Methane Biocatalysis: Paving the Way to Sustainability. Kalyuzhnaya, M.G., Xing, X.H., eds. (Springer, International Publishing AG, Switzerland), pp. 8169–8182. https://doi.org/10.1007/978-3-319-74866-5.

Kameya, M., Arai, H., Ishii, M., and Igarashi, Y. (2010). Purification of three aminotransferases from *Hydrogenobacter* thermophilus TK-6 – novel types of alanine or glycine aminotransferase: enzymes and catalysis. FEBS J. *277*, 1876–1885. https://doi.org/10.1111/j.1742-4658.2010.07604.x

Kelly, D.P., McDonald, I.P., and Wood, A.P. The family *Methylobacteriaceae*. In: Prokaryotes: Alphaproteobacteria and Betaproteobacteria, Rosenberg, E., DeLong, E.F. Lory, S., Stackerbrandt, E., Thomson, F., eds (Springer, Berlin, Heidelberg) pp. 313–340.

Kendziorek, M., and Paszkowski, A. (2008). Properties of serine:glyoxylate aminotransferase purified from *Arabidopsis thaliana* leaves. Acta Biochim. Biophys. Sin. *40*, 102–110.

Khadem, A.F., Pol, A., Jetten, M.S., Op den Camp, H.J. (2010). Nitrogen fixation by the verrucomicrobial methanotroph '*Methylacidiphilum fumariolicum*' SolV. Microbiology *156*, 1052-1059. https://doi.org/10.1099/mic.0.036061-0.

Khadem, A.F., Pol, A., Wieczorek, A., Mohammadi, S.S., Francoijs, K.J., Stunnenberg, H.G., Jetten, M.S., and Op den Camp, H.J. (2011). Autotrophic methanotrophy in Verrucomicrobia: *Methylacidiphilum fumariolicum* SolV uses the Calvin-Benson-Bassham cycle for carbon dioxide fixation. J. Bacteriol. *193*, 4438–4446. https://doi.org/10.1128/JB.00407-11

Khadem, A.F., Wieczorek, A.S., Pol, A., Vuilleumier, S., Harhangi, H.R., Dunfield, P.F., Kalyuzhnaya, M.G., Murrell, J.C., Francoijs, K.J., Stunnenberg, H.G., *et al.* (2012). Draft genome sequence of the volcano-inhabiting thermoacidophilic methanotroph *Methylacidiphilum fumariolicum* strain SolV. J. Bacteriol. *194*, 3729–3730. https://doi.org/10.1128/JB.00501-12

Khmelenina, V.N., Rozova, O.N., and Trotsenko, Y.A. (2011). Characterization of the recombinant pyrophosphate-dependent 6-phosphofructokinases from *Methylomicrobium alcaliphilum* 20Z and *Methylococcus capsulatus* Bath. Meth. Enzymol. *495*, 1–14. https://doi.org/10.1016/B978-0-12-386905-0.00001-2

Khmelenina, V.N., Rozova, O.N., Akberdin, I.R., Kalyuzhnaya, M.G., and Trotsenko, Y.A. (2018). Pyrophosphate-Dependent enzymes in methanotrophs: new findings and views. In Methane Biocatalysis: Paving the Way to Sustainability. Kalyuzhnaya, M.G., Xing, X.H., eds. (Springer, International Publishing AG, Switzerland), pp. 83–98. https://doi.org/10.1007/978-3-319-74866-5.

Matsen, J.B., Yang, S., Stein, L.Y., Beck, D., and Kalyuzhnaya, M.G. (2013). Global molecular analyses of methane metabolism in methanotrophic alphaproteobacterium, *Methylosinus trichosporium* OB3b. Part I: Transcriptomic study. Front. Microbiol. *4*, 40. https://doi.org/10.3389/fmicb.2013.00040

McDonald, I.R., Bodrossy, L., Chen, Y., and Murrell, J.C. (2008). Molecular ecology techniques for the study of aerobic methanotrophs. Appl. Environ. Microbiol. *74*, 1305–1315. https://doi.org/10.1128/AEM.02233-07

Medvedkova, K.A., Khmelenina, V.N., and Trotsenko, I.u.A. (2007). [Sucrose as a factor of thermal adaptation of the thermophilic methanotroph *Methylocaldum szegediense* O-12.] Mikrobiologiia *76*, 567–569.

Mertens, E. (1991). Pyrophosphate-dependent phosphofructokinase, an anaerobic glycolytic enzyme? FEBS Lett. *285*, 1–5.

Mitsui, R., Kusano, Y., Yurimoto, H., Sakai, Y., Kato, N., and Tanaka, M. (2003). Formaldehyde fixation contributes to detoxification for growth of a nonmethylotroph, *Burkholderia cepacia* TM1, on vanillic acid. Appl. Environ. Microbiol. *69*, 6128–6132.

Müller, M., Lee, J.A., Gordon, P., Gaasterland, T., Sensen, C.W., 2001. Presence of prokaryotic and eukaryotic species in all subgroups of the PPi-dependent Group II phosphofructokinase protein family. J. Bacteriol. *183*, 6714-6716.

Murrell, J.C., McDonald, I.R., and Gilbert, B. (2000). Regulation of expression of methane monooxygenases by copper ions. Trends Microbiol. *8*, 221–225.

Mustakhimov, I.I., Rozova, O.N., Reshetnikov, A.S., Khmelenina, V.N., Murrell, J.C., and Trotsenko, Y.A. (2008). Characterization of the recombinant diaminobutyric acid acetyltransferase from *Methylophaga thalassica* and *Methylophaga alcalica*. FEMS Microbiol. Lett. *283*, 91–96. https://doi.org/10.1111/j.1574-6968.2008.01156.x

Mustakhimov, I.I., But, S.Y., Reshetnikov, A.S., Khmelenina, V.N., and Trotsenko, Y.A. (2016). Homo- and heterologous reporter proteins for evaluation of promoter activity in *Methylomicrobium alcaliphilum* 20Z. Appl. Biochem. Microbiol (Moscow) *52*, 279–286.

Mustakhimov, I.I., Rozova, O.N., Solntseva, N.P., Khmelenina, V.N., Reshetnikov, A.S., and Trotsenko, Y.A. (2017). The properties and potential metabolic role of glucokinase in halotolerant obligate methanotroph *Methylomicrobium alcaliphilum* 20Z. Antonie Van Leeuwenhoek *110*, 375–386. https://doi.org/10.1007/s10482-016-0809-z

Ochsner, A.M., Christen, M., Hemmerle, L., Peyraud, R., Christen, B., and Vorholt, J.A. (2017). Transposon sequencing uncovers an essential regulatory function of phosphoribulokinase for methylotrophy. Curr. Biol. *27*, 2579–2588.e6.

Op den Camp, H.J., Islam, T., Stott, M.B., Harhangi, H.R., Hynes, A., Schouten, S., Jetten, M.S., Birkeland, N.K., Pol, A., and Dunfield, P.F. (2009). Environmental, genomic and taxonomic perspectives on methanotrophic Verrucomicrobia. Environ. Microbiol. Rep. *1*, 293–306. https://doi.org/10.1111/j.1758-2229.2009.00022.x

Orita, I., Yurimoto, H., Hirai, R., Kawarabayasi, Y., Sakai, Y., and Kato, N. (2005). The archaeon *Pyrococcus horikoshii* possesses a bifunctional enzyme for formaldehyde fixation via the ribulose monophosphate pathway. J. Bacteriol. *187*, 3636–3642.

Orita, I., Sato, T., Yurimoto, H., Kato, N., Atomi, H., Imanaka, T., and Sakai, Y. (2006). The ribulose monophosphate pathway substitutes for the missing pentose phosphate pathway in the archaeon *Thermococcus kodakaraensis*. J. Bacteriol. *188*, 4698–4704.

Puri, A.W., Mevers, E., Ramadhar, T.R., Petras, D., Liu, D., Piel, J., Dorrestein, P.C., Greenberg, E.P., Lidstrom, M.E., and Clardy, J. (2018). Tundrenone: An Atypical

Secondary Metabolite from Bacteria with Highly Restricted Primary Metabolism. J. Am. Chem. Soc. *140*, 2002–2006. https://doi.org/10.1021/jacs.7b12240

Rasigraf, O., Kool, D.M., Jetten, M.S., Sinninghe Damsté, J.S., and Ettwig, K.F. (2014). Autotrophic carbon dioxide fixation via the Calvin-Benson-Bassham cycle by the denitrifying methanotroph '*Candidatus Methylomirabilis oxyfera*'. Appl. Environ. Microbiol. *80*, 2451–2460.

Reshetnikov, A.S., Rozova, O.N., Khmelenina, V.N., Mustakhimov, I.I., Beschastny, A.P., Murrell, J.C., and Trotsenko, Y.A. (2008). Characterization of the pyrophosphate-dependent 6-phosphofructokinase from *Methylococcus capsulatus* Bath. FEMS Microbiol. Lett. *288*, 202–210. https://doi.org/10.1111/j.1574-6968.2008.01366.x

Reshetnikov, A.S., Khmelenina, V.N., Mustakhimov, I.I., Kalyuzhnaya, M., Lidstrom, M., and Trotsenko, Y.A. (2011). Diversity and phylogeny of the ectoine biosynthesis genes in aerobic, moderately halophilic methylotrophic bacteria. Extremophiles *15*, 653–663. https://doi.org/10.1007/s00792-011-0396-x

Rozova, O.N., Khmelenina, V.N., Vuilleumier, S., and Trotsenko, Y.A. (2010). Characterisation of the recombinant pyrophosphate-dependent 6-phosphofructokinase from halotolerant methanotroph *Methylomicrobium alcaliphilum* 20Z. Res. Microbiol. *161*, 861–868.

Rozova, O.N., Khmelenina, V.N., and Trotsenko, Y.A. (2012). Characterization of recombinant PPi-dependent 6-phosphofructokinases from *Methylosinus trichosporium* OB3b and *Methylobacterium nodulans* ORS 2060. Biochemistry Mosc. 77, 288–295. https://doi.org/10.1134/S0006297912030078

Rozova, O.N., Khmelenina, V.N., Gavletdinova, J.Z., Mustakhimov, I.I., and Trotsenko, Y.A. (2015a). Acetate kinase-an enzyme of the postulated phosphoketolase pathway in *Methylomicrobium alcaliphilum* 20Z. Antonie Van Leeuwenhoek *108*, 965–974. https://doi.org/10.1007/s10482-015-0549-5

Rozova, O.N., Khmelenina, V.N., Bocharova, K.A., Mustakhimov, I.I., and Trotsenko, Y.A. (2015b). Role of NAD+-dependent malate dehydrogenase in the metabolism of *Methylobacterium alcaliphilum* 20Z and *Methylosinus trichosporium* OB3b. Microorganisms 3, 47–59.

Rozova, O.N., But, S.Y., Khmelenina, V.N., Reshetnikov, A.S., Mustakhimov, I.I., and Trotsenko, Y.A. (2017). Characterization of Two recombinant 3-Hexulose-6-Phosphate synthases from the halotolerant obligate methanotroph *Methylomicrobium alcaliphilum* 20Z. Biochemistry Mosc. *82*, 176–185. https://doi.org/10.1134/S0006297917020092

Rozova, O.N., Khmelenina, V.N., Mustakhimov, I.I., But, S.Y., and Trotsenko, Y.A. (2019). Properties of the malic enzyme from aerobic methanotroph Methylosinus trichosporium. Biochemistry Mosc. In press http://www.protein.bio.msu.ru/biokhimiya/inpress/index.html

Sánchez, B., Zúñiga, M., González-Candelas, F., de los Reyes-Gavilán, C.G., and Margolles, A. (2010). Bacterial and eukaryotic phosphoketolases: phylogeny, distribution and evolution. J. Mol. Microbiol. Biotechnol. *18*, 37–51. https://doi.org/10.1159/000274310

Schwibbert, K., Marin-Sanguino, A., Bagyan, I., Heidrich, G., Lentzen, G., Seitz, H., Rampp, M., Schuster, S.C., Klenk, H.P., Pfeiffer, F., *et al.* (2011). A blueprint of ectoine metabolism from the genome of the industrial producer *Halomonas elongata* DSM 2581 T. Environ. Microbiol. *13*, 1973–1994. https://doi.org/10.1111/j.1462-2920.2010.02336.x

Sharp, C.E., Smirnova, A.V., Kalyuzhnaya, M.G., Bringel, F., Hirayama, H., Jetten, M.S., Khmelenina, V.N., Klotz, M.G., Knief, C., Kyrpides, N., *et al.* (2015). Draft genome sequence of the moderately halophilic methanotroph *Methylohalobius crimeensis* strain 10Ki. Genome Announc. *3*, e00644–15. https://doi.org/10.1128/genomeA.00644-15

Slamovits, C.H., and Keeling, P.J. (2006). Pyruvate-phosphate dikinase of oxymonads and parabasalia and the evolution of pyrophosphate-dependent glycolysis in anaerobic eukaryotes. Eukaryotic Cell 5, 148–154.

Stein, L.Y., Yoon, S., Semrau, J.D., Dispirito, A.A., Crombie, A., Murrell, J.C., Vuilleumier, S., Kalyuzhnaya, M.G., Op den Camp, H.J., Bringel, F., *et al.* (2010). Genome sequence of the obligate methanotroph *Methylosinus trichosporium* strain OB3b. J. Bacteriol. *192*, 6497–6498. https://doi.org/10.1128/JB.01144-10

Strong, P.J., Kalyuzhnaya, M., Silverman, J., and Clarke, W.P. (2016). A methanotroph-based biorefinery: potential scenarios for generating multiple products from a single fermentation. Bioresour. Technol. *215*, 314–323.

Tonge, G.M., Drozd, I.W. Higgins, I.I. (1977). Energy coupling in *Methylosinus trichosporium*. J. Gen. Microbiol. *99*, 229–232.

Trotsenko, Y.A., and Murrell, J.C. (2008). Metabolic aspects of aerobic obligate methanotrophy. Adv. Appl. Microbiol. *63*, 183–229. https://doi.org/10.1016/S0065-2164(07)00005-6

Yasueda, H., Kawahara, Y., and Sugimoto, S. (1999). *Bacillus subtilis yckG* and *yckF* encode two key enzymes of the ribulose monophosphate pathway used by methylotrophs, and yckH is required for their expression. J. Bacteriol. *181*, 7154–7160.

Yurimoto, H., Hirai, R., Yasueda, H., Mitsui, R., Sakai, Y., and Kato, N. (2002). The ribulose monophosphate pathway operon encoding formaldehyde fixation in a thermotolerant methylotroph, *Bacillus brevis* S1. FEMS Microbiol. Lett. *214*, 189–193.

van Teeseling, M.C., Pol, A., Harhangi, H.R., van der Zwart, S., Jetten, M.S., Op den Camp, H.J., and van Niftrik, L. (2014). Expanding the verrucomicrobial methanotrophic world: description of three novel species of *Methylacidimicrobium* gen. nov. Appl. Environ. Microbiol. *80*, 6782–6791. https://doi.org/10.1128/AEM.01838-14

Vuilleumier, S., Khmelenina, V.N., Bringel, F., Reshetnikov, A.S., Lajus, A., Mangenot, S., Rouy, Z., Op den Camp, H.J., Jetten, M.S., Dispirito, A.A., *et al.* (2012). Genome sequence of the haloalkaliphilic methanotrophic bacterium *Methylomicrobium alcaliphilum* 20Z. J. Bacteriol. *194*, 551–552. https://doi.org/10.1128/JB.06392-11

Ward, N., Larsen, Ø., Sakwa, J., Bruseth, L., Khouri, H., Durkin, A.S., Dimitrov, G., Jiang, L., Scanlan, D., Kang, K.H., *et al.* (2004). Genomic insights into methanotrophy: the complete genome sequence of

Methylococcus capsulatus (Bath). PLOS Biol. 2, e303. https://doi.org/10.1371/journal.pbio.0020303

Webb, H.K., Ng, H.J., and Ivanova, E.P. (2014). The Family Methylocystaceae. In Prokaryotes: Alphaproteobacteria and Betaproteobacteria, Rosenberg, E., DeLong, E.F., Lory, S., Stackerbrandt, E., Thomson, F., eds (Springer, Berlin Heidelberg), pp. 341–347.

Wu, M.L., van Alen, T.A., van Donselaar, E.G., Strous, M., Jetten, M.S., and van Niftrik, L. (2012a). Co-localization of particulate methane monooxygenase and cd1 nitrite reductase in the denitrifying methanotroph 'Candidatus Methylomirabilis oxyfera'. FEMS Microbiol. Lett. *334*, 49–56. https://doi.org/10.1111/j.1574-6968.2012.02615.x

Wu, M.L., van Teeseling, M.C., Willems, M.J., van Donselaar, E.G., Klingl, A., Rachel, R., Geerts, W.J., Jetten, M.S., Strous, M., and van Niftrik, L. (2012b). Ultrastructure of the denitrifying methanotroph 'Candidatus *Methylomirabilis oxyfera*,' a novel polygon-shaped bacterium. J. Bacteriol. *194*, 284–291. https://doi.org/10.1128/JB.05816-11

Lanthanides in Methylotrophy

5

Elizabeth Skovran[1]*, Charumathi Raghuraman[1] and
Norma Cecilia Martinez-Gomez[2]

[1]Department of Biological Sciences, San Jose State University, San Jose, CA, USA.
[2]Department of Microbiology and Molecular Genetics, Michigan State University, East Lansing, MI, USA.

*Correspondence: elizabeth.skovran@sjsu.edu

https://doi.org/10.21775/9781912530045.05

Abstract

Lanthanides were previously thought to be biologically inert owing to their low solubility; however, they have recently been shown to strongly impact the metabolism of methylotrophic bacteria. Leading efforts in this emergent field have demonstrated far-reaching impacts of lanthanide metabolism in biology; from the identification of novel roles of enzymes and pathways dependent on lanthanide-chemistry to the control of transcriptional regulatory networks to the modification of microbial community interactions. Even further, the recent discovery of lanthanide-dependent enzymes associated with multi-carbon metabolism in both methylotrophs and non-methylotrophs alike suggests that lanthanide biochemistry may be more widespread than initially thought. Current efforts aim to understand how lanthanide chemistry and lanthanide-dependent enzymes affect numerous ecosystems and metabolic functions. These efforts are likely to have a profound impact on biotechnological processes involving methylotrophic communities and the biologically mediated recovery of these critical metals from a variety of waste streams while redefining our understanding of a fundamental set of metals in biology.

Introduction: lanthanides in methylotrophy

In the last few decades, the study of methylotrophy has become an exciting and expanding area of research with applications in carbon capture, chemical production, bioremediation, and biomining (Chistoserdova, 2018). Laboratories across the world are engineering methylotrophic bacteria to produce value-added chemicals from single-carbon compounds, turning atmospheric pollutants like methane and methanol into green chemicals like biofuels, terpenoids, and biodegradable plastics (Höfer *et al.*, 2011; Hwang *et al.*, 2014; Ochsner *et al.*, 2015; Pfeifenschneider *et al.*, 2017). Yet others in the field are investigating the abilities of methylotrophs to promote agricultural growth, degrade toxicants like methylmercury, or recover critical metals for our technologies in environmentally friendly and sustainable ways (Iguchi, 2015; Kumar *et al.*, 2016; Martinez-Gomez, 2016; Lu *et al.*, 2017). It is clear from decades of study that not all methylotrophs operate in the same way in terms of their core central metabolic pathways or their transcriptional regulation. Even among their shared alcohol oxidation enzymes, differences in substrate specificities, cofactor ranges, and catalytic efficiencies are observed. With this complexity comes breadth, flexibility, and challenges in methylotroph bioengineering efforts. By better understanding methylotrophic metabolism and its regulation in a variety of organisms, the applications for methylotrophy can be advanced at a rapid pace. Shared goals may be realized. For example, it may soon become possible to capture methane, produce value-added chemicals, and biomine for critical metals, all in a single process.

A key step in methylotrophy is the oxidation of methanol to either formaldehyde or formate, which is carried out by different enzymes including methanol dehydrogenase (MDH) and alcohol oxidase depending on the specific methylotroph. In 2011, the first evidence was published that suggested that some types of MDHs use lanthanides (Ln) instead of calcium (Ca) as cofactors (Hibi *et al.*, 2011; Fitriyanto *et al.*, 2011) paving the way for novel discoveries surrounding the biochemical use of these elements and allowing cultivation of organisms previously thought uncultivable (Pol *et al.*, 2014; Skovran and Martinez-Gomez, 2015; Vekeman *et al.*, 2016; Del Rocío Bustillos-Cristales *et al.*, 2017; Howat *et al.*, 2018).

The Ln family consists of 15 elements ranging from lanthanum (La) to lutetium (Lu). Along with scandium (Sc) and yttrium (Y), these metals are commonly referred to as rare earth elements or REE. REE have been called 'the seeds of technology' since they are essential components of many technological devices including smartphones, computers, solar cells, batteries, wind turbines, lasers, and optical glasses due to their super magnetic properties and unique chemistries (Zepf *et al.*, 2014; Martinez-Gomez *et al.*, 2016). The term REE is misleading as these metals can be as abundant as other well-known industrial metals such as copper or zinc (Tyler, 2005). However, they are highly insoluble and scarce in pure form. Because of these properties, their inherent use in biological processes was thought implausible. It has now been demonstrated that some methylotrophs are dependent upon REE for growth as it is a cofactor for XoxF-type MDHs and ExaF-type ethanol dehydrogenases (EtDHs) (Pol *et al.*, 2014; Good *et al.*, 2016; Del Rocío Bustillos-Cristales *et al.*, 2017; Howat *et al.*, 2018). And for characterized methylotrophs that contain both Ca- and La-dependent MDHs, the Ln-dependent XoxF-type MDH is often preferred over its Ca counterpart as transcription of the Ca-dependent enzyme is repressed by Ln in multiple studied systems (Farhan Ul Haque *et al.*, 2015; Chu and Lidstrom, 2016; Gu *et al.*, 2016; Vu *et al.*, 2016; Masuda *et al.*, 2018).

In many ecosystems, the concentrations of REE are low; in the pico- to nanomolar range (Byrne and Sholkovitz, 1996; Pol *et al.*, 2014), yet are likely high enough to support methylotrophic growth. For perspective, batch culture growth of Ln-dependent *Methylobacterium extorquens* AM1 strains can reach maximum density (OD_{600} = 1.6 in culture tubes) with 100 nM La, though maximum growth rate and density are achieved with 1 μM La (Vu *et al.*, 2016). Some ecosystems like the acidic volcanic mud pots from the Solfatara contain concentrations of soluble REE in the low micromolar range (Pol *et al.*, 2014). Additionally, environments can be artificially induced to contain increased concentrations of REE. For example, REE are found in different types of fertilizers, and plants have been shown to effectively concentrate these elements (Gao *et al.*, 2003; Hu *et al.*, 2004, Nakagawa *et al.*, 2012). The addition of REE to the soil through fertilization may serve to promote the growth of associated Ln-dependent methylotrophs in the phyllosphere.

One question that remains is how do methylotrophic bacteria obtain these insoluble Ln if they are not located in acidic environments such as volcanic mudpots where soluble Ln are abundant? The capacity for solubilization and transport must exist as methylotrophs such as *M. extorquens* AM1 have even been shown to obtain Ln from computer hard drive magnets for growth (Martinez-Gomez *et al.*, 2016).

With the challenges that methylotrophs face in acquiring Ln, why would they have evolved enzymes that require these elements? Before the discovery of Ln-containing enzymes, it was hypothesized that Ln would have catalytically superior properties compared with Ca when acting as a Lewis acid, as Ln are more potent Lewis acids than Ca (Lim and Franklin, 2004). It has been corroborated that this is the case for some pyrroloquinoline quinone (PQQ)-dependent MDHs (Pol *et al.*, 2014; Lumpe *et al.*, 2018).

In Gram-negative bacteria, methanol utilization begins with oxidation of methanol in the periplasmic space by PQQ-dependent MDHs that are either Ca-dependent (MxaFI), or Ln-dependent (XoxF) enzymes. XoxF-type MDHs are more abundant in characterized environmental communities and often exist in multiple copies within methylotroph genomes (Vorobev *et al.*, 2013; Keltjens *et al.*, 2014; Beck *et al.*, 2015). Intriguingly, XoxF-type MDHs have also been found in non-methylotrophs though their functions are mostly unknown (Vorobev *et al.*, 2013; Chistoserdova, 2016). XoxF has been phylogenetically classified into five major clades

(XoxF1 to XoxF5), but additional metagenomic sequence analyses indicate that *xoxF*-like genes are more widespread than previously thought suggesting the potential for additional XoxF clades (Chistoserdova, 2011; Keltjens *et al.*, 2014; Taubert *et al.*, 2015; Chistoserdova, 2016; Chistoserdova and Kalyuzhnaya, 2018). Detailed biochemical studies that facilitate comparisons between the different types of XoxF MDHs are limited (Pol *et al.*, 2014, Wu *et al.*, 2015; Good *et al.*, 2018; Huang *et al.*, 2018; Jahn *et al.*, 2018) so it is yet unknown if XoxF MDHs from different clades will exhibit distinct and unique characteristics and roles.

Ln such as La and cerium (Ce) are predicted to have similar bond lengths to PQQ and catalytic amino acids when incorporated into MDHs (Pol *et al.*, 2014). However, the biochemical versatility and promiscuity of substrates and cofactors among the XoxF MDHs is slowly emerging (Bogart *et al.*, 2015; Pol *et al.*, 2014; Good *et al.*, 2018; Huang *et al.*, 2018; Jahn *et al.*, 2018; Lumpe *et al.*, 2018). The possibility of different binding and positioning properties for different Ln may allow production of different products. For example, some XoxF-type MDHs have been suggested to produce formate directly from methanol rather than formaldehyde (Pol *et al.*, 2014). Structural, biochemical, and mechanistic differences of the currently characterized PQQ- and Ln-dependent enzymes reveal metabolic flexibility that could explain the diversity of XoxF-type and ExaF-type alcohol dehydrogenases present in methylotrophic and non-methylotrophic bacteria.

Ln-dependent alcohol dehydrogenases are widespread and have ample metabolic flexibility

Overview of Ln-dependent methanol dehydrogenases

So far, the only known Ln-dependent enzymes are PQQ-dependent alcohol dehydrogenases (Chistoserdova, 2016). Though very divergent when compared with each other, they also share common features such as amino acid sequences and spatial orientation in the active site. Specifically, the amino acids that coordinate PQQ are conserved (Pol *et al.*, 2014; Bogart *et al.*, 2015; Jahn *et al.*, 2018; Lumpe *et al.*, 2018; McSkimming *et al.*, 2018). One feature that has been suggested as structurally important for coordinating Ln instead of Ca is an additional Asp (Asp^{301}, numbering according to XoxF from *Methylacidiphilum fumariolicum*) two positions downstream from a catalytic Asp (Asp^{299}) (Keltjens *et al.*, 2014; Bogart *et al.*, 2015) (Fig. 5.1).

Overall, the protein backbone of Ln-MDHs resembles an eight propeller-like peptide with PQQ buried inside its hydrophobic pocket (Pol *et al.*, 2014; Jahn *et al.*, 2018). In addition to the catalytic Asp, two Cys residues that form a disulfide bond reside next to PQQ and are likely essential for electron transfer to cytochrome c_L (XoxG; Ghosh *et al.*, 1994; Yu *et al.*, 2017; Zheng *et al.*, 2018). From the few crystal structures and DFT calculation studies that have been done, there is a consensus that the Ln atom coordinates with PQQ and functions as a Lewis acid to facilitate electron extraction from methanol by PQQ (Bogart *et al.*, 2015; Deng *et*

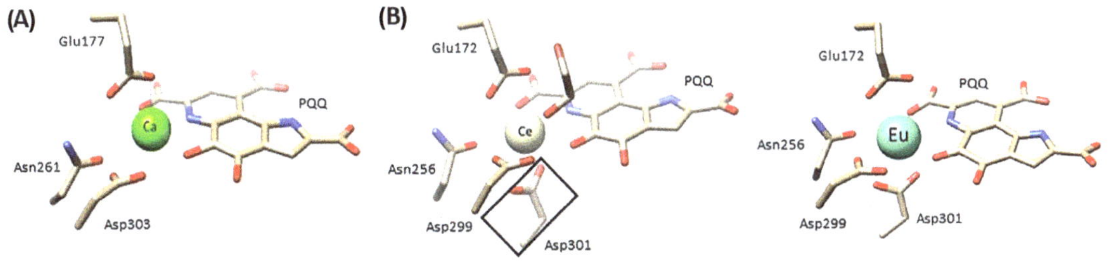

Figure 5.1 Metal coordination for PQQ-dependent alcohol dehydrogenases in the active site. (A) Ca-MxaF from *M. extorquens* AM1 [PDB ID: 1W6S (Williams, *et al.*, 2005)]. (B) Ce-XoxF and Eu-XoxF from the *M. fumariolicum* SolV strain [PDB ID: 4MAE (Pol *et al.*, 2014)]. The additional aspartate that has been suggested to be necessary for Ln coordination is depicted inside a black box. The catalytic aspartate coordinating the Ln is also highlighted (Asp299). In addition, Ce-XoxF shows the catalytic cysteines necessary for electron transfer to the cytochrome.

al., 2018; Lumpe *et al.*, 2018; McSkimming *et al.*, 2018). The mechanism of methanol oxidation also includes a proton abstraction from the alcohol by the catalytic Asp residue where the Ln assists the reaction as a Lewis acid. Some debate exists regarding whether a hydride transfer or an addition/elimination mechanism takes place (Leopoldini *et al.*, 2007; Lumpe *et al.*, 2018; McSkimming *et al.*, 2018). However, two recent studies strongly suggest the hydride transfer mechanism (Lumpe *et al.*, 2018; McSkimming *et al.*, 2018).

Different Ln can be utilized as cofactors for Ln-dependent alcohol dehydrogenases. Owing to Ln contraction throughout the series, there is a decrease in ionic radius that correlates with an increase in Lewis acidity suggesting that smaller, heavier Ln would enhance the catalytic efficiency and oxidation capacity of the enzymes. Currently, only a few kinetic studies of purified MDHs coordinated with different Ln are available, and each demonstrated a correlation of increasing MDH catalytic efficiency with increasing coordination throughout the Ln series up to neodymium (Nd) (Pol *et al.*, 2014; Good *et al.*, 2018; Jahn *et al.*, 2018; Lumpe *et al.*, 2018). However, coordination with Ln heavier than Nd, such as europium (Eu) or Lu, did not provide higher catalytic efficiency (Jahn *et al.*, 2018; Lumpe *et al.*, 2018). These results suggest that the Lewis acidity of the Ln is not the only chemical property influencing activity. It is likely that the Ln size may affect the metal interaction with the catalytic amino acid, PQQ, and/or the substrate (Fig. 5.1).

In *M. extorquens* AM1, incorporation of Ln into XoxF also appears to be less efficient throughout the series. Apo-XoxF (no metal incorporated into XoxF) was not found when purified from cells grown in the presence of La, but in the presence of Nd, ≈ 50% of the recovered XoxF was in the apo-form (Good *et al.*, 2018). Similarly, for *M. fumariolicum*, apo-XoxF is not obtained when grown in the presence of La, but in the presence of Eu, only 70% of XoxF is fully occupied (Jahn *et al.*, 2018). It is yet unknown whether Ln are unequally obtained and transported into the cell, contributing to the decreased occupancy of XoxF with the heavier Ln, or whether decreased occupancy is due to decreased insertion of the Ln into XoxF once inside the cell, or whether this phenomenon is due to a combination of both possibilities.

Numerous reports have shown differences in the distance between the catalytic Asp and metal cation (tested so far with Ca, Ce, and Eu), highlighting the flexibility of this residue (Leopoldini *et al.*, 2007; Jahn *et al.*, 2018; Lumpe *et al.*, 2018). Furthermore, differences in substrate affinities and V_{max} values have also been observed when different Ln were coordinated with the enzyme (Good *et al.*, 2018; Lumpe *et al.*, 2018). It has also been shown that in some cases, Ln can influence the substrate orientation when bound to the protein (McSkimming *et al.*, 2018). The ability of these enzymes to coordinate different Ln provides flexibility in the use of a wide range of alcohols and aldehydes available in the environment for which concentration and availability can fluctuate greatly.

The oxidation capacity of the Ln-dependent MDHs is also relevant. XoxF from *M. fumariolicum* has a higher oxidation capacity than XoxF from *M. extorquens* AM1 such that the final product of methanol oxidation appears to be formate in the former case, while formaldehyde is the product in the latter case (Pol *et al.*, 2014; Good *et al.*, 2018). It has been shown *in vitro* that XoxF from *M. extorquens* AM1 is able to efficiently use formaldehyde as a substrate but *in vivo*, formaldehyde is still being produced and accumulates at similar rates whether Ln are present or not (Good *et al.*, 2018). These findings highlight the fact that the current assays used to define the activities and the kinetic parameters of alcohol/aldehyde dehydrogenases may not reflect physiological realities. *In vivo*, PQQ-dependent MDHs are coupled to natural electron acceptors (typically cytochromes) and other accessory proteins that can influence the oxidation capacities of these enzymes (Huang *et al.*, 2018; Zheng *et al.*, 2018).

Metabolic consequences of differing methanol oxidation capacities

A major consequence of oxidizing methanol directly to formate is the loss of NAD(P)H which is produced to store electrons when oxidizing formaldehyde to formate via either the tetrahydromethanopterin or the glutathione-dependent formaldehyde oxidation pathways (Barber *et al.*, 1996; Vorholt *et al.*, 1998; Hagemeier *et al.*, 2000). This is particularly important for methanotrophs that require reduced quinone equivalents for activation of methane in addition to the requirement for NAD(P)H in biosynthesis (Cook and

Shiemke, 2002; Choi *et al.*, 2003; Shiemke *et al.*, 2004). Methanotrophs may instead need to rely on additional reduced cytochromes or on a reversed electron transport complex (e.g. Complex III bc_1) driven by proton motive force to reduce quinones. Alternatively, redistribution of carbon flux through the pentose phosphate pathway or increasing transhydrogenase activity may serve to balance reducing equivalents.

M. extorquens AM1 Ln-dependent XoxF and ExaF enzymes produce both formaldehyde and formate, respectively, as final products when using methanol as substrate. It is interesting to note that for *M. extorquens* AM1 grown with methanol and exogenous La, the tetrahydromethanopterin pathway is essential, but 2- to 3-fold down-regulation of the genes encoding three of the four formate dehydrogenases occurs, and less CO_2 (around 18%) is released by cultures compared with growth with methanol in the absence of La. This suggests that *M. extorquens* AM1 rewires its metabolic network to reduce formate oxidation when Ln are present and alters NAD(P)H production (Good *et al.*, 2018). Further examples characterizing both final products of different types of XoxF, and how different strains may balance differences in catabolism/anabolism derived from higher methanol oxidation capacities are currently needed.

Ln-dependent ethanol dehydrogenases

Ln dependency for PQQ alcohol dehydrogenases was recently expanded to include EtDHs. The first reported Ln-EtDH was discovered in *M. extorquens* AM1, a homologue 57% identical to ExaA from *Psuedomonas aeruginosa* that has been named ExaF to denote its dependency on Ln as a cofactor for catalytic activity and to distinguish it from Ca-dependent enzymes. (Görisch and Rupp, 1989; Good *et al.*, 2016; Vu *et al.*, 2016). Like XoxF-type MDHs, ExaF has a second Asp residue coordinating the Ln. ExaF is the most efficient PQQ-EtDH described to date; an increase in activity of 1,000-fold was observed when the enzyme was purified with La when compared with Ca-ExaF. Similar to XoxF from *M. fumariolicum*, ExaF oxidizes ethanol and methanol to acetate and formate, respectively (Good *et al.*, 2016).

A second representative of the PQQ Ln-EtDHs, PedH, was discovered in a non-methylotroph, *P. putida* KT2440 (Wehrmann *et al.*, 2017). PedH is 49% identical to ExaA and 63% identical to ExaF. PedH has a relatively high catalytic efficiency for ethanol, but its K_M for ethanol is in the micromolar range, around 200-times higher than that seen for ExaF. Further, when PedH is compared with its Ca-dependent counterpart, PedE, an approximate 19% decrease in catalytic efficiency for ethanol is observed. An even greater reduction in efficiency is observed with acetaldehyde (Wehrmann *et al.*, 2017). Hence, the only two Ln-dependent EtDHs identified as of now show important kinetic differences after Ln coordination. Detailed structural studies and DFT calculations of these two PQQ-dependent EtDHs can highlight the Ln influence on substrate orientation and substrate affinity. Importantly, it has become apparent that Ln-biochemistry and its role in metabolism as a whole is not unique to one-carbon metabolism.

Overview of the lanthanide- or rare earth-switch

In many methylotrophic organisms, the presence of exogenous Ln results in up-regulation of the Ln-dependent MDH, *xoxF*, and repression of the Ca-dependent MDH, *mxaFI* (Farhan Ul Haque *et al.*, 2015; Skovran and Martinez Gomez, 2015; Chu and Lidstrom 2016; Gu *et al.*, 2016; Vu *et al.*, 2016). This differential regulation of the Ca- and Ln-dependent MDHs has been termed the Ln- or rare earth-switch. In multiple methylotrophs, Ln-dependent growth and expression appears limited to La, Ce, Nd, and praseodymium (Pr) with severely limited growth and mediation of the Ln switch also facilitated by samarium (Sm) in some cases (Farhan Ul Haque *et al.*, 2015; Chu and Lidstrom 2016; Gu *et al.*, 2016; Vu *et al.*, 2016; Huang *et al.*, 2018; Lv and Tani, 2018; Masuda *et al.*, 2018). However, more recent studies show that some organisms such as *Methylotenera mobilis* and *M. fumariolicum* SolV are capable of supporting Ln-dependent methanol oxidation with heavier Ln such as europium (Eu) and gadolinium (Gd) (Huang *et al.*, 2018; Jahn *et al.*, 2018). However, kinetic studies using Eu-XoxF from SolV suggest that the catalytic efficiency of methanol oxidation decreases when heavier Ln than Nd are bound in the active site (Jahn *et al.*, 2018).

The study of the regulatory network controlling the Ln switch is complicated by a variety of factors. The regulators that control this switch differ in different organisms lending it difficult to transfer knowledge about this regulation across the methylotrophic community. In some methylotrophs, additional signals or nutrients and metabolic processes affect regulation of the Ln-switch, yet how these signals contribute to changes in gene expression is not yet understood. Direct binding to promoters by regulators involved in the switch has not been demonstrated lending it unclear whether their involvement is direct or indirect. Finally, in many methylotrophs, there are multiple copies of the *xoxF* genes, some of which only appear to be expressed *in situ* or in co-culture. It is not yet clear why different versions of XoxF would be required or preferred under different growth conditions. It has even been suggested that microorganisms in communities can excrete signals that affect the Ln switch in their methanotrophic neighbours causing methanol to be released as a carbon source for the methylotrophs present in the same community (Krause *et al.*, 2017). Understanding how the Ln switch is controlled, the regulators involved in the switch, and the signals that control the switch is a nascent area of study.

ExaF: A new member of the lanthanide switch

Good *et al.* (2016) reported that ExaF from *M. extorquens* is a Ln- and PQQ-dependent ethanol dehydrogenase, but that it can also oxidize methanol, albeit with lesser efficiency. Subsequently, RNAseq results from Okayam University in Japan determined that two putative PQQ-dependent non-XoxF MDHs, including c07235 which shares 37% identity with ExaF, are also differentially regulated by Ln in *Methylobacterium aquaticum* Strain 22A (Masuda *et al.*, 2018). In *M. extorquens* AM1, RNAseq work from Michigan State University showed that expression of *exaF* is also upregulated by La adding new members to the Ln switch (Good *et al.*, 2018).

In *M. extorquens* AM1, *exaF* is encoded downstream of the c-type cytochrome *exaG* (*pedF*), and the periplasmic binding protein *exaJ*, which shares sequence similarity with *xoxJ*. A 292 bp gap exists between *exaJ* and *exaF* suggesting the possibility that two independent promoters may be used to control expression of the *exa* genes. To investigate the expression of the *exa* genes, transcriptional reporter fusion assays were used which fused the fluorescent reporter Venus (Nagai *et al.*, 2002) to the predicted *exaF* and *exaGJ* promoter regions (Fig. 5.2C). Expression was measured in the presence and absence of 2 μM La with either succinate, methanol, or ethanol as carbon sources (Fig. 5.2A and B). Expression from the *exaF* promoter was upregulated by exogenous La in both succinate and methanol media and further upregulated in ethanol medium (15.0 ± 6.2 RFU/OD in succinate, 55.3 ± 6.2 RFU/OD in methanol, and 456.0 ± 31.5 RFU/OD in ethanol in the presence of La). However, expression from the *exaF* promoter was below background levels in the absence of Ln when succinate or methanol were the carbon sources, and 9.4-fold reduced expression from the *exaF* promoter was seen in the absence of La when ethanol was the carbon source compared with when La was present (Fig. 5.2A). Of note, background levels of fluorescence are due to expression read-through of the tetracycline resistance gene encoded on the transcriptional reporter fusion plasmid, pAP5 (Skovran *et al.*, 2011; Vu *et al.*, 2016). Depending on the transcriptional reporter fusion construct, a negative value (below background) may indicate that repression of the cloned promoter region is occurring or that the cloned promoter region contains a transcriptional terminator from an upstream gene. Expression from the *exaGJ* promoter was upregulated by ethanol but exogenous Ln did not appear to have a marked effect on expression (Fig. 5.2B).

These studies support the findings by Good *et al* that ExaF is Ln-dependent and that ethanol is the preferred substrate in *M. extorquens* AM1 (Good *et al.*, 2016). These data also suggest that expression of *exaF* is controlled by more than just Ln as providing ethanol as the carbon source increased expression 8.2-fold compared with when methanol was the carbon source in the presence of La. The presence of exogenous La and ethanol appear to have a synergistic effect as expression of *exaF* is highly induced only when both La and ethanol are present. It is intriguing that expression from the *exaGJ* promoter does not appear to be regulated by Ln and that significant expression occurs under all conditions tested. The fact that expression from the

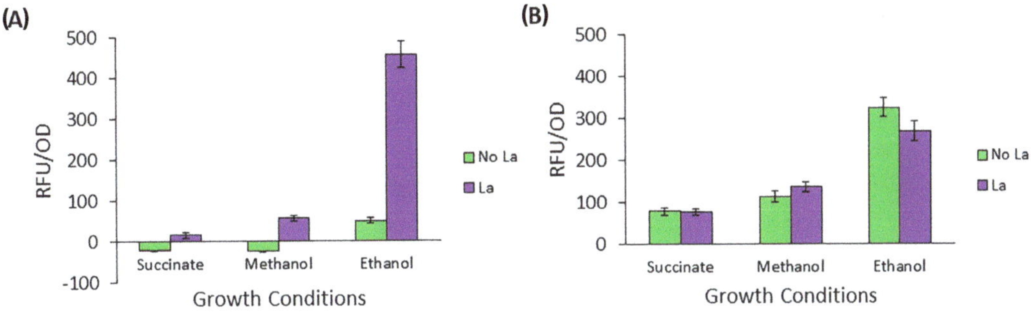

(C)

Plasmid or Primer	Description or Sequence (5'-3')	Reference
pAP5	Promoterless *venus* fusion (from pCM62) (Tc^r)	55
pHV3[a]	pAP5 containing the *exaF* promoter region (Tc^r)	This study
pGS2[b]	pAP5 containing the *exaGJ* promoter region (Tc^r)	This study
p*exaF*_1139acIIbeg	ATCGAACGTTCCGCAACGAGCAGATCTTCCTGA	This study
p*exaF*endEcoRI	GATCGAATTCACCTTGTCCTTGTAGAGGGCGGC	This study
p*exaG*_1137acIIbeg	ACGTAACGTTTCCTCGGCCGGCACATTCAGAAC	This study
p*exaG*endEcoRI	ATGAATTCTAAGGATTCTCCTCGCGCCATGT	This study

Abbreviations: ^{r} = resistant; Tc, tetracycline. [a]Plasmid constructed by Huong N. Vu. [b]Plasmid constructed by Gabriel A. Subuyuj.

Figure 5.2 *exaF* expression is controlled by the Ln switch and expression from the *exaF* and *exaGJ* promoters is upregulated by ethanol. Expression from the *exaF* (A) and *exaGJ* (B) promoters was measured using the modified yellow fluorescence protein, Venus as a fluorescent reporter (Nagai *et al*., 2002). Strains were grown in MP minimal medium with 0.4% succinate, 0.5% methanol, or 0.3% ethanol as the carbon source with (purple bars) and without (green bars) 2 µM lanthanum as previously described (Vu *et al*., 2016). Expression is reported as relative fluorescence units (RFU) per OD at 600 nm. Data are the average of biological triplicates and the error bars represent one standard deviation from the mean. Background expression levels from the promoterless venus fusion vector construct (pAP5) were subtracted from the experimental values for each condition and strain tested. Primers and plasmids used in this study are shown in (C).

exaGJ promoter is upregulated by ethanol suggests that ExaG and ExaJ may work with ExaF during ethanol oxidation but this has not yet been demonstrated.

Master regulators of the lanthanide switch: MxaYB and MxbDM?

Understanding the complexity of factors that control the switch is in its infant stages. Multiple two-component systems have been described that are required for expression of the *mxa* and *xox* genes. In *M. extorquens* AM1, the regulatory network controlling expression of the *mxa* and *xox1* operons involves at least two two-component systems, MxbDM and MxcQE, and an additional orphan response regulator, MxaB. The MxcE, MxaB, and MxbM response regulators are all required for expression of the *mxa* operon while only MxbM is required for the repression of the *xox1* operon (Morris and Lidstrom, 1992; Xu *et al.*, 1995; Springer *et al.*, 1997, 1998; Skovran *et al.*, 2011). In addition, XoxF plays an unknown regulatory role in the network as the *xoxF* genes are required for expression of the *mxa* operon and for repression of the *xox1* operon (Skovran *et al.*, 2011; Vu *et al*, 2016). XoxF is also likely to be required for *mxa* expression in *M. aquaticum* Strain 22A as *xoxF* mutants are unable to grown in methanol medium lacking Ln even though *mxaF* is encoded in the genome and is the predominant enzyme in the absence of Ln (Masuda *et al.*, 2018).

In the methanotroph *Methylomicrobium buryatense* 5GB1C, a functional Ln switch exists yet homologues of the MxcQE and MxbDM two-component systems are absent. Additionally, growth patterns suggest that XoxF is not required for expression of the *mxa* genes in *M. buryatense*, as growth of a *xoxF* mutant is not impaired in methanol medium lacking Ln (Chu and Lidstrom, 2016). These discrepancies highlight the diversity that may exist in the regulatory networks that control the Ln switch in different methylotrophs, as described in more detail below.

Methylomicrobium buryatense 5GB1C

Work from the University of Washington demonstrated that deletion of the *mxaB* response regulator gene results in a growth defect similar to the loss of *mxaF* or *mxaI* when grown in the absence of Ln, suggesting that MxaB is required for expression of the MxaFI-MDH (Chu and Lidstrom, 2016). qRT-PCR data showed that expression of *mxaFI* was reduced ≈1000-fold in an *mxaB* mutant strain while *xoxF* was upregulated 30-fold, signifying that MxaB plays a major role in the Ln switch (Chu and Lidstrom, 2016). *mxaB* expression is also repressed by Ln similar to the two-component systems that are involved in the switch in *M. extorquens* AM1 (Good *et al.*, 2018). In this same study from the University of Washington, second site suppressors of *xoxF* were isolated which restored growth due to induced expression of *mxaFI* in methanol medium containing La. These suppressor mutations were subsequently mapped to the sensor kinase, MxaY (Chu *et al.*, 2016). By changing a Glu to Gly at position 147, the MxaY sensor could be constitutively activated resulting in expression of the *mxaFI* MDH regardless of Ln presence or absence. Intriguingly, both loss of *mxaY* and constitutive activation of MxaY in the presence of La result in a significant growth defect (doubling time ≈7.5 h compared with 2.4 h in WT), suggesting that MxaY has an important regulatory role during Ln growth beyond regulating MDH gene expression.

Methylobacterium extorquens AM1

In *M. extorquens* AM1, the regulatory network controlling differential expression of the *mxa* and *xox1* genes is complex, involving at least two sensor kinases, MxbD and MxcQ, and three response regulators, MxbM, MxcE, MxaB, though neither the phosphorylation state nor direct binding of these regulators to DNA has been shown (Morris and Lidstrom, 1992, Xu *et al.*, 1995; Springer *et al.*, 1997, 1998; Skovran *et al.*, 2011). It has been suggested that the MxcQE and MxbDM two-component systems controlling *mxa* and *xox1* expression may work in a cascade with MxcQE activating expression of *mxbDM* (Springer *et al.*, 1997). To add an additional layer of complexity, the XoxF-type MeDH is required for expression of the MxaFI-type MeDH and the MxbDM two-component system when cells are grown in the absence of La, and also for repression of itself (Skovran *et al.*, 2011; Vu *et al.*, 2016). Vu *et al.* (2016) have hypothesized that Ln sensing may be done through a sensor kinase such as MxcQ or MxbD using apo-XoxF rather than sensing Ln directly. Similar to work with *M. buryatense* 5GB1C, second site suppressors of *xoxF* were isolated (but in the absence of Ln) which restored wild-type levels of expression from the *mxa* promoter (Skovran *et al.*, 2011). Complementation and sequence analyses mapped these suppressor mutations to the MxbD sensor kinase. Strains containing a suppressor mutation in MxbD no longer require XoxF for expression of the *mxa* genes in the absence of Ln (Bang Luong and Elizabeth Skovran, San Jose State University, unpublished data).

Taken together, work with *M. buryatense* 5GB1C and *M. extorquens* AM1 suggests that MxbD and MxaY may serve analogous roles in the Ln switch as the main Ln sensors. If the hypotheses from these authors are correct, that MxaY senses Ln directly, and MxbD may instead sense apo-XoxF, this may help to explain why divergent regulatory systems have evolved as the signal for Ln presence is different. As these theories have yet to be demonstrated via binding interaction studies, it is only possible to speculate at this point.

Regulation of the lanthanide switch in non-methylotrophs

To date, characterization of a Ln switch has only been reported in one non-methylotroph, *P. putidia* KT2440 (Werhmann *et al.*, 2018). In this organism, the Ln switch controls differential expression of *pedH* and *pedE*, which encode the Ln- and Ca-dependent EtDHs that facilitate oxidation of a broad range of volatile alcohols and aldehydes (Werhmann *et al.*, 2017). In the absence of *pedH*, strains are unable to grow in the presence of Ln with phenylethanol as a carbon source. Using an adaptive evolution approach, Werhmann *et al.* (2018) generated and isolated *pedH* suppressor mutant strains that could grow in the presence of Ln. These strains contained a broken Ln switch which allowed expression of *pedE* which encodes the Ca-dependent EtDH enzyme. The suppressor mutations mapped to *pedS2* which encodes a sensor histidine kinase with 25% amino acid identity to MxaY. Adjacent to *pedS* is a LuxR-type response regulator, *pedR2*. Further genetic analyses suggested that PedR2 is the response regulator mediating the Ln switch and identified Asp 53 as the site of phosphorylation. The authors propose a model of Ln-dependent regulation where in the absence of Ln, PedS2 phosphorylates PedR2 which functions as both a strong activator of *pedE* expression and a repressor of *pedH* expression. When Ln are present, PedS2 kinase activity decreases such that *pedH* expression occurs while *pedE* expression is repressed (Werhmann *et al.*, 2018).

Copper as an override control of the lanthanide switch

Copper is an extremely important metal in methanotrophy as it is required in the active site of the membrane-associated or particulate methane monooxygenase (pMMO) (Lieberman and Rosenzweig, 2005; Semrau *et al.*, 2010). Of the methanotrophs that contain pMMO, some contain an additional soluble methane monooxygenase (sMMO) which instead contains a di-iron centre (Sirajuddin and Rosenzweig, 2015; Ross and Rosenzweig, 2017). Organisms that contain both pMMO and sMMO are subject to the Cu-switch which promotes expression of the Cu-containing pMMO over sMMO when Cu is available (Stanley *et al.*, 1983; Nielsen *et al.*, 1997; Murrell *et al.*, 2000).

The effect of Cu on the Ln switch has been studied in Type I and Type II methanotrophs and appears to override the Ln switch in some methanotrophs but not others. In the Type II methanotroph, *M. trichosporium* OB3b, exogenous Cu attenuates the effect of Ln on the switch allowing expression of the Mxa-MDH even when Ln are present (Gu *et al.*, 2016; Farhan Ul Haque *et al.*, 2015). Cu also decreases expression of both *xox* genes encoded in the *M. trichosporium* genome by almost an order of magnitude in the presence of Ln (Farhan Ul Haque *et al.* 2015). This is in stark contrast to the Type I methanotroph, *M. buryatense*, in which exogenous Cu has very little effect on the expression of the *mxa* and *xoxF* genes (Chu and Lidstrom 2016). There are not yet sufficient data to conclude if this difference is generally conserved among the Type I and Type II methanotrophs. Cu control over the Ln switch in some methanotrophs does however lend further support to the idea that the regulatory networks controlling the Ln switch in methylotrophic bacteria are complex and diverse.

The effects of community interactions and substrate availability on the lanthanide switch

Recent work with two species co-cultures as well as with more complex mixed communities suggests that the Ln switch may be controlled by additional factors, among them community interactions, though the nature of those interactions is not yet well understood (Vorobev *et al.*, 2013; Krause *et al.*, 2017; Yu and Chistoserdova, 2017; Yu *et al.*, 2017; Zheng *et al.*, 2018). Krause *et al.* (2017) compared expression of *mxaF* and *xoxF* genes in cultures of a methanotroph, *Methylobacter tundripaludum*, grown with and without non-methane-utilizing methylotrophic bacteria of the genus *Methylotenera*. While the Ln switch responded as expected when *M. tundripaludum* was grown in pure culture, in co-cultures, the *M. tundripaludum mxaFI* genes were predominantly expressed even when La was available. This apparently resulted in methanol excretion into the culture medium which could provide carbon to and perhaps spare Ln for the non-methane utilizing methylotrophs (Krause *et al.*, 2017). Methanol excretion was not observed when *M. tundripaludum* expressed *xoxF* in pure culture possibly due to the proposed superior catalytic properties of the XoxF enzymes (Lim and Franklin, 2004; Pol *et al.*, 2014). Interestingly, in the non-methane-utilizing methylotroph, *Methylotenera mobilis* JLW8 which lacks the *mxaFI* genes but encodes two different XoxF enzymes instead, *xoxF_1770* was highly expressed in pure culture, while *xoxF_2048* was overexpressed in the co-culture. A similar trend was seen for *M. mobilis* 13, in which one *xoxF* gene was preferentially expressed in pure culture while the other *xoxF* gene was preferentially expressed in co-culture (Krause *et al.*, 2017). It is yet unclear why different XoxF enzymes would be preferred over others under different conditions.

Recently, a large-scale OMICS study was completed that followed population and transcriptomic changes that occurred when 50 pure cultures of bacteria representing different functional guilds were combined to create a synthetic methylotrophic community. This synthetic community was subjected to different nitrogen sources (nitrate versus ammonium) and partial pressures of methane and oxygen, in the absence of Ln (Yu *et al.*, 2017). Community population dynamics were followed over 212 days and transcriptomics were analysed at various time points up to 159 days. This study showed that expression of *mxaF* and *xoxF* could be affected by the nitrogen source as well as oxygen and methane partial pressures, and that these responses could differ among species (Fig. 5.3). With nitrate as the nitrogen source, high methane and high oxygen promoted expression of *mxaF* in *Methylomonas* LW13 and *Methylphilus* Q8, which were the dominant species in the community under these conditions. When methane and oxygen were low, expression of *mxaF* was repressed in *Methylomonas* LW13 and *xoxF* was induced. Perhaps when more methane is available to the methanotrophs in the community, production of MxaF may promote sharing of methanol as 'public goods' as carbon is not as precious/scarce. There was more variability in expression of the *xoxF* and *mxaF* genes among the non-methanotrophic methylotrophs which often seemed to favour *xoxF* expression regardless of the condition tested. Production of XoxF may be a way for organisms that have access to methanol (1) to keep the methanol rather than excrete it in the case of a methanotroph; or (2) to efficiently use methanol, if scarce, in the case of a non-methanotroph. This would only make sense however, if Ln were available. As Ln are not present as in this study, the patterns observed only reflected the transcriptional regulation in response to the different factors and not necessarily enzyme activity. Perhaps, XoxF enzymes could serve as methanol sequestration and storage systems by binding the volatile methanol. Perhaps some of the less-characterized XoxF enzymes may have a wider cofactor range. It is also possible that trace Ln contamination may have allowed for expression of *xoxF* in these experiments. Regardless, these results highlight the importance of community studies and again point to the complex nature of the Ln-switch, further suggesting that the Ln switch is affected by more than simple presence or absence of exogenous Ln.

Conclusions

Ln-dependent methylotrophy and its regulation is an emerging field of study. Very little is known about how organisms that use Ln for growth obtain these insoluble micronutrients and incorporate them into the enzymes that utilize them as cofactors. Likely, the importance of Ln will extend beyond the world of methylotrophs. Recently, PedH, a homologue of ExaF was described from non-methylotrophic *Pseudomonas putida* KT2440, and it was shown to also use light Ln and be transcriptionally regulated by their presence (Wehrmann *et al.*, 2017, 2018). It remains to be seen whether these critical metals will have an inherent role in enzymes other than PQQ-dependent alcohol dehydrogenases.

The fact that many methylotrophs have multiple copies of *xoxF* is intriguing. Once thought by some to be duplicate or remnant copies, variants of *xoxF* that have low expression in pure culture appear to have functional roles in communities based upon their gene expression profiles. Biochemically, different types of XoxF MDHs exhibit diverse kinetic parameters for a wide set of substrates. Parsing out the distinct roles for these variant XoxF MDHs is an exciting area of study that is just beginning. Understanding how expression of these variant *xoxF* genes fits into the poorly understood regulatory networks that control the Ln switch will likely add further complexity to an already complex system.

For organisms that contain both Ca- and Ln-dependent MDHs, regulation of the Ln switch appears to be mainly controlled by exogenous Ln when grown in pure culture with the ability of Cu to override the switch in some methanotrophs. However, as there are multiple regulators that contribute to expression of genes under control of the Ln switch in characterized organisms, this suggests that there are likely additional signals that have yet to be identified. In *M. extorquens* AM1 for example,

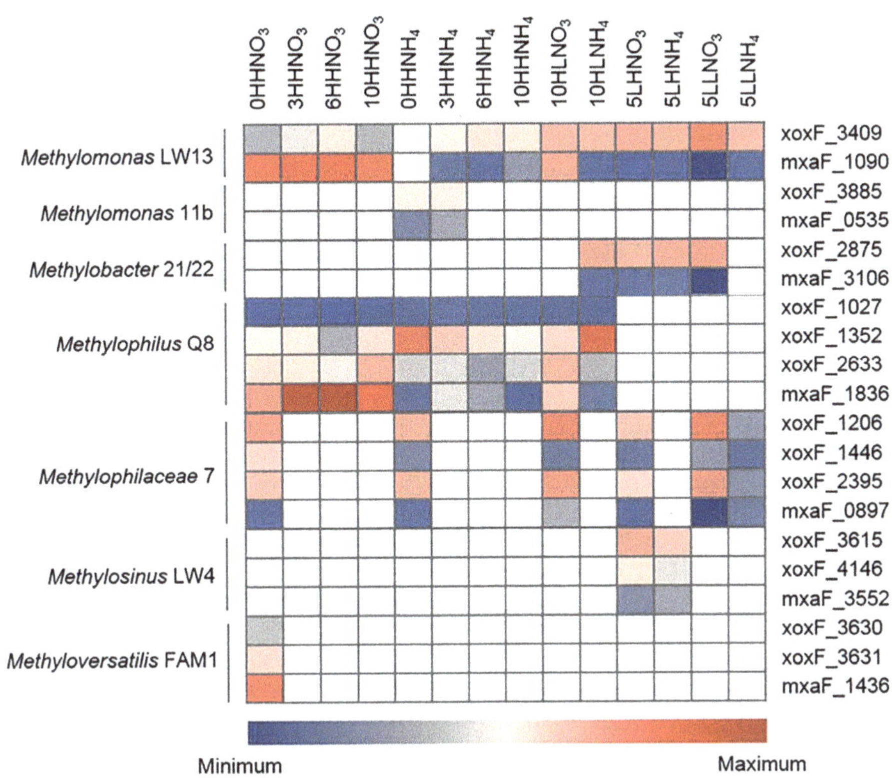

Figure 5.3 Heat map depicting expression of *xoxF* and *mxaF* genes in selected species. Figure and legend are adapted from Yu *et al.* (2017). Sample designations refer to number of transfers (bottom number), followed by methane and O_2 regimens (H for high partial pressure, L for low partial pressure), followed by the nitrogen source. On the left, genus and strain names. On the right, gene name followed by gene number as per genome annotation.

expression of *exaF* is upregulated by Ln, but presence of ethanol rather than methanol or succinate exhibits a larger and synergistic effect on expression (Fig. 5.2A).

Regulation of methanol oxidation is of particular importance as the product of methanol oxidation is either formaldehyde which is highly reactive and toxic, or formate, an acid. It is likely that a variety of signals beyond Ln exert control over the MDHs to help maintain homeostasis in organisms whose central metabolism must balance production and consumption of multiple toxic intermediates while sequestering enough volatile carbon for growth (Skovran *et al.*, 2010). These enzymes controlled by the Ln switch require not only metals for activity but PQQ and cytochromes, therefore also iron and haem. A multitude of processes are required for a single metabolic step and may have as of yet unknown effects on expression of the enzymes that carry out this step. As use of XoxF over MxaFI affects production of nucleotide pools in some organisms either by circumventing NADPH production pathways or by decreasing formate oxidation, it is possible that nucleotide homeostasis may also be sensed by the Ln switch which could help maintain balanced nucleotide pools by shifting the MDH that is used. More studies are needed to go beyond speculation to gain important knowledge about how methylotrophs sense and adapt to Ln availability in the presence and absence of other organisms.

There is much left to learn regarding Ln-dependent enzymes and the regulation of the *mxa*, *xox*, and *exa* genes. The mechanisms of Ln acquisition and transport have yet to be explored. A deeper understanding of these processes may facilitate the engineering of methanotrophs and methylotrophs to better function as environmental platforms for carbon capture, production of value added chemicals, and recovery of REEs from electronic waste and other Ln waste streams. Ln in biology is an emergent and rapidly expanding field full of 'first discoveries.' Researchers have just scratched the surface of novel revelations surrounding Ln in ecology, biochemistry, genetics, bioremediation, biotechnology, and more.

References

Barber, R.D., Rott, M.A., and Donohue, T.J. (1996). Characterization of a glutathione-dependent formaldehyde dehydrogenase from *Rhodobacter sphaeroides*. J. Bacteriol. *178*, 1386–1393.

Beck, D.A., McTaggart, T.L., Setboonsarng, U., Vorobev, A., Goodwin, L., Shapiro, N., Woyke, T., Kalyuzhnaya, M.G., Lidstrom, M.E., and Chistoserdova, L. (2015). Multiphyletic origins of methylotrophy in Alphaproteobacteria, exemplified by comparative genomics of Lake Washington isolates. Environ. Microbiol. *17*, 547–554. https://doi.org/10.1111/1462-2920.12736

Bogart, J.A., Lewis, A.J., and Schelter, E.J. (2015). DFT study of the active site of the XoxF-type natural, cerium-dependent methanol dehydrogenase enzyme. Chemistry *21*, 1743–1748. https://doi.org/10.1002/chem.201405159

Byrne, R.H., and Sholkovitz, E.R. (1996). Marine chemistry and geochemistry of the lanthanides. In Handbook on the Physics and Chemistry of Rare Earths 23, 497–523. https://doi.org/10.1016/S0168-1273(96)23009-0

Chistoserdova, L. (2016). Lanthanides: New life metals? World J. Microbiol. Biotechnol. *32*, 138. https://doi.org/10.1007/s11274-016-2088-2

Chistoserdova, L. (2018). Applications of methylotrophs: can single carbon be harnessed for biotechnology? Curr. Opin. Biotechnol. *50*, 189–194.

Chistoserdova, L., and Kalyuzhnaya, M.G. (2018). Current trends in methyltrophy. Trends Microbiol. *26*, 703–7 14. https://doi.org/10.1016/j.tim.2018.01.011

Choi, D.W., Kunz, R.C., Boyd, E.S., Semrau, J.D., Antholine, W.E., Han, J.I., Zahn, J.A., Boyd, J.M., de la Mora, A.M., and DiSpirito, A.A. (2003). The membrane-associated methane monooxygenase (pMMO) and pMMO-NADH:quinone oxidoreductase complex from *Methylococcus capsulatus* Bath. J. Bacteriol. *185*, 5755–5764.

Chu, F., and Lidstrom, M.E. (2016). XoxF acts as the predominant methanol dehydrogenase in the Type I methanotroph *Methylomicrobium buryatense*. J. Bacteriol. *198*, 1317–1325. https://doi.org/10.1128/JB.00959-15

Chu, F., Beck, D.A., and Lidstrom, M.E. (2016). MxaY regulates the lanthanide-mediated methanol dehydrogenase switch in *Methylomicrobium buryatense*. PeerJ *4*, e2435. https://doi.org/10.7717/peerj.2435

Cook, S.A., and Shiemke, A.K. (2002). Evidence that a type-2 NADH:quinone oxidoreductase mediates electron transfer to particulate methane monooxygenase in methylococcus *capsulatus*. Arch. Biochem. Biophys. *398*, 32–40. https://doi.org/10.1006/abbi.2001.2628

Del Rocío Bustillos-Cristales, M., Corona-Gutierrez, I., Castañeda-Lucio, M., Águila-Zempoaltécatl, C., Seynos-García, E., Hernández-Lucas, I., Muñoz-Rojas, J., Medina-Aparicio, L., and Fuentes-Ramírez, L.E. (2017). culturable facultative methylotrophic bacteria from the cactus *Neobuxbaumia macrocephala* possess the locus xoxF and consume methanol in the presence of Ce3+ and Ca2. Microbes Environ. *32*, 244–251. https://doi.org/10.1264/jsme2.ME17070

Deng, Y.W., Ro, S.Y., and Rosenzweig, A.C. (2018). Structure and function of the lanthanide-dependent methanol dehydrogenase XoxF from the methanotroph *Methylomicrobium buryatense* 5GB1C. J. Biol. Inorg. Chem. [Epub ahead of print]. https://doi.org/10.1007/s00775-018-1604-2

Farhan Ul Haque, M., Kalidass, B., Bandow, N., Turpin, E.A., DiSpirito, A.A., and Semrau, J.D. (2015). Cerium regulates expression of alternative methanol dehydrogenases in *Methylosinus trichosporium* OB3b. Appl. Environ. Microbiol. *81*, 7546–7552. https://doi.org/10.1128/AEM.02542-15

Fitriyanto, N.A., Fushimi, M., Matsunaga, M., Pertiwiningrum, A., Iwama, T., and Kawai, K. (2011). Molecular structure and gene analysis of Ce3+ -induced methanol dehydrogenase of *Bradyrhizobium* sp. MAFF211645. J. Biosci. Bioeng. *111*, 613–617. https://doi.org/10.1016/j.jbiosc.2011.01.015

Gao, Y., Zeng, F., Yi, A., Ping, S., and Jing, L. (2003). Research of the entry of rare earth elements Eu3+ and La3+ into plant cell. Biol. Trace Elem. Res. *91*, 253–265.

Ghosh, M., Avezoux, A., Anthony, C., Harlos, K., and Blake, C.C. (1994). X-ray structure of PQQ-dependent methanol dehydrogenase. EXS *71*, 251–260.

Good, N.M., Vu, H.N., Suriano, C.J., Subuyuj, G.A., Skovran, E., and Martinez-Gomez, N.C. (2016). Pyrroloquinoline Quinone ethanol dehydrogenase in *Methylobacterium extorquens* AM1 extends lanthanide-dependent metabolism to multicarbon substrates. J. Bacteriol. *198*, 3109–3118.

Good, N.M., Walser, O.N., Moore, R.S., Suriano, C.J., Huff, A.F., and Martinez-Gomez N.C. (2018). Investigation of lanthanide-dependent methylotrophy in *Methylobacterium extorquens* AM1 uncovers complementary roles for alcohol dehydrogenases enzymes. (2018). bioRxiv. https://doi.org/100.1101/329011

Görisch, H., and Rupp, M. (1989). Quinoprotein ethanol dehydrogenase from *Pseudomonas*. Antonie Van Leeuwenhoek *56*, 35–45.

Gu, W., Farhan Ul Haque, M., DiSpirito, A.A., and Semrau, J.D. (2016). Uptake and effect of rare earth elements on gene expression in *Methylosinus trichosporium* OB3b. FEMS Microbiol. Lett. *363*, fnw129. https://doi.org/10.1093/femsle/fnw129

Hagemeier, C.H., Chistoserdova, L., Lidstrom, M.E., Thauer, R.K., and Vorholt, J.A. (2000). Characterization of a second methylene tetrahydromethanopterin dehydrogenase from *Methylobacterium extorquens* AM1. Eur. J. Biochem. *267*, 3762–3769.

Hibi, Y., Asai, K., Arafuka, H., Hamajima, M., Iwama, T., and Kawai, K. (2011). Molecular structure of La3+-induced methanol dehydrogenase-like protein in *Methylobacterium radiotolerans*. J. Biosci. Bioeng. *111*, 547–549. https://doi.org/10.1016/j.jbiosc.2010.12.017

Höfer, P., Vermette, P., and Groleau, D. (2011). Introducing a new bioengineered bug: *Methylobacterium extorquens* tuned as a microbial bioplastic factory. Bioeng. Bugs *2*, 71–79. https://doi.org/10.4161/bbug.2.2.15009

Howat, A.M., Vollmers, J., Taubert, M., Grob, C., Dixon, J.L., Todd, J.D., Chen, Y., Kaster, A.K., and Murrell, J.C. (2018). Comparative genomics and mutational analysis reveals a novel XoxF-utilizing methylotroph in the Roseobacter group isolated from the marine environment. Front. Microbiol. *9*, 766. https://doi.org/10.3389/fmicb.2018.00766

Hu, Z.H., Richter, H., Sparovek, G., and Schnug, E. (2004). Physiological and biochemical effects of rare earth elements on plants and their agricultural significance: a review. J. Plant Nutr. *27*, 183–220. https://doi.org/10.1081/PLN-120027555

Huang, J., Yu, Z., and Chistoserdova, L. (2018). Lanthanide-dependent methanol dehydrogenases of XoxF4 and XoxF5 clades are differentially distributed among methylotrophic bacteria and they reveal different biochemical properties. Front. Microbiol. *9*, 1366. https://doi.org/10.3389/fmicb.2018.01366

Hwang, I.Y., Lee, S.H., Choi, Y.S., Park, S.J., Na, J.G., Chang, I.S., Kim, C., Kim, H.C., Kim, Y.H., Lee, J.W., *et al.* (2014). Biocatalytic conversion of methane to methanol as a key step for development of methane-based biorefineries. J. Microbiol. Biotechnol. *24*, 1597–1605.

Iguchi, H., Yurimoto, H., and Sakai, Y. (2015). Interactions of methylotrophs with plants and other heterotrophic bacteria. microorganisms *3*, 137–151. https://doi.org/10.3390/microorganisms3020137

Jahn, B., Pol, A., Lumpe, H., Barends, T.R.M., Dietl, A., Hogendoorn, C., Op den Camp, H.J.M., and Daumann, L.J. (2018). Similar but not the same: First kinetic and structural analyses of a methanol dehydrogenase containing a europium ion in the active site. Chembiochem. *19*, 1147–1153. https://doi.org/10.1002/cbic.201800130

Keltjens, J.T., Pol, A., Reimann, J., and Op den Camp, H.J. (2014). PQQ-dependent methanol dehydrogenases: rare-earth elements make a difference. Appl. Microbiol. Biotechnol. *98*, 6163–6183. https://doi.org/10.1007/s00253-014-5766-8

Krause, S.M., Johnson, T., Samadhi Karunaratne, Y., Fu, Y., Beck, D.A., Chistoserdova, L., and Lidstrom, M.E. (2017). Lanthanide-dependent cross-feeding of methane-derived carbon is linked by microbial community interactions. Proc. Natl. Acad. Sci. U.S.A. *114*, 358–363. https://doi.org/10.1073/pnas.1619871114

Kumar, M., Tomar, R.S., Lade, H., and Paul, D. (2016). Methylotrophic bacteria in sustainable agriculture. World J. Microbiol. Biotechnol. *32*, 120. https://doi.org/10.1007/s11274-016-2074-8

Leopoldini, M., Russo, N., and Toscano, M. (2007). The preferred reaction path for the oxidation of methanol by PQQ-containing methanol dehydrogenase: addition-elimination versus hydride-transfer mechanism. Chemistry *13*, 2109–2117. https://doi.org/10.1002/chem.200601123

Lieberman, R.L., and Rosenzweig, A.C. (2005). Crystal structure of a membrane-bound metalloenzyme that catalyses the biological oxidation of methane. Nature *434*, 177–182.

Lim, S., and Franklin, S.J. (2004). Lanthanide-binding peptides and the enzymes that Might Have Been. Cell. Mol. Life Sci. *61*, 2184–2188. https://doi.org/10.1007/s00018-004-4156-2

Lu, X., Gu, W., Zhao, L., Farhan Ul Haque, M., DiSpirito, A.A., Semrau, J.D., and Gu, B. (2017). Methylmercury uptake and degradation by methanotrophs. Sci. Adv. *3*, e1700041. https://doi.org/10.1126/sciadv.1700041

Lumpe, H., Pol, A., Op den Camp, H., and Daumann, L. (2018). Impact of the lanthanide contraction on the activity of a lanthanide-dependent methanol dehydrogenase – a kinetic and DFT study. Dalton Trans. *47*, 10463–10472. https://doi.org/10.1039/c8dt01238e

Lv, H., and Tani, A. (2018). Genomic characterization of methylotrophy of *Oharaeibacter diazotrophicus* strain SM30T. J. Biosci. Bioeng. [Epub ahead of print].

Martinez-Gomez, N.C., Vu, H.N., and Skovran, E. (2016). Lanthanide chemistry: from coordination in chemical complexes shaping our technology to coordination in enzymes shaping bacterial metabolism. Inorg. Chem. *55*, 10083–10089. https://doi.org/10.1021/acs.inorgchem.6b00919

Masuda, S., Suzuki, Y., Fujitani, Y., Mitsui, R., Nakagawa, T., Shintani, M., and Tani, A. (2018). Lanthanide-dependent regulation of methylotrophy in *Methylobacterium aquaticum* strain 22A. mSphere. *3*, e00462–17. https://doi.org/10.1128/mSphere.00462-17

McSkimming, A., Cheisson, T., Carroll, P.J., and Schelter, E.J. (2018). Functional synthetic model for the lanthanide-dependent quinoid alcohol dehydrogenase active site. J. Am. Chem. Soc. *140*, 1223–1226. https://doi.org/10.1021/jacs.7b12318

Morris, C.J., and Lidstrom, M.E. (1992). Cloning of a methanol-inducible *moxF* promoter and its analysis in *moxB* mutants of *Methylobacterium extorquens* AM1rif. J. Bacteriol. *174*, 4444–4449.

Murrell, J.C., McDonald, I.R., and Gilbert, B. (2000). Regulation of expression of methane monooxygenases by copper ions. Trends Microbiol. *8*, 221–225.

Nagai, T., Ibata, K., Park, E.S., Kubota, M., Mikoshiba, K., and Miyawaki, A. (2002). A variant of yellow fluorescent protein with fast and efficient maturation for cell-biological applications. Nat. Biotechnol. *20*, 87–90. https://doi.org/10.1038/nbt0102-87

Nakagawa, T., Mitsui, R., Tani, A., Sasa, K., Tashiro, S., Iwama, T., Hayakawa, T., and Kawai, K. (2012). A catalytic role of XoxF1 as La3+-dependent methanol dehydrogenase in *Methylobacterium extorquens* strain AM1. PLOS ONE 7, e50480. https://doi.org/10.1371/journal.pone.0050480

Nielsen, A.K., Gerdes, K., and Murrell, J.C. (1997). Copper-dependent reciprocal transcriptional regulation of methane monooxygenase genes in *Methylococcus capsulatus* and *Methylosinus trichosporium*. Mol. Microbiol. *25*, 399–409.

Ochsner, A.M., Sonntag, F., Buchhaupt, M., Schrader, J., and Vorholt, J.A. (2015). *Methylobacterium extorquens*: methylotrophy and biotechnological applications. Appl. Microbiol. Biotechnol. *99*, 517–534. https://doi.org/10.1007/s00253-014-6240-3

Pfeifenschneider, J., Brautaset, T., and Wendisch, V.F. (2017). Methanol as carbon substrate in the bio-economy: Metabolic engineering of aerobic methylotrophic bacteria for production of value-added chemicals. Biofuels Bioprod. Bioref. *11*, 719–731. https://doi.org/10.1002/bbb.1773

Pol, A., Barends, T.R., Dietl, A., Khadem, A.F., Eygensteyn, J., Jetten, M.S., and Op den Camp, H.J. (2014). Rare earth metals are essential for methanotrophic life in volcanic mudpots. Environ. Microbiol. *16*, 255–264. https://doi.org/10.1111/1462-2920.12249

Ross, M.O., and Rosenzweig, A.C. (2017). A tale of two methane monooxygenases. J. Biol. Inorg. Chem. *22*, 307–319. https://doi.org/10.1007/s00775-016-1419-y

Semrau, J.D., DiSpirito, A.A., and Yoon, S. (2010). Methanotrophs and copper. FEMS Microbiol. Rev. *34*, 496–531. https://doi.org/10.1111/j.1574-6976.2010.00212.x

Shiemke, A.K., Arp, D.J., and Sayavedra-Soto, L.A. (2004). Inhibition of membrane-bound methane monooxygenase and ammonia monooxygenase by diphenyliodonium: implications for electron transfer. J. Bacteriol. *186*, 928–937.

Sirajuddin, S., and Rosenzweig, A.C. (2015). Enzymatic oxidation of methane. Biochemistry *54*, 2283–2294. https://doi.org/10.1021/acs.biochem.5b00198

Skovran, E., and Martinez-Gomez, N.C. (2015). Microbiology. Just add lanthanides. Science *348*, 862–863. https://doi.org/10.1126/science.aaa9091

Skovran, E., Crowther, G.J., Guo, X., Yang, S., and Lidstrom, M.E. (2010). A systems biology approach uncovers cellular strategies used by *Methylobacterium extorquens* AM1 during the switch from multi- to single-carbon growth. PLOS ONE *5*, e14091. https://doi.org/10.1371/journal.pone.0014091

Skovran, E., Palmer, A.D., Rountree, A.M., Good, N.M., and Lidstrom, M.E. (2011). XoxF is required for expression of methanol dehydrogenase in *Methylobacterium extorquens* AM1. J. Bacteriol. *193*, 6032–6038. https://doi.org/10.1128/JB.05367-11

Springer, A.L., Morris, C.J., and Lidstrom, M.E. (1997). Molecular analysis of *mxbD* and *mxbM*, a putative sensor-regulator pair required for oxidation of methanol in *Methylobacterium extorquens* AM1. Microbiology *143*, 1737–1744. https://doi.org/10.1099/00221287-143-5-1737

Springer, A.L., Auman, A.J., and Lidstrom, M.E. (1998). Sequence and characterization of *mxaB*, a response regulator involved in regulation of methanol oxidation, and of *mxaW*, a methanol-regulated gene in *Methylobacterium extorquens* AM1. FEMS Microbiol. Lett. *160*, 119–124.

Stanley, S.H., Prior, S.D., Leak, D.J., and Dalton, H. (1983). Copper stress underlies the fundamental change in intracellular location of methane-monooxygenase in methane-oxidizing organisms – studies in batch and continuous cultures. Biotechnol. Lett. *5*, 487–492. https://doi.org/10.1007/BF00132233

Taubert, M., Grob, C., Howat, A.M., Burns, O.J., Dixon, J.L., Chen, Y., and Murrell, J.C. (2015). XoxF encoding an alternative methanol dehydrogenase is widespread in coastal marine environments. Environ. Microbiol. *17*, 3937–3948. https://doi.org/10.1111/1462-2920.12896

Tyler, G. (2005). Rare earth elements in soil and plant systems – A review. Plant Soil *267*, 191–206. https://doi.org/10.1007/s11104-005-4888-2

Yu, Z., and Chistoserdova, L. (2017). Communal metabolism of methane and the rare Earth element switch. J. Bacteriol. [Epub ahead of print].

Yu, Z., Beck, D.A.C., and Chistoserdova, L. (2017). Natural selection in synthetic communities highlights the roles of *Methylococcaceae* and *Methylophilaceae* and suggests differential roles for alternative methanol dehydrogenases in methane consumption. Front. Microbiol. *8*, 2392. https://doi.org/10.3389/fmicb.2017.02392

Vekeman, B., Speth, D., Wille, J., Cremers, G., De Vos, P., Op den Camp, H.J., and Heylen, K. (2016). Genome characteristics of two novel Type I methanotrophs enriched from North Sea Sediments containing exclusively a lanthanide-dependent XoxF5-type methanol dehydrogenase. microb. Ecol. *72*, 503–509. https://doi.org/10.1007/s00248-016-0808-7

Vorholt, J.A., Chistoserdova, L., Lidstrom, M.E., and Thauer, R.K. (1998). The NADP-dependent methylene tetrahydromethanopterin dehydrogenase in *Methylobacterium extorquens* AM1. J. Bacteriol. *180*, 5351–5356.

Vorobev, A., Beck, D.A., Kalyuzhnaya, M.G., Lidstrom, M.E., and Chistoserdova, L. (2013). Comparative transcriptomics in three *Methylophilaceae* species uncover different strategies for environmental adaptation. PeerJ *1*, e115. https://doi.org/10.7717/peerj.115

Vu, H.N., Subuyuj, G.A., Vijayakumar, S., Good, N.M., Martinez-Gomez, N.C., and Skovran, E. (2016). Lanthanide-dependent regulation of methanol oxidation systems in *Methylobacterium extorquens* AM1 and their contribution to methanol growth. J. Bacteriol. *198*, 1250–1259. https://doi.org/10.1128/JB.00937-15

Wehrmann, M., Billard, P., Martin-Meriadec, A., Zegeye, A., and Klebensberger, J. (2017). Functional role of lanthanides in enzymatic activity and transcriptional regulation of pyrroloquinoline quinone-dependent alcohol dehydrogenases in *Pseudomonas putida* KT2440. MBio *8*, e00570–17.

Wehrmann, M., Berthelot, C., Billard, P., and Klebensberger, J. (2018). The PedS2/PedR2 Two-Component System Is Crucial for the Rare Earth Element Switch in Pseudomonas putida KT2440. mSphere 3, e00376–18.

Williams, P.A., Coates, L., Mohammed, F., Gill, R., Erskine, P.T., Coker, A., Wood, S.P., Anthony, C., and Cooper, J.B. (2005). The atomic resolution structure of methanol dehydrogenase from *Methylobacterium extorquens*. Acta Crystallogr. Sect. D Biol. Crystallogr. *61*, 75–79. https://doi.org/10.1107/S0907444904026964

Wu, M.L., Wessels, J.C., Pol, A., Op den Camp, H.J., Jetten, M.S., and van Niftrik, L. (2015). XoxF-type methanol dehydrogenase from the anaerobic methanotroph 'Candidatus Methylomirabilis oxyfera'. Appl. Environ. Microbiol. *81*, 1442–1451.

Xu, H.H., Janka, J.J., Viebahn, M., and Hanson, R.S. (1995). Nucleotide sequence of the *mxcQ* and *mxcE* genes, required for methanol dehydrogenase synthesis in

Methylobacterium organophilum XX: a two-component regulatory system. Microbiology *141*, 2543–2551. https://doi.org/10.1099/13500872-141-10-2543

Zepf, V., Simmons, J., Reller, A., Ashfield, M., and Rennie, C. (2014). Materials critical to the energy sector – An introduction, 2nd edition (BP p.l.c, London). https://www.bp.com/content/dam/bp/pdf/sustainability/group-reports/ESC_Materials_handbook_BP_Apr2014.pdf

Zheng, Y., Huang, J., Zhao, F., and Chistoserdova, L. (2018). Physiological effect of XoxG(4) on lanthanide-dependent methanotrophy. MBio *9*, e02430–17.

Diversity of Methylotrophy Pathways in the Genus *Paracoccus* (*Alphaproteobacteria*)

6

Jakub Czarnecki[1,2]* and Dariusz Bartosik[1]

[1]Department of Bacterial Genetics, Institute of Microbiology, Faculty of Biology, University of Warsaw, Warsaw, Poland.

[2]Bacterial Genome Plasticity, Department of Genomes and Genetics, Institut Pasteur, Paris, France.

*Correspondence: jczarnecki@biol.uw.edu.pl

https://doi.org/10.21775/9781912530045.06

Abstract

Paracoccus denitrificans Pd 1222 is a model methylotrophic bacterium. Its methylotrophy is based on autotrophic growth (enabled by the Calvin cycle) supported by energy from the oxidation of methanol or methylamine. The growing availability of genome sequence data has made it possible to investigate methylotrophy in other *Paracoccus* species. The examination of a large number of *Paracoccus* spp. genomes reveals great variability in C_1 metabolism, which have been shaped by different evolutionary mechanisms. Surprisingly, the methylotrophy schemes of many *Paracoccus* strains appear to have quite different genetic and biochemical bases. Besides the expected 'autotrophic methylotrophs', many strains of this genus possess another C_1 assimilatory pathway, the serine cycle, which seems to have at least three independent origins. Analysis of the co-occurrence of different methylotrophic pathways indicates, on the one hand, evolutionary linkage between the Calvin cycle and the serine cycle, and, on the other hand, that genes encoding some C_1 substrate-oxidizing enzymes occur more frequently in association with one or the other. This suggests that some genetic module combinations form more harmonious enzymatic sets, which act with greater efficiency in the methylotrophic process and thus undergo positive selection.

Introduction

The genus *Paracoccus* (class *Alphaproteobacteria*, order *Rhodobacterales*, family *Rhodobacteraceae*) currently includes around 70 defined species and hundreds of strains whose taxonomic position has yet to be precisely assigned (NCBI Taxonomy, 10 January 2018). Representatives of this genus have been identified in diverse environments. The original isolate and the type species, *P. denitrificans*, was isolated from soil (Beijerinck, 1910), like many other *Paracoccus* spp. (Urakami *et al.*, 1990; Siller *et al.*, 1996; Tsubokura *et al.*, 1999). Numerous strains have been isolated from fresh water (Sheu *et al.*, 2018), seawater (Kim and Lee, 2015), sediments (G. Zhang *et al.*, 2016), activated sludge (Liu *et al.*, 2006), biofilters (Lipski *et al.*, 1998), or from environments linked to higher organisms, such as plant roots (rhizosphere) (Doronina *et al.*, 2002), marine bryzoans (Pukall *et al.*, 2003), insects (S. Zhang *et al.*, 2016), and human tissues (opportunistic pathogens *P. yeei* and *P. sanguinis*) (Funke *et al.*, 2004; McGinnis *et al.*, 2015).

The ubiquity of *Paracoccus* spp. is due to their great metabolic diversity and flexibility. All paracocci have an aerobic respiratory metabolism and utilize multi-carbon compounds. However, many of them can switch between different growth modes, using different carbon and energy sources, and employing various final electron acceptors. In the absence of oxygen, some *Paracoccus* spp. conduct nitrate respiration (Baker *et al.*, 1998; Kelly *et al.*, 2006). This process leads to denitrification and has been applied for the removal of nitrates from wastewater (Liu *et al.*, 2012). In addition to 'standard' carbon sources, like sugars, amino acids and succinate, *Paracoccus* strains isolated from polluted environments can utilize xenobiotics, e.g. polycyclic aromatic hydrocarbons (PAHs), making them useful in bioremediation (Sun *et al.*, 2013). Numerous representatives of the genus can grow chemolithoautotrophically, coupling CO_2 assimilation with the oxidation of inorganic compounds or elements, such as thiosulfate, thiocyanate, elemental sulfur, molecular hydrogen, or ferrous ions (Kelly *et al.*, 2006). Finally, many *Paracoccus* spp. are methylotrophs. Most utilize methanol (MeOH) and methylamine (MA) as sole carbon and energy sources, e.g. *P. denitrificans* and closely related *P. versutus* and *P. kondratievae* (Kelly *et al.*, 2006). However, other strains isolated from environments polluted with C1 compounds are able to metabolize dimethylamine (DMA), trimethylamine (TMA), *N,N*-dimethylformamide (DMF) (Urakami *et al.*, 1990; Kim *et al.*, 2001; Sanjeevkumar *et al.*, 2013) or dichloromethane (Doronina *et al.*, 1998).

The purpose of this study is to describe the diversity of methylotrophy in the genus *Paracoccus* at the biochemical and genetic levels, including an examination of the origin and evolution of C1 metabolism in this group of bacteria.

Paracoccus denitrificans Pd 1222 as an example of an 'autotrophic' methylotroph

Since its isolation at the beginning of the 20th century (Beijerinck, 1910), *P. denitrificans* has been extensively studied and different aspects of its energy metabolism have been revealed. The composition of its core respiratory chain closely resembles that of the classic mitochondrial respiratory chain (unlike respiratory chains of many other bacteria, including *E. coli*), which made it a valuable model for studies on the energetic processes in eukaryotes (John and Whatley, 1975). However, the electron transport chain of *P. denitrificans* also has many branches at both the entrance and exit sides of the core. On the one hand, this allows the bacterium to utilize alternative final electron receptors, namely nitrate, nitrite, nitric oxide and nitrous oxide (which leads to denitrification – *P. denitrificans* is an important model in studies on this process (Baker *et al.*, 1998), thus permitting growth when oxygen is limited. On the other hand, different electron donors may be used. As a consequence, *P. denitrificans* has the ability to grow chemolithoautotrophically on inorganic energy sources such as hydrogen and thiosulphate (Friedrich and Mitrenga, 1981). Its autotrophic growth may also be supported by the oxidation of some organic C1 compounds, namely MeOH, MA and formate (Baker *et al.*, 1998). These compounds are oxidized to CO_2, the released electrons are used for oxidative phosphorylation, and the ATP and CO_2 produced are used in the Calvin cycle for biomass production. Thus, *P. denitrificans* is an example of an 'autotrophic methylotroph', which lacks a 'heterotrophic' pathway dedicated to the assimilation of reduced C_1 units (such as the serine cycle or ribulose monophosphate pathway), but it can assimilate carbon from C_1 compounds after their total oxidation (Baker *et al.*, 1998; Chistoserdova, 2011) (Fig. 6.1).

Since the 1970s, numerous studies have sought to understand the details of C_1 metabolism in *P. denitrificans* (Harms *et al.*, 1985; Baker *et al.*, 1998), especially in strain Pd 1222, which is readily transformed by conjugation to enable genetic manipulation (Devries *et al.*, 1989). The results of these studies have uncovered the properties of many *P. denitrificans* proteins involved in methylotrophy (mainly MeOH and MA dehydrogenases, as well as associated proteins, i.e. those involved in the transfer of electrons from the dehydrogenases to the respiratory chain), and have shed light on the regulation of their expression (Baker *et al.*, 1998).

The whole genome sequence of *P. denitrificans* Pd 1222 was obtained in 2006 (NCBI Genomes). It has an unusual structure consisting of two chromosomes (chromosome 1, 2.8 Mb, and chromosome 2, 1.7 Mb) and one large plasmid (plasmid 1 1650 kb). The availability of this sequence has permitted elucidation of the genetic basis of its methylotrophy. *P. denitrificans* Pd 1222 carries several gene clusters responsible for C_1 metabolism, dispersed across the three replicons. Genes for the enzymes involved in the oxidation of primary C_1 substrates to formaldehyde are located on chromosome 2 (gene cluster encoding MxaFI-type MeOH dehydrogenase and associated proteins) and

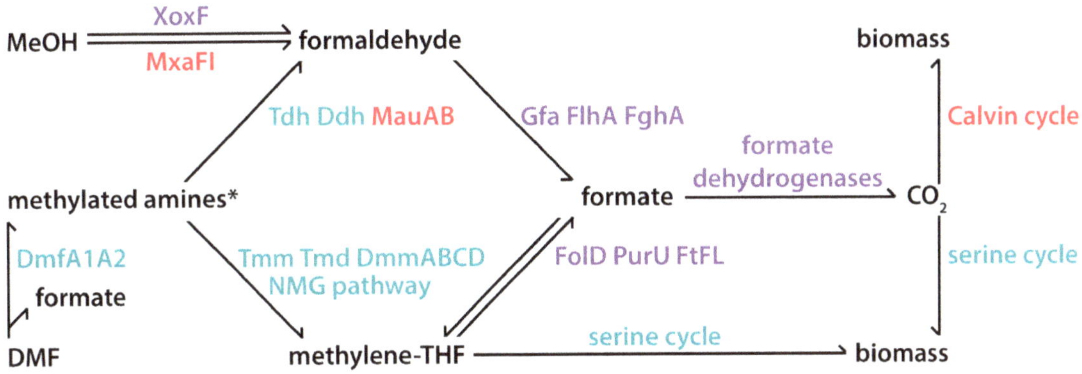

Figure 6.1 Summary of the methylotrophic pathways of *Paracoccus* spp. *P. denitrificans* Pd 1222 and *P. aminovorans* JCM 7685 are used as examples because these strains possess all of the methylotrophic pathways discussed in this study. The enzymes and pathways present in *P. denitrificans* Pd 1222 are shown in red, those present in *P. aminovorans* JCM 7685 are shown in blue, and those present in both strains are shown in violet. Ddh, DMA dehydrogenase; DmfA1A2, DMFase; DmmABCD, DMA monooxygenase; FghA, *S*-formylglutathione hydrolase; FlhA, *S*-(hydroxymethyl)glutathione dehydrogenase; FolD, methylenetetrahydrofolate dehydrogenase (NADP+)/methenyltetrahydrofolate cyclohydrolase; FtfL, formate-tetrahydrofolate ligase; Gfa, *S*-(hydroxymethyl) glutathione synthase; MauAB, MA dehydrogenase; MxaFI, MxaFI-type MeOH dehydrogenase; PurU, formyltetrahydrofolate deformylase; Tdh, TMA dehydrogenase; Tmd, TMA *N*-oxide demethylase; Tmm, TMA monooxygenase; XoxF, XoxF-type MeOH dehydrogenase. *Methylated amines – TMA, DMA and MA.

plasmid 1 (the *mau* genes encoding small and large subunits of MA dehydrogenase and associated proteins). The second step in the methylotrophy of *P. denitrificans* Pd 1222 is oxidation of formaldehyde to formate in the glutathione-dependent pathway, which is essential for growth of this strain on C_1 compounds (Harms *et al.*, 1996). Three enzymes of this pathway, *S*-(hydroxymethyl)glutathione synthase (Gfa), *S*-(hydroxymethyl)glutathione dehydrogenase (FlhA), and *S*-formylglutathione hydrolase (FghA), are encoded within chromosome 1. Interestingly, these glutathione-dependent formaldehyde oxidation genes occur in the immediate vicinity of genes encoding XoxF-type MeOH dehydrogenase and associated proteins. XoxF was recently confirmed as a MeOH-oxidizing enzyme (Keltjens *et al.*, 2014; Chistoserdova, 2016). However, its involvement in MeOH metabolism in *P. denitrificans* was suggested many years before (Harms *et al.*, 1996), although its redundancy with a MxaFI-type system remains unexplained. Formate is oxidized to CO_2 by two multi subunit (encoded in chromosomes 1 and 2) or one single subunit (encoded in chromosome 1) formate dehydrogenase. Finally, the Calvin cycle gene cluster, which is required for assimilation of CO_2, is located on chromosome 1 and consists of genes encoding two subunits of RuBisCO (*rbcL* and *rbcS*), as well as genes for fructose-1,6-bisphosphatase (*fbp*), phosphoribulokinase (*prk*), transketolase (*tkt*), fructose-1,6-bisphosphate aldolase (*fba*), RuBisCO activating protein (*cbbX*), and the Calvin cycle regulator (*cbbR*).

Parallel studies on the methylotrophy of the closely related *P. versutus* have shown that this species utilizes similar routes of C1 metabolism. The assimilatory pathway required for growth on C1 substrates includes highly similar MA dehydrogenase and Calvin cycle enzymes (Karagouni and Kelly, 1989; Baker *et al.*, 1998).

Paracoccus aminophilus JCM 7686 and *Paracoccus aminovorans* JCM 7685 as serine cycle methylotrophs specialized in DMF utilization

In 1990, the isolation of two DMF-degrading strains from a sample of DMF-polluted soil in Japan was reported (Urakami *et al.*, 1990). These strains, designated JCM 7686 and JCM 7685 were recognized as representatives of two new *Paracoccus* species: *P. aminophilus* and *P. aminovorans* (Urakami *et al.*, 1990). The methylotrophic pathways of these isolates were shown to be more complex than those of *P. denitrificans*, because they include enzymes required for the utilization of a wider range of C1 compounds. Both strains are able to decompose DMF to formate and DMA, and oxidize TMA via trimethylamine *N*-oxide (TMAO) to DMA, and then degrade the resultant DMA to MA (Urakami *et al.*, 1990).

The entire genome sequences of *P. aminophilus* JCM 7686 and *P. aminovorans* JCM 7685 were obtained by our group (Dziewit *et al.*, 2014; Czarnecki *et al.*, 2017), facilitating the reconstruction of their methylotrophic pathways. As expected, genes required for the metabolism of additional C_1 substrates were identified in both strains. These encode small and large subunits of DMFase (DmfA1A2), as well as TMA monooxygenase (Tmm), TMAO demethylase (Tmd), and multi-subunit DMA monooxygenase (DmmABCD). Besides TMA and DMA monooxygenases, the *P. aminovorans* JCM 7685 genome also encodes TMA and DMA dehydrogenases (Tdh and Ddh, respectively), which may catalyse the oxidation of TMA and DMA (Fig. 6.1). The role of these genes in the metabolism of specific C1 substrates has been confirmed in both strains (Dziewit *et al.*, 2010, 2015; Czarnecki *et al.*, 2017).

Surprisingly, the enzymes involved in the metabolism of MeOH and MA by *P. aminophilus* and *P. aminovorans* differ from those employed by *P. denitrificans*. In the case of MeOH utilization, *P. aminophilus* and *P. aminovorans* do not possess a MxaFI-type MeOH dehydrogenase, and their growth on this compound relies fully on a XoxF-type dehydrogenase, as has been confirmed by mutational analysis (Dziewit *et al.*, 2015). In the case of MA, both strains have genes for an alternative MA oxidation pathway: the *N*-methylglutamate (NMG) pathway. In *P. aminovorans* the NMG pathway is the only pathway for MA oxidation, while in *P. aminophilus* it co-exists with the MA dehydrogenase pathway, which was previously characterized in *P. denitrificans* (Fig. 6.1). Furthermore, both strains lack RuBisCO genes, so cannot assimilate CO_2. Thus, their methylotrophy has to be supported by another pathway of C1 unit assimilation. A serine cycle gene cluster was found in both genomes; its involvement in methylotrophy has been confirmed in *P. aminovorans*, and the role of transcriptional regulator ScyR in its regulation was revealed (Czarnecki *et al.*, 2017). The serine cycle requires glyoxylate regeneration, which is accomplished by the action of the ethylmalonyl-CoA pathway in *P. aminovorans* JCM 7685 (Czarnecki *et al.*, 2017). All genes required for this pathway are also present in *P. aminophilus*, in the non-serine cycle methylotroph *P. denitrificans* and even in non-methylotrophic strains, as they are used for other purposes, such as growth on C2 compounds (Schneider *et al.*, 2012). The genes required for another glyoxylate-regenerating process, the glyoxylate shunt, are found in *P. aminophilus* and *P. aminovorans*, but, as they were unable to support methylotrophic growth in a strain with a blocked ethylmalonyl-CoA pathway (Czarnecki *et al.*, 2017), their role remains unclear.

Owing to the presence of different enzymes oxidizing primary C1 substrates and the serine cycle, the fate of C1 units released during methylotrophic metabolism is more complex in *P. aminophilus* and *P. aminovorans* than in *P. denitrificans*. In the serine cycle, carbon is assimilated in the form of methylene group bound to tetrahydrofolate (THF) and CO_2 (Fig. 6.1). Methylene-THF may be delivered directly by the C1-substrate-oxidizing enzymes TMAO demethylase and DMA monooxygenase or the NMG pathway, which do not release free formaldehyde, but transfer the C1 unit directly to THF (Fig. 6.1). On the other hand, the XoxF-type MeOH dehydrogenase, TMA dehydrogenase, DMA dehydrogenase and MA dehydrogenase release free formaldehyde, which has to be oxidized to formate by the glutathione-dependent pathway present in *P. aminophilus*, *P. aminovorans* and *P. denitrificans*. To feed the serine cycle, the formate has to be bound to THF, and then it has to be reduced to a methylene group in an energy-requiring process. The THF-dependent formate reduction pathway is found in all three *Paracoccus* spp., and requires the action of two enzymes: formate-THF ligase (FtfL) and 5,10-methylene-tetrahydrofolate dehydrogenase/methenyl-tetrahydrofolate cyclohydrolase (FolD). This pathway may also act in the opposite direction to oxidize the methylene group to formate. In this case, FolD promotes the reverse reaction itself, while the second reaction is catalysed by formyltetrahydrofolate deformylase (PurU), which is also present in all three species (Fig. 6.1). A special situation occurs during growth on DMF, where DMFase releases a C1 unit directly in the form of formate and the second product is DMA. The fate of C1 units has been analysed experimentally in *P. aminovorans*, where growth of a strain lacking the glutathione-dependent pathway for formaldehyde oxidation on different C1 substrates was examined. As expected, this strain was unable to utilize MeOH (it has the XoxF-type MeOH dehydrogenase releasing free formaldehyde, which cannot be further oxidized without the glutathione-dependent pathway), but no effect on growth on MA was detected (it has the NMG pathway which produces methylene-THF, which directly enters the serine cycle or is oxidized in the THF-dependent pathway, without involvement of the glutathione-dependent pathway). Intermediate phenotypes were observed during growth on DMF, TMA or DMA, since monooxygenases can oxidize some portion of these compounds to produce methylene-THF, while another portion is oxidized by dehydrogenases to produce free formaldehyde (Czarnecki *et al.*, 2017) (Fig. 6.1). Like *P. denitrificans*, *P. aminophilus* and *P. aminovorans* have formate dehydrogenases to deal with an excess of formate (Dziewit *et al.*, 2015; Czarnecki *et al.*, 2017).

The genomic localization and clustering of methylotrophy genes of *P aminophilus* and *P. aminovorans* suggests that many of them could have been acquired horizontally to confer increased fitness for growth in DMF-polluted soil. Some DMFase genes of *P. aminophilus* are located on the small plasmid pAMI2 (18.6 kb), which also carries a genetic module putatively involved in its mobilization for conjugal transfer (Dziewit *et al.*, 2010). A closely related DMFase is encoded in the chromosomes of both *P. aminovorans* and *P. aminophilus* (84% aa identity of large subunits and 73% aa identity of small subunits). However, these chromosomal DMFase genes and their adjacent transcriptional regulator genes are surrounded by genes encoding transposases and other proteins typical of mobile genetic elements, which indicates their recent acquisition. Similarly, the TMA and DMA dehydrogenase genes of *P. aminovorans* JCM 7685, that are not found in any other *Paracoccus* strain (Czarnecki *et al.*, 2017), are located on the large extrachromosomal replicon pAMV3 (740 kb), which, like other replicons of this type, seems to be a reservoir of various horizontally transmitted genes.

A notable example of methylotrophy genes acquired by HGT are those clustered within a 40-kb methylotrophy island (MEI) located on the extrachromosomal replicon pAMV1 of *P. aminovorans* (Fig. 6.2). The genes present on this island are involved in all steps of methylotrophy: (i) oxidation of primary C1 substrates with methylene-THF generation (TMA monooxygenase, TMAO demethylase, DMA monooxygenase, the NMG pathway), (ii) oxidation of methylene-THF to formate (FolD and PurU), and assimilation of C1 units in the form of methylene-THF and CO_2 (the serine cycle). The closest homologue of this MEI was identified in *Paracoccus* sp. N5 (Beck *et al.*, 2015; Dziewit *et al.*, 2015; Czarnecki *et al.*, 2017), and similarly clustered genes are located in the chromosomes of many bacteria of the *Roseobacter* clade, including *Ruegeria pomeroyi* (Dziewit *et al.*, 2015). The MEI genes are also present in the *P. aminophilus* genome. However, in this case the island is divided in two, with one part encoding the serine

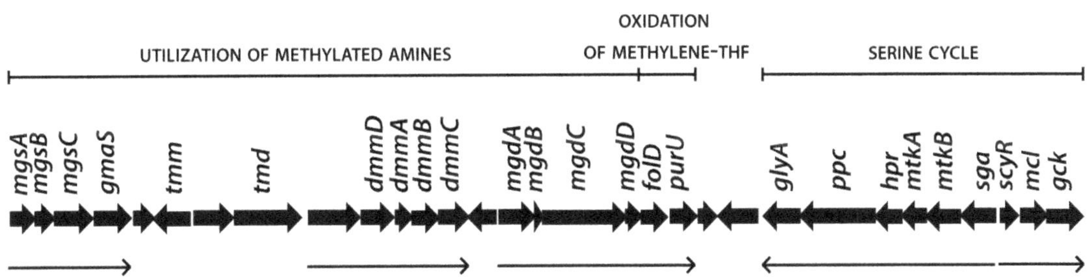

Figure 6.2 Genetic organization of the methylotrophy island (MEI) of *P. aminovorans* JCM 7685. The general functions of genes are indicated above their names. Potential operons are indicated by thin arrows. *dmmABCD*, DMA monooxygenase; *folD*, methylenetetrahydrofolate dehydrogenase (NADP+)/methenyltetrahydrofolate cyclohydrolase; *gck*, glycerate 2-kinase; *glyA*, serine hydroxymethyltransferase; *gmaS*, glutamate-methylamine ligase; *hpr*, hydroxypyruvate reductase; *mcl*, malyl-CoA lyase; *mgdABCD*, *N*-methylglut amate dehydrogenase; *mgsABCD*, *N*-methylglutamate synthase; *mtkAB*, malate-CoA ligase; *scyR*, serine cycle transcriptional regulator; *sga*, serine-glyoxylate aminotransferase; *tmd*, TMA N-oxide demethylase; *tmm*, TMA monooxygenase; *ppc*, phosphoenolpyruvate carboxylase; *purU*, formyltetrahydrofolate deformylase.

cycle enzymes located in the chromosome, and the second part, comprising the rest of the MEI, in extrachromosomal replicon pAMI6 (207 kb) (Dziewit *et al.*, 2015), whose genetic load is 40% homologous to that of *P. aminovorans* pAMV1. It appears that acquisition of the MEI (most probably from representatives of the *Roseobacter* clade (Dziewit *et al.*, 2015) could have enhanced the methylotrophic ability of *P. aminophilus* and *P. aminovorans*, which made them better adapted to living in DMF-polluted soil. To fully reconstruct the evolution of C1 pathways in the genus *Paracoccus*, deeper analyses are required, which consider not only the presence of given pathways (which has already been done (Dziewit *et al.*, 2015; Czarnecki *et al.*, 2017), but also their phylogenetic relationships.

Occurrence and phylogenetic relationships of methylotrophy genes in *Paracoccus* spp.

We previously examined 44 *Paracoccus* spp. genomes to determine the diversity of methylotrophy genes in this genus (Dziewit *et al.*, 2015; Czarnecki *et al.*, 2017). In this study we have broadened this analysis using additional recently deposited *Paracoccus* spp. sequences (in total, 62 *Paracoccus* genomes were available in NCBI GenBank on 11 May 2018) plus 10 *Paracoccus* spp. genomes obtained by our group, which will be fully described in a forthcoming publication (*P. bengalensis* DSM 17099, *P. ferrooxidans* NCCB 1300066, *P. haundaensis* LGM P-21903, *P. kondratievae* NCIBM 13773, *P. pantotrophus* DSM 11072, *P. solventivorans* DSM 11592, *P. sulfuroxidans* JCM 14013, *P. thiocyanatus* JCM 20756, *P. versutus* UW1 and *P. yeei* CCUG 32053). The majority of the studied *Paracoccus* strains (almost 65%), contain sets of genetic modules involved in all three steps of methylotrophy (Chistoserdova, 2011), which are potentially sufficient for methylotrophic growth.

Comparative analysis of this large number of *Paracoccus* genomes allowed us to distinguish two groups of methylotrophy-related genes based on their degree of conservation: (i) highly conserved, vertically transmitted genes, present in almost all of the genomes, and (ii) genes present only in some strains because of selective loss, horizontal gene transfer, or a combination of these two evolutionary mechanisms. The genes of the first group are usually located on chromosomes, while many of the genes of the second group are found on extrachromosomal replicons, or in the company of mobile genetic elements when present on chromosomes.

The first group includes genes comprising fundamental pathways, that are important for non-methylotrophic metabolism, to which further C1 pathways are appended. The first example of this group is a cluster consisting of genes for glutathione-dependent formaldehyde oxidation (*gfa, flhA, fghA*) and for MeOH oxidation by XoxF (Fig. 6.3). The compact and conserved nature of the gene cluster comprising both pathways suggests their cooperation. As mentioned above, both the XoxF and the glutathione-dependent pathways may be essential for some methylotrophic processes in *Paracoccus* spp. (Ras *et al.*, 1995; Harms *et al.*, 1996). However, the ubiquity of this gene cluster in non-methylotrophs indicates that its significance is more general. Its probable function in non-methylotrophic organisms is in the utilization of MeOH as an additional source of energy, without its assimilation into biomass. Despite its overall conservation, the cluster is truncated in some *Paracoccus* spp. and lacks the *xox* genes. For example, *P. halophilus* JCM 14014 does not have any homologues of *xoxF*. In *P. alcaliphilus* JCM 7364 the *xox* genes and the genes for the glutathione-dependent pathway are separated, being located on a large (430 kb) extrachromosomal replicon and on the chromosome, respectively.

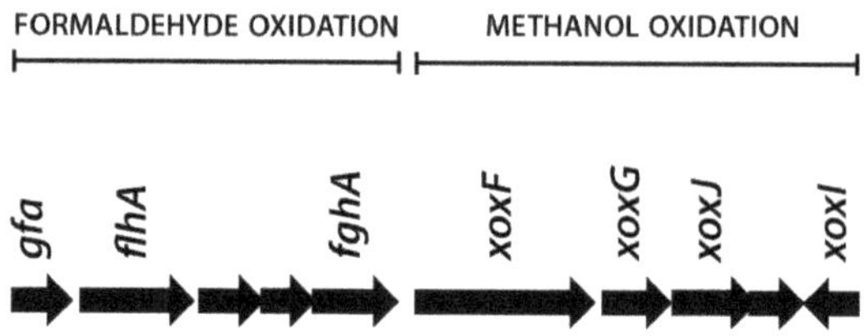

Figure 6.3 Genetic organization of the gene cluster for the glutathione-dependent formaldehyde oxidation pathway and the XoxF-type methanol dehydrogenase of *P. denitrificans* Pd 1222. The general functions of genes are indicated above their names. *fghA*, *S*-formylglutathione hydrolase; *flhA*, *S*-(hydroxymethyl)glutathione dehydrogenase; *gfa*, *S*-(hydroxymethyl)glutathione synthase; *xoxF*, XoxF-type MeOH dehydrogenase; *xoxG*, c-type cytochrome; *xoxI*, SRPBCC family protein; *xoxJ*, quinoprotein dehydrogenase-associated putative ABC transporter substrate-binding protein.

Another example of genes of the conserved group are those encoding the numerous enzymes of the ethylmalonyl-CoA pathway. These do not form a single gene cluster but are scattered throughout *Paracoccus* genomes. Enzymes of the ethylmalonyl-CoA pathway are involved in various processes, such as the utilization of C2 compounds as carbon and energy sources or synthesis of polyhydroxyalkanoates. As mentioned above, in some *Paracoccus* spp., the ethylmalonyl-CoA pathway is responsible for glyoxylate regeneration, which is crucial for functioning of the serine cycle during methylotrophic growth (Chistoserdova, 2011). A few *Paracoccus* spp. lack some enzymes of this pathway, e.g. *P. chinensis* CGMCC 1.7655 or strains of *P. sanguinis*. Nevertheless, a complete set of ethylmalonyl-CoA pathway genes was found in all strains that employ the serine cycle.

The next examples of conserved methylotrophy-linked genes in *Paracoccus* spp. are those involved in THF-dependent transformations of C1 units: *folD*, *purU* and *ftfL*. The encoded enzymes are important for the generation of C1-THF intermediates required by biosynthetic pathways, such as purine synthesis. As in the serine cycle C1 units are incorporated into biomass in the form of methylene-THF, the THF-dependent pathway constitutes a central metabolic process in *Paracoccus* serine cycle methylotrophs (Fig. 6.1). It should be noted that besides the conserved set of *folD*, *purU* and *ftfL* genes, there are also numerous homologues that are horizontally transferred, for example in the company of serine cycle genes or NMG pathway genes. It is possible that the serine cycle-associated and the NMG pathway-associated homologues are better adapted to co-operate with the methylotrophic pathways, while the conserved homologues are mainly responsible for anabolic housekeeping functions, but members of these groups are likely to be interchangeable to some extent.

The last representatives of the group of conserved genes are gene clusters encoding multi-subunit formate dehydrogenases, which are responsible for formate detoxification in both methylotrophs and non-methylotrophs, and form part of the core genome of *Paracoccus* spp.

The second group of genes, which only occurs in some *Paracoccus* strains, includes gene sets involved in C1 unit assimilation pathways, the Calvin cycle and the serine cycle, as well as genes involved in primary oxidation of C1 substrates. The Calvin cycle is widespread among *Paracoccus* species (Fig. 6.4) and a complete set of genes required for this process was identified in 56% of analysed strains. Phylogenetic analysis of the RuBisCO large subunit (RbcL) and phosphoribulokinase (Prk) indicates that there are at least two evolutionarily distinct lineages of Calvin cycle genes in this genus (Figs. 6.5–6.7). This is reflected in variations in the organization of Calvin cycle gene clusters in different *Paracoccus* spp. (Fig. 6.5). The Calvin cycle genes are generally found in chromosomes, with some exceptions, e.g. in *Paracoccus* sp. N5 and *P. kondratievae* they are located on large extrachromosomal replicons (1 Mb and 423 kb, respectively), most probably as a result of translocation from the chromosome.

Compared with the Calvin cycle, the serine cycle is less prevalent in *Paracoccus* spp. and its occurrence was predicted in 17% of the analysed strains. Serine cycle gene sets are present in a few different lineages of the genus (Fig. 6.4), and three types of gene organization were identified (Fig. 6.5). The occurrence of three gene organization schemes is probably an effect of the independent acquisition of the serine cycle from different sources. These separate events are reflected in the phylogenetic tree of phosphoenolpyruvate carboxylase (Ppc), the key enzyme of the cycle (Fig. 6.8). The first type of serine cycle gene organization occurs in the MEI of *P. aminovorans* JCM 7685 and, as mentioned above, is also found in many representatives of the *Roseobacter* clade, in the order *Rhodobacterales*, as are *Paracoccus* spp. The second type of serine cycle gene organization, found on the chromosome of *P. zhejiangensis* J6, for example, is also most similar to arrangements present in members of the *Roseobacter* clade. The third type of serine cycle gene organization, found in *P. denitrificans* ISTOD1, is typical for strains of the *Aminobacter* and *Labrys* genera (Beck *et al.*, 2015), within the order *Rhizobiales*. In some *Paracoccus* spp., two types of serine cycle module coexist in the same genome (Fig. 6.4). The serine cycle clusters may be located within extrachromosomal replicons, as in pAMV1 in *P. aminovorans* JCM 7685 and probably also ISTOD1 in *P. denitrificans* (the unfinished genome sequence of this strain does not allow confirmation of the genomic localization of this gene cluster, although it is present in a sequence scaffold containing a rep*ABC*

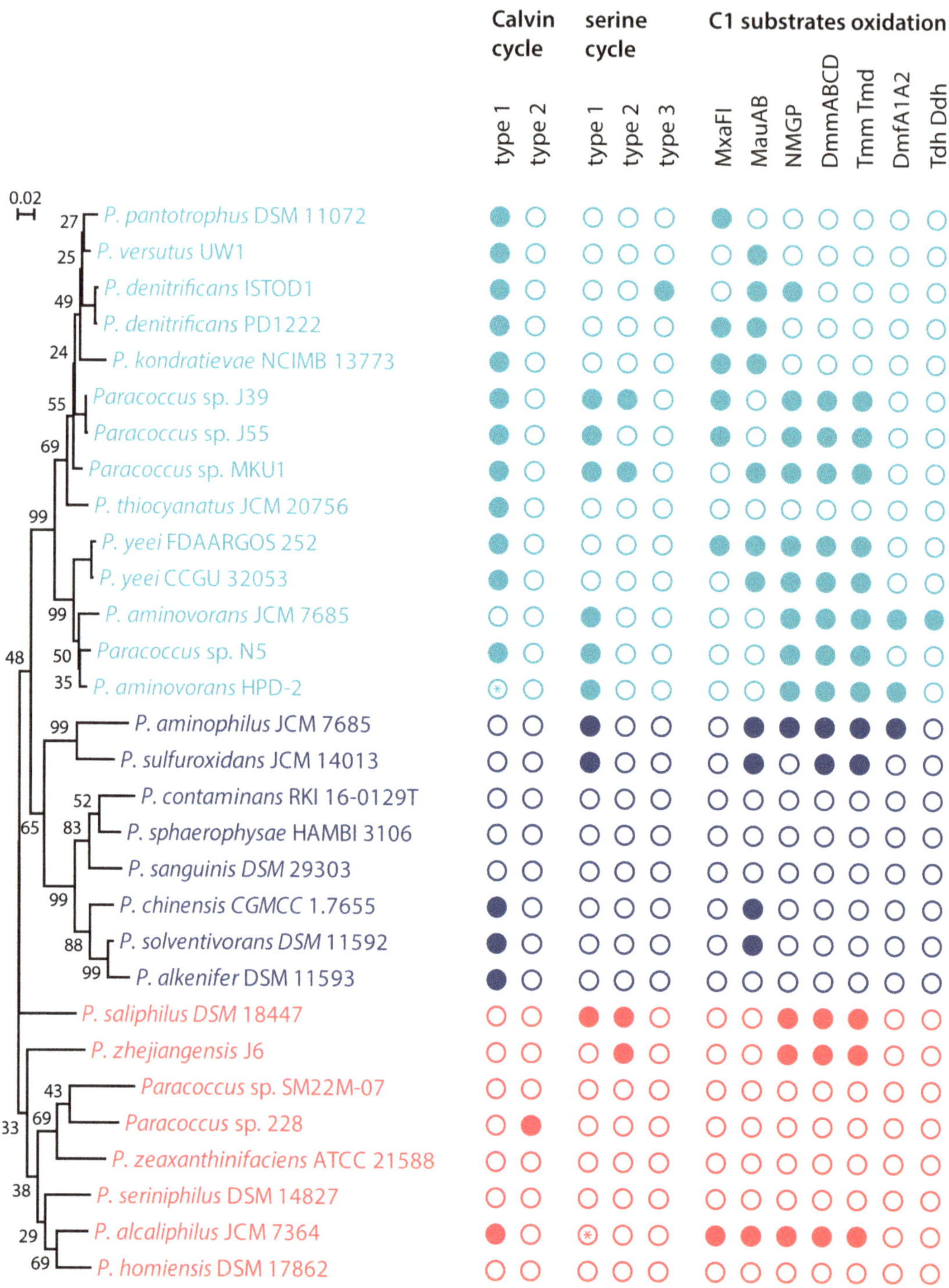

Figure 6.4 Occurrence of methylotrophy genes in *Paracoccus* spp. The maximum likelihood phylogenetic tree of *Paracoccus* spp. is based on the conserved gene for ethylmalonyl-CoA mutase (*ecm*). Where not indicated, the bootstrap support value is 100. Three sub-clades of *Paracoccus* spp., classified on the basis of 16S rDNA, *dnaA* (data not shown) and *ecm* comparisons, are indicated in light blue, dark blue and red. Ddh, DMA dehydrogenase; DmfA1A2, DMFase; DmmABCD, DMA monooxygenase; MauAB, MA dehydrogenase; MxaFI, MxaFI-type MeOH dehydrogenase; NMGP, *N*-methylglutamate pathway; Tdh, TMA dehydrogenase; Tmd, TMA *N*-oxide demethylase; Tmm, TMA monooxygenase. *Truncated gene clusters.

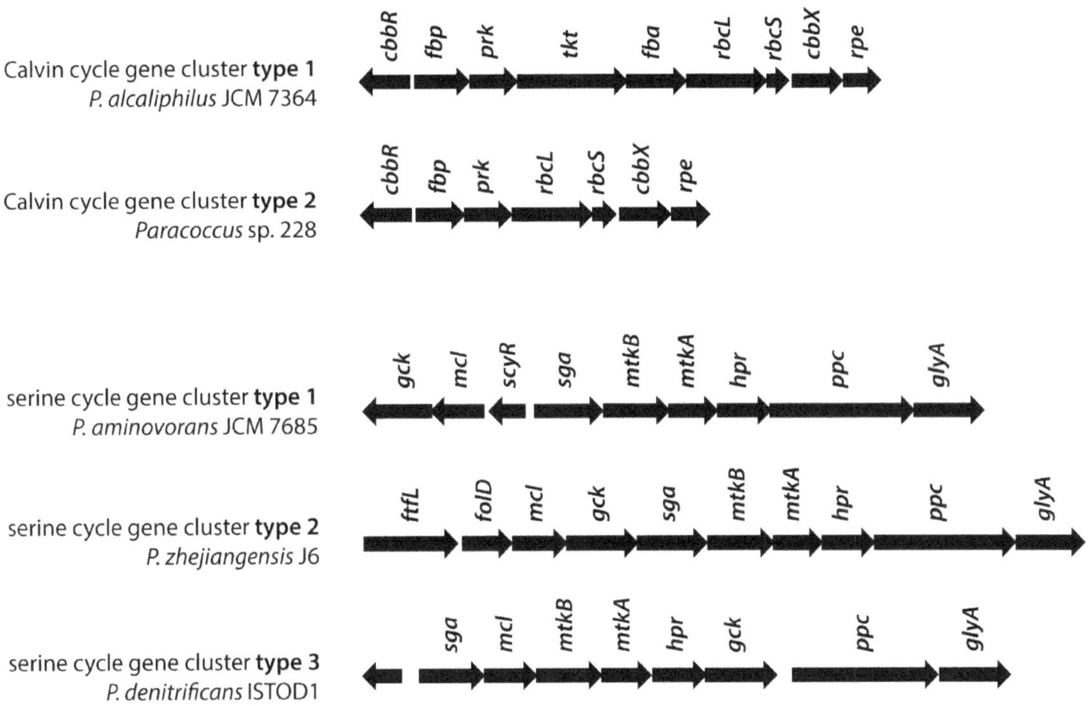

Figure 6.5 Organization of genetic modules for the Calvin cycle and the serine cycle in *Paracoccus* spp. The composition of the Calvin cycle gene cluster type 1 varies in some strains, e.g. it lacks the *rpe* gene in *P. denitrificans, P. pantotrophus, P. versutus, P. kondratievae* and *P. thiocyanatus*. It is also truncated in *Paracoccus* sp. N5 (lacks the *tkt*, *fba* and *rpe* genes) and in *P. alkenifer* DSM 11593 (lacks the *tkt* gene). In all these cases, homologues of the absent genes are present at other locations in the genome. An incomplete Calvin cycle gene cluster type 1 is found in *P. aminovorans* HPD-2 (lacks *rbcS*, pseudogenes of *rbcL* and *cbbX*). The serine cycle gene cluster type 1 occurs without the *gck* gene in *P. aminophilus* JCM 7686 and *P. sulfuroxidans* JCM 14013. In both cases, other homologues of *gck* are present in the genome. *cbbR*, Calvin cycle transcriptional regulator; *cbbX*, RuBisCO activating protein; *fba*, fructose-bisphosphate aldolase; *fbp*, fructose-bisphosphatase; *flfL*, formate-THF ligase; *folD*, 5,10-methylene-tetrahydrofolate dehydrogenase/methenyl-tetrahydrofolate cyclohydrolase (FolD); *gck*, glycerate 2-kinase; *glyA*, serine hydroxymethyltransferase; *hpr*, hydroxypyruvate reductase; *mcl*, malyl-CoA lyase; *mtkA*, malate-CoA ligase subunit alpha; *mtkB*, malate-CoA ligase subunit beta; *ppc*, phosphoenolpyruvate carboxylase; *prk*, phosphoribulokinase; *rbcL*, RuBisCO large subunit; *rbcS*, RuBisCO small subunit; *rpe*, ribulose-phosphate 3-epimerase; *scyR*, serine cycle transcriptional regulator; *sga*, serine-glyoxylate aminotransferase; *tkt*, transketolase.

replication-partitioning module, typical for large plasmids of *Alphaproteobacteria*). The serine cycle genes have a chromosomal location in *P. aminophilus* JCM 7685 or *P. sulfuroxidans* JCM 14013. The distribution of the serine cycle genes indicates that they were acquired horizontally several times in different lineages of the genus. Nevertheless, their prevalence could also have been shaped by gene loss. The best example is the incomplete serine cycle cluster of *P. alcaliphilus* JCM 7364, which lacks the *pcc* gene (there are no *pcc* homologues in the entire *P. alcaliphilus* genome).

One particularly interesting aspect is co-occurrence of the Calvin cycle and the serine cycle, and co-evolution of these two cycles in *Paracoccus* genomes. The phylogeny and distribution patterns of the key enzymes of these cycles (Figs 6.4, 6.6–6.8) indicate that the Calvin cycle is the more ancient in this group of bacteria, and that the serine cycle has been acquired independently in several lineages. In some strains, these two cycles coexist (Fig. 6.4), but in others, the Calvin cycle seems to have been lost after serine cycle acquisition. This phenomenon is perfectly illustrated by a group of three isolates: *Paracoccus* sp. N5, *P. aminovorans* HPD-2 and *P. aminovorans* JCM 7685. The first strain has the intact Calvin cycle gene cluster, the second one possesses an incomplete cluster lacking

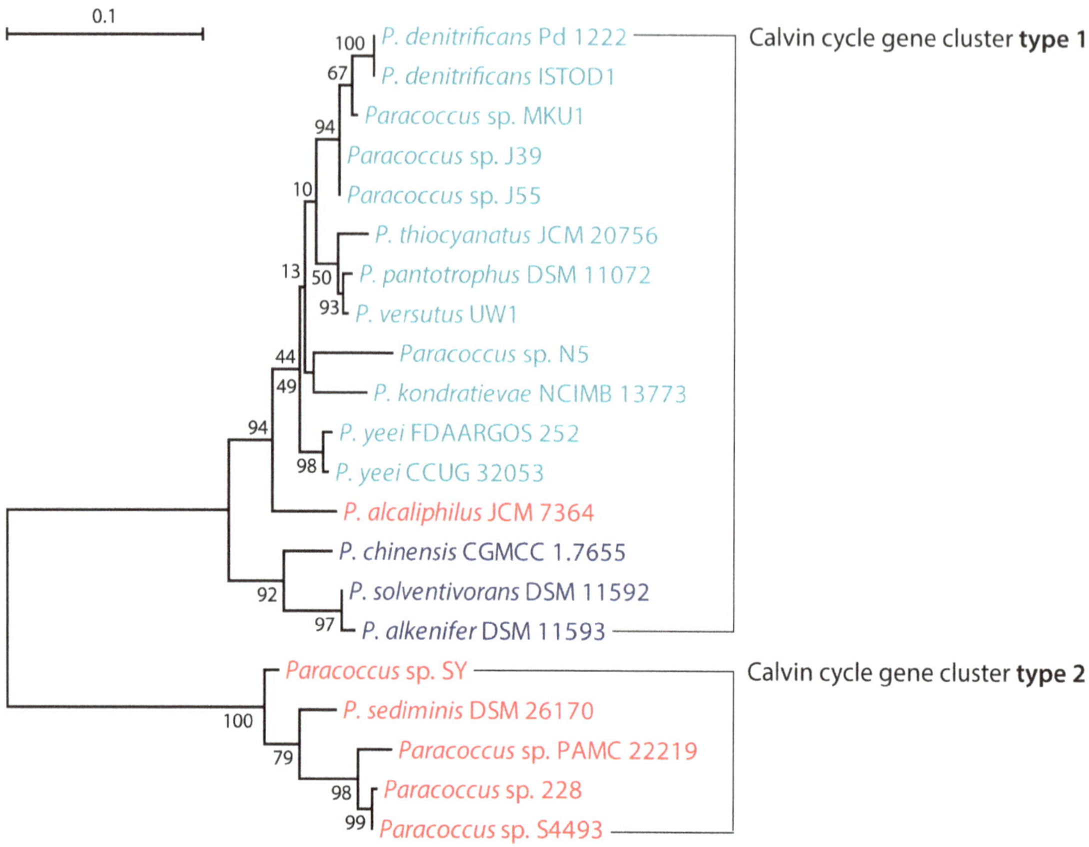

Figure 6.6 Maximum likelihood phylogenetic tree of RuBisCO large subunits (RbcL) of *Paracoccus* spp. Two phylogenetic groups matching different types of genetic organization of the Calvin cycle genes are indicated. Three sub-clades of *Paracoccus* spp. are coloured as in Fig. 6.4.

the *rbcS* gene but containing *rbcL* and *cbbX* pseudogenes, while the third strain does not possess the *rbcS, rbcL* nor *cbbX* genes (Fig. 6.4). However, not all of the Calvin cycle genes are lost in *Paracoccus* serine cycle methylotrophs lacking this cycle. These strains still have a truncated version of the Calvin cycle gene cluster, including its three first genes encoding the Calvin cycle regulator (CbbR), fructose-1,6-bisphosphatase (Fbp) and phosphoribulokinase (Prk). The phylogeny of the *Paracoccus* phosphoribulokinases (Fig. 6.7) indicates that the same reduction of the Calvin cycle gene cluster arose in different lineages where the serine cycle appeared. Moreover, retention of a reduced Calvin cycle gene cluster is typical only for the serine cycle methylotrophs, whereas the *cbbR, fbp* and *prk* genes are not present in *Paracoccus* non-methylotrophs. An identically organized gene cluster was detected in another serine cycle methylotroph, *Methylobacterium extorquens* PA1. This includes the gene encoding QscR, a transcriptional regulator with homology to CbbR, which is a global regulator of the serine cycle genes (Kalyuzhnaya and Lidstrom, 2003, 2005). Recently, QscR was shown to be regulated by phosphoribulokinase, and it was demonstrated that both the *qscR* and *prk* genes are essential for the methylotrophy of *M. extorquens* (Ochsner *et al.*, 2017). Parallel evolution of the same gene set in *Paracoccus* serine cycle methylotrophs (and *M. extorquens*) implies that there is a universal evolutionary linkage between the Calvin cycle and the serine cycle (Ochsner *et al.*, 2017). As the role of the *cbbR–fbp–prk* cluster is only regulatory, some strains may have evolved towards the loss of these genes. In the analysed strain set there is one example where partial loss of this regulatory cluster has occurred: *P. saliphilus* DSM 18447 has two different serine cycle gene sets, plus a remnant of the *cbbR–fbp–prk* cluster (it lacks *cbbR*, has a *fbp* pseudogene and an intact *prk* gene). However, the

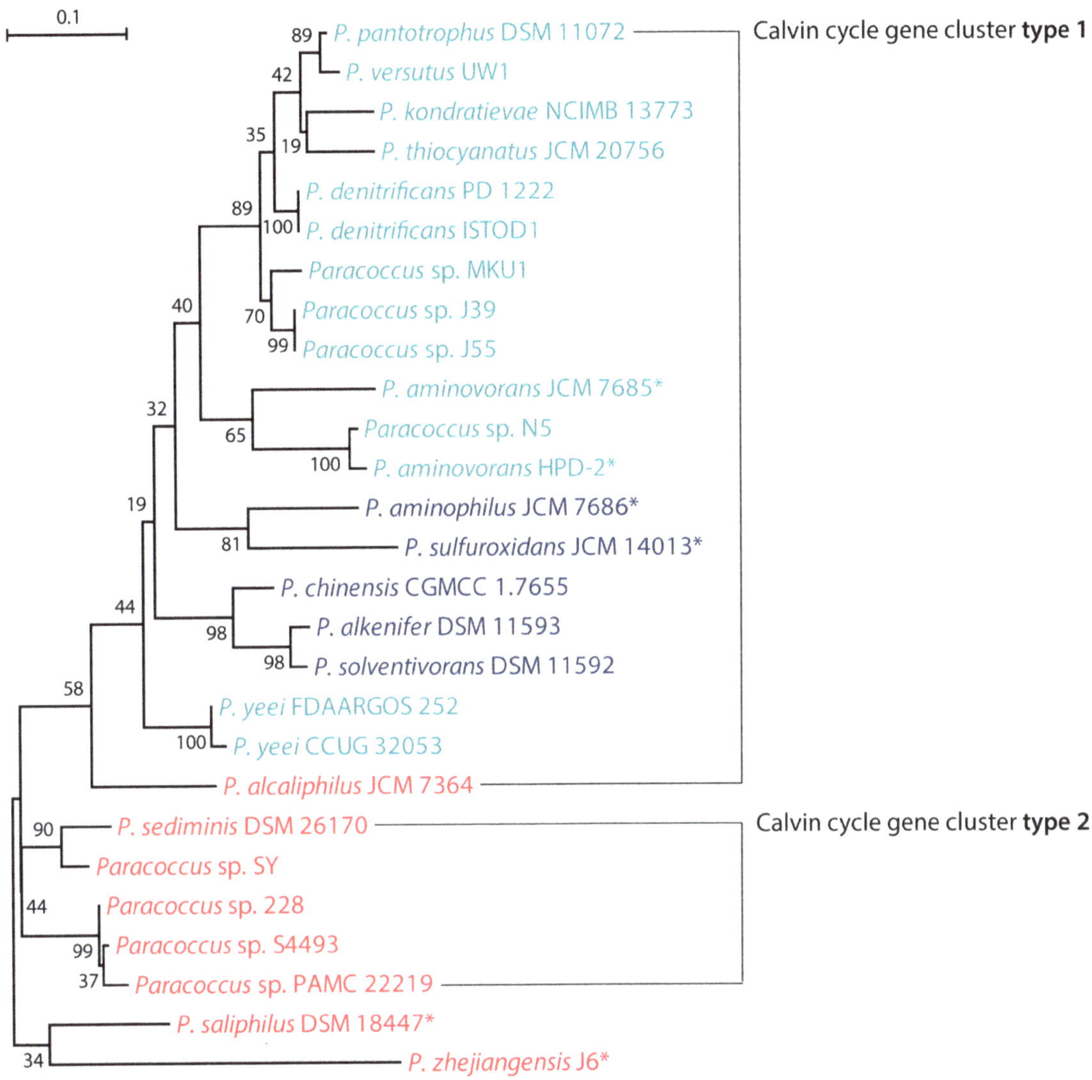

Figure 6.7 Maximum likelihood phylogenetic tree of phosphoribulokinases (Prk) of *Paracoccus* spp. Two phylogenetic groups matching different types of genetic organization of the Calvin cycle genes are indicated. Three sub-clades of *Paracoccus* spp. are coloured as in Fig. 6.4. *Strains with the serine cycle, which have truncated Calvin cycle gene clusters (and lack RuBisCO).

ability of this strain to grow methylotrophically has yet to be tested.

Among the genes involved in C1 metabolism, those encoding enzymes participating in the primary oxidation of C1 substrates, the first stage of methylotrophy, show the greatest variability in their occurrence. One prominent exception is the aforementioned conserved XoxF-type MeOH dehydrogenase. Conversely, a second type of MeOH dehydrogenase, the two subunit enzyme MxaFI, is found only in a few *Paracoccus* spp. (Fig. 6.4). A similarly limited distribution is observed in the case of MA dehydrogenase (Fig. 6.4). Both enzymes, which were thought to be reliable markers of methylotrophy in this genus based on studies on *P. denitrificans* Pd 1222, seem to be rather rare among *Paracoccus* spp. whose genomes have been sequenced. MxaFI-type MeOH dehydrogenase and MA dehydrogenase usually co-exist with the Calvin cycle (Fig. 6.4), which may result from adaptation of their mode of action to interact better with the Calvin cycle than with the serine cycle (release of the free formaldehyde, which is then oxidized to formate in the glutathione-dependent

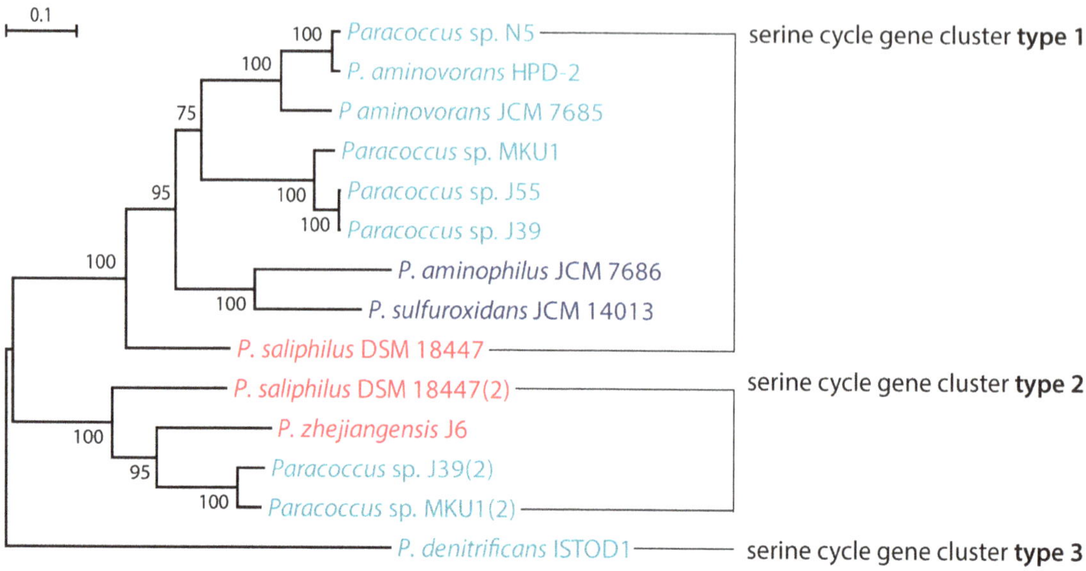

Figure 6.8 Maximum likelihood phylogenetic tree of phosphoenolpyruvate carboxylases (Ppc) of *Paracoccus* spp. Three phylogenetic groups matching different types of genetic organization of the serine cycle genes are indicated. Three sub-clades of *Paracoccus* spp. are coloured as in Fig. 6.4.

pathway, but not direct binding of the methylene group to THF). Genes of the NMG pathway for MA oxidation are mainly found in strains that possess the serine cycle (Fig. 6.4). The NMG pathway enzymes appear to be evolutionarily bound to the serine cycle enzymes, since their phylogenetic trees show a similar topology (Fig. 6.9). While the third type of serine cycle gene cluster (Fig. 6.4) is found in the company of only the NMG pathway genes, the first and the second types co-occur with other methylene-THF-producing enzymes, TMA monooxygenase, TMAO demethylase and DMA monooxygenase (Fig. 6.4), e.g. within the MEI of *P. aminovorans* JCM 7685 (Fig. 6.2) (Czarnecki *et al.*, 2017). The exception are strains of *P. yeei* which lack the serine cycle, but have the NMG pathway accompanied by DMA monooxygenase, TMA monooxygenase and TMA *N*-oxide demethylase genes (Fig. 6.4).

The DMFase genes *dmfA1A2* were only found in three strains, *P. aminophilus* JCM 7686, *P. aminovorans* JCM 7685 and *P. aminovorans* HPD-2, and always in association with serine cycle, NMG pathway, TMA monooxygenase, TMAO demethylase and DMA monooxygenase genes (Fig. 6.4). The methylotrophic potential of *Paracoccus* spp. does not seem to have been fully determined. Genes for TMA and DMA dehydrogenases were found only in *P. aminovorans* JCM 7685, but these enzymatic activities had already been identified in other *Paracoccus* isolates, and they may be important for the utilization of methylated amines when the oxygen concentration is changeable (Kim *et al.*, 2001, 2003). Moreover, one strain whose genome has yet to be sequenced, *P. methylutens* DM 12, expresses dichloromethane halogenase, which is required for dichloromethane utilization. Thus, further investigation of the content of *Paracoccus* spp. genomes may reveal interesting and unexpected features of their C_1 pathways.

Conclusions

From our detailed analysis of numerous *Paracoccus* spp. genome sequences, it is now clear that the archetypal autotrophic methylotroph, *P. denitrificans* Pd 1222, represents only a small part of the methylotrophic capacity present in the genus. Besides some generally conserved features, like the presence of the glutathione-dependent formaldehyde oxidation pathway and the ethylmalonyl-CoA pathway, methylotrophic *Paracoccus* spp. vary greatly in the genetic and biochemical basis of their C_1 metabolism. As anticipated there are differences in the sets of enzymes responsible for the primary oxidation of C_1 compounds, which are located at the periphery

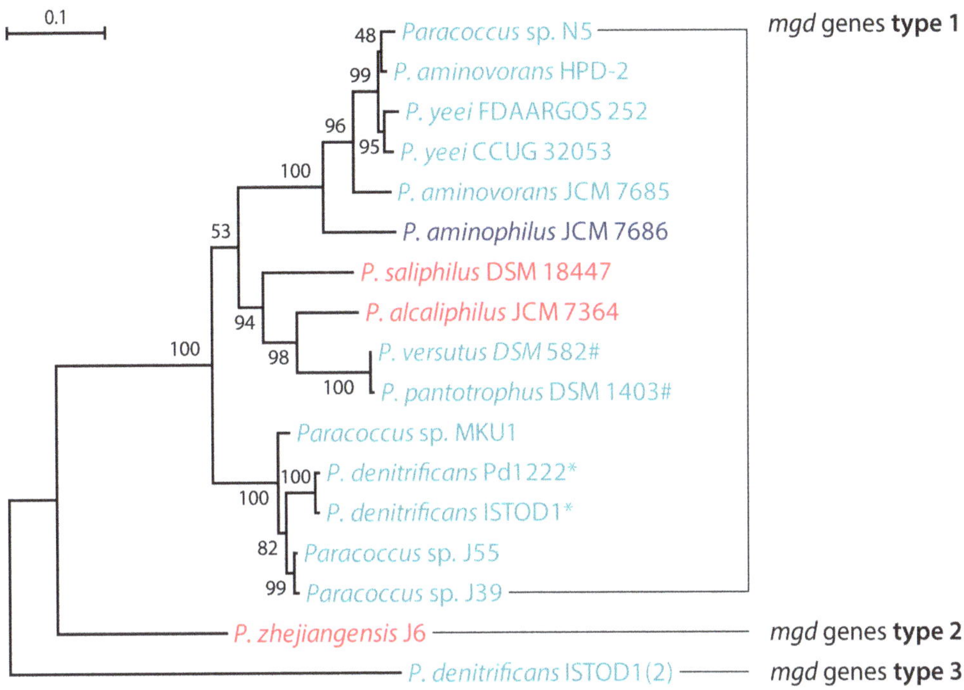

Figure 6.9 Maximum likelihood phylogenetic tree of *N*-methylglutamate dehydrogenase subunits C (MgdC) of *Paracoccus* spp. Three phylogenetic groups corresponding to the different types of serine cycle gene clusters in *Paracoccus* spp. (compare with Fig. 6.8) are indicated. Three sub-clades of *Paracoccus* spp. are coloured as in Fig. 6.4. #Strains with the NMG pathway, which do not have the serine cycle; *strains with orphan *mgd* operons, which occur without other genes of the NMG pathway.

of the C_1 metabolic net, and thus may be most 'exposed' to evolutionary changes. However, there is also considerable variation in the very nucleus of the C_1 metabolism: the C_1 assimilatory pathways. This variability is an effect of different evolutionary mechanisms, which are very hard to retrace. As was proposed previously, genetic modules representing different steps of methylotrophy may be reassembled in different genomes, generating new metabolic capabilities (Chistoserdova, 2011). This appears to be the case in *Paracoccus* spp. The more frequent coexistence of certain modules indicates that they may 'fit' together better, probably due to more efficient cooperation. This explains why the methylene THF-generating NMG pathway for MA oxidation is typically found with the methylene THF-consuming serine cycle, whereas the formaldehyde-producing MA dehydrogenase is found with the Calvin cycle. The opposite combinations of these pathways occur in some *Paracoccus* spp., but they are much less common.

The evolutionary tendencies in methylotrophy of the genus *Paracoccus* are the same as in the class *Alphaproteobacteria* as a whole, where the polyphyletic origins of C_1 metabolism have already been described (Beck *et al.*, 2015). Interestingly, these tendencies may be observed even on a microscale level, within strains of the same species. For example, comparison of two strains of *P. denitrificans*, Pd 1222 and ISTOD1, shows how evolutionary processes can easily rebuild a metabolic net. It is likely that the discovery of additional genetic modules associated with methylotrophy in this species will follow the sampling of new strains, especially those from environments where C_1 compounds are present.

Future trends

Although analysis of the increasing body of genomic data from *Paracoccus* spp. can give many interesting results, there is a need for experimental studies on the C_1 metabolism of these bacteria. The predicted functions of many genes have to be verified, the involvement of others confirmed (e.g. genes encoding putative transporters located in vicinity of methylotrophy genes), and their regulatory mechanisms characterized. It would also be informative to determine whether serine cycle-based methylotrophy can be transmitted to other strains, e.g. via the transfer of the pAMV1 methylotrophy island. *Paracoccus* spp. seem to be ideal models for studying the evolution of methylotrophy. Studies on these bacteria may shed light on some unexplored molecular aspects of C_1 compound utilization, such as the relationship between metabolic routes representing different steps of methylotrophy (e.g. the serine cycle and the NMG pathway), and the regulatory dependencies between the Calvin cycle and the serine cycle. Such studies may also help clarify some ecological aspects, such as differences in fitness in particular niches between 'autotrophic' and 'heterotrophic' methylotrophs. A greater understanding of the *Paracoccus* C_1 metabolism will not only broaden general knowledge on methylotrophy, but may also assist the construction of novel methylotrophic strains that are adapted to perform industrially important processes.

Acknowledgements

This work was supported by the National Science Centre (NCN), Poland, on the basis of decision no. DEC-2013/09/B/NZ1/00133.

References

Baker, S.C., Ferguson, S.J., Ludwig, B., Page, M.D., Richter, O.M., and van Spanning, R.J. (1998). Molecular genetics of the genus *Paracoccus*: metabolically versatile bacteria with bioenergetic flexibility. Microbiol. Mol. Biol. Rev. *62*, 1046–1078.

Beck, D.A., McTaggart, T.L., Setboonsarng, U., Vorobev, A., Goodwin, L., Shapiro, N., Woyke, T., Kalyuzhnaya, M.G., Lidstrom, M.E., and Chistoserdova, L. (2015). Multiphyletic origins of methylotrophy in *Alphaproteobacteria*, exemplified by comparative genomics of Lake Washington isolates. Environ. Microbiol. *17*, 547–554. https://doi.org/10.1111/1462-2920.12736

Beijerinck, M.W. (1910). Bildung und verbrauch von stickoxydul durch bakterien. Zentbl Bakteriol Parasitenkd Infektionskr Hyg Abt II 25, 30–63.

Chistoserdova, L. (2011). Modularity of methylotrophy, revisited. Environ. Microbiol. *13*, 2603–2622. https://doi.org/10.1111/j.1462-2920.2011.02464.x

Chistoserdova, L. (2016). Lanthanides: New life metals? World J. Microbiol. Biotechnol. 32, 138. https://doi.org/10.1007/s11274-016-2088-2

Czarnecki, J., Dziewit, L., Puzyna, M., Prochwicz, E., Tudek, A., Wibberg, D., Schlüter, A., Pühler, A., and Bartosik, D. (2017). Lifestyle-determining extrachromosomal replicon pAMV1 and its contribution to the carbon metabolism of the methylotrophic bacterium *Paracoccus aminovorans* JCM 7685. Environ. Microbiol. *19*, 4536–4550. https://doi.org/10.1111/1462-2920.13901

Devries, G.E., Harms, N., Hoogendijk, J., and Stouthamer, A.H. (1989). Isolation and characterization of *Paracoccus denitrificans* mutants with increased conjugation frequencies and pleiotropic loss of a (nGATCn) DNA-modifying property. Arch. Microbiol. *152*, 52–57. https://doi.org/10.1007/Bf00447011.

Doronina, N.V., Trotsenko, Y.A., Krausova, V.I., and Suzina, N.E. (1998). *Paracoccus methylutens* sp. nov. – a new aerobic facultatively methylotrophic bacterium utilizing dichloromethane. Syst. Appl. Microbiol. *21*, 230–236. https://doi.org/10.1016/S0723-2020(98)80027-1.

Doronina, N.V., Trotsenko, Y.A., Kuznetzov, B.B., and Tourova, T.P. (2002). Emended description of *Paracoccus kondratievae*. Int. J. Syst. Evol. Microbiol. 52, 679–682. https://doi.org/10.1099/00207713-52-2-679.

Dziewit, L., Dmowski, M., Baj, J., and Bartosik, D. (2010). Plasmid pAMI2 of *Paracoccus aminophilus* JCM 7686 carries *N,N*-dimethylformamide degradation-related genes whose expression is activated by a LuxR family regulator. Appl. Environ. Microbiol. *76*, 1861–1869. https://doi.org/10.1128/AEM.01926-09

Dziewit, L., Czarnecki, J., Wibberg, D., Radlinska, M., Mrozek, P., Szymczak, M., Schlüter, A., Pühler, A., and Bartosik, D. (2014). Architecture and functions of a multipartite genome of the methylotrophic bacterium *Paracoccus aminophilus* JCM 7686, containing primary and secondary chromids. BMC Genomics *15*, 124. https://doi.org/10.1186/1471-2164-15-124

Dziewit, L., Czarnecki, J., Prochwicz, E., Wibberg, D., Schlüter, A., Pühler, A., and Bartosik, D. (2015). Genome-guided insight into the methylotrophy of *Paracoccus aminophilus* JCM 7686. Front. Microbiol. 6, 852. https://doi.org/10.3389/fmicb.2015.00852

Friedrich, C.G., and Mitrenga, G. (1981). Oxidation of thiosulfate by *Paracoccus denitrificans* and other hydrogen bacteria. FEMS Microbiol. Lett. *10*, 209-212. https://doi.org/10.1111/j.1574-6968.1981.tb06239.x.

Funke, G., Frodl, R., and Sommer, H. (2004). First comprehensively documented case of *Paracoccus yeei* infection in a human. J. Clin. Microbiol. *42*, 3366–3368. https://doi.org/10.1128/JCM.42.7.3366-3368.2004

Harms, N., de Vries, G.E., Maurer, K., Veltkamp, E., and Stouthamer, A.H. (1985). Isolation and characterization of *Paracoccus denitrificans* mutants with defects in the metabolism of one-carbon compounds. J. Bacteriol. *164*, 1064–1070.

Harms, N., Ras, J., Reijnders, W.N., van Spanning, R.J., and Stouthamer, A.H. (1996). S-formylglutathione hydrolase of *Paracoccus denitrificans* is homologous to human esterase D: a universal pathway for formaldehyde detoxification? J. Bacteriol. *178*, 6296–6299.

John, P., and Whatley, F.R. (1975). *Paracoccus denitrificans* and the evolutionary origin of the mitochondrion. Nature *254*, 495–498.

Kalyuzhnaya, M.G., and Lidstrom, M.E. (2003). QscR, a LysR-type transcriptional regulator and CbbR homolog, is involved in regulation of the serine cycle genes in *Methylobacterium extorquens* AM1. J. Bacteriol. *185*, 1229–1235.

Kalyuzhnaya, M.G., and Lidstrom, M.E. (2005). QscR-mediated transcriptional activation of serine cycle genes in *Methylobacterium extorquens* AM1. J. Bacteriol. *187*, 7511–7517.

Karagouni, A.D., and Kelly, D.P. (1989). Carbon-dioxide fixation by *Thiobacillus versutus* - apparent absence of a CO2-concentrating mechanism in organisms grown under carbon-limitation in the chemostat. FEMS Microbiol. Lett. *58*, 179–182.

Kelly, D.P., Rainey, F.A., and Wood, A.P. (2006). The Genus *Paracoccus*. In Prokaryotes: A Handbook on the Biology of Bacteria, Vol 5, Third Edition, Dworkin, M., Falkow, S., Rosenberg, E., Schleifer, K.-H., and Stackebrandt, E. (Springer Nature, Switzerland), pp. 232–249. https://doi.org/10.1007/0-387-30745-1_12.

Keltjens, J.T., Pol, A., Reimann, J., and Op den Camp, H.J. (2014). PQQ-dependent methanol dehydrogenases: rare-earth elements make a difference. Appl. Microbiol. Biotechnol. *98*, 6163–6183. https://doi.org/10.1007/s00253-014-5766-8

Kim, K., and Lee, S.S. (2015). *Paracoccus aquimaris* sp. nov., isolated from seawater. Antonie van Leeuwenhoek. *108*, 871–877. https://doi.org/10.1007/s10482-015-0541-0.

Kim, S.G., Bae, H.S., and Lee, S.T. (2001). A novel denitrifying bacterial isolate that degrades trimethylamine both aerobically and anaerobically via two different pathways. Arch. Microbiol. *176*, 271–277. https://doi.org/10.1007/s002030100319

Kim, S.G., Bae, H.S., Oh, H.M., and Lee, S.T. (2003). Isolation and characterization of novel halotolerant and/or halophilic denitrifying bacteria with versatile metabolic pathways for the degradation of trimethylamine. FEMS Microbiol. Lett. *225*, 263–269.

Lipski, A., Reichert, K., Reuter, B., Spröer, C., and Altendorf, K. (1998). Identification of bacterial isolates from biofilters as *Paracoccus alkenifer* sp. nov. and *Paracoccus solventivorans* with emended description of *Paracoccus solventivorans*. Int. J. Syst. Bacteriol. *48*, 529–536. https://doi.org/10.1099/00207713-48-2-529

Liu, X.Y., Wang, B.J., Jiang, C.Y., and Liu, S.J. (2006). *Paracoccus sulfuroxidans* sp. nov., a sulfur oxidizer from activated sludge. Int. J. Syst. Evol. Microbiol. *56*, 2693–2695. https://doi.org/10.1099/ijs.0.64548-0.

Liu, Y., Gan, L., Chen, Z., Megharaj, M., and Naidu, R. (2012). Removal of nitrate using *Paracoccus* sp. YF1 immobilized on bamboo carbon. J. Hazard. Mater. 229-230, 419–425. https://doi.org/10.1016/j.jhazmat.2012.06.029

McGinnis, J.M., Cole, J.A., Dickinson, M.C., Mingle, L.A., Lapierre, P., Musser, K.A., and Wolfgang, W.J. (2015). *Paracoccus sanguinis* sp. nov., isolated from clinical specimens of New York State patients. Int. J. Syst. Evol. Microbiol. *65*, 1877–1882. https://doi.org/10.1099/ijs.0.000193.

Ochsner, A.M., Christen, M., Hemmerle, L., Peyraud, R., Christen, B., and Vorholt, J.A. (2017). Transposon sequencing uncovers an essential regulatory function of phosphoribulokinase for methylotrophy. Curr. Biol. *27*, 2579–2588.

Pukall, R., Laroche, M., Kroppenstedt, R.M., Schumann, P., Stackebrandt, E., and Ulber, R. (2003). *Paracoccus seriniphilus* sp. nov., an L-serine-dehydratase-producing coccus isolated from the marine bryozoan *Bugula plumosa*. Int. J. Syst. Evol. Microbiol. *53*, 443–447. https://doi.org/10.1099/ijs.0.02352-0.

Ras, J., Van Ophem, P.W., Reijnders, W.N., Van Spanning, R.J., Duine, J.A., Stouthamer, A.H., and Harms, N. (1995). Isolation, sequencing, and mutagenesis of the gene encoding NAD- and glutathione-dependent formaldehyde dehydrogenase (GD-FALDH) from *Paracoccus denitrificans*, in which GD-FALDH is essential for methylotrophic growth. J. Bacteriol. *177*, 247–251.

Sanjeevkumar, S., Nayak, A.S., Santoshkumar, M., Siddavattam, D., and Karegoudar, T.B. (2013). *Paracoccus denitrificans* SD1 mediated augmentation with indigenous mixed cultures for enhanced removal of *N,N*-dimethylformamide from industrial effluents. Biochem. Eng. J. *79*, 1-6. https://doi.org/10.1016/j.bej.2013.06.016.

Schneider, K., Peyraud, R., Kiefer, P., Christen, P., Delmotte, N., Massou, S., Portais, J.C., and Vorholt, J.A. (2012). The ethylmalonyl-CoA pathway is used in place of the glyoxylate cycle by Methylobacterium extorquens AM1 during growth on acetate. J. Biol. Chem. *287*, 757–766. https://doi.org/10.1074/jbc.M111.305219

Sheu, S.Y., Hsieh, T.Y., Young, C.C., and Chen, W.M. (2018). *Paracoccus fontiphilus* sp. nov., isolated from a freshwater spring. Int. J. Syst. Evol. Microbiol. *68*, 2054–2060. https://doi.org/10.1099/ijsem.0.002793

Siller, H., Rainey, F.A., Stackebrandt, E., and Winter, J. (1996). Isolation and characterization of a new gram-negative, acetone-degrading, nitrate-reducing bacterium from soil, *Paracoccus solventivorans* sp. nov. Int. J. Syst. Bacteriol. *46*, 1125–1130. https://doi.org/10.1099/00207713-46-4-1125

Sun, M., Luo, Y., Teng, Y., Christie, P., Jia, Z., and Li, Z. (2013). Tenax TA extraction to understand the rate-limiting factors in methyl-β-cyclodextrin-enhanced bioremediation of PAH-contaminated soil. Biodegradation *24*, 365–375. https://doi.org/10.1007/s10532-012-9593-2

Tsubokura, A., Yoneda, H., and Mizuta, H. (1999). *Paracoccus carotinifaciens* sp. nov., a new aerobic gram-negative astaxanthin-producing bacterium. Int. J. Syst. Bacteriol. *49*, 277–282. https://doi.org/10.1099/00207713-49-1-277

Urakami, T., Araki, H., Oyanagi, H., Suzuki, K., and Komagata, K. (1990). *Paracoccus aminophilus* sp. nov. and *Paracoccus aminovorans* sp. nov., which utilize *N,N*-dimethylformamide. Int. J. Syst. Bacteriol. *40*, 287–291. https://doi.org/10.1099/00207713-40-3-287

Zhang, G., Xian, W., Yang, J., Liu, W., Jiang, H., and Li, W. (2016). *Paracoccus gahaiensis* sp. nov. isolated from sediment of Gahai Lake, Qinghai-Tibetan Plateau, China. Arch. Microbiol. *198*, 227–232. https://doi.org/10.1007/s00203-015-1184-2

Zhang, S., Gan, L., Qin, Q., Long, X., Zhang, Y., Chu, Y., and Tian, Y. (2016). Paracoccusacridae sp. nov., isolated from the insect *Acrida cinerea* living in deserted cropland. Int. J. Syst. Evol. Microbiol. *66*, 3492–3497. https://doi.org/10.1099/ijsem.0.001222

Microbiology and Ecology of Methylated Amine Metabolism in Marine Ecosystems

7

Michaela A. Mausz* and Yin Chen*

School of Life Sciences, Gibbet Hill Campus, University of Warwick, Coventry, UK.

*Correspondence: m.mausz@warwick.ac.uk and y.chen.25@warwick.ac.uk

https://doi.org/10.21775/9781912530045.07

Abstract

Methylated amines (MAs) are ubiquitous in marine ecosystems, found from surface seawaters to sediment pore waters. These volatile ammonium analogues play important roles in biogeochemical cycles of carbon and nitrogen in the marine water column. They also contribute to the release of climate-active gases, being precursors of the potent greenhouse gas methane through methanogenesis in coastal sediments. Very recently, it also became acknowledged that MAs are important precursors for new particle growth, hence forming cloud condensation nuclei in the marine atmosphere. Microbial metabolism of MAs has been demonstrated in the marine ecosystems for both Archaea and Bacteria. In this chapter, we summarize the latest developments in analytical methods for quantifying MA concentrations in marine surface water and sediments. We discuss the metabolic pathways leading to the formation and degradation of MAs by marine microbes and the novel biochemistry and structural biology of the enzymes for MA transformation. Lastly, we highlight the need for future research towards a better understanding of the microbiology and ecology of oceanic MA cycles.

Introduction

Methylated amines (MAs), namely monomethylamine (MMA), dimethylamine (DMA), trimethylamine (TMA) and its oxidation product TMA *N*-oxide (TMAO), are ubiquitously present throughout marine ecosystems, from surface seawaters to deep sediments (Fitzsimons *et al.*, 1997; Gibb *et al.*, 1999b; Carpenter *et al.*, 2012). Researchers have studied MAs in marine environments for many decades, and various analytical pipelines, such as gas chromatography (GC), high-performance liquid chromatography (HPLC), and ion-exchange chromatography (IC) have been developed for quantifying ambient concentrations of these molecules in the sea, which usually range from nanomolar concentrations (nM) in surface waters to micromolar concentrations (μM) in the sediments. Microbial metabolism of MAs has been of special interest because these compounds can be metabolized by methanogenic Archaea to produce methane. In marine sediments and coastal salt marshes, MAs are the well-known non-competitive substrates for methylotrophic methanogens allowing them to thrive in these high sulfate environments (Oremland *et al.*, 1982; King, 1984). Metabolism of MAs in well-oxygenated marine water columns has been overlooked in the past, however, we now know MAs appear to be important sources of carbon, nitrogen and energy for abundant marine heterotrophic bacteria, particularly members of the *Roseobacter* clade (members of the *Rhodobacteraceae*) and the SAR11 clade (*Pelagibacter* spp.) (Chen *et al.*, 2011; Sun *et al.*, 2011). The recent resurgence of interest in studying the marine MA cycles encompasses efforts towards better understanding of their atmospheric chemistry, particularly their role in catalysing the formation of new particles in ambient air (e.g. Almeida *et al.*, 2013; Schobesberger *et al.*, 2013). The aim of this chapter is to summarize the current knowledge on analytical methods for measuring ambient concentrations of MAs in the sea, discuss recent advances in understanding the microbiology and ecology of MA metabolism by marine microbes, and highlight the critical directions for future research.

Methylated amines in marine systems

Natural concentrations of methylated amines in coastal and open ocean systems

Although MAs are ubiquitous in marine environments and contribute as part of the dissolved organic nitrogen (DON) pool (Gibb *et al.*, 1999b; Carpenter *et al.*, 2012), their *in situ* quantification has proven rather challenging. This is partly due to low natural concentrations of MAs, which typically lie in the nanomolar (nM) range for seawater, while in sediment pore water concentrations can increase to low micromolar (μM) levels (Lee and Olson, 1984; Van Neste *et al.*, 1987; Yang *et al.*, 1993; Gibb *et al.*, 1995a, 1999b; Fitzsimons *et al.*, 1997; Hatton and Gibb, 1999) (Table 7.1). Consequently, studies monitoring natural MA concentrations are rare and were mostly performed by a few qualified groups applying custom-made instrumentation (reviewed by Carpenter *et al.*, 2012). Table 7.1 summarizes reports for MA measurements in marine environments.

Interestingly, many studies reporting MA concentrations from sea- or pore water also determined ammonium (NH_4^+), showing that the latter regularly exceeds the dominating MA species by at least one order of magnitude (Fitzsimons *et al.*, 2006; Carpenter *et al.*, 2012). Among MAs, monomethylamine (MMA) often is the most abundant species in seawater samples, e.g. in the Pacific or Atlantic Ocean, the Arabian Sea, the Mediterranean, the Irish Sea, and the English Channel (Van Neste *et al.*, 1987; Abduhl-Rashid, 1990; Gibb, 1994; Gibb *et al.*, 1995a, 1999a,b). According to a study in the Arabian Sea, MMA concentrations in coastal seawater were ≈12 nM in August to October and ≈22 nM in November to December 1994. In contrast, DMA accounted for ≈ 3.0 nM and ≈ 4.2 nM, while TMA only reached concentrations of ≈ 0.10 and 0.45 nM, respectively (Gibb *et al.*, 1999b). Comparison with cruise samples taken at offshore sites demonstrated that MA concentrations decreased further away from the coastline. MMA was still the most abundant MA, but its concentrations in offshore samples from the Arabian Sea decreased to ≈ 6 nM in August to October and ≈ 12 nM in November to December. DMA accounted for ≈ 2.9 nM during both seasons and TMA reached ≈ 0.05 nM and 0.13 nM, respectively (Gibb *et al.*, 1999b).

In coastal areas of the Mediterranean Sea, concentrations were in a similar range with MMA being the dominant MA species (4–38 nM), although TMA was more prevalent and DMA was the least abundant (TMA: 4–22 nM, DMA: 3–15 nM) (Gibb, 1994; Gibb *et al.*, 1995a). Similarly, in the Arabian Sea, MA concentrations also decreased in offshore samples, with both MMA and DMA accounting for ≤ 9 nM, and TMA reaching up to 7 nM (Gibb, 1994; Gibb *et al.*, 1995a).

Van Neste and co-workers reported similar concentrations from the coastal Pacific Ocean near Oahu, Hawaii (MMA: 52 ± 20 nM, DMA: 1.5 ± 2.0 nM, TMA: 12 ± 3 nM), whereas concentrations were slightly higher in the Atlantic Ocean along the coast of Massachusetts (MMA: 200 ± 58 nM, DMA: 8.9 ± 4.4 nM, TMA: 41 ± 27 nM) (Van Neste *et al.*, 1987). Concentrations reported by Abdul-Rashid (1990) from both coastal (Port Erin Jetty, Isle of Man) and offshore (Cypris Station, Isle of Man) samples collected in the Irish Sea were even higher than in the Atlantic Ocean samples, although concentrations varied with season being highest for MMA from mid-July until November (coastal up to 619 nM, offshore up to 418 nM). Both DMA and TMA levels were much lower throughout most of the year with DMA reaching ≤ 110 nM at the coastal and ≤ 93 nM at the offshore site, and TMA accounting for ≤ 34 nM and ≤ 30 nM, respectively (Abduhl-Rashid, 1990). While MMA dominated in coastal and offshore surface seawater samples, water samples collected from a salt marsh and an anoxic estuary showed highest concentrations of DMA (Flax Pond salt marsh, USA: 15–180 nM; Pettaquamscutt estuary, USA: 10–210 nM) followed by TMA and slightly lower levels of MMA (Flax Pond salt marsh, TMA:

Table 7.1 Methylated amine concentrations in marine environments from seawater or pore water samples as reported in the literature. Numbers in brackets indicate maximum observed concentrations. For explanations cf the methods see the text

Location – sample type	Concentration (nM)					Method	References
	MMA	DMA	TMA	TMAO	EA		
Pacific – Hawaii coastal	52 ± 20	1.5 ± 2.0	12 ± 3	nd	nd	Vacuum distillation + GC-CLD	Van Neste *et al.* (1987)
Atlantic – Massachusetts coastal	200 ± 58	8.9 ± 4.4	41 ± 27	nd	nd		
Irish Sea – coastal	0–619	0–110	0–34	nd	nd	Microdiffusion + GC-NPD	Abduhl-Rashid (1990)
Irish Sea – offshore	0–418	0–93	0–30	nd	nd		
Flax Pond salt marsh, USA – seawater	10–60	15–180	< 3–80	nd	< 3–30	Diffusion cell + GC-NPD	Yang *et al.* (1993)
Flax Pond salt marsh, USA – pore water	587	154	514	nd	98		
Pettaquamscutt, USA – anoxic estuary basin	20–22	10–210	5–76	nd	< 5–30		
Flax Pond salt marsh, US – seawater	5–40	25–180	10–50	nd	nd	Diffusion cell + GC-NPD	Yang *et al.* (1994)
Mediterranean – coastal (Gulf of Lions)	18 ± 10.0	12 ± 11.4	10 ± 6.9	nd	nd	FIGD-IC	Gibb (1994)
Mediterranean – offshore	7.5 ± 5.5	4.6 ± 3.0	1.4 ± 1.6	nd	nd		
NW Mediterranean – coastal	4–38	3–15	4–22	nd	nd	FIGD-IC	Gibb *et al.* (1995a)
NW Mediterranean – offshore	< LoD – 9	< LoD – 9	< LoD – 7	nd	nd		
Tamar and Plym estuary, UK – seawater	37–100	0–49	0–13	nd	nd		
Plymouth Sound, UK – seawater	4–23	13–22	4–17	nd	nd		
Sutton Harbour, UK – seawater	91	< LoD	15	nd	nd		
Arabian Sea – coastal	12 ± 20 (66)	3.0 ± 4.1 (14)	0.10 ± 0.37 (2.0)	nd	nd	FIGD-IC	Gibb *et al.* (1999b)
	22 ± 13 (55)	4.2 ± 2.8 (14)	0.45 ± 0.81 (4.0)	nd	nd		
Arabian Sea – offshore	6 ± 7 (31)	2.9 ± 2.8 (12)	0.05 ± 0.21 (1.8)	nd	nd		
	12 ± 7 (34)	2.9 ± 1.6 (7)	0.13 ± 0.24 (0.8)	nd	nd		
Arabian Sea – Gulf of Oman	11 ± 9 (32)	2.8 ± 3.1 (11)	0.19 ± 0.42 (1.9)	nd	nd		
Arabian Sea – coastal	16.2 (65.8)	4.54 (13.9)	0.044 (0.44)	nd	nd	FIGD-IC	Gibb *et al.* (1999a)
	25.6 (49.8)	4.85 (9.95)	0.14 (0.8)	nd	nd		
Arabian Sea – offshore	5.24 (13.9)	3.85 (11.1)	< LoD	nd	nd		
	8.8 (16.5)	2.04 (5.7)	0.036 (0.3)	nd	nd		
Plymouth Sound, UK – seawater	nd	nd	< LoD	3.2	nd	Enzymatic reduction + FIGD-IC	Hatton and Gibb (1999)
Ryder Bay, Antarctica – seawater	nd	nd	< LoD – 4.1	2.1–64	nd		

Table 7.1 Continued

Location – sample type	Concentration (nM)					Method	References
	MMA	DMA	TMA	TMAO	EA		
Ryder Bay, Antarctica – seawater	12.0 ± 9.1 (36.0)	3.8 ± 3.9 (12.7)	1.6 ± 1.8 (12.0)	15.2 ± 16.6 (76.9)	0.6 ± 0.6 (1.9)	(Enzymatic reduction +) FIGD-IC	Gibb and Hatton (2004)
Western English Channel – coastal	3	6	20	nd	nd	SPME + GC-NPD	Cree *et al.* (2018)
Southern Ocean – offshore	<LoD – 2.6	<LoD – 5.4	1.4–6.9	nd	nd		
Buzzards Bay, USA – coastal pore water	<LoD	48,000	29,000	nd	nd	Steam distillation + GC-CLD	Lee and Olson (1984)
Buzzards Bay, USA – pore water core	<LoD	1100–1900	400–500	nd	nd		
E Tropical N-Pacific Ocean – offshore pore water	<LoD	100–1300	80–140	nd	nd		
Norsminde Fjord estuary, Denmark – pore water	nd	nd	0–10,000	nd	nd	GC-FID	Sørensen and Glob (1987)
Flax Pond salt marsh, USA – pore water	1060	190	nd	nd	nd	Steam distillation + RP-HPLC	Wang and Lee (1990)
Long Island Sound estuary, USA – pore water	870	120	nd	nd	nd		
Peru upwelling region – coastal pore water	650	100	nd	nd	nd		
Flax Pond salt marsh, USA – pore water	100–1700	100–3600	100–600	nd	nd	Diffusion cell + GC-NPD	Wang and Lee (1994)
Oglet Bay salt marsh, UK – pore water	0–319,000	0–9000	0–50,000	nd	nd	Microdiffusion + GC-NPD	Fitzsimons *et al.* (1997)
Burnham Overy Staithe mudflat, UK – pore water	100–1300	100–2800	0–4680	nd	nd	Microdiffusion + GC-NPD	Fitzsimons *et al.* (2001)
Thames Estuary, UK – pore water, July	80–130	310–370	30–60	nd	nd	Microdiffusion + GC-NPD	Fitzsimons *et al.* (2006)
Thames Estuary, UK – pore water, Nov	980–1440	1200–1850	1400–2230	nd	nd		

CLND, chemiluminescent nitrogen detector; EA, ethylamine; FID, flame ionization detector; FIGD, flow injection gas diffusion; GC, gas chromatography; HPLC, high performance liquid chromatography; IC, ion chromatography; LoD, limit of detection; nd, not determined; NPD, nitrogen-phosphorus detector; RP, reverse phase; SPME, solid phase microextraction.

<3–80 nM, MMA: 5–60 nM; Pettaquamscutt estuary, TMA: 5–76 nM, MMA: 20–22 nM) (Yang *et al.*, 1993, 1994). Nevertheless, the authors could not explain the observed pattern of high DMA concentrations in the seawater overlying the sediment, as decomposition of marsh grasses and degradation of quaternary amines primarily release TMA as the key intermediate (Yang *et al.*, 1994).

Natural concentrations of methylated amines in sediment pore water

MA concentrations reported from sediment pore water are at least one order of magnitude higher than those found in seawater (see Table 7.1). Flax Pond salt marsh has been repeatedly sampled for MAs from sediment pore water (Wang and Lee, 1990, 1994; Yang *et al.*, 1993). Concentrations varied depending on sampling depth, and season with the highest concentrations observed in the autumn, which was attributed to the decomposing of the cordgrass *Spartina alterniflora* (Wang and Lee, 1994). DMA was found at highest concentrations in the pore water with concentrations usually <0.5 μM, while average concentrations for MMA and TMA reached <0.2 μM (Wang and Lee, 1994). TMA was the most abundant species among MAs in pore water of the Burnham Overy Staithe mudflat, UK, (up to 4.68 μM) (Fitzsimons *et al.*, 2001) and the Thames Estuary, UK, where also a seasonal pattern was observed (0.03–0.06 μM in July, 1.4–2.23 μM in November) (Fitzsimons *et al.*, 2006). The authors postulated that MAs adsorbed to sediments are released during tidal inundation, which would explain the documented concentrations in the pore water (Fitzsimons *et al.*, 2006).

Analytical methods to quantify methylated amine concentration in marine environments

The difficulties of determining methylated amine concentrations from marine samples

As described above, several studies have measured MA concentrations in marine environments, but their analyses remain challenging and sample pre-treatments as well as the application of specially designed instrumentation impede data comparability. Low compound concentrations in natural seawater near the detection limits of analytical instruments (nM range) require tedious pre-concentration approaches such as steam or vacuum distillation, rotary evaporation, concentration in the oven, or the usage of diffusion techniques (static, circular, membrane or microdiffusion, application of diffusion cells) (e.g. Lee and Olson, 1984; Van Neste *et al.*, 1987; Wang and Lee, 1990; Abdul-Rashid *et al.*, 1991; Yang *et al.*, 1993; Fitzsimons *et al.*, 2001, 2006) (see Table 7.1). Consequently, recovery rates after pre-concentration steps are often low and thus influence reproducibility of the data. In addition, the high concentration of inorganic salts in seawater samples interferes with peak resolution. High water solubility and adsorption of MAs, or their polar, volatile nature add additional difficulties to the successful determination of their natural concentrations (Yang *et al.*, 1993; Gibb *et al.*, 1995b). To overcome these limitations a sensitive and reliable method is required that allows the parallel determination of various amines in natural seawater samples. Since the 1980s, several groups started to develop reproducible techniques for measuring ambient concentrations of MAs in seawater, which are summarized below.

High-performance liquid chromatography (HPLC)-based methods

The application of high-performance liquid chromatography (HPLC) for the measurement of amines requires a proper derivatization approach. Unfortunately, the lack of a suitable derivatization agent usable for primary, secondary, and tertiary amines in parallel reduces the applicability of HPLC-based methods for amine detection (Mopper and Zika, 1987; Wang and Lee, 1990; Gibb *et al.*, 1995b). Thus, most methods were developed to target amino acids or primary amines, or they require a combination of derivatization approaches (Jacobs, 1987; Mopper and Zika, 1987; Carlucci and Karmas, 1988). Historically, a whole range of detection methods were combined with HPLC such as UV, fluorescence, chemiluminescence excitation, and electrochemical detection (Mellbin and Smith, 1984; Jacobs, 1987; Mopper and Zika, 1987; Carlucci and Karmas, 1988; Wang and Lee, 1990), although not all of these approaches were applied to determine MA concentrations in sea- and pore water samples.

Using a pre-column fluorescence derivatization with *O*-phthalaldehyde and the thiols 2-mercapto-ethanol and *N*-acetyl-L-cysteine, Mopper and Zika (1987) measured amino acid and primary amine concentrations in rain and air samples collected during cruises in the Gulf of Mexico, as well as in seawater samples taken in the north-western Atlantic Ocean. Derivatization products were detected by a fluorescence detector after HPLC separation. The authors determined total primary amine concentration to be 4 nM (2-aminoethanol 82.5 mole%, MMA 17.5 mole%) in surface seawater (Mopper and Zika, 1987), but this technique does not allow the detection of secondary and tertiary amines.

In a quantitative derivatization method for amino acids, Heinrikson and Meredith (1984) used the reaction of amino acids with phenylisothiocyanate to form phenylthiocarbamyl-amino acid derivatives (Edman degradation). These derivatives produced a fluorescent complex that, after separation by reverse-phase HPLC, could be detected on a fluorescence detector at 254 nm (Heinrikson and Meredith, 1984). The method also proved suitable to determine primary and secondary amines such as MMA and DMA in sediment pore water samples (Wang and Lee, 1990). Therefore, pore water samples were first concentrated by steam distillation (Lee and Olson, 1984), followed by addition of phenylisothiocyanate to form phenylthiohydantoin derivatives of MMA and DMA (Wang and Lee, 1990). The products were then analysed by reverse-phase HPLC with fluorescence detection (Wang and Lee, 1990).

Gas chromatography (GC)-based methods

After pre-concentration, usually by distillation or a diffusion technique (e.g. Lee and Olson, 1984; Van Neste *et al.*, 1987; Abdul-Rashid *et al.*, 1991; Yang *et al.*, 1993; Fitzsimons *et al.*, 2001), gas chromatography (GC), which exploits the volatility of the target compounds, has been widely used for MA detection from environmental samples. Although suffering from its own limitations (summarized by Gibb *et al.*, 1995a,b), GC separation does not require derivatization, and it can be combined with a range of detectors. A concept generally applied in sample preparation exploits the fact that when MAs are exposed to highly basic conditions, they convert to their volatile gaseous forms. For example, TMA was converted into its gas phase by base addition to specifically analyse TMA in pore water samples, transferred to a cryogenic trap and injected into a gas chromatograph with a flame ionization detector (GC-FID) (Glob and Sørensen, 1987; Sørensen and Glob, 1987).

Alternatively, a chemiluminescent technique originally developed by Lee and Olson (1984) exploits light emission from nitrogen compounds reacting with ozone at high temperature, which can be detected by a chemiluminescent nitrogen detector (CLND) (Lee and Olson, 1984; Van Neste *et al.*, 1987). The method is claimed to be selective for highly volatile nitrogen compounds and was applied to determine MA concentrations in coastal seawaters from Hawaii and Massachusetts, and also in different pore water samples (Lee and Olson, 1984; Van Neste *et al.*, 1987).

One of the more frequently used approaches for the measurement of MAs in seawater samples provides the coupling of a GC to a nitrogen-phosphorus detector (GC-NPD) (Abdul-Rashid *et al.*, 1991; Yang *et al.*, 1993; Fitzsimons *et al.*, 2001). An NPD works similarly to an FID using the electronic conductivity of a hydrogen/air flame combined with an alkali metal salt bead (usually rubidium or caesium chloride). Under heat, ions utilized from the bead interact specifically with nitrogen- or phosphorus-containing compounds and the ion amount is measured at the amplifier (CHROMEDIA Analytical Sciences). Amine determination by GC-NPD requires sample concentration prior to measurement. Dr Cindy Lee's laboratory used a diffusion cell to pre-concentrate amines. Therefore, they combine membrane diffusion with static and circulation diffusion chamber approaches and exploit the ability of amines to protonate or deprotonate with pH changes, which either allows them to diffuse across membranes (deprotonated) or to get trapped (re-protonated) (Yang *et al.*, 1993). Alternatively, Dr Mark Fitzsimons' group uses microdiffusion via a Cavett diffusion flask consisting of a stoppered Quickfit Pyrex Erlenmeyer flask and a small collecting cup at fixed distance from the glass stopper (Abdul-Rashid *et al.*, 1991). The second approach was commonly used to analyse sediment pore waters (Fitzsimons *et al.*, 1997, 2001, 2006).

The newest developments in MA analysis aim to apply more widely available equipment rather than previous custom-made instrumentation. To specifically target TMA in marine sediments, a cryogenic purge and trap system combined with GC coupled to mass spectrometry (P&T-GC-MS) was applied. Using a stream of inert gas, volatile analytes such as TMA were drawn from the liquid sample phase and trapped cryogenically, while in parallel by a headspace method the isotopic composition of stable carbon isotopes were analysed (Zhuang *et al.*, 2017). Whether or not this method is also applicable to the measurement of MMA and DMA awaits experimental validation, because the detection of compounds with such a small mass to charge ratio (m/z) is problematic with most common MS detectors. A suitable alternative approach combines headspace solid phase microextraction (SPME) with GC-NPD (Cree *et al.*, 2018). Here, an SPME fibre was exposed to the headspace above a stirred, basic (pH ≈ 13) seawater sample maintained at 60°C to extract MAs. Upon GC injection, analytes thermally desorbed from the fibre and were detected by an NPD (Cree *et al.*, 2018). MMA, DMA and TMA concentrations from Western English Channel and Southern Ocean seawater samples were thus successfully analysed in lab- and ship-based settings (Cree *et al.*, 2018). As this widely available approach allows the parallel sensitive and robust analysis of MAs and other low-molecular-weight amines, it might encourage further measurements and henceforth improve our understanding of MA cycles in seawater.

Flow injection gas-diffusion coupled to ion chromatography (FIGD–IC)

Some investigators suggested that ion chromatography (IC) is a good approach for detecting MAs. In addition to the sensitivity of this method, the polar nature of MAs and ammonium (NH_4^+) allows their measurements in parallel in low ionic strength samples such as sea- or pore water without resolution problems sometimes associated with GC separation or the requirement for derivatization (Gibb *et al.*, 1995a,b). However, the samples need pre-concentration to prevent column overloading due to large excess of sodium (Na^+) and potassium (K^+) ions in seawater, and the required sensitivity in the nM range. Similarly to the diffusion cell (Yang *et al.*, 1993), an automated flow injection gas diffusion-ion chromatography (FIGD-IC) system exploits that NH_4^+ and MAs deprotonate to their volatile gaseous forms under high pH conditions (> 12). In this state the molecules diffuse from the sample stream through a hydrophobic membrane and get re-protonated and trapped after contact with an acidic acceptor stream. Recycling of the stream in the enrichment loop results in analyte enrichment over time, which can then be transferred to the IC for injection (Gibb *et al.*, 1995a,b).

Combination of FIGD-IC with an enzymatic reaction allowed the additional determination of the nitrogenous osmolyte TMAO from seawater. TMAO was reduced to TMA in the presence of the enzyme dimethyl sulfoxide (DMSO) reductase and reducing solutions consisting of flavin mononucleotide (FMN) and ethylenediaminetetraacitic acid (EDTA) (Hatton and Gibb, 1999; Gibb and Hatton, 2004). The authors successfully quantified concentrations of MAs as well as their potential precursor TMAO from seawater samples (Gibb and Hatton, 2004). Although FIGD-IC is a sensitive and powerful approach for measuring MAs, it is not widely used probably because it is technically challenging to assemble the complex combination of the FIGD preconcentration system to the IC (Gibb *et al.*, 1995a,b).

Microbial metabolism of methylated amines in marine environments

In the marine environments, MAs can be metabolized by at least four biochemical pathways (Fig. 7.1):

1. methylotrophic methanogenesis, which occurs in anaerobic sediments and is conducted by methylotrophic Archaea (Rother, 2010);
2. the dehydrogenase pathway, which uses a series of dehydrogenases to sequentially demethylate each methylated amine species while forming formaldehyde as a by-product (Chistoserdova, 2011);
3. the aerobic TMA oxidation pathway, which forms TMAO as the key intermediate;
4. the TMA-dependent acetogenesis pathway, which is still poorly understood.

We will briefly summarize pathways 1, 2 and 4, but focus primarily on the aerobic TMAO-dependent pathway throughout this section, to reflect its prevalence in marine surface waters.

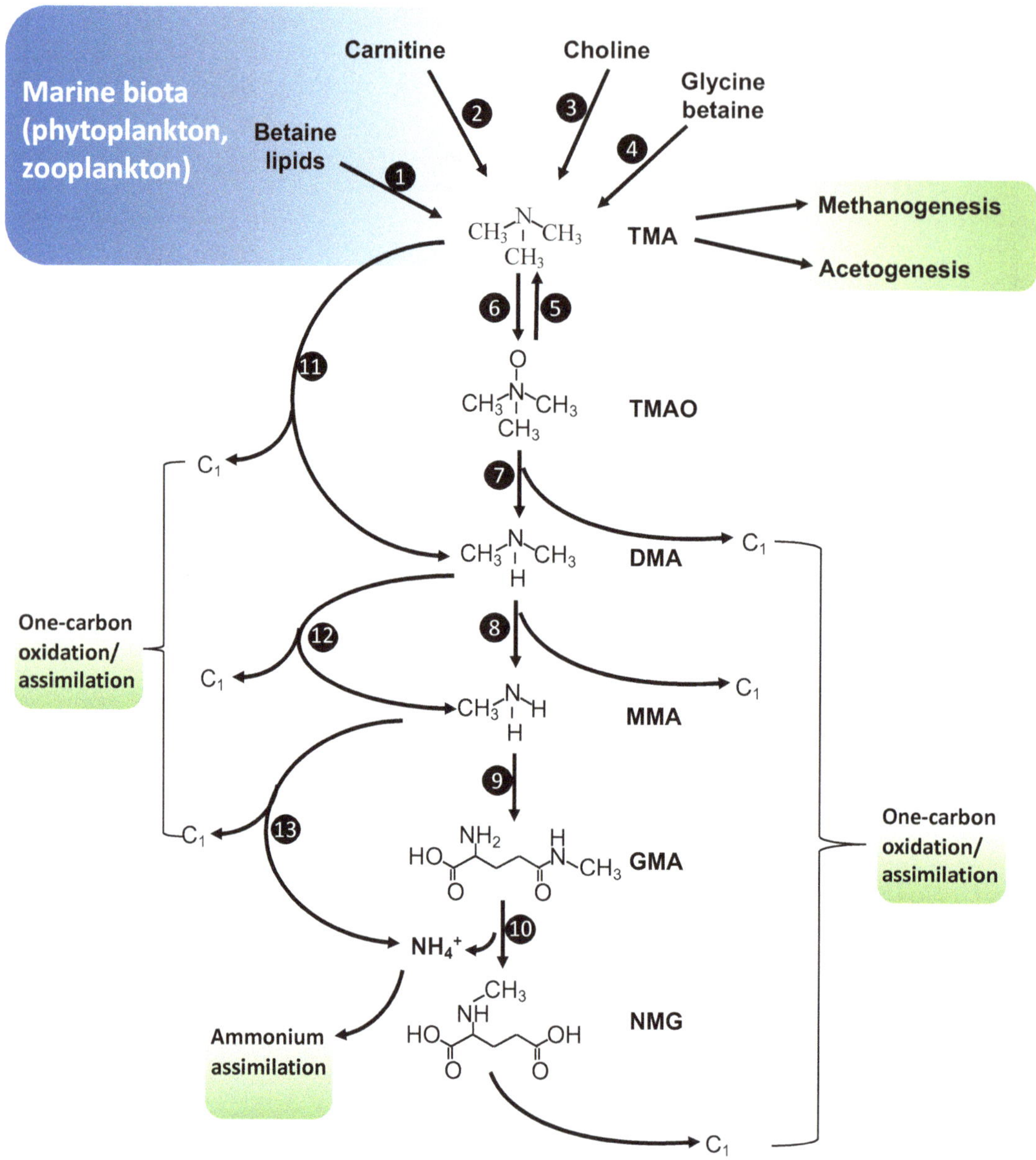

Figure 7.1 Methylated amine formation and degradation in marine environments. (1) degradation of betaine lipids; (2) carnitine monooxygenase; (3) choline –TMA lyase; (4) glycine betaine reductase; (5) TMAO reductase; (6) TMA monooxygenase; (7) TMAO demethylase; (8) DMA monooxygenase; (9) GMA synthetase; (10) NMG synthase; (11) TMA dehydrogenase; (12) DMA dehydrogenase; (13) MMA dehydrogenase. C_1 represents either formaldehyde or its derivatives (e.g. conjugated to tetrahydrofolate).

Anaerobic methylated amine oxidation by methylotrophic methanogens

In anaerobic habitats methanogenesis comprises the final step in the degradation of organic matter resulting in the production of gaseous methane by methanogenic Archaea belonging to the *Euryarchaeota* (reviewed in Rother, 2010). Three different pathways exist: hydrogenotrophic, methylotrophic, and acetotrophic methanogenesis. Methylotrophic methanogens, such as the well-studied order of *Methanosarcinales*, can utilize methylated compounds such as methanol, methylamine (MMA, DMA, TMA), and dimethylsulfide (Hippe *et al.*, 1979; Mah, 1980; van der Maarel and Hansen, 1997). For example, in surface sediment from a salt marsh near Dorset, UK, the turnover time for TMA was 80 days and *Methanosarcinales* were the major methanogens found in the sediment (Parkes *et al.*, 2012). Similarly, we have recently demonstrated that *Methanococcoides* are key TMA utilizers in a saltmarsh sediment (Jameson *et al.*, 2019). More recently, a new group of methylotrophic methanogens was described, *Methanomassiliicoccales*, which represent obligate hydrogen-dependent methanogens that can utilize methanol or MAs (Iino *et al.*, 2013). Their capability to grow on MAs has been originally hypothesized from the metatranscriptomic data, followed by *in vivo* validation in the strain *Methanomassiliicoccus luminyensis* B10 (Poulsen *et al.*, 2013; Brugère *et al.*, 2014; Kröninger *et al.*, 2017). Although the known *Methanomassiliicoccales* were isolated from gut and rumen samples, their occurrence in marine habitats is supported by metagenomic and phylogenomic datasets (Paul *et al.*, 2012; Borrel *et al.*, 2013).

Methylated amine dehydrogenases pathway

Several microbes such as *Hyphomicrobium* and *Methylophaga* use a pathway for sequential demethylation of TMA and DMA through TMA dehydrogenase, DMA dehydrogenase and MMA dehydrogenase. TMA dehydrogenase has been purified from *Methylophaga* sp. strain SK1 and *Methylophilus methylotrophus* (Lim *et al.*, 1986; Kim *et al.*, 2006). TMA dehydrogenase is a homodimeric flavoprotein containing a 4Fe–4S iron–sulfur cluster (Burgess *et al.*, 2008). DMA dehydrogenase is also a homodimeric flavin-containing protein which is closely related to TMA dehydrogenase ($\approx$60% sequence identity in amino acid sequence) (Yang *et al.*, 1995). In contrast, MMA dehydrogenase is a periplasmic enzyme which requires a specific cofactor, tryptophan tryptophylquinone (TTQ). MMA dehydrogenases have been purified from several bacteria, including *Methylobacterium extorquens* (Eady and Large, 1968) and *Paracoccus versutus* (Haywood *et al.*, 1982), and they are composed of two large and two small subunits MauBA.

Oxidation pathway involving TMAO as the key intermediate

The third pathway for MA metabolism, involving the oxidation of TMA to TMAO as the key intermediate, appears to be even more widely distributed across environments (Chen *et al.*, 2011; Nayak and Marx, 2015). In this pathway, TMA is oxidized by a flavin-containing monooxygenase, Tmm, to TMAO, followed by demethylation to DMA by a novel binuclear Zn/Fe metalloprotein, Tdm (Chen *et al.*, 2011; Lidbury *et al.*, 2014; Zhu *et al.*, 2014b, 2016). Both Tmm and Tdm are ubiquitously found in cosmopolitan marine heterotrophs including the *Roseobacter* clade and the SAR11 (*Pelagibacter* spp.) clade (Chen *et al.*, 2011; Lidbury *et al.*, 2014), suggesting a key role of this so-called indirect methylamine oxidation pathway in marine biogeochemical cycles of carbon and nitrogen. We recently also identified the secondary amine oxidase, DMA monooxygenase, in marine *Roseobacter* bacteria (Lidbury *et al.*, 2017). DMA monooxygenase, Dmm, is composed of four subunits, DmmDABC, which catalyse the formation of MMA and formaldehyde from DMA oxidation. Dmm has previously been purified from *Aminobacter aminovorans* (known as *Pseudomonas aminovorans*) by Dr John Dawson's group in the late 1980s (Alberta and Dawson, 1987; Alberta *et al.*, 1989). DMA monooxygenase is a NADPH-dependent haem-containing protein which is sensitive to carbon monoxide (Alberta and Dawson, 1987; Alberta *et al.*, 1989). Recently, the structure of the alpha subunit of Dmm containing the haem-binding domain has been resolved by X-ray crystallography (Ortmayer *et al.*, 2016).

MMA oxidation to formaldehyde and ammonium in this pathway is rather complicated, involving two unusual amino acids, gamma-glutamylmethylamide (GMA) and *N*-methylglutamate (NMG), and three enzymes, GMA synthetase, NMG synthase and NMG dehydrogenase (Anthony, 1982; Chen *et al.*, 2010b; Latypova *et al.*, 2010). GMA synthetase has been purified from *Methylovorus mays* and *Methylophaga* sp. (Kimura *et al.*, 1992; Yamamoto *et al.*, 2008). It is evolutionarily related to bacterial glutamine synthetase and similarly it is likely an oligomer made of 8–12 identical subunits (Kimura *et al.*, 1992). NMG synthase, on the other hand, is still poorly studied in terms of substrate specificity, co-factors and the role of individual subunits, MgsABC (Pollock and Hersh, 1971, 1973; Jorns and Hersh, 1975; Latypova *et al.*, 2010). The last enzyme in this pathway is a multimeric NMG dehydrogenase (MgdABCD), one of which contains a tetrahydrofolate (THF) binding domain involved in formaldehyde conjugation. This enzyme has only been partially purified from *Aminobacter aminovorans* owing to its attachment to cytoplasmic membranes (Boulton *et al.*, 1980; Chen *et al.*, 2010b; Latypova *et al.*, 2010). Interestingly, in marine environments, this indirect methylamine oxidation pathway seems to play a key role for sequestering nitrogen from methylated compounds; the methyl groups being oxidized by Roseobacters and SAR11 as a supplementary energy source (Chen *et al.*, 2010a; Sun *et al.*, 2011; Lidbury *et al.*, 2015).

TMA-dependent acetogenesis pathway

It appears that TMA can be also demethylated for acetate formation through an as-yet uncharacterized acetogenesis pathway found in a halophilic homoacetogenic bacterium (Zhilina and Zavarzin, 1990). This bacterium, *Acetohalobium arabaticum*, was isolated from a lagoon near the Sea of Azov and is capable of methylotrophic growth on TMA as a sole carbon source while producing DMA, MMA and acetate. We speculate that this bacterium may use methylamine methyltransferases to channel C1 carbon to the Wood-Ljungdahl pathway for acetogenesis (Wood and Ljundahl, 1991) but this hypothesis still awaits experimental validation. Interestingly, the methyltransferases encoded by *Acetohalobium arabaticum* are induced by TMA, and they have been shown to contain the unusual amino acid pyrrolysine (Prat *et al.*, 2012). Phylogenetically *Acetohalobium arabaticum* pyrrolysine-containing methylamine methyltransferases emerged within the archaeal groups and it is therefore suggested that these bacterial pyrrolysine-containing methylamine methyltransferases were acquired through horizontal gene transfer (Prat *et al.*, 2012). The distribution of this pathway in marine environments and its ecological significance in biogeochemical cycles of MAs are yet to be carefully studied.

Microbial ecology of the metabolism of methylated amines by marine microbes

Thanks to the biochemical characterization of new enzymes involved in methylated amine metabolism using model bacteria over the past decade, it is now possible to use cultivation-independent approaches to evaluate the distribution, diversity and activity of marine microbes involved in MA metabolism.

Development of PCR primers targeting functional genes

PCR primers targeting functional genes involved in MA metabolism have been developed in several studies. In a study to characterize TMA and MMA utilizers in coastal surface waters, two PCR primers targeting *tmm* and *gmaS* of marine Roseobacters have been developed (Chen, 2012). These primers have been applied successfully to retrieve *tmm* and *gmaS* sequences from marine Roseobacters, including several groups that are associated with pelagic marine *Roseobacter* clades (e.g. strains HTCC2255, HTCC2083). The *gmaS* PCR primers were subsequently optimized in order to retrieve novel methylamine-metabolizing methylotrophs in Movile Cave (Wischer *et al.*, 2015). In this study, two sets of reverse PCR primers targeting *Alphaproteobacteria* and *Beta-/Gammaproteobacteria*, respectively, have been designed and optimized.

In order to better understand the diversity of microbes capable of MA metabolism in salt marsh sediments, we have recently also developed PCR primers targeting *tdm* (Lidbury, 2015). These primers were optimized to detect not only alphaproteobacterial Roseobacters and SAR11, but also *Gammaproteobacteria*. A small clone library was made using DNA extracted from the top few centimetres of a coastal salt marsh sediment located in the UK and we have found several novel *tdm* clades.

PCR primers targeting the large subunit of MMA dehydrogenase (MauA) are also available (Neufeld *et al.*, 2007). These primers were designed based on *mauA* from *Paracoccus, Methylobacillus, Methylophilus, Methylobacterium* and *Methylophaga*. These primers have been used successfully to retrieve *mauA* from *Methylophaga* spp. as well as from novel methylamine utilizers in seawater from Station L4 off the coast of Plymouth, UK.

Omics and single-cell approaches

Taking the advantage of the readily available large-scale oceanic metagenomics/metaproteomics datasets, we have carried out studies to uncover the ecology and to estimate the abundance of MA utilizers in marine ecosystems. In a study by Chen *et al.* (2011), we estimated that ≈20% of the bacteria in the surface waters of the Global Oceanic Sampling (GOS) Expedition datasets have the genetic potential for TMA oxidation by Tmm. Tmm sequences retrieved from the GOS datasets are primarily phylogenetically assigned to the SAR11 clade or the *Roseobacter* clade, reflecting the dominance of these bacteria in marine surface waters. Similarly, we showed that SAR11 and *Roseobacter* clade bacteria are predominant in the oceanic metagenome datasets for Tdm, GmaS as well as TmoX (encoding a substrate binding protein for TMAO uptake) (Lidbury *et al.*, 2014, 2017). Interestingly, genes encoding the DMA monooxygenase (DmmDABC) are absent in the SAR11 clade bacteria, and metagenomics/metatranscriptomics data analyses provide no support for the presence of the Dmm pathway in the SAR11 clade (Lidbury *et al.*, 2017). It therefore remains to be seen, if and how SAR11 clade bacteria metabolize DMA.

Several studies have used metagenomics/metaproteomics or single cell approaches to link the identity of marine microbes and their function in MA metabolism. Taubert *et al.* (2017) used ^{15}N-stable isotope probing in combination with metaproteomics to characterize the flow of nitrogen from ^{15}N-MMA to ^{15}N-ammonium in coastal seawater from Plymouth. They were able to confirm the uptake of ^{15}N by *Methylophaga* spp., *Leisingera* spp. as well as a novel gammaproteobacterium that was closely related to strain IMCC2047. Similarly, in an attempt to characterize MMA utilizers in a deep subseafloor, Trembath-Reichert *et al.* (2017) combined stable isotope labelling and nanometre scale secondary ion mass spectrometry (Nano-SIMS). The authors showed that MMA was primarily metabolized by heterotrophic bacteria but also, to some extent, by methanogenic Archaea.

Conclusions and future directions

Although much progress has been made over the past decades towards a better understanding of MA cycling in marine environments, a number of issues are still remaining unresolved, and these are addressed below.

Biochemistry and structural biology of enzymes involved in methylated amine metabolism

Several key enzymes in the MA degradation pathways remain challenging for purification and/or crystallization for X-ray analysis. NMG synthase is an intriguing enzyme which could potentially use either MMA or GMA as substrates. This enzyme has so far only been partially purified, and recombinant expression of the genes *mgsABC* in *Escherichia coli* remains problematic. Two of the NMG synthase subunits, MgsB and MgsC have a significant similarity to glutamate synthase domains. Further characterization of this enzyme will help to better understand its evolutionary relationship to glutamate synthase. The membrane-bound NMG dehydrogenase has also been proven difficult for isolation and purification. It shows similarity to the multimeric sarcosine oxidase complex and has a characteristic glycine cleavage T-protein domain (Moriguchi *et al.*, 2010). TMAO demethylase (Tdm) is another interesting enzyme involved in MA degradation. This enzyme catalyses an unusual oxygen-independent oxidative demethylation, probably using the substrate TMAO as the oxygen donor (Zhu *et al.*, 2016). Structural characterization of these novel enzymes will aid in better understanding of their catalytic activities and cofactor requirements.

Alternative route of DMA catabolism in SAR11 clade bacteria?

SAR11 clade bacteria are numerically abundant in the marine water column and can make up to 50% of bacterioplankton (Giovannoni, 2017). We have demonstrated that Tmm, Tdm and the TMAO transporter protein (TmoX) are functional in catalysing TMA oxidation, TMAO demethylation and TMAO transport, respectively (Chen *et al.*, 2011; Lidbury *et al.*, 2014). The SAR11 bacterium strain HTCC1062 can actively oxidize ^{14}C-TMA, ^{14}C-TMAO and ^{14}C-MMA to $^{14}CO_2$ (Sun *et al.*, 2011). It is therefore puzzling that the genomes of SAR11 bacteria do not contain the *dmmDABC* genes required for DMA oxidation (Lidbury *et al.*, 2017). It remains to be established whether TMA and TMAO can be completely mineralized by SAR11 bacteria to obtain both energy and ammonium from these methylated amines. Our own unpublished data suggest that DMA can be oxidized by at least one strain of SAR11 bacteria, HTCC7211, suggesting the presence of an uncharacterized DMA oxidation pathway in this bacterium.

Interplay between methylated amine cycles and methylated sulfur cycles

Methylated amine metabolism, in many ways, resembles methylated sulfur (dimethylsulfide-dimethylsulfoniopropionate) cycles in the marine environment. Both methylated amines and methylated sulfurs are likely mainly derived from marine phytoplankton. Of particular interest is the recent finding that TMA monooxygenase from marine bacteria can also oxidize dimethylsulfide (Chen *et al.*, 2011; Lidbury *et al.*, 2016). The apparent K_m of *Roseobacter* and SAR11 Tmm for TMA and dimethylsulfide is comparable and the Tmm enzyme can completely oxidize dimethylsulfide to dimethylsulfoxide. Oxidation of dimethylsulfide to dimethylsulfoxide in the model *Roseobacter* clade bacterium *Ruegeria pomeroyi* DSS-3 is entirely dependent on the presence of MAs, particularly DMA, which often is the most abundant MA species in surface seawaters. This interesting interplay between marine methylamine and dimethylsulfide metabolism highlights the importance of studying these molecules concomitantly.

Oceanic sources of methylated amines

The sources of oceanic MAs, particularly in the well-oxygenated surface seawaters remain undetermined. TMA can be generated via anaerobic microbial metabolism through the degradation of quaternary amines, such as choline, glycine betaine and TMAO (Fig. 7.1). Choline is a common dissolved organic nitrogen species which can be released into the marine environment through degradation of phospholipids or conversion from choline sulfate, an osmolyte used by certain seagrasses and coastal plants (Hutchings and Saenger, 1987). Choline degradation to TMA is carried out by an oxygen-sensitive glycyl radical containing choline-TMA lyase (Bodea *et al.*, 2016). Glycine betaine and TMAO are both common osmolytes used by many marine organisms, which can be converted to TMA via betaine reductase and TMAO reductase, respectively (Oren, 1990; Seibel and Walsh, 2002). However, it remains unestablished whether these largely anaerobic processes occur in the oxygenated marine water column and whether they contribute to the formation of MA in surface waters.

Formation of TMA through aerobic processes does occur in microorganisms. For example, we have recently identified a Rieske-containing oxygenase which can generate TMA for the cleavage of carnitine, a common metabolite in mammals (Zhu *et al.*, 2014a). It is therefore plausible that carnitine oxygenase or other similar Rieske oxygenases may contribute to the formation of TMA in surface seawaters. Interestingly, degradation of betaine-containing diacylglycerol lipids may represent an overlooked source of oceanic TMA. Betaine lipids are common in many marine phytoplankton, synthesized in response to phosphorus deficiency, and at least one species of betaine lipids, 1(3),2-Diacylglyceryl-3(1)-*O*-2′-(hydroxymethyl)(*N,N,N*,-trimethyl)-β-alanine (DGTA), can be spontaneously degraded to TMA (Vogel *et al.*, 1990; Dembitsky, 1996). It therefore remains to be established whether these aerobic processes are mainly responsible for TMA releases into the water column in marine environments.

Acknowledgements

The work on methylated amine metabolism was generously supported by the Natural Environment Research Council (NERC) of the UK, through grants NE/H016236/1, NE/M002233/1 and NE/I027061/1.

References

Abduhl-Rashid, M.K. (1990). Studies of Aliphatic Amines and Other Volatile Organic Compounds in the Marine Environment. PhD.thesis (The University of Liverpool, Liverpool), pp. 326.

Abdul-Rashid, M.K., Riley, J.P., Fitzsimons, M.F., and Wolff, G.A. (1991). Determination of volatile amines in sediment and water samples. Anal. Chim. Acta *252*, 223–226. https://doi.org/10.1016/0003-2670(91)87219-W

Alberta, J.A., and Dawson, J.H. (1987). Purification to homogeneity and initial physical characterization of secondary amine monooxygenase. J. Biol. Chem. *262*, 11857–11863.

Alberta, J.A., Andersson, L.A., and Dawson, J.H. (1989). Spectroscopic characterization of secondary amine mono-oxygenase. Comparison to cytochrome P-450 and myoglobin. J. Biol. Chem. *264*, 20467–20473.

Almeida, J., Schobesberger, S., Kürten, A., Ortega, I.K., Kupiainen-Määttä, O., Praplan, A.P., Adamov, A., Amorim, A., Bianchi, F., Breitenlechner, M., *et al.* (2013). Molecular understanding of sulphuric acid-amine particle nucleation in the atmosphere. Nature *502*, 359–363. https://doi.org/10.1038/nature12663

Anthony, C. (1982). The Biochemistry of Methylotrophs (Academic Press Inc., London).

Bodea, S., Funk, M.A., Balskus, E.P., and Drennan, C.L. (2016). Molecular basis of C-N bond cleavage by the glycyl radical enzyme choline trimethylamine-lyase. Cell Chem. Biol. *23*, 1206–1216.

Borrel, G., O'Toole, P.W., Harris, H.M., Peyret, P., Brugère, J.F., and Gribaldo, S. (2013). Phylogenomic data support a seventh order of methylotrophic methanogens and provide insights into the evolution of methanogenesis. Genome Biol. Evol. *5*, 1769–1780. https://doi.org/10.1093/gbe/evt128

Boulton, C.A., Haywood, G.W., and Large, P.J. (1980). *N*-methylglutamate dehydrogenase, a flavohaemoprotein purified from a new pink trimethylamine-utilizing bacterium. J. Gen. Microbiol. *117*, 293–304.

Brugère, J.F., Borrel, G., Gaci, N., Tottey, W., O'Toole, P.W., and Malpuech-Brugère, C. (2014). Archaebiotics: proposed therapeutic use of archaea to prevent trimethylaminuria and cardiovascular disease. Gut Microbes 5, 5–10. https://doi.org/10.4161/gmic.26749

Burgess, S.G., Messiha, H.L., Katona, G., Rigby, S.E., Leys, D., and Scrutton, N.S. (2008). Probing the dynamic interface between trimethylamine dehydrogenase (TMADH) and electron transferring flavoprotein (ETF) in the TMADH-2ETF complex: role of the Arg-α237 (ETF) and Tyr-442 (TMADH) residue pair. Biochemistry 47, 5168–5181. https://doi.org/10.1021/bi800127d

Carlucci, F.V., and Karmas, E. (1988). Liquid chromatographic determination of some amines and their amino acid precursors in protein foods. J. Assoc. Off. Anal. Chem. *71*, 564–568.

Carpenter, L.J., Archer, S.D., and Beale, R. (2012H). Ocean-atmosphere trace gas exchange. Chem. Soc. Rev. *41*, 6473–6506. https://doi.org/10.1039/c2cs35121h

Chen, Y. (2012). Comparative genomics of methylated amine utilization by marine *Roseobacter* clade bacteria and development of functional gene markers (*tmm, gmaS*). Environ. Microbiol. *14*, 2308–2322. https://doi.org/10.1111/j.1462-2920.2012.02765.x

Chen, Y., McAleer, K.L., and Murrell, J.C. (2010a). Monomethylamine as a nitrogen source for a nonmethylotrophic bacterium, *Agrobacterium tumefaciens*. Appl. Environ. Microbiol. *76*, 4102–4104. https://doi.org/10.1128/AEM.00469-10

Chen, Y., Scanlan, J., Song, L., Crombie, A., Rahman, M.T., Schäfer, H., and Murrell, J.C. (2010b). γ-Glutamylmethylamide is an essential intermediate in the metabolism of methylamine by *Methylocella silvestris*. Appl. Environ. Microbiol. *76*, 4530–4537. https://doi.org/10.1128/AEM.00739-10

Chen, Y., Patel, N.A., Crombie, A., Scrivens, J.H., and Murrell, J.C. (2011). Bacterial flavin-containing monooxygenase is trimethylamine monooxygenase. Proc. Natl. Acad. Sci. U.S.A. *108*, 17791–17796. https://doi.org/10.1073/pnas.1112928108

Chistoserdova, L. (2011). Modularity of methylotrophy, revisited. Environ. Microbiol. *13*, 2603–2622. https://doi.org/10.1111/j.1462-2920.2011.02464.x

Cree, C.H.L., Airs, R., Archer, S.D., and Fitzsimons, M.F. (2018). Measurement of methylamines in seawater using solid phase microextraction and gas chromatography. Limnol. Oceanogr. Meth. *16*, 411–420. https://doi.org/10.1002/lom3.10255

Dembitsky, V.M. (1996). Betaine ether-linked glycerolipids: chemistry and biology. Prog. Lipid Res. *35*, 1–51.

Eady, R.R., and Large, P.J. (1968). Purification and properties of an amine dehydrogenase from *Pseudomonas* AM1 and its role in growth on methylamine. Biochem. J. *106*, 245–255.

Fitzsimons, M.F., Jemmett, A.W., and Wolff, G.A. (1997). A preliminary study of the geochemistry of methylamines in a salt marsh. Org. Geochem. *27*, 15–24. https://doi.org/10.1016/S0146-6380(97)00062-4

Fitzsimons, M.F., Kahni-danon, B., and Dawitt, M. (2001). Distributions and adsorption of the methylamines in the inter-tidal sediments of an East Anglian Estuary. Environ. Exp. Bot. *46*, 225–236. https://doi.org/10.1016/S0098-8472(01)00102-2

Fitzsimons, M.F., Millward, G.E., Revitt, D.M., and Dawit, M.D. (2006). Desorption kinetics of ammonium and methylamines from estuarine sediments: Consequences for the cycling of nitrogen. Mar. Chem. *101*, 12–26. https://doi.org/10.1016/j.marchem.2005.12.006

Gibb, S.W. (1994). Trace Determination of Ammonia and Methylamines by Flow Injection Extraction: Ion Chromatography in Estuarine and Marine Environments. Ph.D. Thesis (The University of East Anglia, Norwich).

Gibb, S.W., and Hatton, A.D. (2004). The occurrence and distribution of trimethylamine-*N*-oxide in Antarctic coastal waters. Mar. Chem. *91*, 65–75. https://doi.org/10.1016/j.marchem.2004.04.005

Gibb, S.W., Mantoura, R.F.C., and Liss, P.S. (1995a). Analysis of ammonia and methylamines in natural waters by flow injection gas diffusion coupled to ion chromatography. Anal. Chim. Acta *316*, 291–304. https://doi.org/10.1016/0003-2670(95)00372-7

Gibb, S.W., Wood, J.W., Fauzi, R., and Mantoura, C. (1995b). Automation of flow injection gas diffusion-ion chromatography for the nanomolar determination of methylamines and ammonia in seawater and atmospheric samples. J. Automat. Chem. *17*, 205–212. https://doi.org/10.1155/S1463924695000320

Gibb, S.W., Mantoura, R.F.C., and Liss, P.S. (1999a). Ocean-atmosphere exchange and atmospheric speciation of ammonia and methylamines in the region of the NW Arabian Sea. Global Biogeochem. Cy. *13*, 161–178. https://doi.org/10.1029/98GB00743

Gibb, S.W., Mantoura, R.F.C., Liss, P.S., and Barlow, R.G. (1999b). Distributions and biogeochemistries of methylamines and ammonium in the Arabian Sea. Deep Sea Res. II *46*, 593–615. https://doi.org/10.1016/S0967-0645(98)00119-2

Giovannoni, S.J. (2017). SAR11 bacteria: the most abundant plankton in the oceans. Ann. Rev. Mar.

Sci. *9*, 231–255. https://doi.org/10.1146/annurev-marine-010814-015934
Glob, E., and Sørensen, J. (1987). Determination of dissolved and exchangeable trimethylamine pools in sediments. J. Microbiol. Methods *6*, 347–355. https://doi.org/10.1016/0167-7012(87)90020-0
Hatton, A.D., and Gibb, S.W. (1999). A technique for the determination of trimethylamine-*N*-oxide in natural waters and biological media. Anal. Chem. *71*, 4886–4891.
Haywood, G.W., Janschke, N.S., Large, P.J., and Wallis, J.M. (1982). Properties and subunit structure of methylamine dehydrogenase from *Thiobacillus* A2 and *Methylophilus methylotrophus*. FEMS Microbiol. Lett. *15*, 79-82. https://doi.org/doi:10.1111/j.1574-6968.1982.tb00042.x
Heinrikson, R.L., and Meredith, S.C. (1984). Amino acid analysis by reverse-phase high-performance liquid chromatography: precolumn derivatization with phenylisothiocyanate. Anal. Biochem. *136*, 65–74.
Hippe, H., Caspari, D., Fiebig, K., and Gottschalk, G. (1979). Utilization of trimethylamine and other *N*-methyl compounds for growth and methane formation by *Methanosarcina barkeri*. Proc. Natl. Acad. Sci. U.S.A. *76*, 494–498.
Hutchings, P., and Saenger, P. (1987). The Ecology of Mangroves (University of Queensland Press, St. Lucia).
Iino, T., Tamaki, H., Tamazawa, S., Ueno, Y., Ohkuma, M., Suzuki, K., Igarashi, Y., and Haruta, S. (2013). *Candidatus* Methanogranum caenicola: a novel methanogen from the anaerobic digested sludge, and proposal of *Methanomassiliicoccaceae* fam. nov. and *Methanomassiliicoccales* ord. nov., for a methanogenic lineage of the class *Thermoplasmata*. Microbes Environ. *28*, 244–250.
Jacobs, W.A. (1987). o-Phthalaldehyde-sulfite derivatization of primary amines for liquid chromatography-electrochemistry. J. Chromatogr. *392*, 435–441.
Jameson, E., Stephenson, J., Jones, H., Millard, A., Kaster, A.K., Purdy, K.J., Airs, R., Murrell, J.C., and Chen, Y. (2019). *Deltaproteobacteria* (*Pelobacter*) and *Methanococcides* are responsible for choline-dependent methanogenesis in a coastal saltmarsh sediment. ISME J. *13*, 277–289. https://doi.org/10.1038s41396-018-0269-8
Jorns, M.S., and Hersh, L.B. (1975). *N*-Methylglutamate synthetase. Substrate-flavin hydrogen transfer reactions probed with deazaflavin mononucleotide. J. Biol. Chem. *250*, 3620–3628.
Kim, H.G., Kim, Y., Lim, H.M., Shin, H.-J., and Kim, S.W. (2006). Purification, characterization, and cloning of trimethylamine dehydrogenase from *Methylophaga* sp. strain SK1. Biotechnol. Bioproc. E. *11*, 337–343. https://doi.org/10.1007/BF03026250
Kimura, T., Sugahara, I., Hanai, K., and Tonomura, Y. (1992). Purification and characterization of γ-glutamylmethylamide synthetase from *Methylophaga* sp. AA-30. Biosci. Biotechnol. Biochem. *56*, 708–711. https://doi.org/10.1271/bbb.56.708
King, G.M. (1984). Metabolism of trimethylamine, choline, and glycine betaine by sulfate-reducing and methanogenic bacteria in marine sediments. Appl. Environ. Microbiol. *48*, 719–725.
Kröninger, L., Gottschling, J., and Deppenmeier, U. (2017). Growth characteristics of *Methanomassiliicoccus luminyensis* and expression of methyltransferase encoding genes. Archaea *2017*, 2756573. https://doi.org/10.1155/2017/2756573
Latypova, E., Yang, S., Wang, Y.S., Wang, T., Chavkin, T.A., Hackett, M., Schäfer, H., and Kalyuzhnaya, M.G. (2010). Genetics of the glutamate-mediated methylamine utilization pathway in the facultative methylotrophic beta-proteobacterium *Methyloversatilis universalis* FAM5. Mol. Microbiol. *75*, 426–439. https://doi.org/10.1111/j.1365-2958.2009.06989.x
Lee, C., and Olson, B.L. (1984). Dissolved, exchangeable and bound aliphatic amines in marine sediments: initial results. Org. Geochem. *6*, 259–263. https://doi.org/10.1016/0146-6380(84)90047-0
Lidbury, I. (2015). Microbial Methylated Amine Metabolism in Marine Surface Waters. PhD. Thesis (University of Warwick, Coventry), pp. 245.
Lidbury, I., Murrell, J.C., and Chen, Y. (2014). Trimethylamine *N*-oxide metabolism by abundant marine heterotrophic bacteria. Proc. Natl. Acad. Sci. U.S.A. *111*, 2710–2715. https://doi.org/10.1073/pnas.1317834111
Lidbury, I.D., Murrell, J.C., and Chen, Y. (2015). Trimethylamine and trimethylamine *N*-oxide are supplementary energy sources for a marine heterotrophic bacterium: implications for marine carbon and nitrogen cycling. ISME J. *9*, 760–769. https://doi.org/10.1038/ismej.2014.149
Lidbury, I., Kröber, E., Zhang, Z., Zhu, Y., Murrell, J.C., Chen, Y., and Schäfer, H. (2016). A mechanism for bacterial transformation of dimethylsulfide to dimethylsulfoxide: a missing link in the marine organic sulfur cycle. Environ. Microbiol. *18*, 2754–2766. https://doi.org/10.1111/1462-2920.13354
Lidbury, I., Mausz, M.A., Scanlan, D.J., and Chen, Y. (2017). Identification of dimethylamine monooxygenase in marine bacteria reveals a metabolic bottleneck in the methylated amine degradation pathway. ISME J. *11*, 1592–1601. https://doi.org/10.1038/ismej.2017.31
Lim, L.W., Xia, Z., Mathews, F.S., and Davidson, V.L. (1986). Preliminary X-ray crystallographic study of methanol dehydrogenase from *Methylophilus methylotrophus*. J. Mol. Biol. *191*, 141–142.
Mah, R.A. (1980). Isolation and characterization of *Methanococcus mazei*. Curr. Microbiol. *3*, 321–326. https://doi.org/10.1007/BF02601895
Mellbin, G., and Smith, B.E.F. (1984). Trace determination of aliphatic amines using high-performance liquid chromatography with chemiluminescence excitation and photon counting. J. Chromatogr. A *312*, 203–210. https://doi.org/10.1016/S0021-9673(01)92775-X
Mopper, K., and Zika, R.G. (1987). Free amino acids in marine rains: evidence for oxidation and potential role in nitrogen cycling. Nature *325*, 246–249. https://doi.org/10.1038/325246a0
Moriguchi, T., Ida, K., Hikima, T., Ueno, G., Yamamoto, M., and Suzuki, H. (2010). Channeling and conformational changes in the heterotetrameric sarcosine oxidase from *Corynebacterium* sp. U-96. J. Biochem. *148*, 491–505. https://doi.org/10.1093/jb/mvq083

Nayak, D.D., and Marx, C.J. (2015). Experimental horizontal gene transfer of methylamine dehydrogenase mimics prevalent exchange in nature and overcomes the methylamine growth constraints posed by the sub-optimal *N*-methylglutamate pathway. Microorganisms *3*, 60–79. https://doi.org/10.3390/microorganisms3010060

Neufeld, J.D., Schäfer, H., Cox, M.J., Boden, R., McDonald, I.R., and Murrell, J.C. (2007). Stable-isotope probing implicates *Methylophaga* spp and novel *Gammaproteobacteria* in marine methanol and methylamine metabolism. ISME J. *1*, 480–491.

Oremland, R.S., Marsh, L.M., and Polcin, S. (1982). Methane production and simultaneous sulphate reduction in anoxic, salt marsh sediments. Nature *296*, 143–145. https://doi.org/10.1038/296143a0

Oren, A. (1990). Formation and breakdown of glycine betaine and trimethylamine in hypersaline environments. Antonie Van Leeuwenhoek *58*, 291–298.

Ortmayer, M., Lafite, P., Menon, B.R., Tralau, T., Fisher, K., Denkhaus, L., Scrutton, N.S., Rigby, S.E., Munro, A.W., Hay, S., *et al.* (2016). An oxidative *N*-demethylase reveals PAS transition from ubiquitous sensor to enzyme. Nature *539*, 593–597. https://doi.org/10.1038/nature20159

Parkes, J.R., Brock, F., Banning, N., Hornibrook, E.R.C., Roussel, E.G., Weightman, A.J., and Fry, J.C. (2012). Changes in methanogenic substrate utilization and communities with depth in a salt-marsh, creek sediment in southern England. Estuar. Coast. Shelf Sci. *96*, 170–178. https://doi.org/10.1016/j.ecss.2011.10.025

Paul, K., Nonoh, J.O., Mikulski, L., and Brune, A. (2012). '*Methanoplasmatales*,' *Thermoplasmatales*-related archaea in termite guts and other environments, are the seventh order of methanogens. Appl. Environ. Microbiol. *78*, 8245–8253. https://doi.org/10.1128/AEM.02193-12

Pollock, R.J., and Hersh, L.B. (1971). *N*-methylglutamate synthetase. Purification and properties of the enzyme. J. Biol. Chem. *246*, 4737–4743.

Pollock, R.J., and Hersh, L.B. (1973). *N*-methylglutamate synthetase. The use of flavin mononucleotide in oxidative catalysis. J. Biol. Chem. *248*, 6724–6733.

Poulsen, M., Schwab, C., Jensen, B.B., Engberg, R.M., Spang, A., Canibe, N., Højberg, O., Milinovich, G., Fragner, L., Schleper, C., *et al.* (2013). Methylotrophic methanogenic Thermoplasmata implicated in reduced methane emissions from bovine rumen. Nat. Commun. *4*, 1428. https://doi.org/10.1038/ncomms2432

Prat, L., Heinemann, I.U., Aerni, H.R., Rinehart, J., O'Donoghue, P., and Söll, D. (2012). Carbon source-dependent expansion of the genetic code in bacteria. Proc. Natl. Acad. Sci. U.S.A. *109*, 21070–21075. https://doi.org/10.1073/pnas.1218613110

Rother, M. (2010). Methanogensis. In Handbook of Hydrocarbon and Lipid Microbiology, K.N. Timmis, ed. (Springer, Berlin, Heidelberg), pp. 481-499. https://0-doi-org.pugwash.lib.warwick.ac.uk/10.1007/978-3-540-77587-4_36

Schobesberger, S., Junninen, H., Bianchi, F., Lönn, G., Ehn, M., Lehtipalo, K., Dommen, J., Ehrhart, S., Ortega, I.K., Franchin, A., *et al.* (2013). Molecular understanding of atmospheric particle formation from sulfuric acid and large oxidized organic molecules. Proc. Natl. Acad. Sci. U.S.A. *110*, 17223–17228. https://doi.org/10.1073/pnas.1306973110

Seibel, B.A., and Walsh, P.J. (2002). Trimethylamine oxide accumulation in marine animals: relationship to acylglycerol storage. J. Exp. Biol. *205*, 297–306.

Sørensen, J., and Glob, E. (1987). Influence of benthic fauna on trimethylamine concentrations in coastal marine sediments. Mar. Ecol. Prog. Ser. *39*, 15–21. http://www.jstor.org/stable/24825641

Sun, J., Steindler, L., Thrash, J.C., Halsey, K.H., Smith, D.P., Carter, A.E., Landry, Z.C., and Giovannoni, S.J. (2011). One carbon metabolism in SAR11 pelagic marine bacteria. PLOS ONE *6*, e23973. https://doi.org/10.1371/journal.pone.0023973

Taubert, M., Grob, C., Howat, A.M., Burns, O.J., Pratscher, J., Jehmlich, N., von Bergen, M., Richnow, H.H., Chen, Y., and Murrell, J.C. (2017). Methylamine as a nitrogen source for microorganisms from a coastal marine environment. Environ. Microbiol. *19*, 2246–2257. https://doi.org/10.1111/1462-2920.13709

Trembath-Reichert, E., Morono, Y., Ijiri, A., Hoshino, T., Dawson, K.S., Inagaki, F., and Orphan, V.J. (2017). Methyl-compound use and slow growth characterize microbial life in 2-km-deep subseafloor coal and shale beds. Proc. Natl. Acad. Sci. U.S.A. *114*, E9206–E9215. https://doi.org/10.1073/pnas.1707525114

van der Maarel, M.J.E.C., and Hansen, T.A. (1997). Dimethylsulfoniopropionate in anoxic intertidal sediments: a precursor of methanogenesis via dimethyl sulfide, methanethiol, and methiolpropionate. Mar. Geol. *137*, 5–12. https://doi.org/10.1016/S0025-3227(96)00074-6

Van Neste, A., Duce, R.A., and Lee, C. (1987). Methylamines in the marine atmosphere. Geophys. Res. Lett. *14*, 711–714. https://doi.org/10.1029/GL014i007p00711

Vogel, G., Woznica, M., Gfeller, H., Müller, C., Stämpfli, A.A., Jenny, T.A., and Eichenberger, W. (1990). 1(3),2-Diacylglyceryl-3(1)-*O*-2′-(hydroxymethyl) (*N*,*N*,*N*,-trimethyl)-β-alanine (DGTA): A novel betaine lipid from *Ochoromonas danica* (Chrysophyceae). Chem. Phys. Lipids *52*, 99–109. https://doi.org/10.1016/0009-3084(90)90154-J

Wang, X.-C., and Lee, C. (1990). The distribution and adsorption behavior of aliphatic amines in marine and lacustrine sediments. Geochim. Cosmochim. Acta *54*, 2759–2774. https://doi.org/10.1016/0016-7037(90)90010-I

Wang, X.-C., and Lee, C. (1994). Sources and distribution of aliphatic amines in salt marsh sediment. Org. Geochem. *22*, 1005–1021. https://doi.org/10.1016/0146-6380(94)90034-5

Wischer, D., Kumaresan, D., Johnston, A., El Khawand, M., Stephenson, J., Hillebrand-Voiculescu, A.M., Chen, Y., and Colin Murrell, J. (2015). Bacterial metabolism of methylated amines and identification of novel methylotrophs in Movile Cave. ISME J. *9*, 195–206. https://doi.org/10.1038/ismej.2014.102

Wood, H.G., and Ljundahl, L. (1991). Autotrophic Character of the Acetogenic Bacteria. In Variations in Autotrophic Life, J.M. Shively, and L. Barton, eds. (Academic Press Inc., San Diego).

Yamamoto, S., Wakayama, M., and Tachiki, T. (2008). Cloning and expression of *Methylovorus mays* No. 9 gene encoding gamma-glutamylmethylamide synthetase: an enzyme usable in theanine formation by coupling with the alcoholic fermentation system of baker's yeast. Biosci. Biotechnol. Biochem. 72, 101–109.

Yang, C.C., Packman, L.C., and Scrutton, N.S. (1995). The primary structure of *Hyphomicrobium* X dimethylamine dehydrogenase. Relationship to trimethylamine dehydrogenase and implications for substrate recognition. Eur. J. Biochem. *232*, 264–271.

Yang, X.-H., Lee, C., and Scranton, M.I. (1993). Determination of nanomolar concentrations of individual dissolved low molecular weight amines and organic acids in seawater. Anal. Chem. *65*, 572–576. https://doi.org/10.1021/ac00053a014

Yang, X.-H., Scranton, M.I., and Lee, C. (1994). Seasonal variations in concentration and microbial uptake of methylamines in estuarine waters. Mar. Ecol. Prog. Ser. *108*, 303–312.

Zhilina, T.N., and Zavarzin, G.A. (1990). Extremely halophilic, methylotrophic, anaerobic bacteria. FEMS Microbiol. Lett. *87*, 315–322. https://doi.org/10.1111/j.1574-6968.1990.tb04930.x

Zhu, Y., Jameson, E., Crosatti, M., Schäfer, H., Rajakumar, K., Bugg, T.D., and Chen, Y. (2014a). Carnitine metabolism to trimethylamine by an unusual Rieske-type oxygenase from human microbiota. Proc. Natl. Acad. Sci. U.S.A. *111*, 4268–4273. https://doi.org/10.1073/pnas.1316569111

Zhu, Y., Jameson, E., Parslow, R.A., Lidbury, I., Fu, T., Dafforn, T.R., Schäfer, H., and Chen, Y. (2014b). Identification and characterization of trimethylamine *N*-oxide (TMAO) demethylase and TMAO permease in *Methylocella silvestris* BL2. Environ. Microbiol. *16*, 3318–3330. https://doi.org/10.1111/1462-2920.12585

Zhu, Y., Ksibe, A.Z., Schäfer, H., Blindauer, C.A., Bugg, T.D., and Chen, Y. (2016). O2 -independent demethylation of trimethylamine *N*-oxide by Tdm of *Methylocella silvestris*. FEBS J. *283*, 3979–3993. https://doi.org/10.1111/febs.13902

Zhuang, G.-C., Lin, Y.-S., Bowles, M.W., Heuer, V.B., Lever, M.A., Elvert, M., and Hinrichs, K.-U. (2017). Distribution and isotopic composition of trimethylamine, dimethylsulfide and dimethylsulfoniopropionate in marine sediments. Mar. Chem. *196*, 35–46. https://doi.org/10.1016/j.marchem.2017.07.007

Methylotrophs and Methylotroph Populations for Chloromethane Degradation

8

Françoise Bringel[1*], Ludovic Besaury[2], Pierre Amato[3], Eileen Kröber[4], Steffen Kolb[4], Frank Keppler[5,6], Stéphane Vuilleumier[1] and Thierry Nadalig[1]

[1]Université de Strasbourg UMR 7156 UNISTRA CNRS, Molecular Genetics, Genomics, Microbiology (GMGM), Strasbourg, France.
[2]Université de Reims Champagne-Ardenne, Chaire AFERE, INRA, FARE UMR A614, Reims, France.
[3] Université Clermont Auvergne, CNRS, SIGMA Clermont, ICCF, Clermont-Ferrand, France.
[4]Microbial Biogeochemistry, Research Area Landscape Functioning – Leibniz Centre for Agricultural Landscape Research – ZALF, Müncheberg, Germany.
[5]Institute of Earth Sciences, Heidelberg University, Heidelberg, Germany.
[6]Heidelberg Center for the Environment HCE, Heidelberg University, Heidelberg, Germany.

*Correspondence: francoise.bringel@unistra.fr

https://doi.org/10.21775/9781912530045.08

Abstract

Chloromethane is a halogenated volatile organic compound, produced in large quantities by terrestrial vegetation. After its release to the troposphere and transport to the stratosphere, its photolysis contributes to the degradation of stratospheric ozone. A better knowledge of chloromethane sources (production) and sinks (degradation) is a prerequisite to estimate its atmospheric budget in the context of global warming. The degradation of chloromethane by methylotrophic communities in terrestrial environments is a major underestimated chloromethane sink. Methylotrophs isolated from soils, marine environments and more recently from the phyllosphere have been grown under laboratory conditions using chloromethane as the sole carbon source. In addition to anaerobes that degrade chloromethane, the majority of cultivated strains were isolated in aerobiosis for their ability to use chloromethane as sole carbon and energy source. Among those, the Proteobacterium *Methylobacterium* (recently reclassified as *Methylorubrum*) harbours the only characterized 'chloromethane utilization' (*cmu*) pathway, so far. This pathway may not be representative of chloromethane-utilizing populations in the environment as *cmu* genes are rare in metagenomes. Recently, combined 'omics' biological approaches with chloromethane carbon and hydrogen stable isotope fractionation measurements in microcosms, indicated that microorganisms in soils and the phyllosphere (plant aerial parts) represent major sinks of chloromethane in contrast to more recently recognized microbe-inhabited environments, such as clouds. Cultivated chloromethane-degraders lacking *cmu* genes display a singular isotope fractionation signature of chloromethane. Moreover, $^{13}CH_3Cl$ labelling of active methylotrophic communities by stable isotope probing in soils identify taxa that differ from those known for chloromethane degradation. These observations suggest that new biomarkers for detecting active microbial chloromethane-utilizers in the environment are needed to assess the contribution of microorganisms to the global chloromethane cycle.

Introduction

Chloromethane and stratospheric ozone depletion

Chloromethane (methyl chloride, CH_3Cl) is the most abundant organohalogen in the Earth atmosphere. Its global production is estimated at 4–5 megatons per year, with main sources stemming from terrestrial vegetation (Keppler, 2005). Photolytic degradation of chloromethane releases a halogen radical, which catalyses the destruction of ozone. Thus, chloromethane contributes to depletion of the stratospheric ozone layer (altitude of approximately 20 to 30 km), which constitutes the Earth's natural protective shield that absorbs the solar UVC and partially UVB radiation dangerous for living organisms. Until the 1990s, chloromethane was used as a refrigerant under the name Freon 40. Chloromethane has a stratospheric lifetime of about one year, much shorter than most other chlorofluorocarbons (CFCs), solvents and halons also banned by the international agreement of 1987 known as the Montreal Protocol on substances that deplete the ozone layer (see web resource section). There are still some uncertainties about chloromethane anthropogenic emissions (e.g. coal combustion, feedstock for chemical industries) (Li *et al.*, 2017). Chloromethane is responsible alone for approximately 16% of stratospheric chlorine-catalysed ozone destruction (Carpenter *et al.*, 2014). A detailed understanding of its sources and sinks will be essential to predict changes in atmospheric chloromethane fluxes in the context of global climate change.

Chloromethane sources and sinks

Chloromethane formation in plants and soil involves biotic and abiotic processes. Chloride ion can be alkylated during the abiotic oxidation of organic matter by an electron acceptor such as Fe(III) in soils and sediments (Keppler *et al.*, 2000). Abiotic chloromethane mainly results from the conversion of plant methoxyl groups (ether- or ester-bonded methyl groups) and their reaction with chloride ion (Keppler *et al.*, 2000; Hamilton *et al.*, 2003; Sailaukhanuly *et al.*, 2014). This process occurs in terrestrial ecosystems at ambient temperatures (Derendorp *et al.*, 2012; Keppler *et al.*, 2014), but it is much more efficient at higher temperatures such as those reached during pyrolysis and biomass burning (Hamilton *et al.*, 2003; Keppler, 2005; McRoberts *et al.*, 2015). Chloromethane release was also detected during thermal conversion of Martian soils by the Mars landers Viking (Biemann *et al.*, 1976) and Curiosity (Ming *et al.*, 2014), indicating the presence of endogenous organic matter on Mars. Chloromethane emission profiles during thermal treatment of soils sampled from other hyperarid environments hostile microbial life such as the Atacama desert were almost identical to those recorded by the Curiosity rover on Mars (Schulze-Makuch *et al.*, 2018). Furthermore, chloromethane formation was observed from thermal conversion of extraterrestrial material such as carbonaceous meteorites (Keppler *et al.*, 2014) and in protostellar environments (Fayolle *et al.*, 2017). These recent observations contributed to the emergence of an 'astronomical' fundamental interest in the understanding of chloromethane's cycle in sun-like stars and on Earth.

On Earth, plants (alive or decaying) are a major biotic source of chloromethane. Chloromethane is produced by enzymatic chloride ion methylation, as shown in higher plants affiliated to the *Brassicaceae* family (Attieh *et al.*, 1995; Rhew *et al.*, 2003) and wood-degrading fungi (Harper *et al.*, 1990). The S-adenosyl-L-methionine-dependent halide ion methyltransferase is encoded by gene *HOL1* (Harmless to Ozone Layer) in *Arabidopsis thaliana* (Nagatoshi and Nakamura, 2009). *HOL1* gene disruption correlates with decreased pathogen defence, possibly due to reduced production of methyl thiocyanate (CH_3SCN) from glucosinolate-derived thiocyanate by HOL1 (Manley, 2002; Rhew *et al.*, 2003; Nagatoshi and Nakamura, 2009). Chloromethane production is considered a by-product of plant thiocyanate metabolism. Such methyltransferases have also been detected in other crop and seaside plants (Itoh *et al.*, 2009) including marine algae (Wuosmaa and Hager, 1990; Ohsawa *et al.*, 2001; Toda and Itoh, 2011).

Identified chloromethane sinks are dominated by abiotic loss processes in the atmosphere involving reaction with OH radicals, or via chlorine radicals in the marine atmospheric boundary layer (see Web resources). The extent of consumption of chloromethane under the control of biological processes, especially by microorganisms, constitutes one of the largest uncertainties regarding the global budget of chloromethane (Harper and Hamilton, 2003; Keppler, 2005).

Microbial degradation: an underestimated sink in the global chloromethane budget

The study of ecology and diversity of chloromethane and other methyl halide-degrading microorganisms remain a challenging field of environmental biology. Key microbial enzymes of (de)halogenation activity and active microorganisms in chlorine and other halogens (fluorine, bromine, iodine) cycling remain largely unknown (Weigold *et al.*, 2016). The global impact of microorganisms on the chloromethane sink is difficult to quantify due to concomitant processes of production and degradation of chloromethane in the environment. For instance, highly fluctuating chloromethane emissions occur in fern plants as recently discussed (Jaeger *et al.*, 2018b). Both production and degradation of chloromethane may coexist within a single organism as reported for lignin-degrading fungi. Unlike most other wood-rotting fungi, in *Phanerochaete chrysosporium*, *Phlebia radiata*, and *Coriolus versicolor*, chloromethane serves as an endogenous methyl donor in veratryl alcohol biosynthesis so that no chloromethane is emitted during lignin degradation and growth (Harper *et al.*, 1990; Coulter *et al.*, 1993). However, so far, methyl halide (CH_3X; chloromethane, bromomethane and iodomethane) cycling studies focused on chloromethane degradation by bacteria rather than by fungi (Leisinger and Braus-Stromeyer, 1995; Harper, 2000; McDonald *et al.*, 2002; Schäfer *et al.*, 2007; Cox *et al.*, 2012). While the possible implications of fungal populations on the chloromethane cycle remain to be assessed using 'omics' approaches, the aim of this review is primarily to discuss recent investigations of bacterial populations associated with chloromethane sinks in contrasting terrestrial environments, these populations potentially differing from the so far characterized microbes with known pathways for chloromethane degradation. This possibility raises questions about the composition, distribution, functioning and evolution of chloromethane-utilizing populations in response to highly fluctuating emissions of this volatile halogenated compound at the soil-plant-atmosphere interfaces.

Assessing bacterial chloromethane sinks

Recent investigations of plant and soil samples suggest that chloromethane-degraders represent a minor fraction of microbial communities (Chaignaud *et al.*, 2018; Jaeger *et al.*, 2018a,b) that are estimated as 10^7 cells per cm^2 of leaf surface to 10^9 cells per g of soil (Vorholt, 2012). In air, microbial concentrations range from a few tens to billions per cubic metre depending on location, altitude above ground, and time of the day and year (Amato *et al.*, 2017a; DasSarma and DasSarma, 2018), representing a total of 10^{19} bacterial cells in the atmosphere (Whitman *et al.*, 1998). Microorganisms collected from aerosols, cloud and rain waters retain metabolic activity under cloud-like conditions and are able to utilize atmospheric organic compounds such as formaldehyde and formate present at relatively high concentrations in cloud droplets (≈ 1–$100\,\mu M$) (Amato *et al.*, 2007; Delort *et al.*, 2010; Vaïtilingom *et al.*, 2013; Šantl-Temkiv *et al.*, 2017). While cloud-borne microbial assemblages include active facultative methylotrophs (Amato *et al.*, 2017b), no isolate grew on chloromethane, no chloromethane consumption was observed (Fig. 8.1), and, consistently, no *cmuA* gene was detected using a highly sensitive capture hybridization method (see below). In contrast, viable methanotrophs (*Methylosinus* and *Methylocystis*) isolated from atmospheric samples (aerosol and rain) (Šantl-Temkiv *et al.*, 2013) were able to use methane under cloud-like physicochemical conditions at concentrations relevant for the atmospheric environment ($\approx 1.5\,ppm$). We concluded that the microbiota thriving in the investigated clouds do not represent a significant sink for chloromethane, unlike soils, aerial parts of plants (phyllosphere), or marine environments (Goodwin *et al.*, 2005; Schäfer *et al.*, 2007; Chaignaud *et al.*, 2018; Jaeger *et al.*, 2018a,b).

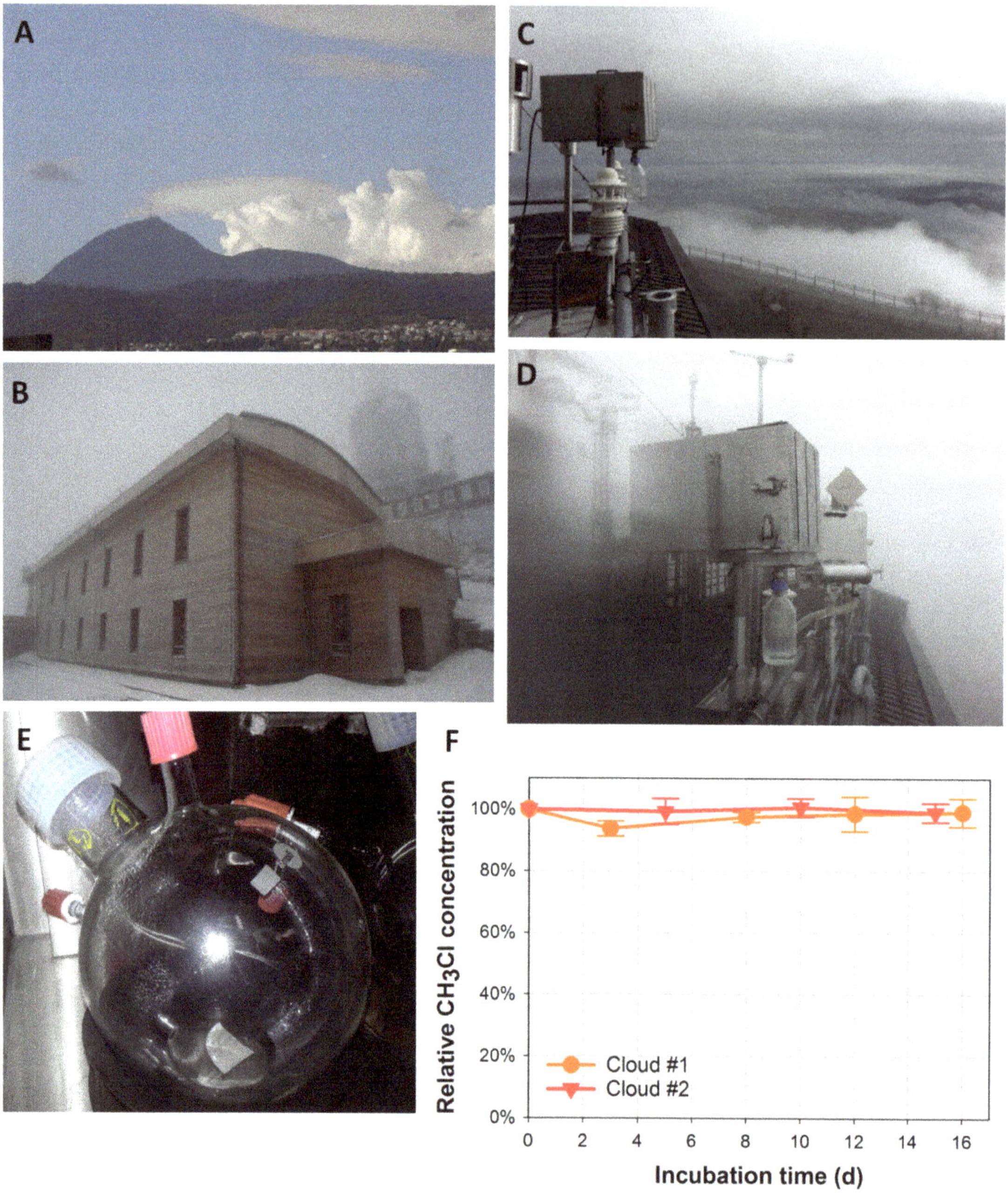

Figure 8.1 Assessing cloud water microorganisms for chloromethane degradation. Cloud water samples were collected under sterile conditions from the atmospheric station at Puy de Dôme mountain summit (1465 m above see level, France) (A) at the roof platform of a historical meteorological station (B) using cloud droplet impactors (C and D) as previously described (Amato *et al.*, 2017b). Airtight microcosms used for testing chloromethane (≈10 ppm) degradation of the collected cloud water samples (E). Two independent cloud water samples (626 ml and 740 ml, respectively, containing $\approx 10^6$ cells pre-concentrated into 20 ml) showed no detectable temporal decrease of chloromethane under the tested conditions at 17°C (F). *M. extorquens* CM4 added in sterile rainwater spiked with 10 mM of chloromethane was used as positive control (data not shown). No isolate grew on chloromethane supplied as the sole source of carbon and energy (data not shown) in enrichment cultures of cloud and rain samples, performed as previously described (Nadalig *et al.*, 2011).

Cultivated chloromethane-degrading methylotrophs

Methylotrophic bacteria capable of utilizing chloromethane as the only source of carbon and/or energy have been isolated (Schaefer *et al.*, 2002; Nadalig *et al.*, 2014) from contrasting environments such as soil, plants, freshwater and marine environments (Table 8.1). Chloromethane degraders are found among facultative aerobes, obligate anaerobes, obligate and facultative methylotrophs, and conditional utilizers of chloromethane. This diversity is encompassed by organisms possessing genomes ranging in size from very small (1.3 Mb) to relatively large (6.2 Mb), G+C contents ranging from 38% to 68%, and chloromethane utilization genes can be chromosome- or plasmid-borne (Table 8.1).

Under strictly anoxic conditions, the only known pure culture that uses chloromethane as the sole energy source is the Gram-positive homoacetogenic autotrophic bacterium, *Acetobacterium dehalogenans* MC (Traunecker *et al.*, 1991). In this strain, chloromethane dehalogenation releases methyl-tetrahydrofolate (H_4F) that is further metabolized to acetate (Meßmer *et al.*, 1993, 1996). *Pseudomonas aeruginosa* NB1 uses chloromethane as the sole source of carbon and energy under either anoxic or aerobic conditions upon switching from nitrate to oxygen as the terminal electron acceptor (Freedman *et al.*, 2004). Another case of anoxic chloromethane-utilization was recently reported and provides the first example of co-utilization of two chlorinated methane substrates in a mixed culture. In the absence of dichloromethane, chloromethane was not degraded as the chloromethane-degrader partner is thought to consume H_2 generated by *Candidatus* Dichloromethanomonas elyunquensis upon dichloromethane degradation (Chen *et al.*, 2017) (Fig. 8.2A).

Under aerobic conditions, cometabolic degradation of chloromethane was observed by nitrifying bacteria (Rasche *et al.*, 1991) and methanotrophs (Han and Semrau, 2000). This is attributed to the activity of ammonium- or methane monooxygenase, respectively. When chloromethane is provided at low concentration in presence of another C_1 substrate, the growth of the facultative methanotroph *Methylomicrobium album* BG8 is enhanced (Fig. 8.2C). Another case of conditional degradation of chloromethane was found for the growth of the abundant coastal Betaproteobacterium strain HTCC2181. This obligate methylotroph utilizes chloromethane for energy only when methanol is supplied as the carbon source (Halsey *et al.*, 2012) by a yet uncharacterized methyltransferase, with possible methyl transfer from chloromethane to H_4F (Fig. 8.2B; discussed in substrate co-utilization section).

Microbial utilization of chloromethane as the sole carbon and energy source (Fig. 8.2A) was first described in *Hyphomicrobium* sp. strain MC1, isolated from a sewage treatment plant in Switzerland in 1986 (Hartmans *et al.*, 1986). Strains affiliated with other taxonomic groups were subsequently recovered from soils (*Aminobacter, Hyphomicrobium, Methylorubrum*), marine environments (*Roseovarius, Leisingera*) and plants (*Hyphomicrobium*) (Table 8.1). Measured doubling time during aerobic growth at 30°C with chloromethane (10–15 mM) ranges from ≈5 hours (*Hyphomicrobium* strains CM2, MC1 and AT2) to ≈19 hours (*Hyphomicrobium* strains AT3 and AT4) (Nadalig *et al.*, 2011). *M. extorquens* strain CM4 became the reference organism of chloromethane utilization (see below).

The *cmu* pathway for chloromethane degradation

Biochemistry and genetics of aerobic chloromethane utilization have been elucidated in detail in strain CM4 affiliated to *Methylobacterium extorquens*, recently reclassified as *Methylorubrum extorquens* (Green and Ardley, 2018). Using random mutagenesis of strain CM4 with a miniTn5 transposon, mutants unable to grow with chloromethane were used to identify chloromethane utilization (*cmu*) genes (Vannelli *et al.*, 1998, 1999) (Fig. 8.2A). The first step involves chloromethane dehalogenation with the transfer of its methyl group to the C_1 carrier tetrahydrofolate (H_4F) to produce methyl-H_4F, and one chloride and one proton. The fact that the primary product of chloromethane demethylation is methyl-H_4F rather than formaldehyde differs this pathway from C_1 oxidation pathways, including the methanol oxidation pathway (Chistoserdova, 2011; Studer *et al.*, 2002). Chloromethane dehalogenase consists of the corrinoid methyltransferase CmuA and the H_4F-dependent methyltransferase CmuB (Studer *et al.*, 1999, 2001). It transforms bromomethane

Table 8.1 Chloromethane-degrading bacteria isolated from contrasting environments

Name (taxonomical class)	Origin	Metabolism/trophic type	Genome	Comments
Acetobacterium dehalogenans MC (Clostridia)	Activated sludge (Traunecker *et al.*, 1991)	Anaerobic homoacetogenic	ns[a]	Chloromethane dehalogenation forms methyl-H_4F (Meßmer *et al.*, 1996)
Acetobacterium sp. in mixed culture RM (with *Candidatus* Dichloromethanomonas elyunquensis)	Pristine river sediment (Justicia-Leon *et al.*, 2012)	Anaerobic hydrogenotrophic chloromethane-degradation	ns	H_2-dependent chloromethane degradation (Chen *et al.*, 2017)
Aminobacter ciceronei IMB1 (Alphaproteobacteria)	Fumigated strawberries (Hancock *et al.*, 1998)	Aerobic facultative methylotroph	ns	*cmuA*[b]
Aminobacter lissarensis CC495 (Alphaproteobacteria)	Beech woodland soil (Coulter *et al.*, 1999)	Aerobic facultative methylotroph	ns	*cmuA*[b]
Celeribacter indicus P73 (Alphaproteobacteria)	Deep-sea sediment (Lai *et al.*, 2014)	Aerobic facultative methylotroph (genome-based)[c]	4.5 Mb, five plasmids; G+C content 66% (Cao *et al.*, 2015)	Genome-based detection of *cmu* genes; 3 *folD-purU* copies (This work)
Hyphomicrobium sp. strains AT2, AT3, AT4 (Alphaproteobacteria)	Phyllosphere of *Arabidopsis thaliana* (Nadalig *et al.*, 2011)	Aerobic facultative methylotroph	ns	*cmuA*[b]
Hyphomicrobium sp. MC1 (Alphaproteobacteria)	Industrial sewage plan (Hartmans *et al.*, 1986)	Aerobic facultative methylotroph	4.7 Mb; G+C content 59% (Vuilleumier *et al.*, 2011)	Complete *cmu* pathway
Hyphomicrobium sp. MC2 (Alphaproteobacteria)	Soil from a petrochemical factory (Doronina *et al.*, 1996)	Aerobic facultative methylotroph	ns	*cmuA*[b]
Leisingera methylohalidovorans MB2 (Alphaproteobacteria)	Marine tide pool (Schaefer *et al.*, 2002)	Aerobic facultative methylotroph	4.6 Mb, two plasmids; G+C content 62% (Buddruhs *et al.*, 2013)	Uncharacterized *cmu*-independent chloromethane degradation
Methylomicrobium album BG8 (Gammaproteobacteria)	Freshwater (Whittenbury *et al.*, 1970)	Aerobic facultative methanotroph, obligatory methylotroph	4.5 Mb, one plasmid; G+C content 56% (Kits *et al.*, 2013)	Low chloromethane concentrations enhance growth on methanol (Han and Semrau, 2000). No *cmu* gene
Methylorubrum[d] *extorquens* CM4 (Alphaproteobacteria)	Soil from a petrochemical factory (Doronina *et al.*, 1996)	Aerobic facultative methylotroph	6.2 Mb, two plasmids; G+C content 68% (Marx *et al.*, 2012)	Plasmid pCMU01-encoded *cmu* (Roselli *et al.*, 2013)
Nocardioides sp. SAC-4 (Actinobacteria)	Topsoil (McAnulla *et al.*, 2001a)	Aerobic facultative methylotroph[e]	ns	
OM43 bacterioplankton clade, strain HTCC2181 (Betaproteobacteria)	Seawater (Giovannoni *et al.*, 2008)	Obligate methylotroph	1.3 Mb (three contigs); G+C content 38% (Giovannoni *et al.*, 2008)	*cmu*-independent chloromethane use as energy source if methanol is the C source (Halsey *et al.*, 2012)

Table 8.1 Continued

Name (taxonomical class)	Origin	Metabolism/trophic type	Genome	Comments
Pseudomonas aeruginosa NB1 (Gammaproteobacteria)	Anoxic activated sludge enrichment culture (Freedman *et al.*, 2004)	Aerobic facultative methylotroph	ns	Nitrate respiration, chloromethane used as C and energy source (Freedman *et al.*, 2004)
Roseovarius pacificus 81-2 (Alphaproteobacteria)	Deep-sea sediment (Wang *et al.*, 2009)	Aerobic facultative methylotroph[c]	4.5 Mb (52 contigs, one plasmid); G+C content 62% (Wang *et al.*, 2009)	Genome-based detection of *cmu* genes (this work)
Roseovarius sp. strain 217 (Alphaproteobacteria)	Seawater (Schäfer *et al.*, 2005)	Aerobic facultative methylotroph	4.6 Mb (37 contigs); G+C content 61% (Moore Foundation project NZ_AAMV00000000)	Uncharacterized *cmu*-independent chloromethane degradation pathway
Roseovarius sp. strains 179, 198 (Alphaproteobacteria)	Seawater (Schäfer *et al.*, 2005)	Aerobic facultative methylotroph	ns	*cmuA*[b]

[a]ns, not sequenced; [b]gene *cmuA* was PCR amplified and sequenced; [c]chloromethane has not been tested to our knowledge; [d]*Methylobacterium* genus reclassified (Green and Ardley, 2018); [e]methylotrophic growth on chloromethane but no other tested C_1 compound (formate, methanol, methane, methylamine) (McAnulla *et al.*, 2001a).

and iodomethane as well as chloromethane (Vannelli *et al.*, 1998). Both proteins CmuB (Studer *et al.*, 1999) and CmuA (Studer *et al.*, 2001) are required for growth-supporting dehalogenation of chloromethane. CmuA acts as a bifunctional enzyme (possessing methyltransferase and the corrinoid-binding domains), and it also serves as an intermediate methyl carrier. In a second reaction, CmuB methyltransferase transfers the methyl group from CmuA to H_4F (Studer *et al.*, 2001). Thus, besides H_4F, chloromethane dehalogenation requires a corrinoid cofactor, which explains an absolute need for cobalt for growth with methyl halides (Studer *et al.*, 2001). This uncharacterized corrinoid cofactor possibly differs from the chromosome-encoded cobalamin compound (Roselli *et al.*, 2013). In *M. extorquens* strain CM4, all genes of the *cmu* pathway and *cob* genes are present on a 380 kb plasmid called pCMU01. Comparison of complete genome sequences of strain CM4 with those of other *M. extorquens* strains unable to grow with chloromethane showed that plasmid pCMU01 harbours unique genes without homologues in the compared genomes (*bluB2, btuB, cobA, cbiD*), as well as 13 duplicated genes with homologues of chromosome-borne genes involved in cobalamin-associated biosynthesis and transport, or in H_4F-dependent metabolism (*folC2*) (Roselli *et al.*, 2013). The clustering of genes coding for dehalogenase enzymes and for biosynthesis of associated cofactors suggests a history of coupled enzyme/coenzyme gene acquisition related to chloromethane utilization in strain CM4.

Coping with chloromethane upon methylotrophic growth

Bacteria that grow on chloromethane need to accommodate multiple stresses such as intracellular production of protons and chloride, adjustment of metabolism towards higher demand for dehalogenase-requiring cofactors (H_4F and cobalamin-related compounds), and the necessity to regulate their metabolism for effective chloromethane utilization (Roselli *et al.*, 2013; Michener *et al.*, 2016; Chaignaud *et al.*, 2017). The limiting factors were inferred from experimental observations: (i) growth with chloromethane is slower than with methanol (Chaignaud *et al.*, 2017); (ii) difference in growth rates under chloromethane utilization are not due to differences in chloromethane dehalogenase activity in cell-free extracts of taxonomically-related strains (Nadalig *et al.*, 2011); and (iii) transfer of a *cmu* catabolic pathway gene cassette (plasmid pJM105) in 'naïve'

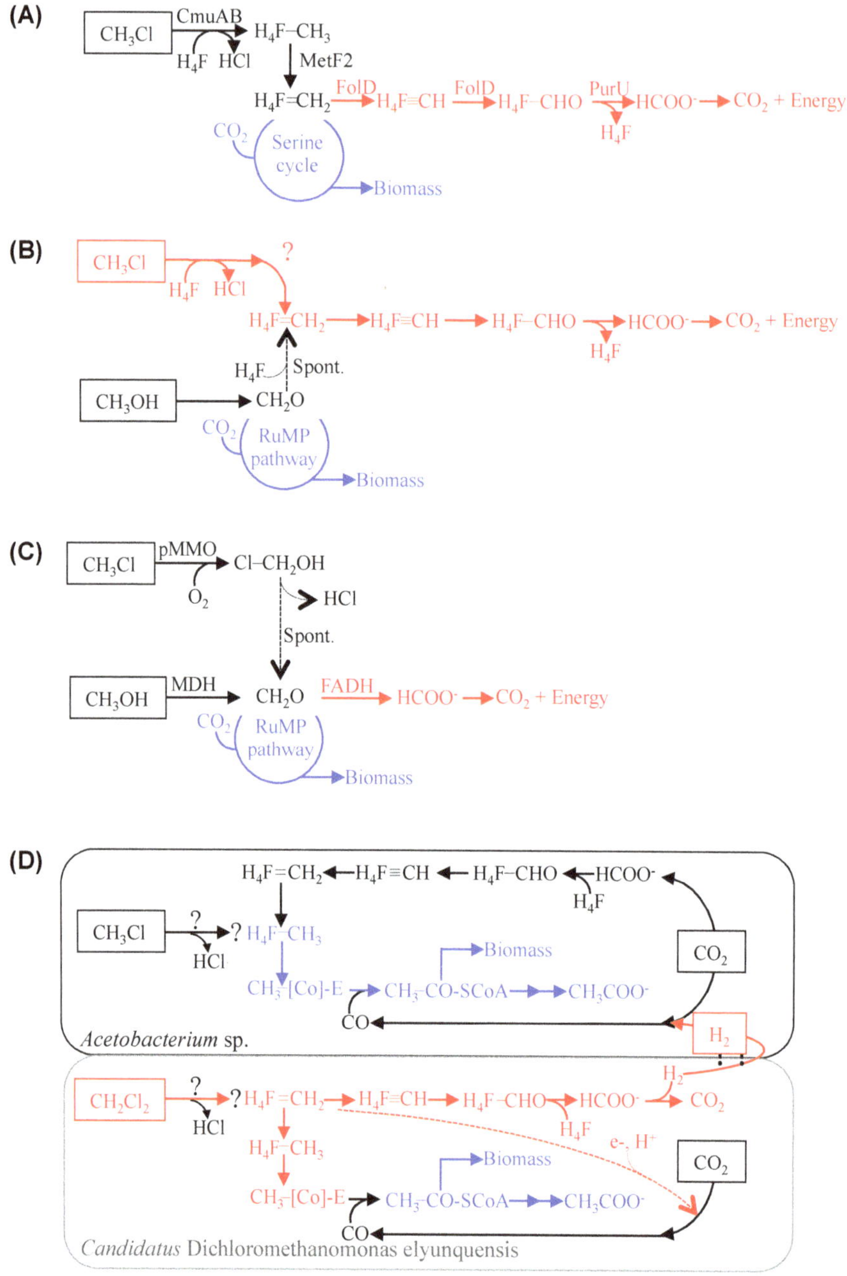

Figure 8.2 Examples of methylotrophic pathways for chloromethane dissimilation (source of energy, in red) and assimilation (source of carbon for biomass, in blue) in cultivated strains and mixed cultures. Bacteria cultivated under aerobic (A to C) and anaerobic (D) conditions. Abbreviations: FADH, formaldehyde dehydrogenase; H_4F, tetrahydrofolate; MDH, methanol dehydrogenase; pMMO, particulate methane monooxygenase; RuMP, ribulose monophosphate pathway. (A) The *cmu* pathway characterized in details in *M. extorquens* CM4 and *Hyphomicrobium* strains, adapted from (Vannelli *et al*., 1999; McAnulla *et al*., 2001b); (B) Bacterioplankton HTCC2181, adapted from (Giovannoni *et al*., 2008); (C) *Methylomicrobium album* BG8, adapted from (Han and Semrau, 2000); and (D) Culture RM, adapted from (Chen *et al*., 2017). Chloromethane and dichloromethane are each degraded by a different organism in a mutualism interaction. Chloromethane could not substitute for dichloromethane as the sole source of energy. It remains unknown how chloromethane enters the C_1 cycle of the chloromethane-degrading homoacetogen strain performing H_2/CO_2 reductive acetogenesis (Kleindienst *et al*., 2017). For *Candidatus* Dichloromethanomonas elyunquensis, dichloromethane is a source of energy. No dichloromethane degradation occurs in medium without CO_2 (bicarbonate) supplementation (Kleindienst *et al*., 2017). Enzymes involved in dichloromethane degradation and CO_2 assimilation have not been characterized.

Methylorubrum strains conferred poor growth ability on chloromethane supplied as the sole energy and carbon source (Michener *et al.*, 2016). Among potential limiting factors, intracellular release of protons and chloride ions is likely, since dehalogenation of chloromethane, as of dichloromethane, initiates growth-supporting degradation. Nonetheless, each of these chlorinated methanes has specific adaptive responses for chloride export. For example, in *Methylorubrum*, the chromosome-encoded proton/chloride antiporter ClcA had a significant fitness impact for growth with dichloromethane but not with chloromethane (Michener *et al.*, 2014a, 2016). Adaptive responses to growth with chloromethane compared with growth with methanol were detected using complementary approaches of random mutagenesis (Vannelli *et al.*, 1999), genome sequencing (Marx *et al.*, 2012), comparative genomics, proteomics and transcriptomics (Roselli *et al.*, 2013; Chaignaud *et al.*, 2017), and experimental evolution (Michener *et al.*, 2016). A summary of the global response to growth with chloromethane in *M. extorquens* strain CM4 is proposed in Fig. 8.3.

According to this model, the adaptive response to chloromethane is complex and includes general oxidative stress response functions (Fig. 8.3) such as molecular chaperones, DNA repair and reactive oxygen species (ROS) removal functions, as well as a dehalogenation-specific response to intracellular hydrochloric acid production (membrane-bound proton translocating pyrophosphatase HppA and a putative chloride/proton antiporter ClcA2 encoded by plasmid pCMU01). The *pnt* gene cluster encoding a NADH/NADPH transhydrogenase with cross-membrane proton translocation activity, most likely plays a role in maintaining both the intracellular pools of NAD^+ and $NADP^+$, and the internal cellular pH with the efflux of protons during chloromethane dehalogenation. Higher expression of genes (enzymes and transporters) for biosynthesis of methyltransferase cofactors were detected, especially *bluB2* (sharing 38% amino acid identity with the chromosomal-encoded *bluB* gene). BluB

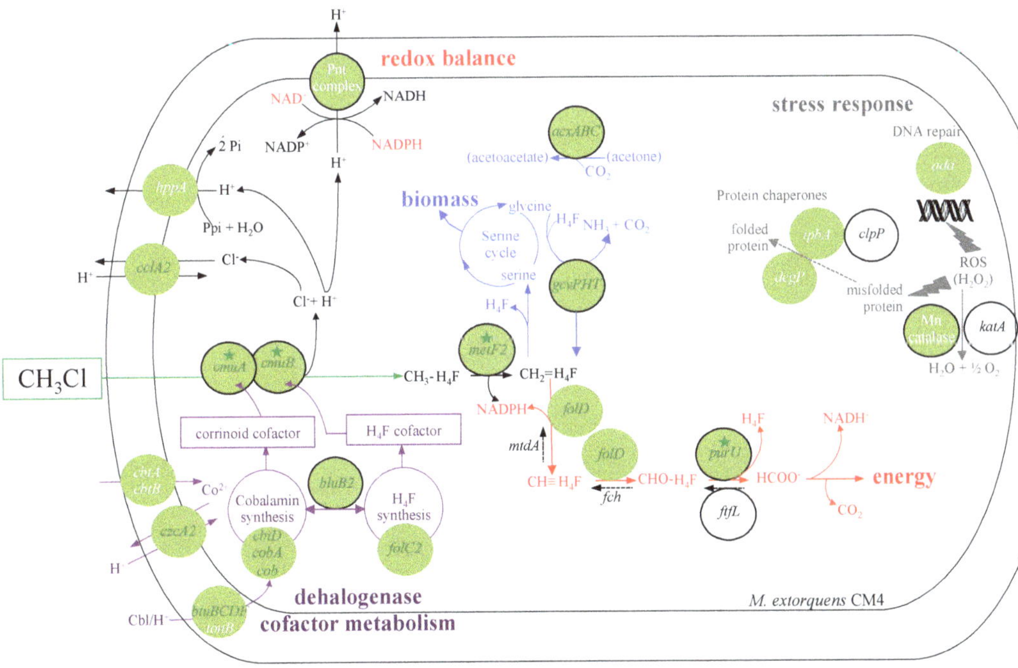

Figure 8.3 Coping with chloromethane upon methylotrophic growth. In *M. extorquens* CM4, the *cmu* pathway-encoding genes and associated genes are plasmid pCMU01-borne (gene names in green). In cultures grown on chloromethane compared with methanol, more abundant RNAs (green bullets) (Chaignaud *et al.*, 2017) and proteins (black-circled bullets) (Roselli *et al.*, 2013) were found. Essential genes in chloromethane assimilation or dechlorination are marked with a star (Vannelli *et al.*, 1999; Michener *et al.*, 2016). For the putative role of the acetone carboxylase components encoded by the plasmid-borne *acxABC* gene cluster chloromethane assimilation, refer to Roselli *et al.* (2013).

catalyses both the production of a lower ligand of cobalamin (5,6-dimethylbenzimidazole) and of a H_4F precursor (erythrose 4P), thus linking together the biosynthesis of the two cofactors of chloromethane dehalogenation (Roselli *et al.*, 2013). In conclusion, the ability to utilize chloromethane likely involves adaptation to oxidative stress, production of reducing equivalents, conversion of H_4F-bound C_1 units, and central metabolism. The adaptive response to growth with chloromethane has little overlap with other toxic halogenated compounds (Roselli *et al.*, 2013; Michener *et al.*, 2014b, 2016; Chaignaud *et al.*, 2017), and possibly with chloromethane degradation by *cmu*-independent pathways.

Probing the diversity of chloromethane-degrading bacteria by comparative genomics and isotopic fractionation

The strategy of combining stable isotope approaches (Compound-Specific Isotope Analysis; CSIA) with comparative genomics helps to explore the diversity of microbial degradation of chloromethane (Nadalig *et al.*, 2014). CSIA is a chemical method to determine the ratio of isotopes of different elements (e.g. C, N, H, O, Cl) in environmental samples (Nijenhuis and Richnow, 2016). This ratio can change during the course of microbial degradation. For example, carbon large fractionation shifts (δ ^{13}C from an initial value of −60‰ to −30‰) were observed when 90% of the chloromethane was degraded by high-cell-density suspensions of methylotrophic bacteria (Miller *et al.*, 2001). Combined measurements of carbon and hydrogen isotope fractionation of the remaining untransformed chloromethane provide further evidence of biotic degradation and clues about the metabolic pathway and enzymes involved (Nadalig *et al.*, 2013) (Table 8.2). When genomes of experimentally validated chloromethane-degrading methylotrophs were searched for the presence of *cmuA*, two groups of bacteria were detected: (i) *M. extorquens* CM4 and *Hyphomicrobium* sp. MC1 isolated from soils share the *cmu* pathway genes; (ii) *Leisingera methylohalidovorans* MB2 and *Roseovarius*

Table 8.2 Degradation of chloromethane and its isotopic enrichment factor for carbon (ε_C) and hydrogen (ε_H) of pure cultures and environmental samples

	Stable isotope enrichment factor for chloromethane upon bacterial degradation[a]			
	Test condition	ε_C (‰)	ε_H (‰)	Reference
Pure bacterial cultures (*cmu* pathway)				
Aminobacter ciceronei IMB1	Resting cells, 26°C	−47 ± 4	Nd[b]	Miller *et al.* (2001)
Aminobacter lissarensis CC495	Resting cells, 26°C	−42 ± 2	Nd	Miller *et al.* (2001)
Methylorubrum extorquens CM4	Dividing cells, 30°C	−42	−39	Nadalig *et al.* (2014)
	Resting cells, 30°C	−41 ± 5	−29 ± 6	Nadalig *et al.* (2013)
Hyphomicrobium sp. MC1	Dividing cells, 30°C	−54	−51	Nadalig *et al.* (2014)
	Resting cells, 30°C	−38 ± 3	−27 ± 10	Nadalig *et al.* (2013)
Pure bacterial cultures (*cmu*-independent pathway)				
Leisingera methylohalidovorans MB2	Dividing cells, 30°C	−76	0	Nadalig *et al.* (2014)
	Resting cells, 26°C	−44 ± 4	Nd	Miller *et al.* (2001)
Environmental samples				
Phyllosphere (fern: *Cyathea cooperi*)	Microcosm, 20–30°C	−43 ± 12	−8 ± 19	Jaeger *et al.* (2018b)
Soils (agricultural, grassland, forest soils)	Microcosm, 20–30°C	−38 to −11 ± 3	−50 ± 19	Jaeger *et al.* (2018a)

[a]Increased carbon isotope fractionation compared with hydrogen isotope fractionation most likely results from the primary isotope effect in cleavage of the carbon-halogen bound during chloromethane dehalogenation, as previously suggested (Elsner *et al.*, 2005); [b]Nd, not determined.

sp. 217 isolated from marine environments harbour no *cmu* genes and degrade chloromethane by a yet uncharacterized mechanism (Table 8.1). Cultures of *L. methylohalidovorans* MB2 displayed clear differences of both carbon and hydrogen isotope signatures upon chloromethane dehalogenation compared with *cmu*-containing strain cultures of *Aminobacter*, *Hyphomicrobium* and *Methylorubrum* (Nadalig *et al.*, 2014).

The *cmu* gene organization is unique in strain CM4 (Roselli *et al.*, 2013), where the genes are plasmid-borne and located in two distant regions (Vannelli *et al.*, 1998). In other strains, when detected by comparative genomics and targeted PCR amplifications, *cmu* genes (Fig. 8.4) were conserved on a single chromosomal cluster (McAnulla *et al.*, 2001b; Nadalig *et al.*, 2014). In particular, a trio of adjacent genes (*cmuBCA*) is conserved in genomes of aerobic chloromethane-degraders isolated from soils and plants, genomes of Alphaproteobacteria isolated from deep-sea (*Celeribacter indicus* P73 and *Roseovarius pacificus* 81-2), as well as of obligate anaerobes (*Desulfobacula* sp. TS; *Desulfomonile tiedjei* DSM 6799, *Desulfurispora thermophila* DSM 16022, *Desulfotomaculum alcoholivorax* DSM 16058, *Thermincola potens* JR and *Thermosediminibacter oceani* DSM 16646). These anaerobic strains harbour *cmuA* homologues closely related to each other (identities at the protein level between 84–93%) (Nadalig *et al.*, 2014) but none of these strains has yet been reported to degrade chloromethane. The overall conserved *cmuBCA* trio suggests a functional link between the encoded CmuA, CmuB and CmuC. While CmuA and CmuB are two methyltransferases essential for chloromethane dehalogenation (Studer *et al.*, 2001), the function of CmuC remains uncharacterized. CmuC, a putative third methyltransferase-like protein most similar to MtaA (corrinoid:coenzyme M methyltransferase), is dispensable for dehalogenation but needed for chloromethane assimilation (Vannelli *et al.*, 1999). Mobile genetic elements (black arrows in Fig. 8.4) in the vicinity of conserved *cmu* clusters potentially suggest their acquisition by horizontal gene transfer. Duplicated *cmu* genes found in strain CM4 (two homologous *cmuC* copies adjacent to either *cmuB* or *cmuA*) and in *D. tiedjei* DSM 6799 (a *cmuAB* copy in addition to the conserved *cmuBCA*) suggest specific strain adaptation. An 'orphan' *cmuA*-like gene (encoding for a protein sharing 43% amino acid identity with *M. extorquens* CM4 CmuA) with no other detected genes of the *cmu* pathway in its genome, raises the question of whether the marine Gammaproteobacterium *Vibrio orientalis* CIP 102891 in able to degrade chloromethane. No chloromethane degradation was observed for *V. orientalis* CIP 102891 resting cells (Fig. 8.5) or by actively growing cell cultures supplied with chloromethane in addition to another carbon source (data not shown). This suggests that the *V. orientalis* CmuA-like homolog might play a yet unknown function, possibly different from chloromethane dehalogenation.

Methylotrophic populations of the chloromethane sink

Chloromethane-degrading populations reduce chloromethane emissions to the atmosphere

In contrast to live soils, sterilized soils or plants do not degrade chloromethane (Chaignaud *et al.*, 2017; Jaeger *et al.*, 2018a,b) suggesting a microbial sink of chloromethane. Compared to chloromethane degrading methylotrophs (Table 8.2), the carbon isotope enrichment factor (ε_C) of remaining chloromethane was similar for fern leaves (mean of −43 ± 12‰) and lower for soils (ranging from −38 to 11 ±3‰ in agricultural, grassland and forest soils, respectively). Stable isotope hydrogen fractionation (ε_H) observed during degradation experiments with fern leaves (−8 ± 19‰), and soils (−50 ± 19‰) was similar to reported ε_H for chloromethane-degrading bacterial strains (Table 8.2). No change in stable isotope fractionation signature was observed when the investigated environmental samples were sterilized, suggesting a biotic microbially driven process (Chaignaud *et al.*, 2017; Jaeger *et al.*, 2018a,b). This hypothesis is in agreement with the recent finding that the occurrence and the diversity of bacterial chloromethane degradation (*cmu* gene expression) is correlated with differential expression of the plant gene *HOL1* involved in the production of chloromethane (Farhan Ul Haque *et al.*, 2017). Thus, a relation was established between chloromethane production by plants and relative abundance of leave-associated chloromethane-degrading bacteria. This suggests that chloromethane-degrading bacteria co-determine the extent of chloromethane

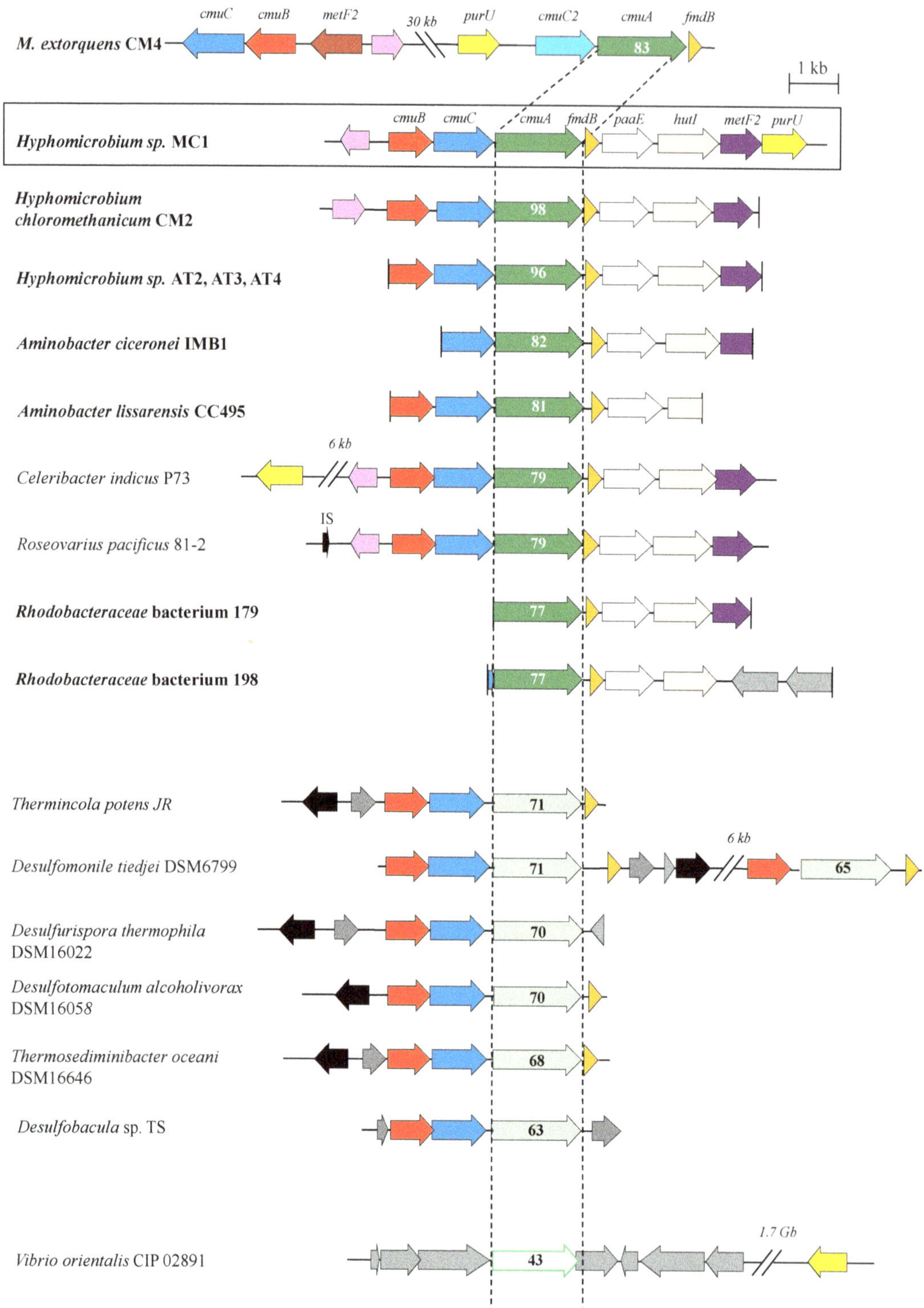

Figure 8.4 Organisation of *cmu* genes in bacteria. Names of experimentally validated chloromethane-degrading bacteria are written in bold. Arrows represent protein-coding genes, and annotations above arrows indicate gene names. Homologous genes are given with identical colour except for arrows in black, pink and grey, which indicate mobile elements, transcriptional regulators, and genes not involved in chloromethane degradation, respectively. The percentage of amino acid sequence identity with the reference *Hyphomicrobium* sp. MC1 CmuA protein is indicated. Gene clusters are drawn to scale.

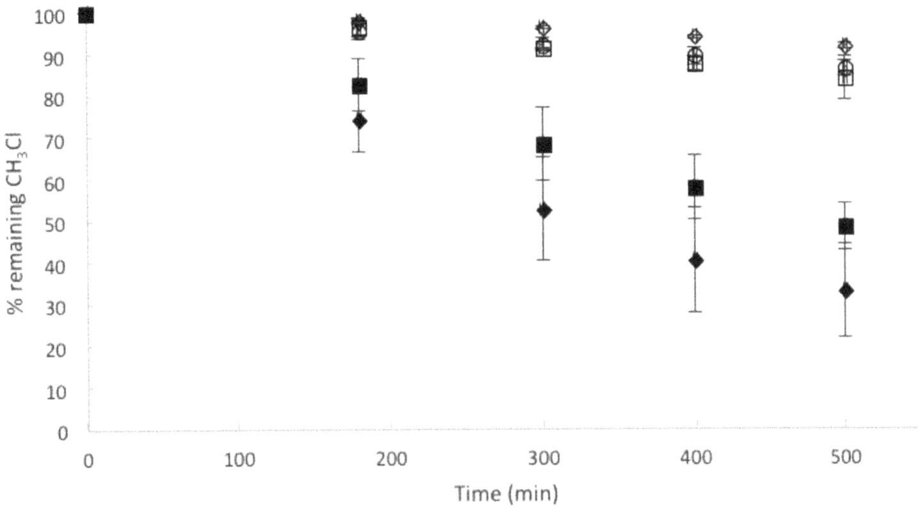

Figure 8.5 Assessing the ability of *Vibrio orientalis* CIP 102891 to degrade chloromethane. Triplicates of resting cells were exposed to chloromethane in gastight flasks at room temperature. The remaining chloromethane was quantified by gas chromatography (GC-FID) at different time points. Cell suspensions of *V. orientalis* CIP 102891 (empty square) were harvested from exponential-grown phase as described for *L. methylohalidovorans* MB2 saline cultures (Nadalig *et al.*, 2014). For *V. orientalis* CIP 102891, the amount of chloromethane remaining was similar to that of the negative controls without cells (empty circle, saline phosphate buffer; empty diamond, phosphate buffer) but significantly different from the positive degraders *M. extorquens* CM4 in (black diamond) and *L. methylohalidovorans* MB2 (black square).

emissions by plants to the atmosphere, and that methylotrophic microorganisms are involved in the terrestrial chloromethane sink.

Rare occurrence of *cmuA* in the environment

Of the genes of the *cmu* pathway, the *cmuA* gene is the most conserved (Farhan Ul Haque *et al.*, 2017; Nadalig *et al.*, 2011). Thus, *cmuA* has been used to study the diversity of methyl halide-degrading organisms in the environment (Miller *et al.*, 2004; Schäfer *et al.*, 2005; Nadalig *et al.*, 2011; Cox *et al.*, 2012; Chaignaud *et al.*, 2018). With increasing numbers of *cmuA* sequences in databases, oligonucleotide primers have been modified for more efficient PCR amplification and improved detection of *cmuA* (see details in Chaignaud *et al.*, 2018, supplemental material). No PCR products were retrieved from environmental DNA prior to enrichment steps of forest soil (Chaignaud *et al.*, 2018) and plant leaves (Jaeger *et al.*, 2018b). When amplified, a low diversity of *cmuA* sequences was found in samples of soils, plant leaves and large volumes of oceanic and coastal waters (Cox *et al.*, 2012; Farhan Ul Haque *et al.*, 2017; Chaignaud *et al.*, 2018). This suggested that *cmuA*-harbouring bacteria might represent a small component of the investigated environmental microbiota.

Recently, an extremely sensitive molecular tool for *cmuA* detection by 'targeted gene capture' was developed (Gasc *et al.*, 2016; Ribière *et al.*, 2016). Its detection limit of 0.00006% makes it particularly adapted to investigate rare genes and biological functions in the environment (Gasc and Peyret, 2018). It consists of utilizing metal beads for recovering the sequences of interest within a community (or within nucleic acid extracts) by hybridization. The beads are coated with biotinylated RNA sequences of ≈ 80 bp, complementary to the targeted gene sequence and potential orthologs. Using KASpOD software (Parisot *et al.*, 2016), such 80-mer sequences have been successfully used to design specific probes based on large sequence datasets to reveal microbial diversity missed using current profiling methods. Based on known *cmuA* gene sequences, gene capture method was attempted in different environments [e.g. soil (Jaeger *et al.*, 2018a), plant leaves (Jaeger *et al.*, 2018b) and cloud microbial communities (L. Besaury, P. Amato, P. Peyret, C. Gasc, unpublished data)] with no success. When homologous sequences were searched in metagenomic databases in a wider range of

Table 8.3 Occurrence of *cmuA* sequences in metagenomes available from public databases

Database	Environment	Metagenome numbers	Total size (Mb)[a]	*cmuA* hit[b]
IMG	**Engineered**	**2332**	**254,327**	
	Bioreactor	108	35,565	0
	Bioremediation	68	12,879	0
	Biotransformation	24	5411	0
	Build environment	1231	31,275	0
	Food production	3	6,600	0
	Lab enrichment	218	11,002	0
	Lab synthesis	4	401	0
	Modelled	65	2804	0
	Solid waste	69	22,575	0
	Wastewater	542	132,349	0
	Aquatic	**9322**	**1,968,340**	
	Freshwater	3912	755,498	0
	Marine	3034	1,049,257	1 (wetland)
	Non-marine saline and alkaline	702	86,625	0
	Sediment	271	34,754	0
	Thermal spring	1403	42,206	0
	Terrestrial	**2915**	**1,508,938**	
	Deep subsurface	183	13,202	0
	Geologic	4	170	0
	Oil reservoir	3	283	0
	Peat	56	10,558	0
	Plant litter	47	2663	0
	Rock-dwelling	8	3220	0
	Soil	2609	1,481,714	0
	Volcanic	5	26	0
	Host-associated	**4106**	**876,365**	
	Algae	64	44,092	0
	Animal	25	6426	0
	Annelida	91	43,120	0
	Arthropoda	147	73,213	0
	Birds	18	5505	0
	Cnidaria	36	723	0
	Fish	4	489	0
	Fungi	102	19,978	0
	Human	2489	139,540	0
	Insecta	34	3153	0
	Invertebrate	15	8056	0
	Mammal	267	104,135	0
	Microbial	24	2132	0

Table 8.3 Continued

Database	Environment	Metagenome numbers	Total size (Mb)[a]	*cmuA* hit[b]
	Mollusca	11	3247	0
	Plants	761	581,931	3 (*Arabidopsis thaliana*)
	Porifera	10	1642	0
	Tunicates	8	3377	0
NCBI	**Diverse**	Metagenome BLAST database		
	Marine		81,542	7 (sediment)
	Freshwater, hydrothermal vent, human gut		94,265	0

[a]Total of 4,783,777 Mb publicly available in May 2018; [b]Blast searches were performed using a *cmuA* consensus template from 122 nucleic acid sequences downloaded from GenBank. Hits displaying an E-value lower than 10^{-5} were kept, and the corresponding proteins aligned with characterized CmuA sequences to only retain members closely related to one of three clades shown in Figure 8.6B.

environments (Table 8.3), only 11 *cmuA* sequences were retrieved, of which three originated from the rhizosphere of the plant *A. thaliana*, and the others from marine environments (Table 8.3). Basing on these results, *cmuA* is estimated to represent around one DNA sequence out of 3 billion, and can be considered as a rare gene (Kislyuk *et al.*, 2011). It may be necessary for future sampling campaigns to assess if environmental niche partitioning of microbial minorities consuming chloromethane is found as for methane (Bodelier *et al.*, 2013).

Using Blast searches, *cmuA* DNA sequences were retrieved from environmental metagenomes, translated to proteins and aligned to reference '*sensu stricto*' proteins of functionally characterized CmuA proteins of Alphaproteobacteria cultivated under aerobic conditions, to uncharacterized proteins of the clade called 'CmuA-anaerobe' found in strict anaerobes and of '*CmuA*-like' proteins found in bacteria lacking *cmuBC* (as defined in Fig. 8.6). Of the 11 metagenomic *cmuA* sequences, one was closely related to the '*sensu stricto* CmuA' clade, one to the 'CmuA-anaerobe', and the remaining belonged to the most-distant CmuA-like clade. Remarkably, chloromethane dehalogenation by '*sensu stricto* CmuA' does not require aerobic conditions and is even sensitive to oxygen (Studer *et al.*, 2001). Thus, anaerobic bacteria with *cmu* genes may be able to use chloromethane as a source of carbon and energy, as previously discussed (Nadalig *et al.*, 2014). Nonetheless, the ability to degrade methyl halides under environmental conditions remains to be confirmed for bacterial populations coding for 'CmuA-anaerobe' and CmuA-like proteins.

Substrate co-utilization as a strategy to sustain growth and resilience of chloromethane-utilizing methylotrophic populations

In the troposphere, chloromethane is a trace gas with a global mean concentration of ≈ 500 to 600 parts per trillion by volume (pptv). In environments with low nutrient levels, methylotroph populations growing on different C_1 substrates (Neufeld *et al.*, 2008; Gifford *et al.*, 2016) would benefit from simultaneous utilization of several C_1 compounds as a strategy to enhance their growth. In coastal ecosystems, methanol is released during phytoplankton blooms (Heikes *et al.*, 2002). When methanol is present in seawater, the methanol-assimilating methylotroph HTCC 2181 uses chloromethane as source of energy (Halsey *et al.*, 2012). Its genome, one of the smallest known for bacteria, lacks the tetrahydromethanopterin (H_4MPT) C_1 transfer module, so C_1 transfer can only rely on H_4F (Giovannoni *et al.*, 2008). Strain HTCC 2181 may have evolved by reducing its C_1 abilities (loss of the genes for the C_1 carrier, H_4MPT synthesis) and thus reducing its metabolic versatility in favour of specialization on carbon compounds at low amounts (Giovannoni *et al.*, 2008).

To our knowledge, no data are available on chloromethane concentration in upper soil of terrestrial

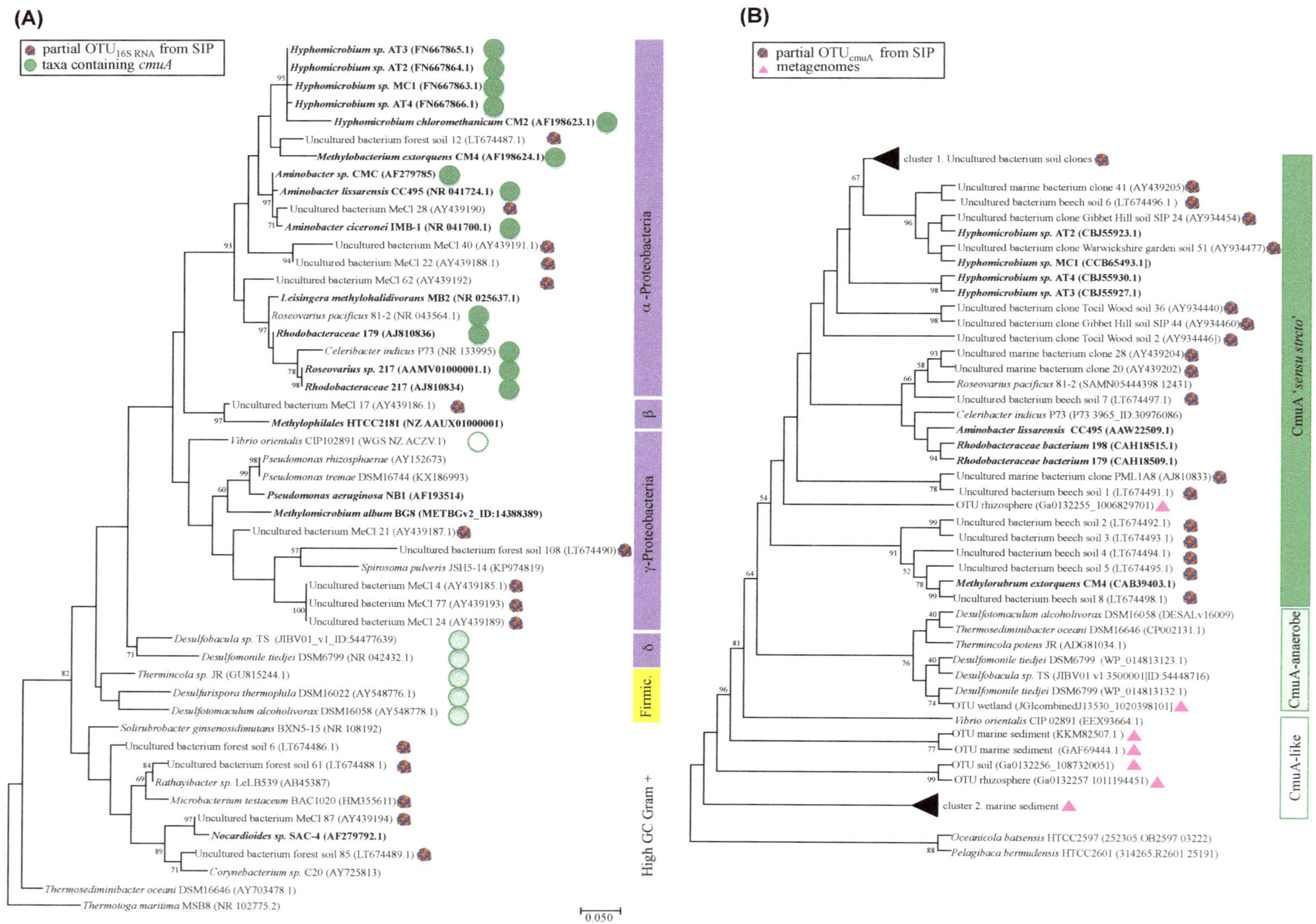

(A)
partial $OTU_{16S\ RNA}$ from SIP
taxa containing cmuA
Hyphomicrobium sp. AT3 (FN667865.1)
Hyphomicrobium sp. AT2 (FN667864.1)
Hyphomicrobium sp. MC1 (FN667863.1)
Hyphomicrobium sp. AT4 (FN667866.1)
Hyphomicrobium chloromethanicum CM2 (AF198623.1)
Uncultured bacterium forest soil 12 (LT674487.1)
Methylobacterium extorquens CM4 (AF198624.1)
Aminobacter sp. CMC (AF279785)
Aminobacter lissarensis CC495 (NR 041724.1)
Uncultured bacterium MeCl 28 (AY439190)
Aminobacter ciceronei IMB-1 (NR 041700.1)
Uncultured bacterium MeCl 40 (AY439191.1)
Uncultured bacterium MeCl 22 (AY439188.1)
Uncultured bacterium MeCl 62 (AY439192)
Leisingera methylohalidivorans MB2 (NR 025637.1)
Roseovarius pacificus 81-2 (NR 043564.1)
Rhodobacteraceae 179 (AJ810836)
Celeribacter indicus P73 (NR 133995)
Roseovarius sp. 217 (AAMV01000001.1)
Rhodobacteraceae 217 (AJ810834)
Uncultured bacterium MeCl 17 (AY439186.1)
Methylophilales HTCC2181 (NZ AAUX01000001)
Vibrio orientalis CIP102891 (WGS NZ ACZV.1)
Pseudomonas rhizosphaerae (AY152673)
Pseudomonas tremae DSM16744 (KX186993)
Pseudomonas aeruginosa NB1 (AF193514)
Methylomicrobium album BG8 (METBGv2_ID:14388389)
Uncultured bacterium MeCl 21 (AY439187.1)
Uncultured bacterium forest soil 108 (LT674490)
Spirosoma pulveris JSH5-14 (KP974819)
Uncultured bacterium MeCl 4 (AY439185.1)
Uncultured bacterium MeCl 77 (AY439193)
Uncultured bacterium MeCl 24 (AY439189)
Desulfobacula sp. TS (JIBV01_v1_ID:54477639)
Desulfomonile tiedjei DSM6799 (NR 042432.1)
Thermincola sp. JR (GU815244.1)
Desulfurispora thermophila DSM16022 (AY548776.1)
Desulfotomaculum alcoholivorax DSM16058 (AY548778.1)
Solirubrobacter ginsenosidimutans BXN5-15 (NR 108192)
Uncultured bacterium forest soil 6 (LT674486.1)
Uncultured bacterium forest soil 61 (LT674488.1)
Rathayibacter sp. LeLB539 (AB45387)
Microbacterium testaceum BAC1020 (HM355611)
Uncultured bacterium MeCl 87 (AY439194)
Nocardioides sp. SAC-4 (AF279792.1)
Uncultured bacterium forest soil 85 (LT674489.1)
Corynebacterium sp. C20 (AY725813)
Thermosediminibacter oceani DSM16646 (AY703478.1)
Thermotoga maritima MSB8 (NR 102775.2)
α-Proteobacteria
β
γ-Proteobacteria
δ
Firmic.
High GC Gram +
0.050
(B)
partial OTU_{cmuA} from SIP
metagenomes
cluster 1. Uncultured bacterium soil clones
Uncultured marine bacterium clone 41 (AY439205)
Uncultured bacterium beech soil 6 (LT674496.1)
Uncultured bacterium clone Gibbet Hill soil SIP 24 (AY934454)
Hyphomicrobium sp. AT2 (CBJ55923.1)
Uncultured bacterium clone Warwickshire garden soil 51 (AY934477)
Hyphomicrobium sp. MC1 (CCB65493.1])
Hyphomicrobium sp. AT4 (CBJ55930.1)
Hyphomicrobium sp. AT3 (CBJ55927.1)
Uncultured bacterium clone Tocil Wood soil 36 (AY934440)
Uncultured bacterium clone Gibbet Hill soil SIP 44 (AY934460)
Uncultured bacterium clone Tocil Wood soil 2 (AY934446])
Uncultured marine bacterium clone 28 (AY439204)
Uncultured marine bacterium clone 20 (AY439202)
Roseovarius pacificus 81-2 (SAMN05444398 12431)
Uncultured bacterium beech soil 7 (LT674497.1)
Celeribacter indicus P73 (P73 3965_ID:30976086)
Aminobacter lissarensis CC495 (AAW22509.1)
Rhodobacteraceae bacterium 198 (CAH18515.1)
Rhodobacteraceae bacterium 179 (CAH18509.1)
Uncultured marine bacterium clone PML1A8 (AJ810833)
Uncultured bacterium beech soil 1 (LT674491.1)
OTU rhizosphere (Ga0132255_1006829701)
Uncultured bacterium beech soil 2 (LT674492.1)
Uncultured bacterium beech soil 3 (LT674493.1)
Uncultured bacterium beech soil 4 (LT674494.1)
Uncultured bacterium beech soil 5 (LT674495.1)
Methylorubrum extorquens CM4 (CAB39403.1)
Uncultured bacterium beech soil 8 (LT674498.1)
Desulfotomaculum alcoholivorax DSM16058 (DESALv16009)
Thermosediminibacter oceani DSM16646 (CP002131.1)
Thermincola potens JR (ADG81034.1)
Desulfomonile tiedjei DSM6799 (WP_014813123.1)
Desulfobacula sp. TS (JIBV01 v1 3500001|ID:54448716)
Desulfomonile tiedjei DSM6799 (WP_014813132.1)
OTU wetland (JGIcombinedJ13530_1020398101]
Vibrio orientalis CIP 02891 (EEX93664.1)
OTU marine sediment (KKM82507.1)
OTU marine sediment (GAF69444.1)
OTU soil (Ga0132256_1087320051)
OTU rhizosphere (Ga0132257 1011194451)
cluster 2. marine sediment
Oceanicola batsensis HTCC2597 (252305.OB2597 03222)
Pelagibaca bermudensis HTCC2601 (314265.R2601 25191)
CmuA 'sensu stricto'
CmuA-anaerobe
CmuA-like

Figure 8.6 (see p.164) Phylogenetic trees of partial 16S rRNA and partial protein CmuA sequences associated with chloromethane-degradation of cultivated strains and environmental samples. The evolutionary history was inferred by using the Maximum Likelihood method based on the JTT matrix-based model (Jones *et al.*, 1992). Bootstrap values were calculated from 1000 replicates,% values ≥50 are presented. Analyses were conducted using the software MEGA7 (Kumar *et al.*, 2016). Sequences of experimentally validated chloromethane-degraders are written in bold. Sequences marked with a ^{13}C icon result from DNA-SIP ($^{13}CH_3Cl$) studies in soil microcosms from a deciduous German forest dominated by beech (16S RNA: LT674486.1 to LT674490.1; CmuA: LT674491.1 to LT674498.1 in Chaignaud *et al.* (2018)), clone libraries from Californian agricultural soils (16S RNA: AY439185.1 to AY439194.1; CmuA: AY439202 to AY439204 in Miller *et al.* (2004)), and CmuA from clone libraries from English woodland soil (AY934428 to AY934484) (Borodina *et al.*, 2005). (A) 16S rRNA tree based on 197 nucleotide sequences. *Thermotoga maritima* sequence served as outgroup. (B) CmuA tree based on 60 amino acid sequences. *Oceanicola batsensis* HTCC2597 and *Pelagibaca bermudensis* HTCC2601 sequences served as outgroup. Pink triangles indicate sequences retrieved from metagenome JGI database. Condensed branch of 11 sequences in cluster 1 (uncultured bacterium clone Warwickshire enrichment 3 (AY934432), 25 (AY934433), 36 (AY934430) and 42 (AY934428); Uncultured bacterium clone Warwickshire garden soil 37 (AY934481); Uncultured bacterium clone Gibbet Hill soil 9 (AY934448), 16 (AY934452), and 37 (AY93448); Uncultured bacterium clone Tocil Wood soil 17 (AY934484), 25 (AY934444), and 38 (AY934439)). Condensed branch of 4 marine sediment metagenome sequences in cluster 2 (GAF77597.1, GAF84365.1, GAG35382.1, GAG79179.1 and GAI71391.1).

ecosystems, which are highly active sinks (Borodina *et al.*, 2005; Chaignaud *et al.*, 2018). In the laboratory, high chloromethane ratios (in the percent range) have been used to isolate and grow pure chloromethane-degrading cultures that originated from soil (Schäfer *et al.*, 2007) in order to obtain enough biomass for growth and chloromethane degradation studies. The identification of natural microbial populations actively assimilating carbon from chloromethane was first achieved using stable isotope probing (DNA SIP) (Miller *et al.*, 2004). Studies in soil revealed that bacterial population members of soil microbial communities take up carbon from amended chloromethane (Miller *et al.*, 2004; Borodina *et al.*, 2005; Chaignaud *et al.*, 2017). Moreover, it has been demonstrated that methanotrophs, as investigated with a pure culture of *Methylomicrobium album* BG8, can assimilate carbon from chloromethane when a surplus of methanol occurs, although chloromethane alone was not assimilated (Han and Semrau, 2000). These findings suggest a possible co-metabolic lifestyle of some methanotrophs regarding chloromethane degradation. In a forest soil, the rate of chloromethane degradation was recently shown to be enhanced when methanol was present (Chaignaud *et al.*, 2018). In this study, *Beijerinckiceae* (i.e. phylotypes affiliating closely with *Methylovirgula ligni*), assimilated carbon from methanol or from chloromethane as suggested by SIP experiments of soil microcosms amended with labelled or unlabelled methanol, chloromethane or both simultaneously (Chaignaud *et al.*, 2017; Morawe *et al.*, 2017). Methylotrophic co-utilization of chloromethane with other C_1 compounds such as methanol, which is found in larger amounts in the environment and utilized by most methylotrophs, may thus represent an important component of chloromethane sink activity. The utilization of methanol as an alternate substrate to chloromethane may favour the persistence of chloromethane-degrading populations in the context of shortage of chloromethane in natural habitats. The presence of pollutants such as toluene in ocean water was found to inhibit chloromethane and bromomethane uptake, thereby challenging the persistence of methyl halide-degrading populations and subsequent methyl halide oceanic uptake, which would in turn disturb an important component of the biogeochemical cycle of these ozone-depleting compounds (Goodwin *et al.*, 2005).

Future trends

Low- and high-affinity degraders

One of the key aspects of future explorations of the methylotrophic chloromethane sink is the need to unravel the enzymes and pathways that define the global sink of halocarbons not only in bacteria but also in fungi. Future research needs to grasp the diversity of chloromethane-degrading-associated metabolisms and taxa associated with the microbial chloromethane sink. Much effort has focused on molecular processes for methylotrophic growth under laboratory conditions. Owing to technical limitations, so far biological chloromethane sinks have been assessed using higher than natural concentrations. Future investigations should focus on environmentally relevant chloromethane levels (picomolar). High variations in chloromethane flux emissions have been reported (Jaeger *et al.*, 2018) so it can be postulated that low and high affinity chloromethane-degrading populations may be found in the environment, as for methane (Maxfield *et al.*, 2009).

Unknowns of the chloromethane cycle

How plant chloromethane emissions vary during their developmental stage or their physiological states (chloride stress, pathogen invasion, exposure to chlorinated pesticides), and the role of chloromethane production by plants, is unknown (Bringel and Couée, 2018). Host-associated microbiota may elicit a response that modulates the interactive interplay between host/microbial cells/insects as found for other plant-produced VOCs (review by Bringel and Couée, 2015). For example, chloromethane regenerates veratryl alcohol during degradation of lignin by some white-rot fungi (Harper *et al.*, 1990), and one may speculate that associated bacterial communities may interfere with this process. Similarly, methylotroph populations may interfere with chlorine output through volatilization via chloromethane emissions, as white-rot fungi and higher plants methylate chloride (Oberg, 2002).

On the bacterial side, even with *M. extorquens* strain CM4, in the *in-labo* best-characterized chloromethane degrader, our current understanding is not sufficient to predict its behaviour under environmental conditions. Key actors remain uncharacterized: the regulator of the *cmu* pathway, the chemical nature of the native corrinoid cofactor of the CmuA methyltransferase from chloromethane to H_4F, many genes involved in corrinoid cofactor metabolism (Roselli *et al.*, 2013), and conserved genes adjacent to *cmuA* such as *cmuC* (Fig. 8.4) (Michener *et al.*, 2016). In fact, all the details of the dehalogenation reactions and carbon funnelling to central metabolism remain to be investigated physiologically, biochemically on wild-type and mutants. Unfortunately, reverse genetics for site-directed mutagenesis has not been successful so far in *M. extorquens* strain CM4 (Michener *et al.*, 2016) and *Hyphomicrobium* sp. Nonetheless, complementary 'omics' technologies such metabolomics, fluxomics and experimental evolution are available for *Methylorubrum* studies (Ochsner *et al.*, 2015).

Biomarkers of chloromethane sinks

Gene *cmuA* is the only known biomarker for bacterial chloromethane utilization. It is rarely found in metagenomes (Table 8.2) (Jaeger *et al.*, 2018a,b), and is absent in the genome of some chloromethane utilizers (Buddruhs *et al.*, 2013) (Table 8.1). Together, these observations reveal the need to define *cmuA*-independent biomarkers of the chloromethane microbial sinks. In open oceans (Goodwin *et al.*, 2005) and soils (Rasche *et al.*, 1991; Chaignaud *et al.*, 2018), co-substrate consumption may be a major driver of chloromethane microbial sinks. Thus, new strategies should be explored for investigating microbial *cmu*-independent chloromethane-degrading pathways, such as experiments on co-consumption of chloromethane with mixtures of other environmentally relevant carbon sources, co-cultures of mixed populations (Nai and Meyer, 2018), and cultivation-independent approaches using single-cell labelling and identification, such as microfluidic cell sorting combined with 'omics' methods and Raman-FISH (Musat *et al.*, 2012; Hol and Dekker, 2014). This quest for new biomarkers will help explore the dynamics and diversity of chloromethane-degrading microorganisms active *in situ* in natural and anthropogenic environments.

Web resources

Data reporting on production and consumption of ozone depleting substances

The Montreal Protocol on Substances that Deplete the Ozone Layer:
http://ozone.unep.org/en/treaties-and-decisions/montreal-protocol-substances-deplete-ozone-layer
Scientific Assessments of Ozone Depletion conducted under the auspices of the World Meteorological Organization (WMO) and the United Nations Environment Programme (UNEP): https://www.esrl.noaa.gov/csd/assessments/ozone/and http://www.wmo.int/pages/prog/arep/gaw/ozone/

Genome of chloromethane-degrading strains

http://www.genoscope.cns.fr/agc/microscope opens the MicroScope platform that hosts *M. extorquens* CM4 genome and provides annotation results relative to BLAST similarities, COG assignations, enzymatic function prediction (PRIAM software), TMHMM and SignalP predictions, and synteny conservation (Syntonizer software). Complete *Hyphomicrobium* sp. MC1 and *Vibrio orientalis* CIP 102891 genomes available at https://www.ncbi.nlm.nih.gov/nuccore/FQ859181 (Vuilleumier *et al.*, 2011) and (http://www.ebi.ac.uk/ena/data/view/Project:40487) (Yang *et al.*, 1983), respectively. Comparative RNA-Seq data of *M. extorquens* CM4 grown with chloromethane or with methanol available at: https://www.genoscope.cns.fr/agc/microscope/transcriptomic/NGSProjectRNAseq.php?projType=RNAseq

References

Amato, P., Demeer, F., Melaouhi, A., Fontanella, S., Martin-Biesse, A.S., Sancelme, M., Laj, P., and Delort, A.M. (2007). A fate for organic acids, formaldehyde and methanol in cloud water: their biotransformation by micro-organisms. Atmos. Chem. Phys. *7*, 4159–4169. https://doi.org/10.5194/acp-7-4159-2007

Amato, P., Brisebois, E., Draghi, M., Duchaine, C., Fröhlich-Nowoisky, J., Huffman, J.A., Mainelis, G., Robine, E., and Thibaudon, M. (2017a). Main biological aerosols, specificities, abundance, and diversity. In Microbiology of Aerosols, A.M. Delort, and P. Amato, eds. (John Wiley & Sons, Inc., Hoboken, NJ), pp. 1–21.

Amato, P., Joly, M., Besaury, L., Oudart, A., Taib, N., Moné, A.I., Deguillaume, L., Delort, A.M., and Debroas, D. (2017b). Active microorganisms thrive among extremely diverse communities in cloud water. PLOS ONE *12*, e0182869. https://doi.org/10.1371/journal.pone.0182869

Attieh, J.M., Hanson, A.D., and Saini, H.S. (1995). Purification and characterization of a novel methyltransferase responsible for biosynthesis of halomethanes and methanethiol in *Brassica oleracea*. J. Biol. Chem. *270*, 9250–9257. https://doi.org/10.1074/jbc.270.16.9250

Biemann, K., Oro, J., Toulmin, P., Orgel, L.E., Nier, A.O., Anderson, D.M., Simmonds, P.G., Flory, D., Diaz, A.V., Rushneck, D.R., *et al.* (1976). Search for organic and volatile inorganic compounds in two surface samples from the chryse planitia region of Mars. Science *194*, 72–76. https://doi.org/10.1126/science.194.4260.72

Bodelier, P.L., Meima-Franke, M., Hordijk, C.A., Steenbergh, A.K., Hefting, M.M., Bodrossy, L., von Bergen, M., and Seifert, J. (2013). Microbial minorities modulate methane consumption through niche partitioning. ISME J. *7*, 2214–2228. https://doi.org/10.1038/ismej.2013.99

Borodina, E., Cox, M.J., McDonald, I.R., and Murrell, J.C. (2005). Use of DNA-stable isotope probing and functional gene probes to investigate the diversity of methyl chloride-utilizing bacteria in soil. Environ. Microbiol. *7*, 1318–1328. https://doi.org/10.1111/j.1462-5822.2005.00819.x

Bringel, F., and Couée, I. (2015). Pivotal roles of phyllosphere microorganisms at the interface between plant functioning and atmospheric trace gas dynamics. Front. Microbiol. *6*, 486. https://doi.org/10.3389/fmicb.2015.00486

Bringel, F., and Couée, I. (2018). Plant–pesticide interactions and the global chloromethane budget. Trends Plant Sci. *23*, 95–99. https://doi.org/10.1016/j.tplants.2017.12.001

Buddruhs, N., Chertkov, O., Petersen, J., Fiebig, A., Chen, A., Pati, A., Ivanova, N., Lapidus, A., Goodwin, L.A., Chain, P., *et al.* (2013). Complete genome sequence of the marine methyl-halide oxidizing *Leisingera methylohalidivorans* type strain (DSM 14336(T)), a representative of the *Roseobacter* clade. Stand. Genomic Sci. *9*, 128–141. https://doi.org/10.4056/sigs.4297965

Cao, J., Lai, Q., Yuan, J., and Shao, Z. (2015). Genomic and metabolic analysis of fluoranthene degradation pathway in *Celeribacter indicus* P73^T. Sci. Rep. *5*, 7741. https://doi.org/10.1038/srep07741

Carpenter, L.J., Reimann, S., Burkholder, J.B., Clerbaux, C., Hall, B.D., Hossaini, R., Laube, J.C., and Yvon-Lewis, S.A. (2014). Update on Ozone-Depleting Substances (ODSs) and other gases of interest to the Montreal protocol. In Scientific Assessment of Ozone Depletion, Global Ozone Research and Monitoring Project Report, A. Engel, and S.A. Montzka, eds. (World Meteorological Organization (WMO), Geneva, Switzerland), pp. 21–125. http://hdl.handle.net/2268/175647

Chaignaud, P., Maucourt, B., Weiman, M., Alberti, A., Kolb, S., Cruveiller, S., Vuilleumier, S., and Bringel, F. (2017). Genomic and transcriptomic analysis of growth-supporting dehalogenation of chlorinated methanes in *Methylobacterium*. Front. Microbiol. *8*, 1600. https://doi.org/10.3389/fmicb.2017.01600

Chaignaud, P., Morawe, M., Besaury, L., Kröber, E., Vuilleumier, S., Bringel, F., and Kolb, S. (2018). Methanol consumption drives the bacterial chloromethane sink in a forest soil. ISME J. *12*, 2681–2693. https://doi.org/10.1038/s41396-018-0228-4

Chen, G., Kleindienst, S., Griffiths, D.R., Mack, E.E., Seger, E.S., and Löffler, F.E. (2017). Mutualistic interaction between dichloromethane- and chloromethane-degrading bacteria in an anaerobic mixed culture. Environ. Microbiol. *19*, 4784–4796. https://doi.org/10.1111/1462-2920.13945

Chistoserdova, L. (2011). Modularity of methylotrophy, revisited. Environ. Microbiol. *13*, 2603–2622. https://doi.org/10.1111/j.1462-2920.2011.02464.x

Coulter, C., Hamilton, J.T., and Harper, D.B. (1993). Evidence for the existence of independent chloromethane- and S-adenosylmethionine-utilizing systems for methylation in *Phanerochaete chrysosporium*. Appl. Environ. Microbiol. *59*, 1461–1466.

Coulter, C., Hamilton, J.T., McRoberts, W.C., Kulakov, L., Larkin, M.J., and Harper, D.B. (1999). Halomethane:bisulfide/halide ion methyltransferase, an unusual corrinoid enzyme of environmental significance isolated from an aerobic methylotroph using chloromethane as the sole carbon source. Appl. Environ. Microbiol. *65*, 4301–4312.

Cox, M.J., Schäfer, H., Nightingale, P.D., McDonald, I.R., and Murrell, J.C. (2012). Diversity of methyl halide-degrading microorganisms in oceanic and coastal waters. FEMS Microbiol. Lett. *334*, 111–118. https://doi.org/10.1111/j.1574-6968.2012.02624.x

DasSarma, P., and DasSarma, S. (2018). Survival of microbes in Earth's stratosphere. Curr. Opin. Microbiol. *43*, 24–30. https://doi.org/10.1016/j.mib.2017.11.002

Delort, A.-M., Vaïtilingom, M., Amato, P., Sancelme, M., Parazols, M., Mailhot, G., Laj, P., and Deguillaume, L. (2010). A short overview of the microbial population in clouds: potential roles in atmospheric chemistry and nucleation processes. Atmos. Res. *98*, 249–260. https://doi.org/10.1016/j.atmosres.2010.07.004

Derendorp, L., Wishkerman, A., Keppler, F., McRoberts, C., Holzinger, R., and Röckmann, T. (2012). Methyl chloride emissions from halophyte leaf litter: dependence on temperature and chloride content. Chemosphere *87*, 483–489. https://doi.org/10.1016/j.chemosphere.2011.12.035

Doronina, N.V., Sokolov, A.P., and Trotsenko, Y.A. (1996). Isolation and initial characterization of aerobic chloromethane-utilizing bacteria. FEMS Microbiol. Lett. *142*, 179–183. https://doi.org/10.1111/j.1574-6968.1996.tb08427.x

Elsner, M., Zwank, L., Hunkeler, D., and Schwarzenbach, R.P. (2005). A new concept linking observable stable isotope fractionation to transformation pathways of organic pollutants. Environ. Sci. Technol. *39*, 6896–6916. https://doi.org/10.1021/es0504587

Farhan Ul Haque, M., Besaury, L., Nadalig, T., Bringel, F., Mutterer, J., Schaller, H., and Vuilleumier, S. (2017). Correlated production and consumption of chloromethane in the *Arabidopsis thaliana* phyllosphere. Sci. Rep. *7*, 17589. https://doi.org/10.1038/s41598-017-17421-y

Fayolle, E.C., Öberg, K.I., Jørgensen, J.K., Altwegg, K., Calcutt, H., Müller, H.S.P., Rubin, M., van der Wiel, M.H.D., Bjerkeli, P., Bourke, T.L., *et al.* (2017). Protostellar and cometary detections of organohalogens. Nat. Astronomy *1*, 703–708. https://doi.org/10.1038/s41550-017-0237-7

Freedman, D.L., Swamy, M., Bell, N.C., and Verce, M.F. (2004). Biodegradation of chloromethane by *Pseudomonas aeruginosa* strain NB1 under nitrate-reducing and aerobic conditions. Appl. Environ. Microbiol. *70*, 4629–4634. https://doi.org/10.1128/AEM.70.8.4629-4634.2004

Gasc, C., and Peyret, P. (2018). Hybridization capture reveals microbial diversity missed using current profiling methods. Microbiome *6*, 61. https://doi.org/10.1186/s40168-018-0442-3

Gasc, C., Peyretaillade, E., and Peyret, P. (2016). Sequence capture by hybridization to explore modern and ancient genomic diversity in model and nonmodel organisms. Nucleic Acids Res. *44*, 4504–4518. https://doi.org/10.1093/nar/gkw309

Gifford, S.M., Becker, J.W., Sosa, O.A., Repeta, D.J., and DeLong, E.F. (2016). Quantitative transcriptomics reveals the growth- and nutrient-dependent response of a streamlined marine methylotroph to methanol and naturally occurring dissolved organic matter. MBio *7*, e01279-16. https://doi.org/10.1128/mBio.01279-16

Giovannoni, S.J., Hayakawa, D.H., Tripp, H.J., Stingl, U., Givan, S.A., Cho, J.C., Oh, H.M., Kitner, J.B., Vergin, K.L., and Rappé, M.S. (2008). The small genome of an abundant coastal ocean methylotroph. Environ. Microbiol. *10*, 1771–1782. https://doi.org/10.1111/j.1462-2920.2008.01598.x

Goodwin, K.D., Tokarczyk, R., Stephens, F.C., and Saltzman, E.S. (2005). Description of toluene inhibition of methyl bromide biodegradation in seawater and isolation of a marine toluene oxidizer that degrades methyl bromide. Appl. Environ. Microbiol. *71*, 3495–3503. https://doi.org/10.1128/AEM.71.7.3495-3503.2005

Green, P.N., and Ardley, J.K. (2018). Review of the genus *Methylobacterium* and closely related organisms: a proposal that some *Methylobacterium* species be reclassified into a new genus, *Methylorubrum* gen. nov. Int. J. Syst. Evol. Microbiol. *68*, 2727–2748. https://doi.org/10.1099/ijsem.0.002856

Halsey, K.H., Carter, A.E., and Giovannoni, S.J. (2012). Synergistic metabolism of a broad range of C1 compounds in the marine methylotrophic bacterium HTCC2181. Environ. Microbiol. *14*, 630–640. https://doi.org/10.1111/j.1462-2920.2011.02605.x

Hamilton, J.T., McRoberts, W.C., Keppler, F., Kalin, R.M., and Harper, D.B. (2003). Chloride methylation by plant pectin: an efficient environmentally significant process. Science *301*, 206–209. https://doi.org/10.1126/science.1085036

Han, J.I., and Semrau, J.D. (2000). Chloromethane stimulates growth of *Methylomicrobium album* BG8 on methanol. FEMS Microbiol. Lett. *187*, 77–81. https://doi.org/10.1111/j.1574-6968.2000.tb09140.x

Hancock, T.L., Costello, A.M., Lidstrom, M.E., and Oremland, R.S. (1998). Strain IMB-1, a novel bacterium for the removal of methyl bromide in fumigated agricultural soils. Appl. Environ. Microbiol. *64*, 2899–2905.

Harper, D.B. (2000). The global chloromethane cycle: biosynthesis, biodegradation and metabolic role. Nat. Prod. Rep. *17*, 337–348.

Harper, D.B., and Hamilton, J.T.G. (2003). The global cycles of the naturally-occurring monohalomethanes. In Natural Production of Organohalogen Compounds, G. Gribble, ed. (Springer, Berlin, Heidelberg), pp. 17–41.

Harper, D.B., Buswell, J.A., Kennedy, J.T., and Hamilton, J.T. (1990). Chloromethane, methyl donor in veratryl alcohol biosynthesis in *Phanerochaete chrysosporium* and other lignin-degrading fungi. Appl. Environ. Microbiol. *56*, 3450–3457.

Hartmans, S., Schmuckle, A., Cook, A.M., and Leisinger, T. (1986). Methyl chloride: naturally occurring toxicant and C-1 growth substrate. Microbiology *132*, 1139–1142. https://doi.org/10.1099/00221287-132-4-1139

Heikes, B.G., Chang, W., Pilson, M.E.Q., Swift, E., Singh, H.B., Guenther, A., Jacob, D.J., Field, B.D., Fall, R., Riemer, D., *et al.* (2002). Atmospheric methanol budget and ocean implication. Global Biogeochem. Cycles *16*, 80-1–80-13. https://doi.org/10.1029/2002GB001895

Hol, F.J., and Dekker, C. (2014). Zooming in to see the bigger picture: microfluidic and nanofabrication tools to study bacteria. Science *346*, 1251821. https://doi.org/10.1126/science.1251821

Itoh, N., Toda, H., Matsuda, M., Negishi, T., Taniguchi, T., and Ohsawa, N. (2009). Involvement of S-adenosylmethionine-dependent halide/thiol methyltransferase (HTMT) in methyl halide emissions from agricultural plants: isolation and characterization of an HTMT-coding gene from *Raphanus sativus* (daikon radish). BMC Plant Biol. *9*, 116. https://doi.org/10.1186/1471-2229-9-116

Jaeger, N., Besaury, L., Kröber, E., Delort, A.M., Greule, M., Lenhart, K., Nadalig, T., Vuilleumier, S., Amato, P., Kolb, S., *et al.* (2018a). Chloromethane degradation in soils: a combined microbial and two-dimensional stable isotope approach. J. Environ. Qual. *47*, 254–262. https://doi.org/10.2134/jeq2017.09.0358

Jaeger, N., Besaury, L., Röhling, A.N., Koch, F., Delort, A.M., Gasc, C., Greule, M., Kolb, S., Nadalig, T., Peyret, P., *et al.* (2018b). Chloromethane formation and degradation in the fern phyllosphere. Sci. Total Environ. *634*, 1278–1287. https://doi.org/10.1016/j.scitotenv.2018.03.316

Jones, D.T., Taylor, W.R., and Thornton, J.M. (1992). The rapid generation of mutation data matrices from protein sequences. Comput. Appl. Biosci. *8*, 275–282. https://doi.org/10.1093/bioinformatics/8.3.275

Justicia-Leon, S.D., Ritalahti, K.M., Mack, E.E., and Löffler, F.E. (2012). Dichloromethane fermentation by a *Dehalobacter* sp. in an enrichment culture derived from pristine river sediment. Appl. Environ. Microbiol. *78*, 1288–1291. https://doi.org/10.1128/AEM.07325-11

Keppler, F., Eiden, R., Niedan, V., Pracht, J., and Schöler, H.F. (2000). Halocarbons produced by natural oxidation processes during degradation of organic matter. Nature *403*, 298–301. https://doi.org/10.1038/35002055

Keppler, F., Harper, D.B., Röckmann, T., Moore, R.M., and Hamilton, J.T.G. (2005). New insight into the atmospheric chloromethane budget gained using stable carbon isotope ratios. Atmos. Chem. Phys. *5*, 2403–2411. https://doi.org/10.5194/acp-5-2403-2005

Keppler, F., Harper, D.B., Greule, M., Ott, U., Sattler, T., Schöler, H.F., and Hamilton, J.T. (2014). Chloromethane release from carbonaceous meteorite affords new insight into Mars lander findings. Sci. Rep. *4*, 7010. https://doi.org/10.1038/srep07010

Kislyuk, A.O., Haegeman, B., Bergman, N.H., and Weitz, J.S. (2011). Genomic fluidity: an integrative view of gene diversity within microbial populations. BMC Genomics *12*, 32. https://doi.org/10.1186/1471-2164-12-32

Kits, K.D., Kalyuzhnaya, M.G., Klotz, M.G., Jetten, M.S., Op den Camp, H.J., Vuilleumier, S., Bringel, F., Dispirito, A.A., Murrell, J.C., Bruce, D., *et al.* (2013). Genome sequence of the obligate gammaproteobacterial methanotroph *Methylomicrobium album* strain BG8. Genome Announc. *1*, e0017013. https://doi.org/10.1128/genomeA.00170-13

Kleindienst, S., Higgins, S.A., Tsementzi, D., Chen, G., Konstantinidis, K.T., Mack, E.E., and Löffler, F.E. (2017). '*Candidatus* Dichloromethanomonas elyunquensis' gen. nov., sp. nov., a dichloromethane-degrading anaerobe of the *Peptococcaceae* family. Syst. Appl. Microbiol. *40*, 150–159. https://doi.org/10.1016/j.syapm.2016.12.001

Kumar, S., Stecher, G., and Tamura, K. (2016). MEGA7: Molecular evolutionary genetics analysis version 7.0 for bigger datasets. Mol. Biol. Evol. *33*, 1870–1874. https://doi.org/10.1093/molbev/msw054

Lai, Q., Cao, J., Yuan, J., Li, F., and Shao, Z. (2014). *Celeribacter indicus* sp. nov., a polycyclic aromatic hydrocarbon-degrading bacterium from deep-sea sediment and reclassification of *Huaishuia halophila* as *Celeribacter halophilus* comb. nov. I. J. Syst. Evol. Microbiol. *64*, 4160–4167. https://doi.org/10.1099/ijs.0.069039-0

Leisinger, T., and Braus-Stromeyer, S.A. (1995). Bacterial growth with chlorinated methanes. Environ. Health Perspect. 103 (Suppl. 5), 33–36. https://doi.org/10.1289/ehp.95103s433

Li, S., Park, M.-K., Jo, C.O., and Park, S. (2017). Emission estimates of methyl chloride from industrial sources in China based on high frequency atmospheric observations. J. Atmos. Chem. *74*, 227–243. https://doi.org/10.1007/s10874-016-9354-4

Manley, S.L. (2002). Phytogenesis of halomethanes: A product of selection or a metabolic accident? Biogeochemistry *60*, 163–180. https://doi.org/10.1023/A:1019859922489

Marx, C.J., Bringel, F., Chistoserdova, L., Moulin, L., Farhan Ul Haque, M., Fleischman, D.E., Gruffaz, C., Jourand, P., Knief, C., Lee, M.C., *et al.* (2012). Complete genome sequences of six strains of the genus *Methylobacterium*. J. Bacteriol. *194*, 4746–4748. https://doi.org/10.1128/JB.01009-12

Maxfield, P.J., Hornibrook, E.R., and Evershed, R.P. (2009). Substantial high-affinity methanotroph populations in Andisols effect high rates of atmospheric methane oxidation. Environ. Microbiol. Rep. *1*, 450–456. https://doi.org/10.1111/j.1758-2229.2009.00071.x

McAnulla, C., McDonald, I.R., and Murrell, J.C. (2001a). Methyl chloride utilising bacteria are ubiquitous in the natural environment. FEMS Microbiol. Lett. *201*, 151–155. https://doi.org/10.1111/j.1574-6968.2001.tb10749.x

McAnulla, C., Woodall, C.A., McDonald, I.R., Studer, A., Vuilleumier, S., Leisinger, T., and Murrell, J.C. (2001b). Chloromethane utilization gene cluster from *Hyphomicrobium chloromethanicum* strain CM2^T and development of functional gene probes to detect halomethane-degrading bacteria. Appl. Environ. Microbiol. *67*, 307–316. https://doi.org/10.1128/AEM.67.1.307-316.2001

McDonald, I.R., Warner, K.L., McAnulla, C., Woodall, C.A., Oremland, R.S., and Murrell, J.C. (2002). A review of bacterial methyl halide degradation: biochemistry, genetics and molecular ecology. Environ. Microbiol. *4*, 193–203. https://doi.org/10.1046/j.1462-2920.2002.00290.x

McRoberts, W.C., Keppler, F., Harper, D.B., and Hamilton, J.T.G. (2015). Seasonal changes in chlorine and methoxyl content of leaves of deciduous trees and their impact on release of chloromethane and methanol at elevated temperatures. Environ. Chem. *12*, 426–437. https://doi.org/10.1071/EN14208

Meßmer, M., Wohlfarth, G., and Diekert, G. (1993). Methyl chloride metabolism of the strictly anaerobic, methyl chloride-utilizing homoacetogen strain MC. Arch.

Microbiol. *160*, 383–387. https://doi.org/10.1007/BF00252225

Meßmer, M., Reinhardt, S., Wohlfarth, G., and Diekert, G. (1996). Studies on methyl chloride dehalogenase and O-demethylase in cell extracts of the homoacetogen strain MC based on a newly developed coupled enzyme assay. Arch. Microbiol. *165*, 18–25. https://doi.org/10.1007/s002030050291

Michener, J.K., Camargo Neves, A.A., Vuilleumier, S., Bringel, F., and Marx, C.J. (2014a). Effective use of a horizontally-transferred pathway for dichloromethane catabolism requires post-transfer refinement. Elife 3. https://doi.org/10.7554/eLife.04279

Michener, J.K., Vuilleumier, S., Bringel, F., and Marx, C.J. (2014b). Phylogeny poorly predicts the utility of a challenging horizontally transferred gene in *Methylobacterium* strains. J. Bacteriol. *196*, 2101–2107. https://doi.org/10.1128/JB.00034-14

Michener, J.K., Vuilleumier, S., Bringel, F., and Marx, C.J. (2016). Transfer of a catabolic pathway for chloromethane in *Methylobacterium* strains highlights different limitations for growth with chloromethane or with dichloromethane. Front. Microbiol. *7*, 1116. https://doi.org/10.3389/fmicb.2016.01116

Miller, L.G., Kalin, R.M., McCauley, S.E., Hamilton, J.T., Harper, D.B., Millet, D.B., Oremland, R.S., and Goldstein, A.H. (2001). Large carbon isotope fractionation associated with oxidation of methyl halides by methylotrophic bacteria. Proc. Natl. Acad. Sci. U.S.A. *98*, 5833–5837. https://doi.org/10.1073/pnas.101129798

Miller, L.G., Warner, K.L., Baesman, S.M., Oremland, R.S., McDonald, I.R., Radajewski, S., and Murrell, J.C. (2004). Degradation of methyl bromide and methyl chloride in soil microcosms: Use of stable C isotope fractionation and stable isotope probing to identify reactions and the responsible microorganisms. Geoch. Cosmoch. Acta *68*, 3271–3283. https://doi.org/10.1016/j.gca.2003.11.028

Ming, D.W., Archer, P.D., Glavin, D.P., Eigenbrode, J.L., Franz, H.B., Sutter, B., Brunner, A.E., Stern, J.C., Freissinet, C., McAdam, A.C., *et al.* (2014). Volatile and organic compositions of sedimentary rocks in Yellowknife Bay, Gale crater, Mars. Science *343*, 1245267. https://doi.org/10.1126/science.1245267

Morawe, M., Hoeke, H., Wissenbach, D.K., Lentendu, G., Wubet, T., Kröber, E., and Kolb, S. (2017). Acidotolerant bacteria and fungi as a sink of methanol-derived carbon in a deciduous forest soil. Front. Microbiol. *8*, 1361. https://doi.org/10.3389/fmicb.2017.01361

Musat, N., Foster, R., Vagner, T., Adam, B., and Kuypers, M.M. (2012). Detecting metabolic activities in single cells, with emphasis on nanoSIMS. FEMS Microbiol. Rev. *36*, 486–511. https://doi.org/10.1111/j.1574-6976.2011.00303.x

Nadalig, T., Farhan Ul Haque, M., Roselli, S., Schaller, H., Bringel, F., and Vuilleumier, S. (2011). Detection and isolation of chloromethane-degrading bacteria from the *Arabidopsis thaliana* phyllosphere, and characterization of chloromethane utilization genes. FEMS Microbiol. Ecol. *77*, 438–448. https://doi.org/10.1111/j.1574-6941.2011.01125.x

Nadalig, T., Greule, M., Bringel, F., Vuilleumier, S., and Keppler, F. (2013). Hydrogen and carbon isotope fractionation during degradation of chloromethane by methylotrophic bacteria. MicrobiologyOpen *2*, 893–900. https://doi.org/10.1002/mbo3.124

Nadalig, T., Greule, M., Bringel, F., Keppler, F., and Vuilleumier, S. (2014). Probing the diversity of chloromethane-degrading bacteria by comparative genomics and isotopic fractionation. Front. Microbiol. *5*, 523. https://doi.org/10.3389/fmicb.2014.00523

Nagatoshi, Y., and Nakamura, T. (2009). *Arabidopsis* HARMLESS TO OZONE LAYER protein methylates a glucosinolate breakdown product and functions in resistance to *Pseudomonas syringae* pv. *maculicola*. J. Biol. Chem. *284*, 19301–19309. https://doi.org/10.1074/jbc.M109.001032

Nai, C., and Meyer, V. (2018). From axenic to mixed cultures: technological advances accelerating a paradigm shift in microbiology. Trends Microbiol. *26*, 538–554 https://doi.org/10.1016/j.tim.2017.11.004.

Neufeld, J.D., Boden, R., Moussard, H., Schäfer, H., and Murrell, J.C. (2008). Substrate-specific clades of active marine methylotrophs associated with a phytoplankton bloom in a temperate coastal environment. Appl. Environ. Microbiol. *74*, 7321–7328. https://doi.org/10.1128/AEM.01266-08

Nijenhuis, I., and Richnow, H.H. (2016). Stable isotope fractionation concepts for characterizing biotransformation of organohalides. Curr. Opin. Biotechnol. *41*, 108–113. https://doi.org/10.1016/j.copbio.2016.06.002

Oberg, G. (2002). The natural chlorine cycle – fitting the scattered pieces. Appl. Microbiol. Biotechnol. *58*, 565–581. https://doi.org/10.1007/s00253-001-0895-2

Ochsner, A.M., Sonntag, F., Buchhaupt, M., Schrader, J., and Vorholt, J.A. (2015). *Methylobacterium extorquens*: methylotrophy and biotechnological applications. Appl. Microbiol. Biotechnol. *99*, 517–534. https://doi.org/10.1007/s00253-014-6240-3

Ohsawa, N., Tsujita, M., Morikawa, S., and Itoh, N. (2001). Purification and characterization of a monohalomethane-producing enzyme S-adenosyl-L-methionine: halide ion methyltransferase from a marine microalga, *Pavlova pinguis*. Biosci. Biotechnol. Biochem. *65*, 2397–2404. https://doi.org/10.1271/bbb.65.2397

Parisot, N., Peyretaillade, E., Dugat-Bony, E., Denonfoux, J., Mahul, A., and Peyret, P. (2016). Probe design strategies for oligonucleotide microarrays. Methods Mol. Biol. *1368*, 67–82. https://doi.org/10.1007/978-1-4939-3136-1_6

Rasche, M.E., Hyman, M.R., and Arp, D.J. (1991). Factors limiting aliphatic chlorocarbon degradation by *Nitrosomonas europaea*: cometabolic inactivation of ammonia monooxygenase and substrate specificity. Appl. Environ. Microbiol. *57*, 2986–2994.

Rhew, R.C., Østergaard, L., Saltzman, E.S., and Yanofsky, M.F. (2003). Genetic control of methyl halide production in *Arabidopsis*. Curr. Biol. *13*, 1809–1813. https://doi.org/10.1016/j.cub.2003.09.055

Ribière, C., Beugnot, R., Parisot, N., Gasc, C., Defois, C., Denonfoux, J., Boucher, D., Peyretaillade, E., and Peyret, P. (2016). Targeted gene capture by hybridization to illuminate ecosystem functioning. Methods Mol. Biol. *1399*, 167–182. https://doi.org/10.1007/978-1-4939-3369-3_10

Roselli, S., Nadalig, T., Vuilleumier, S., and Bringel, F. (2013). The 380 kb pCMU01 plasmid encodes chloromethane utilization genes and redundant genes for vitamin B12- and tetrahydrofolate-dependent chloromethane metabolism in *Methylobacterium extorquens* CM4: a proteomic and bioinformatics study. PLOS ONE *8*, e56598. https://doi.org/10.1371/journal.pone.0056598

Sailaukhanuly, Y., Sárossy, Z., Carlsen, L., and Egsgaard, H. (2014). Mechanistic aspects of the nucleophilic substitution of pectin. On the formation of chloromethane. Chemosphere *111*, 575–579. https://doi.org/10.1016/j.chemosphere.2014.05.001

Šantl-Temkiv, T., Finster, K., Hansen, B.M., Pašić, L., and Karlson, U.G. (2013). Viable methanotrophic bacteria enriched from air and rain can oxidize methane at cloud-like conditions. Aerobiologia *29*, 373–384. https://doi.org/10.1007/s10453-013-9287-1

Šantl-Temkiv, T., Amato, P., Gosewinkel, U., Thyrhaug, R., Charton, A., Chicot, B., Finster, K., Bratbak, G., and Löndahl, J. (2017). High-flow-rate impinger for the study of concentration, viability, metabolic activity, and ice-nucleation activity of airborne bacteria. Environ. Sci. Technol. *51*, 11224–11234. https://doi.org/10.1021/acs.est.7b01480

Schaefer, J.K., Goodwin, K.D., McDonald, I.R., Murrell, J.C., and Oremland, R.S. (2002). *Leisingera methylohalidivorans* gen. nov., sp. nov., a marine methylotroph that grows on methyl bromide. I. J. Syst. Evol. Microbiol. *52*, 851–859. https://doi.org/10.1099/00207713-52-3-851

Schäfer, H., McDonald, I.R., Nightingale, P.D., and Murrell, J.C. (2005). Evidence for the presence of a CmuA methyltransferase pathway in novel marine methyl halide-oxidizing bacteria. Environ. Microbiol. *7*, 839–852. https://doi.org/10.1111/j.1462-2920.2005.00757.x

Schäfer, H., Miller, L.G., Oremland, R.S., and Murrell, J.C. (2007). Bacterial cycling of methyl halides. Adv. Appl. Microbiol. *61*, 307–346. https://doi.org/10.1016/S0065-2164(06)61009-5

Schulze-Makuch, D., Wagner, D., Kounaves, S.P., Mangelsdorf, K., Devine, K.G., de Vera, J.P., Schmitt-Kopplin, P., Grossart, H.P., Parro, V., Kaupenjohann, M., *et al.* (2018). Transitory microbial habitat in the hyperarid Atacama Desert. Proc. Natl. Acad. Sci. U.S.A. *115*, 2670–2675. https://doi.org/10.1073/pnas.1714341115

Studer, A., Vuilleumier, S., and Leisinger, T. (1999). Properties of the methylcobalamin:H_4folate methyltransferase involved in chloromethane utilization by *Methylobacterium* sp. strain CM4. Eur. J. Biochem. *264*, 242–249.

Studer, A., Stupperich, E., Vuilleumier, S., and Leisinger, T. (2001). Chloromethane: tetrahydrofolate methyl transfer by two proteins from *Methylobacterium chloromethanicum* strain CM4. Eur. J. Biochem. *268*, 2931–2938. https://doi.org/10.1046/j.1432-1327.2001.02182.x

Studer, A., McAnulla, C., Büchele, R., Leisinger, T., and Vuilleumier, S. (2002). Chloromethane-induced genes define a third C1 utilization pathway in *Methylobacterium chloromethanicum* CM4. J. Bacteriol. *184*, 3476–3484. https://doi.org/10.1128/JB.184.13.3476-3484.2002

Toda, H., and Itoh, N. (2011). Isolation and characterization of a gene encoding a S-adenosyl-l-methionine-dependent halide/thiol methyltransferase (HTMT) from the marine diatom *Phaeodactylum tricornutum*: Biogenic mechanism of CH_3I emissions in oceans. Phytochem. *72*, 337–343. https://doi.org/10.1016/j.phytochem.2010.12.003

Traunecker, J., Preuß, A., and Diekert, G. (1991). Isolation and characterization of a methyl chloride utilizing, strictly anaerobic bacterium. Arch. Microbiol. *156*, 416–421. https://doi.org/10.1007/BF00248720

Vaïtilingom, M., Deguillaume, L., Vinatier, V., Sancelme, M., Amato, P., Chaumerliac, N., and Delort, A.M. (2013). Potential impact of microbial activity on the oxidant capacity and organic carbon budget in clouds. Proc. Natl. Acad. Sci. U.S.A. *110*, 559–564. https://doi.org/10.1073/pnas.1205743110

Vannelli, T., Studer, A., Kertesz, M., and Leisinger, T. (1998). Chloromethane metabolism by *Methylobacterium* sp. strain CM4. Appl. Environ. Microbiol. *64*, 1933–1936.

Vannelli, T., Messmer, M., Studer, A., Vuilleumier, S., and Leisinger, T. (1999). A corrinoid-dependent catabolic pathway for growth of a *Methylobacterium* strain with chloromethane. Proc. Natl. Acad. Sci. U.S.A. *96*, 4615–4620. https://doi.org/10.1073/pnas.96.8.4615

Vorholt, J.A. (2012). Microbial life in the phyllosphere. Nat. Rev. Microbiol. *10*, 828–840. https://doi.org/10.1038/nrmicro2910

Vuilleumier, S., Nadalig, T., Ul Haque, M.F., Magdelenat, G., Lajus, A., Roselli, S., Muller, E.E., Gruffaz, C., Barbe, V., Médigue, C., *et al.* (2011). Complete genome sequence of the chloromethane-degrading *Hyphomicrobium* sp. strain MC1. J. Bacteriol. *193*, 5035–5036. https://doi.org/10.1128/JB.05627-11

Wang, B., Tan, T., and Shao, Z. (2009). *Roseovarius pacificus* sp. nov., isolated from deep-sea sediment. Int. J. Syst. Evol. Microbiol. *59*, 1116–1121. https://doi.org/10.1099/ijs.0.002477-0

Weigold, P., El-Hadidi, M., Ruecker, A., Huson, D.H., Scholten, T., Jochmann, M., Kappler, A., and Behrens, S. (2016). A metagenomic-based survey of microbial (de)halogenation potential in a German forest soil. Sci. Rep. *6*, 28958. https://doi.org/10.1038/srep28958

Whitman, W.B., Coleman, D.C., and Wiebe, W.J. (1998). Prokaryotes: the unseen majority. Proc. Natl. Acad. Sci. U.S.A. *95*, 6578–6583. https://doi.org/10.1073/pnas.95.12.6578

Whittenbury, R., Phillips, K.C., and Wilkinson, J.F. (1970). Enrichment, isolation and some properties of methane-utilizing bacteria. J. Gen. Microbiol. *61*, 205–218. https://doi.org/10.1099/00221287-61-2-205

Wuosmaa, A.M., and Hager, L.P. (1990). Methyl chloride transferase: a carbocation route for biosynthesis of halometabolites. Science *249*, 160–162. https://doi.org/10.1126/science.2371563

Yang, Y., Yeh, L.p., Cao, Y., Baumann, L., Baumann, P., Tang, J.S.-e., and Beaman, B. (1983). Characterization of marine luminous bacteria isolated off the coast of China and description of *Vibrio orientalis* sp. nov. Curr. Microbiol. *8*, 95–100. https://doi.org/10.1007/BF01566965

Microbial Cycling of Methanethiol

9

Hendrik Schäfer[1]* and Özge Eyice[2]

[1] School of Life Sciences, University of Warwick, Coventry, UK.
[2] School of Biological and Chemical Sciences, Queen Mary University of London, London, UK.

*Correspondence: h.schaefer@warwick.ac.uk

https://doi.org/10.21775/9781912530045.09

Abstract

Methanethiol (MT) is an organic sulfur compound with a strong and disagreeable odour. It has biogeochemical relevance as an important compound in the global sulfur cycle, where it is produced as a reactive intermediate in a number of different pathways for synthesis and degradation of other globally significant sulfur compounds such as dimethylsulfoniopropionate, dimethylsulfide and methionine. With its low odour threshold and unpleasant smell, MT can be a significant cause of malodour originating from animal husbandry, composting, landfill operations, and wastewater treatment and is also associated with faeces, flatus and oral malodour (halitosis). A diverse range of microorganisms drives the production and degradation of MT, including its aerobic and anaerobic metabolism. MT producing and degrading organisms are known to be present in terrestrial, freshwater and marine environments but may also be important in association with plant and animal (including human) hosts. This chapter considers the role of MT as an intermediate of the global sulfur cycle and discusses current knowledge of microbial pathways of MT production and degradation.

Introduction

Methanethiol (MT), also known as methylmercaptan, is a one-carbon organic sulfur compound with the formula CH_3SH. With a boiling point of 6°C it is a colourless gas at room temperature characterized by a pungent and disagreeable odour that has been likened to rotten cabbage. As a product of decaying biomass, it can be found widely in the environment, for instance as a compound emitted from rotting fruit and vegetables but it is also found in association with humans and animals where it is a component of the smell of faeces, flatus and can be associated with bad breath (halitosis) (NCBI Pubchem database, n.d.). The human sense of smell has a low threshold for detection of MT at 1–2 ppb (Devos *et al.*, 1990), which is exploited in its use as an additive to natural gas distribution systems in order to facilitate the detection of leaks. A leak of an unspecified quantity of MT from a chemical factory in northern France in 2013 illustrated the potential of MT as a significant malodour. A bad smell was reported in parts of northern France and as far away as across the Channel in parts of Southern England, due to the chemical having been dispersed by the wind (Reuters, 2013).

Environmental concentrations, production and degradation of MT

Environmental concentrations of MT

Only relatively few studies have measured MT concentrations in seawater, anoxic environments and industrial settings. A detailed study of production of volatile sulfur compounds in anoxic freshwater sediments in a peat bog by Lomans *et al.* (1997) showed that MT was one of the dominant volatile sulfur compounds in anoxic sediments. MT concentrations ranged from 3–76 nM and its production pathway was biological as shown by heat killed controls. Accumulation of MT in sediment slurries to which bromoethanesulfonate had been added, suggested that methanogens were mainly responsible for MT removal in these freshwater sediments. Surface freshwater MT concentrations in the peat bog were 1–8 nM and in a similar range to those of H_2S, DMS and CS_2, demonstrating that MT contributes to emission of S from such freshwater environments (Lomans *et al.*, 1997).

In marine environments, MT concentrations have been reported to be in the range of 0.02–2 nM. In a study primarily reporting carbonyl sulfide (COS) emissions from the Aegean Sea, MT concentrations detected at time zero in incubation experiments with natural seawater samples were in a range of 50–500 pM and it was suggested that MT was subject to photodegradation and could potentially be a precursor for COS photoproduction (Ulshöfer *et al.*, 1996). Measurements of MT were also reported for a sample transect of the Atlantic by Kettle and colleagues, who reported average MT concentrations in surface seawater of 420 pM, with localized higher concentrations up to ≈ 1700 pM in the North African upwelling area, around ≈ 1500 pM in a coastal region close to Montevideo and up to ≈ 1000 pM on a transect between Montevideo and the Falkland Islands (Kettle *et al.*, 2001). In addition to above mentioned photodegradation, the reaction of MT with sulfate, dissolved organic matter (DOM) and trace metals to form sulfate–DOM–metal complexes has also been suggested as an abiotic degradation pathway (Kiene *et al.*, 2000).

Biological Production of MT

Biological MT production is well known as an intermediate in metabolism of dimethylsulfoniopropionate (DMSP) (Kiene and Taylor, 1988; Kiene *et al.*, 2000; Reisch *et al.*, 2011a,b) and DMS (De Bont *et al.*, 1981; Suylen and Kuenen, 1986; Pol *et al.*, 1994; Borodina *et al.*, 2000; Schäfer, 2007) which itself can be produced through microbial degradation of DMSP (Curson *et al.*, 2011b). In marine, estuarine and salt marsh environments, DMSP, an abundant metabolite released by phytoplankton and macroalgae, acts as a key precursor of dimethylsulfide (DMS). Degradation of dissolved DMSP releases DMS by the activity of DMSP-lyases (Johnston *et al.*, 2008), which are found in many aerobic and anaerobic organisms (Todd *et al.*, 2009; Todd *et al.*, 2011; Curson *et al.*, 2011a,b, 2012; Sun *et al.*, 2016). Once produced, DMS is degraded to MT by a DMS monooxygenase, which was reported in some strains of *Hyphomicrobium* and *Arthrobacter* (De Bont *et al.*, 1981; Borodina *et al.*, 2000; Boden *et al.*, 2011).

MT production also occurs as an intermediate of microbial DMSP degradation via the demethylation pathway. In a study of the sulfur metabolism of several isolates from the abundant marine roseobacter clade, Gonzales and colleagues noted the ability of several strains to produce MT from DMSP and other precursors including DMS, dimethyl sulfoxide (DMSO), 3-methylmercaptopropionate (MMPA) and α-ketomethiol-butyrate (González *et al.*, 1999). Subsequent characterization of the metabolic pathway producing MT from DMSP showed that, in this pathway, DMSP is initially demethylated to MMPA by DMSP demethylase encoded by *dmdA,* which was first discovered in the marine roseobacter *Ruegeria pomeroyi* DSS-3 (Howard *et al.*, 2006). Subsequently, Reisch *et al.* (2011) demonstrated that after the demethylation step, MMPA is catabolized first to methylthioacryloyl-CoA (MTA-CoA) and then to MT via the demethiolation pathway (Reisch *et al.*, 2011b). The genes designated as *dmdB, dmdC* and *dmdD* were shown to encode the enzymes catalysing the transformation from MMPA to MT in *R. pomeroyi.* The presence of these genes (with the exception of *dmdD*) in the genome of ubiquitous marine bacterium *Pelagibacter ubique, Pelagibacter* HTCC1062 and *Ruegeria lacuscaerulensis* as well as in marine metagenomes reiterate the significance of this pathway in the marine environment (Reisch *et al.*, 2011b; Sun *et al.*, 2016). Sun and colleagues also demonstrated that eight *Pelagibacterales* genomes contain homologues of the *dmdABC* genes, but not *dmdD*, but *Pelagibacter* is still able to produce MT from DMSP. This suggests that the MT formation from MMPA is widespread in marine ecosystems, however a novel enzyme may be catalysing the last step of MT production from DMSP in some marine bacteria.

It was also noted that several aerobic bacteria from the genera *Corynebacterium, Rhizobium, Flavobacterium, Erwinia, Aeromonas, Pseudomonas* and *Yersinia* isolated from soil, sediment and marine algae cultures have the capacity to methylate hydrogen sulfide to produce MT. The activity of a thiol

S-methyltransferases was demonstrated in distinct fractions of crude cell free extracts of *Pseudomonas fluorescens* PF4 subjected to gel-filtration and ion-exchange chromatography; S-adenosylmethionine was identified as a methyl donor (Drotar *et al.*, 1987). The authors speculated that further methylation of the product MT to DMS would be a possibility and might be carried out by the same enzyme, but they did not observe DMS production in their activity assays, noting that this would depend on the K_m of the second methyl transfer reaction (Drotar *et al.*, 1987).

Anaerobic mechanisms

In anaerobic soils and sediments, MT is primarily produced via the degradation of sulfur containing amino acids and methylation of sulfide (Lomans *et al.*, 2002). Degradation of methionine was shown to lead to formation of MT in anoxic lake sediments (Zinder and Brock, 1978). Similarly, in anoxic salt marsh sediments, addition of methionine and S-methyl cysteine led to production of MT, less MT production was noted from DMSP (Kiene and Capone, 1988). The methyl-thiol group of methionine is cleaved (demethiolation) via the methionine-gamma-lyase enzyme (MegL), which has been purified and characterised from various bacteria such as *Pseudomonas, Brevibacter* and *Trichomonas* species (Bentley and Chasteen, 2004). Another route to anaerobic MT production is by the activity of thiol S-methyltransferases, which transfer methyl groups from S-adenosylmethionine to sulfide resulting in MT formation or methylate MT to form DMS (Bentley and Chasteen, 2004).

Methoxylated aromatic compounds are another precursor from which MT is produced in soil and sediments. Bak and colleagues (1992) isolated two anaerobic homoacetogenic species from the genus *Pelobacter* that produce MT during growth on trimethoxybenzoate or syringate by transferring the methyl group of the aromatic ring to hydrogen sulfide via thiol S-methyltransferase (Bak *et al.*, 1992). Several anaerobic isolates have been described including members of the genera *Holophaga, Sporobacter, Sporobacterium* and *Parasporobacterium* that can methylate sulfide to MT during the degradation of aromatic methoxylated compounds (Kreft and Schink, 1993; Grech-Mora *et al.*, 1996; Mechichi *et al.*, 1999; Lomans *et al.*, 2001). This process is suggested to take place at aerobic/anaerobic interfaces of organic rich freshwater sediments as methoxylated aromatic compounds are produced during the degradation of lignin, an abundant biopolymer (Lomans *et al.*, 2002).

Microbial degradation of MT

The main sink for MT in the environment is its degradation by microorganisms. Owing to the toxicity and foul odour of MT, only few studies have actually attempted to enrich and grow microorganisms on MT as a sole source of carbon and energy source. Therefore, most of what is known about microbial MT degradation is based on isolates in which MT is degraded as an intermediate of the metabolism of other organic sulfur compounds.

Aerobic mechanisms

Aerobic degradation of MT was shown in *Hyphomicrobium* sp. S, which was obtained from soil using DMSO as the enrichment substrate (De Bont *et al.*, 1981). Following this, a wide range of aerobic bacteria that degrade MT have been isolated from several environments including soil, peat biofilter, lake and marine sediments, seawater and marine algal cultures. These methylotrophic and sulfur-oxidizing species were affiliated with the genera *Hyphomicrobium, Thiobacillus, Rhodococcus* and *Methylophaga* (De Bont *et al.*, 1981; Suylen *et al.*, 1987; Cho *et al.*, 1991; Gould and Kanagawa, 1992; Visscher and Taylor, 1993a,b; Pol *et al.*, 1994; Borodina *et al.*, 2000; Schäfer, 2007; Boden *et al.*, 2010). Recently, *Methylotenera mobilis* JLW8 was shown to degrade MT as the sole carbon and energy source (Carrión *et al.*, 2017).

Aerobic bacteria degrade MT by methanethiol oxidase (MTO). This enzyme has been purified from *Hyphomicrobium* and *Thiobacillus* species and shown to degrade MT to formaldehyde, hydrogen sulfide and hydrogen peroxide (Suylen *et al.*, 1987; Gould and Kanagawa, 1992). Recently, the MTO enzyme in *Hyphomicrobium* sp. VS has been characterised in more detail, showing that it requires Cu for activity and suggesting presence of a tryptophan-tryptophylquinone (TTQ) co factor. The identification of the encoding gene revealed that MTO it is a homologue of the so-called selenium-binding protein family whose function had previously been poorly constrained (Eyice *et al.*,

2018). However, identification and characterisation of the human form of selenium-binding protein SELENBP1 by Pol and colleagues demonstrated that the human homologue is also a methanethiol oxidase and that a genetic defect in this gene is the underlying cause of extra-oral, or blood-borne, halitosis (Pol *et al.*, 2018).

Genes encoding MTO are present in genomes of a wide range of microorganisms known to degrade DMS (e.g. *Hyphomicrobium* and *Thiobacillus* spp.), DMSP (*Ruegeria pomeroyi* and other roseobacter clade bacteria), and indeed a number of methanotrophic and methylotrophic bacteria. Detection of *mtoX* in metagenomes as well as the application of specific PCR primers for *mtoX* demonstrated that *mtoX* and thus MT-degrading bacteria are present in a wide range of marine and terrestrial environments. Application of the stable isotope probing method with ^{13}C-labelled DMS has also been used indirectly to identify active MT-degraders in soil and lake sediment samples (Eyice *et al.*, 2018). MT oxidases with different molecular weights to that found in *Hyphomicrobium* sp. VS have been reported in *Rhodococcus* and *Thiobacillus* strains, suggesting that other methanethiol oxidases may yet have to be characterized in detail at the biochemical and genetic level (Kim *et al.*, 2000; Lee *et al.*, 2002).

An alternative sink for MT removal in aerobic environments is MT-dependent DMS production. A recent study demonstrated that MT can be methylated to DMS in aerobic bacteria through the activity of a membrane bound methyltransferase encoded by the gene *mddA* (Carrion *et al.*, 2015). The *mddA* gene was found widely distributed in phylogenetically diverse bacteria and several isolates tested showed that the presence of the *mddA* gene correlated with the ability to form DMS from MT. Based on metagenomic datasets, it was estimated that the *mddA* gene may be present in 5–76% of soil bacteria (Carrión *et al.*, 2015). In a subsequent study, Carrion and colleagues showed that although only a small proportion of MT (≈0.1%) added to a grassland soil was converted to DMS via this pathway, the soil microbial community contained a phylogenetically diverse group of bacteria, mainly *Pseudomonas, Acinetobacter, Gemmobacter, Phyllobacterium, Rhizobium, Ensifer* and *Sinorhizobium* that encoded *mddA*. (Carrión *et al.*, 2017). It is notable that the gene encoding the MddA enzyme can be found abundantly in metagenomes, particularly in soils. The relatively low conversion of MT to DMS observed in a grassland soil in the study could be due to competing pathways of MT and DMS removal and the environmental significance of this pathway requires additional analysis (Carrión *et al.*, 2017).

Anaerobic mechanisms

DMS and MT degradation by microorganisms has been studied in several ecosystems, yet, our knowledge on microbial populations that degrade MT in anoxic environments is very limited.

MT is primarily used by methane-producing archaea (methanogens) and sulfate-reducing bacteria (SRB) in anoxic marine, freshwater and terrestrial ecosystems (Fig. 9.1; Zinder and Brock, 1978; Lomans *et al.*, 1999b). The first study that showed that MT is degraded to methane and carbon dioxide was carried out on samples from Lake Mendota (Wisconsin, USA) using radiolabelled ^{14}C-methyl-methionine (Zinder and Brock, 1978). However, pure methanogenic species that

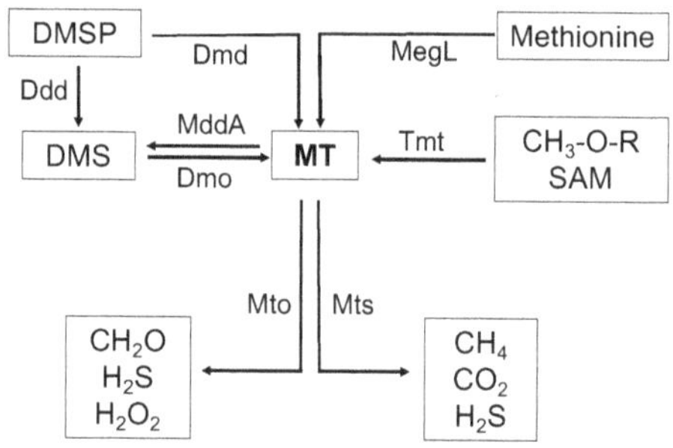

Figure 9.1 Simplified MT cycle and the key synthesis/degradation enzymes/pathways identified. Main MT sources include degradation of DMSP by the demethylation/demethiolation pathway (Dmd), degradation of methionine by methionine-gamma-lyase (MegL) or methyl transfer to sulfide by thiol methyltransferases (Tmt) from methoxylated aromatic compounds (CH_3-O-R) or S-adenosylmethionine (SAM). Cleavage of DMSP by DMSP-lyases (Ddd) produces DMS which can be oxidized to MT by DMS monooxygenase (Dmo). Sinks include methanethiol oxidase (Mto) which degrades MT to formaldehyde (CH_2O), hydrogen sulfide (H_2S) and hydrogen peroxide (H_2O_2), methylation of MT to DMS by methyltransferase (Mdd), and degradation in methanogens to methane (CH_4), carbon dioxide (CO_2) and hydrogen sulfide (H_2S) via activity of Mts methyltransferases.

could utilize MT as the carbon and energy source were not isolated from the samples. Later, Kiene *et al.* (1986) demonstrated that methane is produced in sediment samples from a variety of habitats including freshwater, alkaline and hypersaline lakes as well as estuarine salt marshes. They also obtained, from an estuarine salt marsh sediment, the first methanogenic strain, which was capable of metabolizing MT as the sole source of carbon and energy, yet did not identify this strain (Kiene *et al.*, 1986). Following this, Ni and Boone (1991) identified the first pure methanogenic strain (*Methanolobus siciliae*) from oilfield water samples using DMS as the substrate (Ni and Boone, 1991). The cultures were subsequently shown to use MT as the catabolic substrate (Ni and Boone, 1993; Table 9.1). Another methanogenic strain, *Methanosarcina* sp. MTP4, was isolated from marine sediment using MT as the sole carbon and energy source (Finster *et al.*, 1992; Table 9.1). The biochemical and genetic basis of methanethiol-dependent methanogenesis of *Methanosarcina* spp. has recently been identified. Three fused methyltransferase-corrinoid enzymes (MtsD, MtsF and MtsH) were shown by mutational analysis to be involved in formation of DMS as a metabolic intermediate during carboxidotrophic growth of *Methanosarcina acetivorans*. Further characterization indicated that these were also required for methylotrophic growth of *M. acetivorans* on DMS (Oelgeschläger and Rother, 2009). Work by Fu and Metcalf (2015) further showed that MtsF, which was highly up-regulated during growth on MT, and MtsH were capable of transferring the methyl group from MT to coenzyme M, the latter also seemed to accept DMS as a substrate (Fu and Metcalf, 2015). Other methanogens that can grow on methylated sulfur compounds have been described which appear to be obligately methylotrophic. *Methanomethylovorans hollandica* DMS1 is a MT-degrading methanogen that has been isolated from the sediment of a eutrophic freshwater pond in Nijmegen, The Netherlands, using a chemostat, which enabled high DMS degradation rates by removing inhibitory metabolites (i.e. hydrogen sulfide) (Lomans *et al.*, 1999a). A closely related strain, *Methanomethylovorans uponensis*, was obtained from a wetland sediment and shared the ability to grow using methylated sulfur compounds MT and DMS (Cha *et al.*, 2013). Methanogens related to *Methanomethylovorans hollandica* as well as *Methanolobus* were also present in a lab-scale bioreactor able degrade 6mM MT in the inflowing

Table 9.1 Anaerobic archaea and bacteria that can utilize MT as a carbon and energy source

Isolate	Isolation source	Reference
Archaea		
Methanolobus siciliae HI350	Oil field water	Ni and Boone (1991)
Methanolobus bombayensis B-1[a]	Marine sediment	Kadam *et al.* (1994)
Methanolobus taylorii GS-16	Estuarine sediment	Oremland and Boone (1994)
Methanolobus sp. strain SODA	Bioreactor treating MT at pH 10	van Leerdam *et al.* (2008a)
Methanolobus sp. strain WR1	Bioreactor treating MT at pH ≥8	van Leerdam *et al.* (2008b)
Methanosarcina sp. MTP4	Marine sediment	Finster *et al.* (1992)
Methanosarcina semesiae MD1	Mangrove sediment	Lyimo *et al.* (2000)
Methanomethylovorans hollandica DMS1	Lake sediment	Lomans *et al.* (1999a)
Methanohalophilus zhilinae WeN5T[a]	Alkaline lake sediment	Mathrani *et al.* (1988)
Methanohalophilus oregonense WAL1[a]	Alkaline saline aquifer	Liu *et al.* (1990)
Bacteria		
Desulfotomaculum sp. MTS5	Anaerobic fermentor	Tanimoto and Bak (1994)
Desulfotomaculum sp. SDN4	Anaerobic fermentor	Tanimoto and Bak (1994)
Desulfosarcina sp. SD1	Mangrove sediment	Lyimo *et al.* (2009)
Thiobacillus sp. ASN1	Salt marsh	Visscher and Taylor (1993b)

[a]These strains were isolated using DMS as the carbon and energy source but not tested directly for their ability to grow on MT.

medium which had been inoculated with sludge from a paper pulp mill wastewater treatment plant (de Bok *et al.*, 2006). Further *Methanolobus* isolates (strains WR1 and SODA) were obtained by van Leerdam and colleagues from bioreactors treating MT under alkaline conditions at pH of 8 and above and pH10, respectively, and were shown to grow on MT as sole carbon source (van Leerdam *et al.*, 2008a,b). Additionally, a number of methanogens from the genera *Methanosarcina*, *Methanohalophilus* and *Methanolobus*, which can transform DMS to methane were isolated from anoxic environments including alkaline lake and marine sediments (Table 9.1). Some of these were not tested for their growth on MT directly; however, it appears likely that because of MT being an intermediate of DMS metabolism in other methanogens, that these species can also catabolize MT.

In high-sulfate ecosystems such as marine and salt marsh sediments, SRB degrade MT to hydrogen sulfide and carbon dioxide using sulfate as the final electron acceptor (Kiene and Visscher, 1987; Kiene and Capone, 1988; Tanimoto and Bak, 1994). A relatively small number of SRB that grow on MT and sulfate have been isolated so far. These belong to the genera *Desulfotomaculum* and *Desulfosarcina*, which were obtained from laboratory scale methanogenic fermenters (Tanimoto and Bak, 1994) and mangrove sediments (Lyimo *et al.*, 2009).

A number of studies have been carried out to understand the interaction between methanogens and SRB using 2-bromoethanesulfonic acid (BES) and molybdate as specific inhibitors of methanogenesis and sulfate reduction, respectively (Kiene *et al.*, 1986; Lomans *et al.*, 1997). MT was reported to be a non-competitive substrate for methanogens although competition between methanogens and SRB was observed for DMS at low substrate concentrations (< 10 μM DMS) (Kiene *et al.*, 1986). This microbial interaction may have significant impact on the fate of MT in the environment, particularly in marine sediments which have high sulfate concentrations.

In addition to sulfate, nitrate is also used as an electron acceptor by MT-degrading microorganisms. One example for this metabolism is *Desulfotomaculum* sp. SDN4 that was isolated from a methanogenic thermophilic fermentor and which was able to use nitrate as terminal electron acceptor (Tanimoto and Bak, 1994). DMS-grown cells of *Thiobacillus* ASN-1, which was isolated from a *Spartina*-dominated salt marsh, were also shown to metabolize MT with nitrate and nitrite as electron acceptor (Taylor and Visscher, 1993b).

In a study that used cultivation-independent methods to identify methanogen and SRB populations degrading MT (and DMS) in the environment, Lomans and colleagues found methanogens closely related to *Methanomethylovorans hollandica* to be the dominant MT-degraders in freshwater sediments (Lomans *et al*, 2001). This suggests that *M. hollandica* might be a major player in the MT cycle in freshwater habitats. To our knowledge, there are as yet no studies focusing on the characterization of anaerobic MT-degrading populations using post-genomic approaches, which limits our understanding of the identity and distribution of MT-degrading microorganisms in anoxic environments.

Conclusions

The biogeochemical cycling of methylated sulfur compounds is brought about by a wide range of interconnected and interacting metabolic pathways and microorganisms. The intense study of the metabolism of DMSP and DMS during recent years has brought the role of MT as an intermediate of their microbial degradation into focus. MT is relevant as a malodourous compound in a range of industries but also in a medical context, therefore a more detailed understanding of the biochemistry, genetics and ecology of MT degrading microorganisms has considerable benefit to aid in the exploitation of the properties of microorganisms for removal of MT, and even to help understand the role of organic sulfur metabolism in plants, animals and humans. The identification of several key genes that encode enzymes of MT metabolism facilitates a more holistic analysis of the role of diverse bacteria and archaea using metagenomics and related approaches in the future and will lead to a better understanding of MT cycling in the environment. At the same time, advances in uncovering the role of MT metabolism in human disease may only represent the beginning of a better understanding how sulfur metabolism affects human health, providing an important scope to explore both host and microbiome associated pathways of MT production and degradation in more detail.

References

Bak, F., Finster, K., and Rothfuß, F. (1992). Formation of dimethylsulfide and methanethiol from methoxylated aromatic compounds and inorganic sulfide by newly isolated anaerobic bacteria. Arch. Microbiol. *157*, 529–534.

Bentley, R., and Chasteen, T.G. (2004). Environmental VOSCs – formation and degradation of dimethyl sulfide, methanethiol and related materials. Chemosphere *55*, 291–317. https://doi.org/10.1016/j.chemosphere.2003.12.017

Boden, R., Kelly, D.P., Murrell, J.C., and Schäfer, H. (2010). Oxidation of dimethylsulfide to tetrathionate by *Methylophaga thiooxidans* sp. nov.: a new link in the sulfur cycle. Environ. Microbiol. *12*, 2688–2699. https://doi.org/10.1111/j.1462-2920.2010.02238.x

Boden, R., Borodina, E., Wood, A.P., Kelly, D.P., Murrell, J.C., and Schäfer, H. (2011). Purification and characterization of dimethylsulfide monooxygenase from *Hyphomicrobium sulfonivorans*. J. Bacteriol. *193*, 1250–1258. https://doi.org/10.1128/JB.00977-10

Borodina, E., Kelly, D.P., Rainey, F.A., Ward-Rainey, N.L., and Wood, A.P. (2000). Dimethylsulfone as a growth substrate for novel methylotrophic species of *Hyphomicrobium* and *Arthrobacter*. Arch. Microbiol. *173*, 425–437.

Carrión, O., Curson, A.R., Kumaresan, D., Fu, Y., Lang, A.S., Mercadé, E., and Todd, J.D. (2015). A novel pathway producing dimethylsulphide in bacteria is widespread in soil environments. Nat. Commun. *6*, 6579. https://doi.org/10.1038/ncomms7579

Carrión, O., Pratscher, J., Curson, A.R.J., Williams, B.T., Rostant, W.G., Murrell, J.C., and Todd, J.D. (2017). Methanethiol-dependent dimethylsulfide production in soil environments. ISME J. *11*, 2379–2390. https://doi.org/10.1038/ismej.2017.105

Cha, I.T., Min, U.G., Kim, S.J., Yim, K.J., Roh, S.W., and Rhee, S.K. (2013). *Methanomethylovorans uponensis* sp nov., a methylotrophic methanogen isolated from wetland sediment. Antonie Van Leeuwenhoek Int. J. Gen. Mol. Microbiol. *104*, 1005–1012.

Cho, K.S., Hirai, M., and Shoda, M. (1991). Degradation characteristics of hydrogen sulfide, methanethiol, dimethyl sulfide and dimethyl disulfide by *Thiobacillus thioparus* DW44 isolated from peat biofilter. J. Ferment. Bioeng. *71*, 384–389.

Curson, A.R., Sullivan, M.J., Todd, J.D., and Johnston, A.W. (2011a). DddY, a periplasmic dimethylsulfoniopropionate lyase found in taxonomically diverse species of Proteobacteria. ISME J. *5*, 1191–1200. https://doi.org/10.1038/ismej.2010.203

Curson, A.R., Todd, J.D., Sullivan, M.J., and Johnston, A.W. (2011b). Catabolism of dimethylsulphoniopropionate: microorganisms, enzymes and genes. Nat. Rev. Microbiol. *9*, 849–859. https://doi.org/10.1038/nrmicro2653

Curson, A.R.J., Fowler, E.K., Dickens, S., Johnston, A.W.B., and Todd, J.D. (2012). Multiple DMSP lyases in the gamma-proteobacterium *Oceanimonas doudoroffii*. Biogeochemistry *110*, 109–119.

de Bok, F.A., van Leerdam, R.C., Lomans, B.P., Smidt, H., Lens, P.N., Janssen, A.J., and Stams, A.J. (2006). Degradation of methanethiol by methylotrophic methanogenic archaea in a lab-scale upflow anaerobic sludge blanket reactor. Appl. Environ. Microbiol. *72*, 7540–7547.

De Bont, J.A.M., van Dijken, J.P., and Harder, W. (1981). Dimethyl sulphoxide and dimethyl sulphide as a carbon, sulphur and energy source for growth of *Hyphomicrobium* S. J. Gen. Microbiol. *127*, 315–323.

Devos, M., Patte, F., Rouault, J., Laffort, P., and van Gemert, L.J. (1990). Standardized human olfactory thresholds. In Standardized Human Olfactory Thresholds (IRL Press, Oxford).

Drotar, A., Burton, G.A., Tavernier, J.E., and Fall, R. (1987). Widespread occurrence of bacterial thiol methyltransferases and the biogenic emission of methylated sulfur gases. Appl. Environ. Microbiol. *53*, 1626–1631.

Eyice, Ö., Myronova, N., Pol, A., Carrión, O., Todd, J.D., Smith, T.J., Gurman, S.J., Cuthbertson, A., Mazard, S., Mennink-Kersten, M.A., *et al.* (2018). Bacterial SBP56 identified as a Cu-dependent methanethiol oxidase widely distributed in the biosphere. ISME J. *12*, 145–160. https://doi.org/10.1038/ismej.2017.148

Finster, K., Tanimoto, Y., and Bak, F. (1992). Fermentation of methanethiol and dimethylsulfide by a newly isolated methanogenic bacterium. Arch. Microbiol. *157*, 425–430.

Fu, H., and Metcalf, W.W. (2015). Genetic basis for metabolism of methylated sulfur compounds in *Methanosarcina* species. J. Bacteriol. *197*, 1515–1524. https://doi.org/10.1128/JB.02605-14

González, J.M., Kiene, R.P., and Moran, M.A. (1999). Transformation of sulfur compounds by an abundant lineage of marine bacteria in the alpha-subclass of the class Proteobacteria. Appl. Environ. Microbiol. *65*, 3810–3819.

Gould, W.D., and Kanagawa, T. (1992). Purification and properties of methyl mercaptan oxidase from *Thiobacillus thioparus* TK-m. J. Gen. Microbiol. *138*, 217–221.

Grech-Mora, I., Fardeau, M.-L., Patel, B., Ollivier, B., Rimbault, A., Prensier, G., Garcia, J.-L., and Garnier-Zarli, E. (1996). Isolation and characterization of *Sporobacter termitidis* gen. nov., sp. nov., from the digestive tract of the wood-feeding termite Nasutitermes lujae, *46*. 512–518

Howard, E.C., Henriksen, J.R., Buchan, A., Reisch, C.R., Bürgmann, H., Welsh, R., Ye, W., González, J.M., Mace, K., Joye, S.B., *et al.* (2006). Bacterial taxa that limit sulfur flux from the ocean. Science *314*, 649–652.

Johnston, A.W.B., Todd, J.D., Sun, L., Nikolaidou-Katsaridou, M.N., Curson, A.R.J., and Rogers, R. (2008). Molecular diversity of bacterial production of the climate-changing gas, dimethyl sulphide, a molecule that impinges on local and global symbioses. J. Exp. Bot. *59*, 1059–1067.

Kadam, P.C., Ranade, D.R., Mandelco, L., and Boone, D.R. (1994). Isolation and characterization of *Methanolobus bombayensis* sp. nov., a methylotrophic methanogen that requires high concentrations of divalent cations. Int. J. Syst. Evol. Microbiol. *44*, 603–607.

Kettle, A.J., Rhee, T.S., von Hobe, M., Poulton, A., Aiken, J., and Andreae, M.O. (2001). Assessing the flux of different volatile sulfur gases from the ocean to the atmosphere. J. Geophys. Res. Atmos. *106*, 12193–12209.

Kiene, R.P., and Capone, D.G. (1988). Microbial transformations of methylated sulfur compounds in anoxic salt marsh sediments. Microb. Ecol. *15*, 275–291. https://doi.org/10.1007/BF02012642

Kiene, R.P., and Taylor, B.F. (1988). Biotransformations of organosulphur compounds in sediments via 3-mercaptopopionate. Nature *332*, 148–152.

Kiene, R.P., and Visscher, P.T. (1987). Production and fate of methylated sulfur compounds from methionine and dimethylsulfoniopropionate in anoxic salt marsh sediments. Appl. Environ. Microbiol. *53*, 2426–2434.

Kiene, R.P., Oremland, R.S., Catena, A., Miller, L.G., and Capone, D.G. (1986). Metabolism of reduced methylated sulfur compounds in anaerobic sediments and by a pure culture of an estuarine methanogen. Appl. Environ. Microbiol. *52*, 1037–1045.

Kiene, R.P., Linn, L.J., and Bruton, J.A. (2000). New and important roles for DMSP in marine microbial communities. J. Sea Res. *43*, 209–224.

Kim, S.-J., Shin, H.-J., Kim, Y.-C., Lee, D.-S., and Yang, J.-W. (2000). Isolation and purification of methyl mercaptan oxidase from *Rhodococcus rhodochrous* for mercaptan detection. Biotechnol. Bioprocess Eng. *5*, 465–468.

Kreft, J.-U., and Schink, B. (1993). Demethylation and degradation of phenylmethylethers by the sulfide-methylating homoacotogenic bacterium strain TMBS 4. Arch. Microbiol. *159*, 308–315.

Lee, H.-H., Kim, S.-J., Shin, H.-J., Park, J.-Y., and Yang, J.-W. (2002). Purification and characterisation of methyl mercaptan oxidase from *Thiobacillus thioparus* for mercaptan detection. Biotechnol. Bioprocess Eng. *7*, 375–379.

Liu, Y., Boone, D.R., and Choy, C. (1990). *Methanohalophilus oregonense* sp. nov., a methylotrophic methanogen from an alkaline, saline aquifer. Int. J. Syst. Evol. Microbiol. *40*, 111–116.

Lomans, B.P., Smolders, A., Intven, L.M., Pol, A., Op, D., and Van Der Drift, C. (1997). Formation of dimethyl sulfide and methanethiol in anoxic freshwater sediments. Appl. Environ. Microbiol. *63*, 4741–4747.

Lomans, B.P., Maas, R., Luderer, R., Op den Camp, H.J., Pol, A., van der Drift, C., and Vogels, G.D. (1999a). Isolation and characterization of *Methanomethylovorans hollandica* gen. nov., sp. nov., isolated from freshwater sediment, a methylotrophic methanogen able to grow on dimethyl sulfide and methanethiol. Appl. Environ. Microbiol. *65*, 3641–3650.

Lomans, B.P., Op den Camp, H.J., Pol, A., van der Drift, C., and Vogels, G.D. (1999b). Role of methanogens and other bacteria in degradation of dimethyl sulfide and methanethiol in anoxic freshwater sediments. Appl. Environ. Microbiol. *65*, 2116–2121.

Lomans, B.P., Leijdekkers, P., Wesselink, J.J., Bakkes, P., Pol, A., van der Drift, C., and den Camp, H.J. (2001). Obligate sulfide-dependent degradation of methoxylated aromatic compounds and formation of methanethiol and dimethyl sulfide by a freshwater sediment isolate, *Parasporobacterium paucivorans* gen. nov., sp. nov. Appl. Environ. Microbiol. *67*, 4017–4023.

Lomans, B.P., van der Drift, C., Pol, A., and Op den Camp, H.J.M. (2002). Microbial cycling of volatile organosulfur compounds. Cell. Mol. Life Sci. *59*, 575–588.

Lyimo, T.J., Pol, A., Op den Camp, H.J., Harhangi, H.R., and Vogels, G.D. (2000). *Methanosarcina semesiae* sp. nov., a dimethylsulfide-utilizing methanogen from mangrove sediment. Int. J. Syst. Evol. Microbiol. *50*, 171–178. https://doi.org/10.1099/00207713-50-1-171

Lyimo, T.J., Pol, A., Harhangi, H.R., Jetten, M.S., and Op den Camp, H.J. (2009). Anaerobic oxidation of dimethylsulfide and methanethiol in mangrove sediments is dominated by sulfate-reducing bacteria. FEMS Microbiol. Ecol. *70*, 483–492. https://doi.org/10.1111/j.1574-6941.2009.00765.x

Mathrani, I.M., Boone, D.R., Mah, R.A., Fox, G.E., and Lau, P.P. (1988). *Methanohalophilus zhilinae* sp. nov., an alkaliphilic, halophilic, methylotrophic methanogen. Int. J. Syst. Bacteriol. *38*, 139–142

Mechichi, T., Labat, M., Garcia, J.L., Thomas, P., and Patel, B.K. (1999). *Sporobacterium olearium* gen. nov., sp. nov., a new methanethiol-producing bacterium that degrades aromatic compounds, isolated from an olive mill wastewater treatment digester. Int. J. Syst. Bacteriol. *49*, 1741–1748. https://doi.org/10.1099/00207713-49-4-1741

NCBI (n.d). Methyl Mercaptan. PubChem Compound Database; CID=878. Available online: https://pubchem.ncbi.nlm.nih.gov/compound/878. Accessed 16 September 2018.

Ni, S.S., and Boone, D.R. (1991). Isolation and characterization of a dimethyl sulfide-degrading methanogen, *Methanolobus siciliae* HI350, from an oil well, characterization of *M. siciliae* T4/MT, and emendation of *M. siciliae*. Int. J. Syst. Bacteriol. *41*, 410–416. https://doi.org/10.1099/00207713-41-3-410

Ni, S.S., and Boone, D.R. (1993). Catabolism of dimethylsulfide and methane thiol by methylotrophic methanogens (Chapman & Hall, London).

Oelgeschläger, E., and Rother, M. (2009). In vivo role of three fused corrinoid/methyl transfer proteins in *Methanosarcina acetivorans*. Mol. Microbiol. *72*, 1260–1272. https://doi.org/10.1111/j.1365-2958.2009.06723.x

Oremland, R.S., and Boone, D.R. (1994). *Methanolobus taylorii* sp. nov., a new methylotrophic, estuarine methanogen. Int. J. Syst. Evol. Microbiol. *44*, 573–575.

Pol, A., Op den Camp, H.J., Mees, S.G., Kersten, M.A., and van der Drift, C. (1994). Isolation of a dimethylsulfide-utilizing *Hyphomicrobium* species and its application in biofiltration of polluted air. Biodegradation *5*, 105–112.

Pol, A., Renkema, G.H., Tangerman, A., Winkel, E.G., Engelke, U.F., de Brouwer, A.P.M., Lloyd, K.C., Araiza, R.S., van den Heuvel, L., Omran, H., *et al.* (2018). Mutations in SELENBP1, encoding a novel human methanethiol oxidase, cause extraoral halitosis. Nat. Genet. *50*, 120–129. https://doi.org/10.1038/s41588-017-0006-7

Reisch, C.R., Moran, M.A., and Whitman, W.B. (2011a). Bacterial catabolism of dimethylsulfoniopropionate (DMSP). Front. Microbiol. *2*, 172. https://doi.org/10.3389/fmicb.2011.00172

Reisch, C.R., Stoudemayer, M.J., Varaljay, V.A., Amster, I.J., Moran, M.A., and Whitman, W.B. (2011b).

Novel pathway for assimilation of dimethylsulphoniopropionate widespread in marine bacteria. Nature *473,* 208–211. https://doi.org/10.1038/nature10078

Reuters (2013). Rotten eggs stench reaches UK after French gas leak. Available online: https://www.reuters.com/article/us-france-gasleak/rotten-eggs-stench-reaches-uk-after-french-gas-leak-idUSBRE90L03M20130122. Accessed 16 September 2019.

Schäfer, H. (2007). Isolation of *Methylophaga* spp. from marine dimethylsulfide-degrading enrichment cultures and identification of polypeptides induced during growth on dimethylsulfide. Appl. Environ. Microbiol. 73, 2580–2591.

Sun, J., Todd, J.D., Thrash, J.C., Qian, Y., Qian, M.C., Temperton, B., Guo, J., Fowler, E.K., Aldrich, J.T., Nicora, C.D., *et al.* (2016). The abundant marine bacterium *Pelagibacter* simultaneously catabolizes dimethylsulfoniopropionate to the gases dimethyl sulfide and methanethiol. Nat. Microbiol. *1,* 16065. https://doi.org/10.1038/nmicrobiol.2016.65

Suylen, G.M., and Kuenen, J.G. (1986). Chemostat enrichment and isolation of *Hyphomicrobium* EG. A dimethyl-sulphide oxidizing methylotroph and reevaluation of *Thiobacillus* MS1. Antonie Van Leeuwenhoek 52, 281–293.

Suylen, G.M.H., Large, P.J., Vandijken, J.P., and Kuenen, J.G. (1987). Methyl mercaptan oxidase, a key enzyme in the metabolism of methylated sulfur-compounds by *Hyphomicrobium* EG. J. Gen. Microbiol. *133,* 2989–2997.

Tanimoto, Y., and Bak, F. (1994). Anaerobic degradation of methylmercaptan and dimethyl sulfide by newly isolated thermophilic sulfate-reducing bacteria. Appl. Environ. Microbiol. *60,* 2450–2455.

Todd, J.D., Curson, A.R., Dupont, C.L., Nicholson, P., and Johnston, A.W. (2009). The dddP gene, encoding a novel enzyme that converts dimethylsulfoniopropionate into dimethyl sulfide, is widespread in ocean metagenomes and marine bacteria and also occurs in some Ascomycete fungi. Environ. Microbiol. *11,* 1376–1385. https://doi.org/10.1111/j.1462-2920.2009.01864.x

Todd, J.D., Curson, A.R., Kirkwood, M., Sullivan, M.J., Green, R.T., and Johnston, A.W. (2011). DddQ, a novel, cupin-containing, dimethylsulfoniopropionate lyase in marine roseobacters and in uncultured marine bacteria. Environ. Microbiol. *13,* 427–438. https://doi.org/10.1111/j.1462-2920.2010.02348.x

Ulshöfer, V.S., Flock, O.R., Uher, G., and Andreae, M.O. (1996). Photochemical production and air-sea exchange of carbonyl sulfide in the eastern Mediterranean Sea. Mar. Chem. 53, 25–39.

van Leerdam, R.C., Bonilla-Salinas, M., de Bok, F.A., Bruning, H., Lens, P.N., Stams, A.J., and Janssen, A.J. (2008a). Anaerobic methanethiol degradation and methanogenic community analysis in an alkaline (pH 10) biological process for liquefied petroleum gas desulfurization. Biotechnol. Bioeng. *101,* 691–701. https://doi.org/10.1002/bit.21933

van Leerdam, R.C., de Bok, F.A., Bonilla-Salinas, M., van Doesburg, W., Lomans, B.P., Lens, P.N., Stams, A.J., and Janssen, A.J. (2008b). Methanethiol degradation in anaerobic bioreactors at elevated pH (8): reactor performance and microbial community analysis. Bioresour. Technol. 99, 8967–8973. https://doi.org/10.1016/j.biortech.2008.05.007

Visscher, P.T., and Taylor, B.F. (1993a). A new mechanism for the aerobic catabolism of dimethyl sulfide. Appl. Environ. Microbiol. 59, 3784–3789.

Visscher, P.T., and Taylor, B.F. (1993b). Aerobic and anaerobic degradation of a range of alkyl sulfides by a denitrifying marine bacterium. Appl. Environ. Microbiol. 59, 4083–4089.

Zinder, S.H., and Brock, T.D. (1978). Methane, carbon dioxide, and hydrogen sulfide production from the terminal methiol group of methionine by anaerobic lake sediments. Appl. Environ. Microbiol. 35, 344–352.

Systems Biology Meets Enzymology: Recent Insights into Communal Metabolism of Methane and the Role of Lanthanides

10

Zheng Yu[1], Yue Zheng[1,2,3], Jing Huang[1] and Ludmila Chistoserdova[1*]

[1]Department of Chemical Engineering, University of Washington, Seattle, WA, USA.
[2]CAS Key Laboratory of Urban Pollutant Conversion, Institute of Urban Environment, Chinese Academy of Sciences, Xiamen, China.
[3]University of Chinese Academy of Sciences, Beijing, China.

*Correspondence: milachis@uw.edu

https://doi.org/10.21775/9781912530045.10

Abstract

In this review article, we cover the recent developments in understanding the principles and the mechanisms by which microbial communities participating in methane consumption in natural environmental niches are assembled, and the physiological and biochemical mechanisms and regulators that allow efficient carbon transfer within the communities. We first give a brief overview of methanotrophy. We then describe the recent evidence on non-random assembly of bacterial communities that utilize carbon from methane, based on stable isotope probing experiments as well as on results from natural community manipulations followed by metagenomic analysis. We follow up by highlighting results from synthetic methanotrophic community manipulations identifying the importance of a lanthanide switch that regulates alternative methanol dehydrogenase enzymes in these communities. We further expand on the recently uncovered significance of lanthanides in methylotrophy and review data on the biochemical properties of representatives of two different clades of lanthanide-dependent enzymes. We also provide an overview of the occurrence and the distribution of the lanthanide-dependent alcohol dehydrogenases in the bacterial domain, these data strongly suggesting significance of these metals beyond methylotrophy.

Introduction

A historical outlook on methane oxidation

Methane is an abundant hydrocarbon on Earth. While braking the strong bond between carbon and hydrogen in this molecule is thermodynamically difficult, this reaction is readily catalysed by specialized bacteria, the methanotrophs, at ambient temperatures (Lawton and Rosenzweig, 2016). The ability of microbes to utilize methane as a substrate has been discovered over a century ago, giving rise to the field of science known as the Methylotrophy field (reviewed in Chistoserdova, 2018). It has been especially active since the early 1970s, after multiple methanotrophs have been cultivated in pure cultures (Whittenbury *et al.*, 1970). Those were the so-called aerobic methanotrophs, belonging to the Proteobacteria Phylum of Bacteria, of which originally two types were recognized, gammaproteobacterial (Type I) and alphaproteobacterial

(Type II) methanotrophs (Trotsenko and Murrell, 2008). Key enzymes involved in the primary oxidation of methane in these bacteria have been characterized as two alternative methane monooxygenases, particulate (pMMO) and soluble (sMMO), respectively, the former reliant on copper-mediated catalysis and the latter dependent of iron-mediated catalysis (Ross and Rosenzweig, 2017). The enzyme involved in oxidation of methanol, which is a product of methane oxidation, methanol dehydrogenase (MDH) of the MxaFI-type, has also been thoroughly studied over years, uncovering its calcium dependence and regulation by the presence of methanol (Anthony and Zatman, 1967a,b; Anthony, 2004). However, very recently, an alternative MDH has been identified (see below), suggesting metabolic flexibility for this oxidative step as well. Downstream of methanol, formaldehyde and, sequentially, formate can also be oxidized via alternative enzymes and pathways, further expanding the flexible metabolic scheme of methylotrophy (Chistoserdova, 2011a). Carbon assimilation by proteobacterial methanotrophs has been originally attributed to two specific biochemical pathways, the ribulose monophosphate cycle (in gammaproteobacterial methanotrophs) and the serine cycle (in alphaproteobacterial methanotrophs; Anthony 1982). From recent genomic studies, the gammaproteobacterial methanotrophs also encode at least parts of the serine cycle, but its role in carbon assimilation by these bacteria is still not well understood (But *et al.*, 2017). Some methanotrophs also encode and express genes for the Calvin-Benson-Bassham cycle (Baxter *et al.*, 2002; Kao *et al.*, 2004), this cycle potentially serving as an additional pathway for carbon assimilation. Carbon in the form of CO_2 is also assimilated through the ethylmalonyl-CoA pathway (Anthony, 2011) by the alphaproteobacterial methanotrophs (Chistoserdova *et al.*, 2009).

Two novel taxa within Bacteria capable of methanotrophy have been discovered relatively recently. These are the Verrucomicrobia (Sharp *et al.*, 2014) and the yet unnamed phylum NC10 (Ettwig *et al.*, 2010; Versantvoort *et al.*, 2018). In both of these novel methanotroph taxa, the traditional pathways for methane oxidation are recognized, while carbon assimilation takes place at the level of CO_2 (Khadem *et al.*, 2011; Rasigraf *et al.*, 2014). While the verrucomicrobial methanotrophs are aerobes, the NC10 methanotrophs are considered to be anaerobes (Ettwig *et al.*, 2010; Welte *et al.*, 2016). The latter, so far, have not been isolated in pure cultures, as they potentially require functional partners to carry out methane oxidation anaerobically.

Anaerobic oxidation of methane can also be carried out by the archaeal ANME-types (Hinrichs *et al.*, 1999; Boetius *et al.*, 2000; Orphan *et al.*, 2001). In these species, methanotrophy is carried out through a reverse methanogenesis pathway (McGlynn, 2017), and carbon from methane is assimilated via the Wood–Ljungdahl pathway (Meyerdierks *et al*, 2010; Haroon *et al.*, 2013). Thus, the metabolic scheme employed by methanotrophic archaea is radically different from the one employed by bacteria. These archaeal species are typically engaged in syntrophic co-metabolism involving bacterial species, and, like anaerobic NC10 bacteria, they have not yet been isolated as pure cultures (Knittel and Boetius, 2009; Trembath-Reichert *et al.*, 2016).

Lanthanides in methanotrophy

Over two decades ago, a homologue of the mxaF gene has been identified in several methylotroph genomes (Chistoserdova and Lidstrom, 1997), but it's true function, in encoding an alternative MDH, XoxF, has been revealed only recently, by uncovering its reliance on the so-called rare earth elements lanthanides (Ln; Chistoserdova, 2016; Chistoserdova and Kalyuzhnaya, 2018). The discovery of the role of the mysterious homologue of MxaF (Chistoserdova, 2011a), as an alternative MDH requiring Ln for both expression and activity has stirred-up the field of methylotrophy (Chistoserdova, 2016), raising numbers of questions fundamental not only to microbial physiology and biochemistry, but also to the broader fields of basic chemistry and geobiology: why Ln, and how? Since their discovery, Ln were deemed biologically inert due to their very low solubility (Lim and Franklin, 2004). However, the methylotrophs apparently possess mechanisms for solubilizing Ln and employing them in their central metabolism not only in environments in which Ln concentrations are relatively high (Pol *et al.*, 2014), but also in environments in which Ln concentrations are very low (Chistoserdova, 2016; Shiller *et al.*, 2017).

Laboratory experiments with Ln have mostly employed unnaturally high concentrations of Ln, in the form easily dissolved salts (Fitryanto *et al.*,

2011; Farhan Ul Haque *et al.*, 2015; Gu *et al.*, 2016; Chu and Lidstrom, 2016; Chu *et al.*, 2016). Thus, one needs to consider these results while exercising caution that, in real life, similar conditions probably do not exist. Whenever tested, the addition of Ln ions to growth media has caused a dramatic, arresting effect on the expression of the genes for the classic, calcium-dependent MDH (MxaFI), while positively effecting transcription of *xoxF* (Gu *et al.*, 2016; Chu and Lidstrom, 2016; Chu *et al.*, 2016; Zheng *et al.*, 2018), the mechanism that became known as the Ln switch (Chu and Lidstrom, 2016; Chu *et al.*, 2016). In the presence of Ln, the MxaFI MDH could be removed without a significant effect on growth, while XoxF could be removed when Ln were not present (Farhan Ul Haque *et al.*, 2015; Chu and Lidstrom, 2016; Chu *et al.*, 2016; Zheng *et al.*, 2018). However, the Ln switch was proven to be very fragile. When XoxF mutants were grown in the presence of Ln, suppressor mutants appeared at high frequency, suggesting deregulation and suggesting that some type of selective pressure for such deregulation exists, at least under the conditions of high Ln concentrations (Chu *et al.*, 2016; Zheng *et al.*, 2018). On another hand, there are many examples of methylotrophs that only contain XoxF types of MDH, thus demonstrating that the MxaFI type is an auxiliary enzyme while the XoxF type is essential (Hou *et al.*, 2008; Vekeman *et al.*, 2016). Moreover, many genomes contain multiple, sometimes divergent copies of xoxF genes, suggesting further complexity in the alcohol oxidation potential (Mustakhimov *et al.*, 2013; Huang *et al.*, 2018).

When both the Ca and the Ln forms of MDH are encoded, it is possible that the Ln switch may be a subject to different settings, as opposed to being simply 'on' or 'off'. For example, when assessed in a species encoding both types, both were enzymatically active over a range of Ln concentrations, despite differences in the transcriptional responses (Zheng *et al.*, 2018). Simultaneous expression of both proteins has also been observed in natural settings such as the plant phyllosphere (Delmotte *et al.*, 2009). It is thus likely that each enzyme may play a specialized role in the methylotrophy metabolism, by exploiting not only different metal dependencies, but, potentially, different redox properties, and the presence of both types of enzymes likely confers enhanced environmental fitness.

Interspecies interactions in metabolism of methane

Community function, (re)discovered?

While the process of anaerobic methane oxidation appears to be a communal function (Trembath-Reichert *et al.*, 2016), likely due to the energetic or the metabolic constraints of such a metabolism (Thauer and Shima, 2008), the aerobic methanotrophs have been cultivated as pure cultures for decades, and this is how their metabolism and their physiology have been studied over years (Trotsenko and Murrell, 2008; Chistoserdova and Lidstrom, 2013). However, a possibility of community function in aerobic methane oxidation has also been suggested, based on the observations of the tight associations between the methanotrophs and the generalist heterotrophs during the attempts of obtaining methanotrophs in pure cultures (Dalton, 2005; Dedysh and Dunfield, 2017). The potential of methanotrophs in engaging in communal behaviour has been rediscovered more recently, and, unlike the early notions of methanotrophs supporting random communities of generalist heterotrophs, the recent observations suggest that the communal relationships may not be random after all. For example, stable isotope probing experiments with a lake sediment community responding to methane stimulus have produced interesting and not entirely expected results: the label from methane appeared to be split between a major primary methane consumer, a *Methylobacter* species, and a non-methanotrophic methylotroph, a *Methylotenera* species (Kalyuzhnaya *et al.*, 2008; Fig. 10.1). This result was puzzling because methylotroph communities have been known to be very diverse in the study site, and alternative methanotroph species such as *Methylomonas* and *Methylosinus*, have been readily cultivated from the site, while the *Methylobacter* species remained overlooked (Chistoserdova, 2011b). As to the partner species, many species have been identified in the site that showed robust methylotrophic or growth in laboratory (Beck *et al.*, 2015), but only *Mathylotenera* species were able to partner with *Methylobacter* species in consuming carbon from methane (Kalyuzhnaya *et al.*, 2008). Reconstruction of the metabolisms of the major species involved in methane consumption in these experiments has uncovered a potential for a link between methane oxidation and

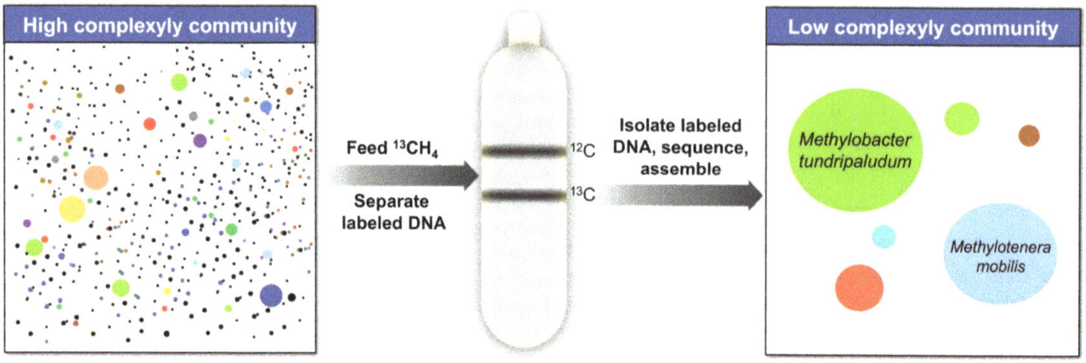

Figure 10.1 Summary of the stable isotope probing experiment followed by metagenomic sequencing. *Methylobacter* and *Methylotenera* were identified as the most competitive consumers of carbon from methane, based on sequence read abundance in the metagenome. See Kalyuzhnaya *et al*. (2008).

denitrification, and thus a potential for adaptation to the conditions of low molecular oxygen availability (Kalyuzhnaya *et al.*, 2008). This hypothesis was tested in additional experiments with labelled methane, under conditions of variable O_2 and nitrate concentrations (Beck *et al.*, 2013). In these experiments, once again, the *Methylobacter* types and the *Methylotenera* types were the species most actively consuming methane in both oxic and hypoxic conditions, and nitrate had a stimulatory effect on both species, further supporting the notion that the two species may be engaged in communal behaviour (Beck *et al.*, 2013). Co-occurrence of the *Methylococcaceae* types along with the *Methylophilaceae* types has been also noted in natural freshwater and soil environments (Martineau *et al.*, 2010; He *et al.*, 2012; Karwautz *et al.*, 2018; Biderre-Petit *et al.*, 2018), while in marine environments, the *Methylococcaceae* type primary methane oxidizers appear to cooperate with *Methylophaga* species as the major carbon sharing partners (Kessler *et al.*, 2011; Paul *et al.*, 2017).

The hypothesis of the non-random partner selection has been tested more directly by manipulating natural communities, as parts of microcosms stimulated with methane (Oshkin *et al.*, 2015). When natural sediment communities were fed methane, in the presence of nitrate, in conditions of either 'high' or 'low' dioxygen partial pressures, very rapid reduction in community complexity occurred, as judged by 16S rRNA gene marker surveys, selecting again for the *Methylococcaceae* and the *Methylophilaceae* types, along with select representatives of non-methylotrophic heterotrophs such as Burkholderiales and Bacteroidetes (Oshkin *et al.*, 2015). Further microcosm experiments that imposed a stricter control of dioxygen concentrations in the headspace allowed for distinguishing genus-specific selection for both major partners: the 'high' oxygen conditions selecting for the *Methylosarcina/Methylophilus* partnerships, and the 'low' oxygen conditions selecting for the *Methylobacter/Methylotenera* partnerships (Hernandez *et al.*, 2015). Thus, it is possible that discrete niche conditions, such as the different ratios of O_2/methane along their counter-gradients typically observed in methane-producing environments such as lake sediments, may be fine-tuning the specific methanotroph–non-methanotroph partnerships, selecting for organisms with optimized metabolic schemes for most successful partnerships.

Insights from synthetic community manipulations

Synthetic communities of organisms relevant to communal methane oxidation have been employed as a simplified and a better controlled model to test the findings from the natural community experiments. Some of the questions addressed via the synthetic community model were whether the dynamics of the natural communities were determined by the initial titres of species present in native conditions, or whether the species most successful in methane consumption were biochemically equipped to be more competitive, under specific conditions created as parts of a specific experiment.

Complex synthetic communities were mixed of 50 species representing bona fide methanotrophs, including the *Methylobacter* and *Methylosarcina* species closely related to the ones identified via stable isotope probing, as well as via microcosm manipulations (Kalyuzhnaya *et al.*, 2008; Oshkin *et al.*, 2015; Hernandez *et al.*, 2015), the *Methylotenera* and *Methylophilus* species identified in prior experiments, and of species that have been isolated from the native environment and grown methylotrophically in laboratory, but not detected in the microcosm experiments (Beck *et al.*, 2015). In addition, a handful of non-methylotrophic heterotrophs were employed to test whether they would persist in communities fed exclusively methane as a carbon source, at variable concentrations of methane and dioxygen, in the presence of alternative (nitrate versus ammonium) nitrogen sources. These synthetic communities were treated similarly to the natural communities (Oshkin *et al.*, 2015; Hernandez *et al.*, 2015), and their dynamics were followed through 16S rRNA gene profiling and through the analysis of select metatranscriptomes, matched to the respective 50 genomes (Yu *et al.*, 2017; Fig. 10.2). The synthetic communities, in general, followed the pattern of the natural communities in terms of the *Methylococcaceae* and the *Methylophilaceae* establishing themselves as the most abundant species. However, when considered at the genus level, the major species differed from the ones in natural community microcosms. The species known for most robust growth in laboratory, the *Methylomonas* species among the *Methylococcaceae* and the *Methylophilus* species among the *Methylophilaceae* gradually outcompeted other species in most microcosms, suggesting that some of the controls present in natural communities, such as, for example, any predatory species, have not been recaptured in the synthetic communities (Yu *et al.*, 2017). Thus, while providing more straightforward and more easily controlled models, synthetic communities function somewhat differently, favouring fast-growing species. This conclusion has been further tested in a simplified model that competed thee major *Methylococcaceae* species against each other under a variety of conditions, confirming that growth rate played a role in the outcomes of the competition experiments (Yu *et al.*, 2016).

One advantage of the synthetic communities is the resolution they offer for the analysis of the metatranscriptomic data, allowing for matching transcript reads to the well assembled and annotated genomic scaffolds (Yu *et al.*, 2017). The strain-resolved transcript abundance analyses produced interesting results: in each of the actively expressing methylotrophs, some of the genes most responsive to the specific experimental conditions were the genes encoding alternative MDH

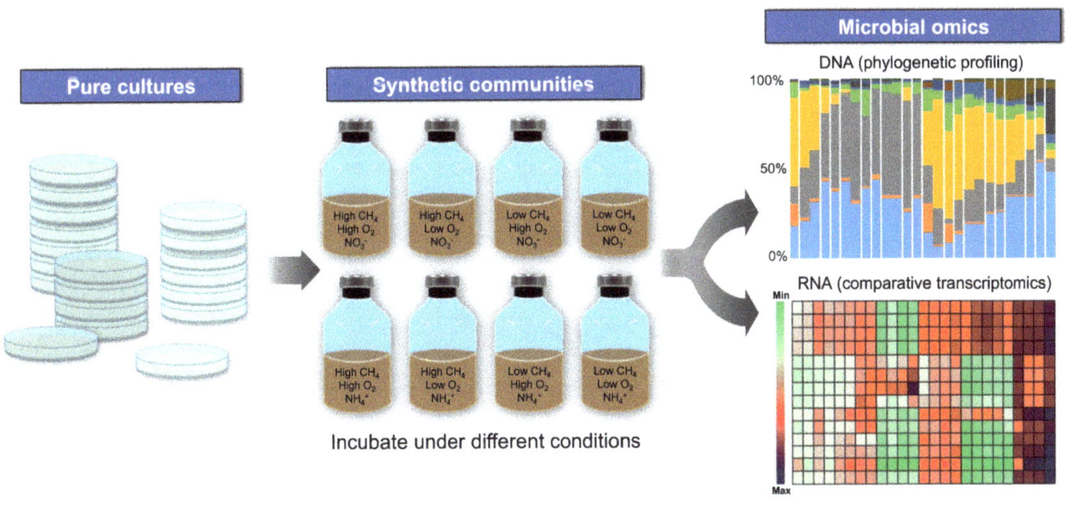

Figure 10.2 Schematic of the synthetic community experiment. Pure cultures were mixed and incubated under different conditions with methane as the only carbon source. Competitive species were identified via iTag analysis, and metabolic features were identified via comparative transcriptomics. See Yu *et al.* (2016, 2017).

enzymes, MxaFI and XoxF. The expression patterns were found very distinct dependent on whether nitrate or ammonium were supplied, or whether 'low' or 'high' concentrations of O_2 and methane were present. Moreover, expression patterns differed among different species and among different MDH enzymes. While the *Methylococcaceae, Methylocystaceae, Rhodocyclaceae,* and the *Methylophilus* species expressed *mxaFI* genes at high levels when 'high' O_2, 'high' methane, and nitrate were supplied, they expressed *xoxF* genes at higher levels when ammonium was supplied instead, or when methane supply was limited. Under the 'high' methane, 'low' O_2, NO_3^- regimen, genes for both enzymes were expressed at approximately equal levels in these species. In contrast, *Methylophilaceae* other than *Methylophilus* (mostly represented by unclassified *Methylophilaceae*) displayed higher expression of *xoxF* genes under all regimens (Yu *et al.*, 2017). It is important to note that the *xoxF* genes that were highly and differentially expressed by major community partners belonged to two main phylogenetic clades (Chistoserdova, 2011a), as follows. While the *Methylococcaceae, Methylocystaceae,* and *Rhodocyclaceae* possessed and express the *xoxF5* clade genes, the *Methylophilaceae* possessed and expressed the *xoxF4* clade genes. Proteins classified as XoxF4 and XoxF5 share approximately 50% amino acid identity, similarly to the identity divergencies between MxaF and XoxF4 and XoxF5 proteins, respectively (Huang *et al.*, 2018, 2019).

The complexity of the lanthanide-dependent methanol oxidation system

The Ln switch mechanism has been proposed as the result of the experiments with pure cultures of methanotrophs, using typical laboratory conditions, i.e. high methane and high dioxygen concentrations and nitrate as the source of nitrogen (Farhan Ul Haque *et al.*, 2015; Gu *et al.*, 2016; Chu and Lidstrom; 2016; Chu *et al.*, 2016; Zheng *et al.*, 2018). Even in these conditions, in some studies, the switch was shown to act in this manner only in the absence of copper, suggesting that, at least in some organisms, the Cu switch may override the Ln switch (Gu *et al.*, 2016). Sampling of the communal transcript response, under a variety of experimental conditions, produced a much more complex picture of how the Ln switch regulates transcription of alternative MDH genes. It appeared that the Ln switch is not only responsive to the presence of Ln, but also to the partial pressures of both methane and dioxygen, as well to the nitrogen source. In addition, a response has been recorded to both the phylogenetic positioning of an organism, and to the phylogenetic affiliation of a XoxF enzyme, providing a very complex picture of how different XoxF enzymes may contribute to the communal function in methane oxidation and, potentially, to interspecies communications (Yu *et al.*, 2017; Fig. 10.3).

Further insights into the Ln switch were gained by manipulating synthetic bacterial communities in their simplest, two-species form, combining a *Methylobacter* type methanotroph

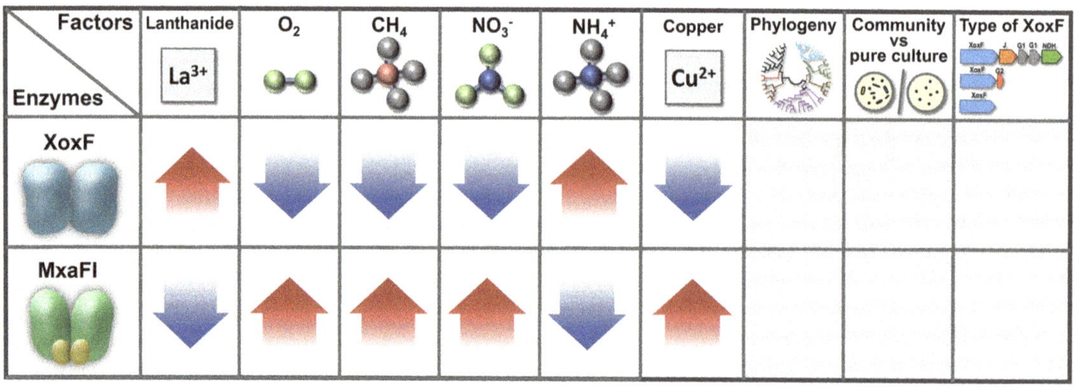

Figure 10.3 Cartoon depicting factors regulating expression of genes encoding XoxF and MxaFI methanol dehydrogenases. Red arrow, up-regulated; blue arrow, down-regulated. See Krause *et al.* (2017), Yu *et al.* (2017) and Zheng *et al.* (2018).

with two different *Methylotenera* partners, into functional methane-oxidizing communities, in separate two-species community experiments (Krause *et al.*, 2017). These were grown in the presence of Ln, either as communities or as axenic cultures, and their transcriptomes were compared (Krause *et al.*, 2017). Pure cultures of the *Methylobacter* species demonstrated the predicted pattern for MDH gene expression: the *xoxF* gene being induced and the *mxa* genes being repressed. Remarkably, in cocultures, this pattern was reversed, and *Methylobacter* cultures expressed the *mxa* genes (Krause *et al.*, 2017). In this case, it appeared that community living has overruled the Ln switch, even in the presence of Ln (Fig. 10.3). It has been proposed that the reversal of the Ln switch was a result of interspecies interactions, the *Methylotenera* species potentially releasing a signal to upregulate the *mxa* machinery in *Methylobacter*, resulting in methanol release (Krause *et al.*, 2017). However, this conclusion on the requirement of the *mxa* system for methanol release needs further investigation. In a more recent study, knockout mutants of a *Methylomonas* species separately expressing either MxaFI or XoxF both appeared to maintain populations of the satellite *Methylophilaceae* of similar size, while the sizes of the satellite populations were correlated with the concentrations of the methanol released (Zheng *et al.*, 2018).

When multiple *xoxF* genes are present in the genomes, these genes are also subjects to differential regulation. For example, the *Methylotenera* species employed in the two-species communities showed reversed transcriptional regulation of the alternative *xoxF* genes, in response to communal living (Krause *et al.*, 2017).

Insights into lanthanide-dependent catalysis

The complexity of the Ln switch has highlighted the necessity of a better understanding of the biochemical properties of XoxF enzymes. It needs to be noted that the diversity of enzymes referred to XoxF has been appreciated for a long time, and, as early as 2011, they have been classified into five divergent phylogenetic clades (Chistoserdova, 2011a). More recent phylogenetic inferences suggested that Ln dependencies may extend beyond these five clusters, and, perhaps, even beyond primary alcohol/aldehyde dehydrogenase enzymes (Keltjens *et al.*, 2014; Chistoserdova and Kalyuzhnaya, 2018). A second aspartate has been proposed to be essential in the active site of XoxF-type enzymes (Pol *et al.*, 2014; Jahn *et al.*, 2018; Lumpe *et al.*, 2018), and this aspartate appears to be conserved in many divergent clades of XoxF and XoxF-like enzymes (Keltjens *et al.*, 2014; Chistoserdova and Kalyuzhnaya, 2018). Two clades, XoxF4 and XoxF5 appear to be especially widely spread among the proteobacterial methylotrophs, and intriguingly, these two types appear to be mutually exclusive: while a variety of Alpha-, Beta- and Gammaproteobacteria encode, sometimes multiple, XoxF5 types, the XoxF4 types are only encoded by the members of *Methylophilaceae* (Betaproteobacteria; Huang *et al.*, 2018). Thus, among the main partners in methane oxidation, as described above, the *Methylococcaceae* encode the XoxF5 type and the *Methylophilaceae* encode the XoxF4 type. The uncompetitive methanol consumers, such as *Rhodocyclaceae, Methylobacterium, Hyphomicrobium, Methylopila* types, etc. based on the synthetic community experiments, (Yu *et al.*, 2017) also only encode and express XoxF5 types, raising a question whether the XoxF4 type, possessed by the *Methylophilaceae* gives them advantage in competing for methanol spilled by the *Methylococcaceae*? This hypothesis was tested by purifying and evaluating the catalytic properties of XoxF5 and XoxF4 enzymes from select organisms (Fitriyanto *et al.*, 2011; Huang *et al.*, 2018, 2019). Both types revealed similarities in terms of Ln dependences and in terms of a broad range of substrates. However, the catalytic efficiencies of the XoxF5 enzymes appeared to be generally higher than the catalytic efficiencies of the XoxF4 enzymes (Fitriyanto *et al.*, 2011; Huang *et al.*, 2018, 2019; Fig. 10.4), thus not supporting the original hypothesis. What does this result mean for understanding the relationships among partner species sharing carbon from methane (Krause *et al.*, 2017; Yu and Chistoserdova, 2017)? The discrepancies between the measured kinetic constants and the observed communal behaviour further point to the limitations of our understanding of the metabolic step mediated by alternative MDH enzymes. One obvious problem is the reliance on the artificial dye assay, practised since the 1960s, for measuring the activity of MDH enzymes (Anthony and Zatman, 1967a,b). However, more likely, enzyme

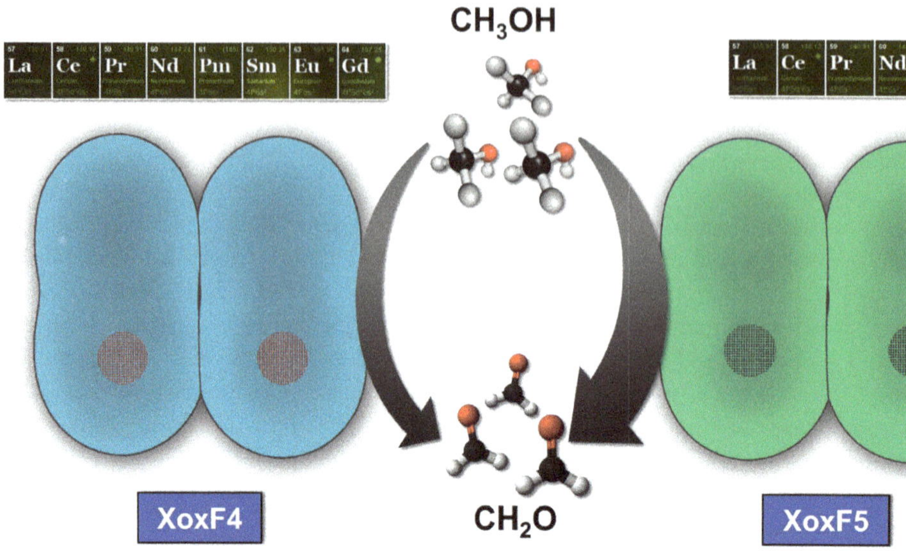

Figure 10.4 Cartoon depicting catalytic properties of XoxF4 versus XoxF5. While XoxF4 enzymes reveal activity with a broader range of lanthanides, XoxF5 enzymes reveal higher catalytic efficiency (size of arrow). The range of metals supporting activity is shown for each enzyme. See Fitriyanto *et al*. (2011) and Huang *et al*. (2018, 2019).

efficiencies and their differential functions could only be understood when considered in the context of natural redox metabolic complexes, which would include the primary alcohol-oxidizing enzyme, their natural electron acceptors (dedicated or promiscuous cytochromes), and the respective cytochrome oxidases, and these additional electron transfer chain components remain poorly understood for XoxF type MDH enzymes. Recently, complex relationships for XoxF enzymes with the downstream redox enzymes have been suggested, with different types of cytochromes potentially metabolically coupling with different types of XoxF enzymes (Yu *et al*., 2017; Zheng *et al*., 2018), and these metabolic connections and their metabolic efficiencies need to be further characterized, to advance our understanding of Ln-linked versus Ca-linked methanol oxidation in methanotrophs and non-methane-oxidizing methylotrophs, as well as their potential involvement in the community function.

The inferred diversity and distribution of lanthanide-dependent alcohol dehydrogenases

Besides XoxF-type MDH enzymes, two enzymes have been recently biochemically characterized, named ExaF/PedH, which also depend on Ln for activity, revealing highest affinities towards ethanol, thus these enzymes were assumed to be functional ethanol dehydrogenases (Good *et al*., 2016; Wehrmann *et al*., 2017). In *Pseudomonas putida*, which also encodes a Ca version of this enzyme (PedE/QedH), expression of the two enzymes is regulated by a Ln switch (Wehrmann *et al*., 2018), a mechanism that is similar to the one described for XoxF (Chu and Lidstrom, 2016; Chu *et al*., 2016). How frequently do Ln-dependent enzymes occur in nature? Based on the conclusion that the Ln MDH enzymes are more widely spread among the methylotrophs, and based on their ancestral positioning in phylogenetic trees (Chistoserdova, 2011a; Keltjens *et al*., 2014; Chistoserdova and Kalyuzhnaya, 2018), they should be expected to be abundant beyond organisms with a demonstrated methylotrophic life style. Indeed, close homologues of XoxF5 are found among a variety of Proteobacteria, belonging, so far, to 12 orders, of which only one, the *Methylococcales*, is represented exclusively by the methylotrophs (Huang *et al*., 2019). Five of the remaining orders are represented by taxa that include methylotroph species but also species not known for methylotrophic life style, and the remaining six orders are only represented by non-methylotrophs, including such important orders as

Vibrionales and *Pseudomonadales* (Fig. 10.5; Huang *et al.*, 2019). While XoxF4 type Ln-MDH enzymes have only been identified so far in members of *Methylophilales*, these represent abundant environmental populations that play important functions in global carbon cycling (Sowell *et al.*, 2011; Salcher *et al.*, 2015; Gifford *et al.*, 2016). ExaF/PedH enzymes appear to be even more broadly spread among the Proteobacteria Phylum microbes, and these are found in bacteria belonging to 14 different orders (Fig. 10.5; Huang *et al.*, 2019), compared with 16 orders possessing the Ca-dependent homologues of ExaF/PedH, named ExaA/PedE (Fig. 10.5). The two counterparts co-occur in 14 orders

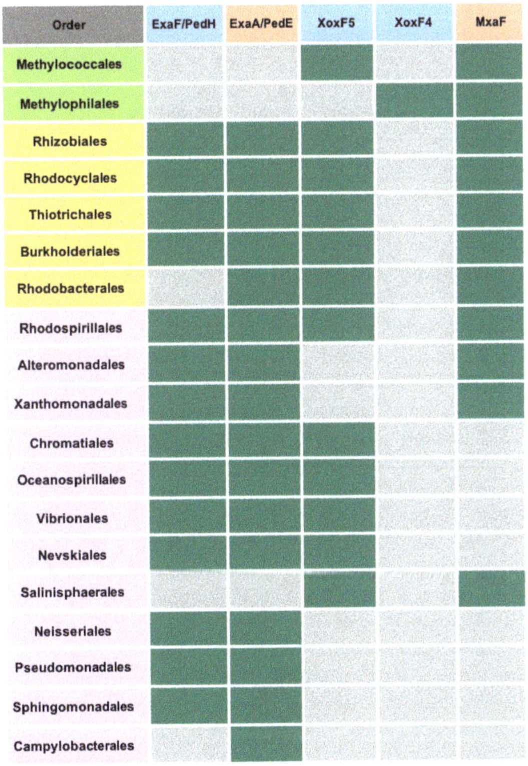

Figure 10.5 Occurrence and distribution of lanthanide-dependent alcohol dehydrogenases among Proteobacteria, compared with the occurrence and distribution of their calcium-dependent counterparts. Taxa are shown at the Order level. In green, taxa that only contain methylotroph species. In yellow, taxa that contain both methylotroph and non-methylotroph species. In purple, taxa that are not known to contain any methylotroph species. Enzymes highlighted in blue, Ln-dependent; in pink, Ca-dependent. Green, gene present, grey, gene absent. See Huang *et al.* (2019).

(Huang *et al.*, 2019). Note that Ln enzymes are also common in Verrucomicrobia and NC10 phylum microbes. However, the databases of the relevant genomes remain limited, while most of the proteins possessing the Ln-binding motif (Keltjens *et al.*, 2014) have not yet been biochemically characterized. However, the currently available genomic and biochemical data strongly suggest that Ln should be considered as true life metals, to join other metals recognized for their importance in biological functions, such as iron, calcium, copper, magnesium, zinc, cobalt, manganese, potassium. Based on the wide distribution of Ln-dependent counterparts of the Ca-dependent alcohol dehydrogenases, and based on existence of special cellular mechanisms of Ln acquisition and the Ln-dependent regulatory mechanisms, it is reasonable to expect that enzymatic reactions beyond alcohol oxidation, along with specific transport and regulation mechanisms, and potentially community interactions involving these reactions, may also be reliant on Ln. Thus further discoveries of Ln-dependent functions are expected.

Conclusions and future perspectives

The communal metabolism of methane, while not an entirely new concept, has regained interest, as the recent meta-omics data suggest that environmental co-occurrences of the methanotroph species along with non-methanotrophic methylotroph species may not be random, and, instead, these partnerships may be specific. While data are available on methanol being the one carbon source shared among the community members, it remains unknown why would the methanotrophs release methanol, and whether this carbon sharing is accidental or is specifically designed to feed populations of the satellite species. The feedbacks that these methanol-consuming populations may be providing to the methanotrophs remain unknown. However, evidence is also available that alternative enzymes involved in methanol oxidation in both the methane-oxidizing species and in the methanol-consuming species are some of the major metabolic sites for regulation, which follows a rather complex pattern, involving, as the major controlling factors, the availability of Ln, the partial pressures of O_2 and methane, the phylogenetic makeups of the communities, and the types of MDH enzymes encoded in the genomes. Deciphering the details of this complex regulation presents a significant challenge for furthering our understanding of the communal methanotrophy under aerobic conditions. The potentially discrete functions of the MDH enzymes with alternative metal specificities, and, further, potentially discrete functions of the multiple Ln-dependent MDH enzymes encoded in many methylotroph genomes present another challenge. While a collection of the Ln-dependent alcohol dehydrogenases have been already characterized, more data are needed in order to decipher the differences among the distinct clades of Ln enzymes, at higher confidence. With this respect, the emergent problem is the paucity of the data on the classic, Ca-dependent MDH enzymes, despite the long history of these enzymes and the intense efforts on their investigation over decades. Next, the functions of the non-methylotrophic heterotrophs in the communities, while they typically are present at minor populations, also need to be addressed, and most parsimonious interspecies interaction scenarios elucidated: whether such heterotrophs are directly fed by the methanotrophs, and whether alternative carbon sources are consumed by them, such as acetate (Kalyuzhnaya *et al.*, 2013), or whether the carbon flow also involves the major methylotroph partners. Further sampling of the potentially Ln-dependent enzymes is necessary, across the phylogenetic trees of the homologues, and their functions within their hosts or potentially within communities need to be revealed. Perhaps, deployment of Ln-binding motifs could identify Ln-dependent enzymes beyond alcohol dehydrogenases, furthering insights into Ln-dependent biological functions. Likely, the insights into the communal function in the aerobic methanotrophy, as a momentum of the ongoing effort, will identify and highlight similar patterns in communal behaviours relevant to other important microbially driven biogeochemical processes.

Acknowledgements

This material is based upon work supported by the U.S. Department of Energy, Office of Science, Office of Biological and Environmental Research under Award Number DE-SC-0016224. YZ acknowledges support from the China Scholarship Council programme under Award Number 20160491078.

References

Anthony, C. (1982). The Biochemistry of Methylotrophs (Academic Press, London, UK).

Anthony, C. (2004). The quinoprotein dehydrogenases for methanol and glucose. Arch. Biochem. Biophys. *428*, 2–9. https://doi.org/10.1016/j.abb.2004.03.038

Anthony, C. (2011). How half a century of research was required to understand bacterial growth on C1 and C2 compounds; the story of the serine cycle and the ethylmalonyl-CoA pathway. Sci. Prog. *94*, 109–137.

Anthony, C., and Zatman, L.J. (1967a). The microbial oxidation of methanol. Purification and properties of the alcohol dehydrogenase of *Pseudomonas* sp. M27. Biochem. J. *104*, 953–959.

Anthony, C., and Zatman, L.J. (1967b). The microbial oxidation of methanol. The prosthetic group of the alcohol dehydrogenase of *Pseudomonas* sp. M27: a new oxidoreductase prosthetic group. Biochem. J. *104*, 960–969.

Baxter, N.J., Hirt, R.P., Bodrossy, L., Kovacs, K.L., Embley, T.M., Prosser, J.I., and Murrell, J.C. (2002). The ribulose-1,5-bisphosphate carboxylase/oxygenase gene cluster of *Methylococcus capsulatus* (Bath). Arch. Microbiol. *177*, 279–289. https://doi.org/10.1007/s00203-001-0387-x

Beck, D.A., Kalyuzhnaya, M.G., Malfatti, S., Tringe, S.G., Glavina Del Rio, T., Ivanova, N., Lidstrom, M.E., and Chistoserdova, L. (2013). A metagenomic insight into freshwater methane-utilizing communities and evidence for cooperation between the *Methylococcaceae* and the *Methylophilaceae*. PeerJ *1*, e23. https://doi.org/10.7717/peerj.23

Beck, D.A., McTaggart, T.L., Setboonsarng, U., Vorobev, A., Goodwin, L., Shapiro, N., Woyke, T., Kalyuzhnaya, M.G., Lidstrom, M.E., and Chistoserdova, L. (2015). Multiphyletic origins of methylotrophy in Alphaproteobacteria, exemplified by comparative genomics of Lake Washington isolates. Environ. Microbiol. *17*, 547–554. https://doi.org/10.1111/1462-2920.12736

Biderre-Petit, C. Taib, N., Gardon, H., Hochart, C., and Debroas, D. (2018). New insights into the pelagic microorganisms involved in the methane cycle in the meromictic Lake Pavin through metagenomics. FEMS Microbiol. Ecol. https://doi.org/10.1093/femsec/fiy183 [Epub ahead of print].

Boetius, A., Ravenschlag, K., Schubert, C.J., Rickert, D., Widdel, F., Gieseke, A., Amann, R., Jørgensen, B.B., Witte, U., and Pfannkuche, O. (2000). A marine microbial consortium apparently mediating anaerobic oxidation of methane. Nature *407*, 623–626. https://doi.org/10.1038/35036572

But, S.Y., Egorova, S.V., Khmelenina, V.N., and Trotsenko, Y.A. (2017). Biochemical properties and phylogeny of hydroxypyruvate reductases from methanotrophic bacteria with different C1-assimilation pathways. Biochemistry Mosc. *82*, 1295–1303. https://doi.org/10.1134/S0006297917110074

Chistoserdova, L. (2011a). Modularity of methylotrophy, revisited. Environ. Microbiol. *13*, 2603–2622. https://doi.org/10.1111/j.1462-2920.2011.02464.x

Chistoserdova, L. (2011b). Methylotrophy in a lake: from metagenomics to single-organism physiology. Appl.

Environ. Microbiol. 77, 4705–4711. https://doi.org/10.1128/AEM.00314-11

Chistoserdova, L. (2016). Lanthanides: New life metals? World J. Microbiol. Biotechnol. *32*, 138. https://doi.org/10.1007/s11274-016-2088-2

Chistoserdova, L. (2018). Methanotrophy: an evolving field. In Methane Biocatalysis: Paving the Way to Sustainability. Kalyuzhnaya, M., Xing, X.-H., eds (Springer, Springer-Verlag, Berlin Heidelberg), pp. 1–15.

Chistoserdova, L., and Lidstrom, M.E. (1997). Molecular and mutational analysis of a DNA region separating two methylotrophy gene clusters in *Methylobacterium extorquens* AM1. Microbiology *143*, 1729–1736. https://doi.org/10.1099/00221287-143-5-1729

Chistoserdova, L., and Lidstrom, M.E. (2013). Aerobic methylotrophic prokaryotes. In The Prokaryotes, Fourth Edition. Rosenberg, E., DeLong, E.F., Thompson, F., Lory, S., Stackebrandt, E., eds. (Springer, Berlin, Heidelberg), pp. 267–285.

Chistoserdova, L., and Kalyuzhnaya, M.G. (2018). Current trends in methylotrophy. Trends Microbiol. *26*, 703–714.

Chistoserdova, L., Kalyuzhnaya, M.G., and Lidstrom, M.E. (2009). The expanding world of methylotrophic metabolism. Annu. Rev. Microbiol. *63*, 477–499. https://doi.org/10.1146/annurev.micro.091208.073600

Chu, F., and Lidstrom, M.E. (2016). XoxF acts as the predominant methanol dehydrogenase in the Type I methanotroph *Methylomicrobium buryatense*. J. Bacteriol. *198*, 1317–1325. https://doi.org/10.1128/JB.00959-15

Chu, F., Beck, D.A., and Lidstrom, M.E. (2016). MxaY regulates the lanthanide-mediated methanol dehydrogenase switch in *Methylomicrobium buryatense*. PeerJ *4*, e2435. https://doi.org/10.7717/peerj.2435

Dalton, H. (2005). The Leeuwenhoek Lecture 2000 the natural and unnatural history of methane-oxidizing bacteria. Philos. Trans. R. Soc. Lond., B, Biol. Sci. *360*, 1207–1222.

Dedysh, S.N., and Dunfield, P.F. (2017). Cultivation of methanotrophs. In Hydrocarbon and Lipid Microbiology Protocols, McGenity, T.J., Timmis, K.N., Nogales Fernandez, B.N., eds (Springer-Ferlag, Berlin, Heidelberg), pp. 231–247.

Delmotte, N., Knief, C., Chaffron, S., Innerebner, G., Roschitzki, B., Schlapbach, R., von Mering, C., and Vorholt, J.A. (2009). Community proteogenomics reveals insights into the physiology of phyllosphere bacteria. Proc. Natl. Acad. Sci. U.S.A. *106*, 16428–16433. https://doi.org/10.1073/pnas.0905240106

Ettwig, K.F., Butler, M.K., Le Paslier, D., Pelletier, E., Mangenot, S., Kuypers, M.M., Schreiber, F., Dutilh, B.E., Zedelius, J., de Beer, D., *et al.* (2010). Nitrite-driven anaerobic methane oxidation by oxygenic bacteria. Nature *464*, 543–548. https://doi.org/10.1038/nature08883

Farhan Ul Haque, M., Gu, W., DiSpirito, A.A., and Semrau, J.D. (2015). Marker exchange mutagenesis of *mxaF*, encoding the large subunit of the Mxa methanol dehydrogenase, in *Methylosinus trichosporium* OB3b. Appl. Environ. Microbiol. *82*, 1549–1555. https://doi.org/10.1128/AEM.03615-15

Fitriyanto, N.A., Fushimi, M., Matsunaga, M., Pertiwiningrum, A., Iwama, T., and Kawai, K. (2011). Molecular structure and gene analysis of Ce3+ -induced methanol dehydrogenase of *Bradyrhizobium* sp. MAFF211645. J. Biosci. Bioeng. *111*, 613–617. https://doi.org/10.1016/j.jbiosc.2011.01.015

Gifford, S.M., Becker, J.W., Sosa, O.A., Repeta, D.J., and DeLong, E.F. (2016). Quantitative transcriptomics reveals the growth- and nutrient-dependent response of a streamlined marine methylotroph to methanol and naturally occurring dissolved organic matter. MBio 7, e01279–16.

Good, N.M., Vu, H.N., Suriano, C.J., Subuyuj, G.A., Skovran, E., and Martinez-Gomez, N.C. (2016). Pyrroloquinoline quinone ethanol dehydrogenase in *Methylobacterium extorquens* AM1 extends lanthanide-dependent metabolism to multicarbon substrates. J. Bacteriol. *198*, 3109–3118.

Gu, W., Farhan Ul Haque, M., DiSpirito, A.A., and Semrau, J.D. (2016). Uptake and effect of rare earth elements on gene expression in *Methylosinus trichosporium* OB3b. FEMS Microbiol. Lett. *363*, fnw129. https://doi.org/10.1093/femsle/fnw129

Haroon, M.F., Hu, S., Shi, Y., Imelfort, M., Keller, J., Hugenholtz, P., Yuan, Z., and Tyson, G.W. (2013). Anaerobic oxidation of methane coupled to nitrate reduction in a novel archaeal lineage. Nature *500*, 567–570. https://doi.org/10.1038/nature12375

He, R., Wooller, M.J., Pohlman, J.W., Catranis, C., Quensen, J., Tiedje, J.M., and Leigh, M.B. (2012). Identification of functionally active aerobic methanotrophs in sediments from an arctic lake using stable isotope probing. Environ. Microbiol. *14*, 1403–1419. https://doi.org/10.1111/j.1462-2920.2012.02725.x

Hernandez, M.E., Beck, D.A., Lidstrom, M.E., and Chistoserdova, L. (2015). Oxygen availability is a major factor in determining the composition of microbial communities involved in methane oxidation. PeerJ 3, e801. https://doi.org/10.7717/peerj.801

Hou, S., Makarova, K.S., Saw, J.H., Senin, P., Ly, B.V., Zhou, Z., Ren, Y., Wang, J., Galperin, M.Y., Omelchenko, M.V., *et al.* (2008). Complete genome sequence of the extremely acidophilic methanotroph isolate V4, *Methylacidiphilum infernorum*, a representative of the bacterial phylum Verrucomicrobia. Biol. Direct 3, 26. https://doi.org/10.1186/1745-6150-3-26

Huang, J., Yu, Z., and Chistoserdova, L. (2018). Lanthanide-dependent methanol dehydrogenases of XoxF4 and XoxF5 clades are differentially distributed among methylotrophic bacteria and they reveal different biochemical properties. Front. Microbiol. *9*, 1366. https://doi.org/10.3389/fmicb.2018.01366

Huang, J., Yu, Z., Groom, J., Cheng, J.-F., Tarver, A., Yoshikunin, Y., and Chistoserdova, L. (2019). Rare earth element alcohol dehydrogenases widely occur among globally distributed, numerically abundant and environmentally important microbes. ISME J. In Press.

Jahn, B., Pol, A., Lumpe, H., Barends, T.R.M., Dietl, A., Hogendoorn, C., Op den Camp, H.J.M., and Daumann, L.J. (2018). Similar but not the same: first kinetic and structural analyses of a methanol dehydrogenase containing a europium Ion in the active site. Chembiochem. 2018 Mar 11. doi: 10.1002/cbic.201800130.

Kalyuzhnaya, M.G., Lapidus, A., Ivanova, N., Copeland, A.C., McHardy, A.C., Szeto, E., Salamov, A., Grigoriev, I.V., Suciu, D., Levine, S.R., *et al.* (2008). High-resolution metagenomics targets specific functional types in complex microbial communities. Nat. Biotechnol. *26*, 1029–1034. https://doi.org/10.1038/nbt.1488

Kalyuzhnaya, M.G., Yang, S., Rozova, O.N., Smalley, N.E., Clubb, J., Lamb, A., Gowda, G.A., Raftery, D., Fu, Y., Bringel, F., *et al.* (2013). Highly efficient methane biocatalysis revealed in a methanotrophic bacterium. Nat. Commun. *4*, 2785. https://doi.org/10.1038/ncomms3785

Kao, W.C., Chen, Y.R., Yi, E.C., Lee, H., Tian, Q., Wu, K.M., Tsai, S.F., Yu, S.S., Chen, Y.J., Aebersold, R., *et al.* (2004). Quantitative proteomic analysis of metabolic regulation by copper ions in *Methylococcus capsulatus* (Bath). J. Biol. Chem. *279*, 51554–51560. https://doi.org/10.1074/jbc.M408013200

Karwautz, C., Kus, G., Stöckl, M., Neu, T.R., and Lueders, T. (2018). Microbial megacities fueled by methane oxidation in a mineral spring cave. ISME J. *12*, 87–100. https://doi.org/10.1038/ismej.2017.146

Keltjens, J.T., Pol, A., Reimann, J., and Op den Camp, H.J. (2014). PQQ-dependent methanol dehydrogenases: rare-earth elements make a difference. Appl. Microbiol. Biotechnol. *98*, 6163–6183. https://doi.org/10.1007/s00253-014-5766-8

Kessler, J.D., Valentine, D.L., Redmond, M.C., Du, M., Chan, E.W., Mendes, S.D., Quiroz, E.W., Villanueva, C.J., Shusta, S.S., Werra, L.M., *et al.* (2011). A persistent oxygen anomaly reveals the fate of spilled methane in the deep Gulf of Mexico. Science *331*, 312–315. https://doi.org/10.1126/science.1199697

Khadem, A.F., Pol, A., Wieczorek, A., Mohammadi, S.S., Francoijs, K.J., Stunnenberg, H.G., Jetten, M.S., and Op den Camp, H.J. (2011). Autotrophic methanotrophy in verrucomicrobia: *Methylacidiphilum fumariolicum* SolV uses the Calvin-Benson-Bassham cycle for carbon dioxide fixation. J. Bacteriol. *193*, 4438–4446. https://doi.org/10.1128/JB.00407-11

Knittel, K., and Boetius, A. (2009). Anaerobic oxidation of methane: progress with an unknown process. Annu. Rev. Microbiol. *63*, 311–334. https://doi.org/10.1146/annurev.micro.61.080706.093130

Krause, S.M., Johnson, T., Samadhi Karunaratne, Y., Fu, Y., Beck, D.A., Chistoserdova, L., and Lidstrom, M.E. (2017). Lanthanide-dependent cross-feeding of methane-derived carbon is linked by microbial community interactions. Proc. Natl. Acad. Sci. U.S.A. *114*, 358–363. https://doi.org/10.1073/pnas.1619871114

Lawton, T.J., and Rosenzweig, A.C. (2016). Biocatalysts for methane conversion: big progress on breaking a small substrate. Curr. Opin. Chem. Biol. 35, 142–149.

Lim, S., and Franklin, S.J. (2004). Lanthanide-binding peptides and the enzymes that Might Have Been. Cell. Mol. Life Sci. *61*, 2184–2188. https://doi.org/10.1007/s00018-004-4156-2

Lumpe, H., Pol, A., Op den Camp, H.J.M., and Daumann, L.J. (2018). Impact of the lanthanide contraction on the activity of a lanthanide-dependent methanol dehydrogenase - a kinetic and DFT study. Dalton Trans. *47*, 10463–10472. https://doi.org/10.1039/c8dt01238e

Martineau, C., Whyte, L.G., and Greer, C.W. (2010). Stable isotope probing analysis of the diversity and activity of methanotrophic bacteria in soils from the Canadian high Arctic. Appl. Environ. Microbiol. *76*, 5773–5784. https://doi.org/10.1128/AEM.03094-09

McGlynn, S.E. (2017). Energy metabolism during anaerobic methane oxidation in ANME Archaea. Microbes Environ. *32*, 5–13. https://doi.org/10.1264/jsme2.ME16166

Meyerdierks, A., Kube, M., Kostadinov, I., Teeling, H., Glöckner, F.O., Reinhardt, R., and Amann, R. (2010). Metagenome and mRNA expression analyses of anaerobic methanotrophic archaea of the ANME-1 group. Environ. Microbiol. *12*, 422–439. https://doi.org/10.1111/j.1462-2920.2009.02083.x

Mustakhimov, I., Kalyuzhnaya, M.G., Lidstrom, M.E., and Chistoserdova, L. (2013). Insights into denitrification in *Methylotenera mobilis* from denitrification pathway and methanol metabolism mutants. J. Bacteriol. *195*, 2207–2211. https://doi.org/10.1128/JB.00069-13

Orphan, V.J., House, C.H., Hinrichs, K.U., McKeegan, K.D., and DeLong, E.F. (2001). Methane-consuming archaea revealed by directly coupled isotopic and phylogenetic analysis. Science *293*, 484–487. https://doi.org/10.1126/science.1061338

Oshkin, I.Y., Beck, D.A., Lamb, A.E., Tchesnokova, V., Benuska, G., McTaggart, T.L., Kalyuzhnaya, M.G., Dedysh, S.N., Lidstrom, M.E., and Chistoserdova, L. (2015). Methane-fed microbial microcosms show differential community dynamics and pinpoint taxa involved in communal response. ISME J. *9*, 1119–1129. https://doi.org/10.1038/ismej.2014.203

Paul, B.G., Ding, H., Bagby, S.C., Kellermann, M.Y., Redmond, M.C., Andersen, G.L., and Valentine, D.L. (2017). Methane-oxidizing bacteria shunt carbon to microbial mats at a marine hydrocarbon seep. Front. Microbiol. *8*, 186. https://doi.org/10.3389/fmicb.2017.00186

Pol, A., Barends, T.R., Dietl, A., Khadem, A.F., Eygensteyn, J., Jetten, M.S., and Op den Camp, H.J. (2014). Rare earth metals are essential for methanotrophic life in volcanic mudpots. Environ. Microbiol. *16*, 255–264. https://doi.org/10.1111/1462-2920.12249

Rasigraf, O., Kool, D.M., Jetten, M.S., Sinninghe Damsté, J.S., and Ettwig, K.F. (2014). Autotrophic carbon dioxide fixation via the Calvin-Benson-Bassham cycle by the denitrifying methanotroph '*Candidatus* Methylomirabilis oxyfera'. Appl. Environ. Microbiol. *80*, 2451–2460. https://doi.org/10.1128/AEM.04199-13

Ross, M.O., and Rosenzweig, A.C. (2017). A tale of two methane monooxygenases. J. Biol. Inorg. Chem. *22*, 307–319. https://doi.org/10.1007/s00775-016-1419-y

Salcher, M.M., Neuenschwander, S.M., Posch, T., and Pernthaler, J. (2015). The ecology of pelagic freshwater methylotrophs assessed by a high-resolution monitoring and isolation campaign. ISME J. *9*, 2442–2453. https://doi.org/10.1038/ismej.2015.55

Sharp, C.E., Smirnova, A.V., Graham, J.M., Stott, M.B., Khadka, R., Moore, T.R., Grasby, S.E., Strack, M., and Dunfield, P.F. (2014). Distribution and diversity of

Verrucomicrobia methanotrophs in geothermal and acidic environments. Environ. Microbiol. *16*, 1867–1878. https://doi.org/10.1111/1462-2920.12454

Shiller, A.M., Chan, E.W., Joung, D.J., Redmond, M.C., and Kessler, J.D. (2017). Light rare earth element depletion during Deepwater Horizon blowout methanotrophy. Sci. Rep. *7*, 10389. https://doi.org/10.1038/s41598-017-11060-z

Sowell, S.M., Abraham, P.E., Shah, M., Verberkmoes, N.C., Smith, D.P., Barofsky, D.F., and Giovannoni, S.J. (2011). Environmental proteomics of microbial plankton in a highly productive coastal upwelling system. ISME J. *5*, 856–865. https://doi.org/10.1038/ismej.2010.168

Thauer, R.K., and Shima, S. (2008). Methane as fuel for anaerobic microorganisms. Ann. N. Y. Acad. Sci. *1125*, 158–170.

Trembath-Reichert, E., Case, D.H., and Orphan, V.J. (2016). Characterization of microbial associations with methanotrophic archaea and sulfate-reducing bacteria through statistical comparison of nested Magneto-FISH enrichments. PeerJ *4*, e1913. https://doi.org/10.7717/peerj.1913

Trotsenko, Y.A., and Murrell, J.C. (2008). Metabolic aspects of aerobic obligate methanotrophy. Adv. Appl. Microbiol. *63*, 183–229. https://doi.org/10.1016/S0065-2164(07)00005-6

Vekeman, B., Speth, D., Wille, J., Cremers, G., De Vos, P., Op den Camp, H.J., and Heylen, K. (2016). Genome characteristics of two novel Type I methanotrophs enriched from north sea sediments containing exclusively a lanthanide-dependent xoxf5-Type Methanol Dehydrogenase. Microb. Ecol. *72*, 503–509. https://doi.org/10.1007/s00248-016-0808-7

Versantvoort, W., Guerrero-Cruz, S., Speth, D.R., Frank, J., Gambelli, L., Cremers, G., van Alen, T., Jetten, M.S.M., Kartal, B., Op den Camp, H.J.M., *et al.* (2018). Comparative genomics of *Candidatus Methylomirabilis* Species and description of *Ca. Methylomirabilis Lanthanidiphila*. Front. Microbiol. 9, 1672. https://doi.org/10.3389/fmicb.2018.01672

Vu, H.N., Subuyuj, G.A., Vijayakumar, S., Good, N.M., Martinez-Gomez, N.C., and Skovran, E. (2016). Lanthanide-dependent regulation of methanol oxidation systems in *Methylobacterium extorquens* AM1 and their contribution to methanol growth. J. Bacteriol. *198*, 1250–1259. https://doi.org/10.1128/JB.00937-15

Wehrmann, M., Billard, P., Martin-Meriadec, A., Zegeye, A., and Klebensberger, J. (2017). Functional Role of lanthanides in enzymatic activity and transcriptional regulation of pyrroloquinoline quinone-dependent alcohol dehydrogenases in *Pseudomonas putida* KT2440. MBio *8*, e00570–17.

Wehrmann, M., Berthelot, C., Billard, P., and Klebensberger, J. (2018). The PedS2/PedR2 two-component system is crucial for the rare earth element switch in *Pseudomonas putida* KT2440. mSphere *3*, e00376–18.

Welte, C.U., Rasigraf, O., Vaksmaa, A., Versantvoort, W., Arshad, A., Op den Camp, H.J., Jetten, M.S., Lüke, C., and Reimann, J. (2016). Nitrate- and nitrite-dependent anaerobic oxidation of methane. Environ. Microbiol. Rep. *8*, 941–955. https://doi.org/10.1111/1758-2229.12487

Whittenbury, R., Phillips, K.C., and Wilkinson, J.F. (1970). Enrichment, isolation and some properties of methane-utilizing bacteria. J. Gen. Microbiol. *61*, 205–218. https://doi.org/10.1099/00221287-61-2-205

Yu, Z., and Chistoserdova, L. (2017). Communal metabolism of methane and the rare Earth element switch. J. Bacteriol. *19*, pii JB.00323-17. https://doi.org/10.1128/JB.00328-17

Yu, Z., Krause, S.M., Beck, D.A., and Chistoserdova, L. (2016). A synthetic ecology perspective: how well does behavior of model organisms in the laboratory predict microbial activities in natural habitats? Front. Microbiol. 7, 946.

Yu, Z., Beck, D.A.C., and Chistoserdova, L. (2017). Natural selection in synthetic communities highlights the roles of *Methylococcaceae* and *Methylophilaceae* and suggests differential roles for alternative methanol dehydrogenases in methane consumption. Front. Microbiol. *8*, 2392. https://doi.org/10.3389/fmicb.2017.02392

Zheng, Y., Huang, J., Zhao, F., and Chistoserdova, L. (2018). Physiological effect of XoxG(4) on lanthanide-dependent methanotrophy. MBio 9, e02430–17.

Methylotrophic Yeasts: Current Understanding of Their C1-Metabolism and its Regulation by Sensing Methanol for Survival on Plant Leaves

11

Hiroya Yurimoto and Yasuyoshi Sakai*

Division of Applied Life Sciences, Graduate School of Agriculture, Kyoto University, Kitashirakawa-Oiwake, Sakyo-ku, Kyoto, Japan.

*Correspondence: ysakai@kais.kyoto-u.ac.jp

https://doi.org/10.21775/9781912530045.11

Abstract

Methylotrophic yeasts, which are able to utilize methanol as the sole carbon and energy source, have been intensively studied in terms of physiological function and practical applications. When these yeasts grow on methanol, the genes encoding enzymes and proteins involved in methanol metabolism are strongly induced. Simultaneously, peroxisomes, organelles that contain the key enzymes for methanol metabolism, massively proliferate. These characteristics have made methylotrophic yeasts efficient hosts for heterologous protein production using strong and methanol-inducible gene promoters and also model organisms for the study of peroxisome dynamics. Much attention has been paid to the interaction between methylotrophic microorganisms and plants. In this chapter, we describe how methylotrophic yeasts proliferate and survive on plant leaves, focusing on their physiological functions and lifestyle in the phyllosphere. Our current understanding of the molecular basis of methanol-inducible gene expression, including methanol-sensing and its applications, is also summarized.

Introduction

Methylotrophic yeasts are capable of utilizing methanol as the sole source of carbon and energy. These yeasts belong to a restricted number of genera, including *Komagataella*, *Ogataea*, *Kuraishia*, and *Candida* (Kurtzman, 2005; Péter *et al.*, 2005; Suh *et al.*, 2006; Limtong *et al.*, 2008), while methylotrophic bacteria belong to diverse genera and subclasses (Kolb, 2009). Methylotrophic yeasts can also use another one-carbon (C1) compound, methylamine, as a nitrogen source, but not as the sole carbon source.

Since the first isolation of the methylotrophic yeast *Kloeckera* sp. (later identified as *Candida boidinii*) in 1969 (Ogata *et al.*, 1969), both their physiological functions and their applications have been intensively studied. During the 1970s, the metabolic pathways for methanol assimilation and dissimilation were elucidated mainly with *Candida boidinii* and *Hansenula polymorpha* (reclassified as *Ogataea polymorpha*), and methylotrophic yeasts were revealed to have a common methanol-utilizing pathway (Anthony, 1982; Veenhuis *et al.*, 1983; Tani, 1984; Yurimoto *et al.*, 2002). One of the differences between bacterial and yeast methanol metabolism is in the initial oxidation reaction of methanol. While methylotrophic bacteria oxidize methanol to formaldehyde by using a pyrroloquinoline quinone (PQQ)- or NAD^+-dependent dehydrogenase, methylotrophic yeasts oxidize methanol with an alcohol oxidase (AOD) using molecular oxygen as an electron acceptor. When methylotrophic yeasts grow on methanol, genes encoding a number of enzymes and proteins involved in methanol metabolism are strongly induced. This strong inducibility by methanol together with the ability to grow to extremely high cell density have enabled methylotrophic yeasts to become promising hosts for high-level heterologous protein production. During the mid-1980s to early 1990s, heterologous gene expression systems driven by strong methanol-inducible gene promoters were developed in various methylotrophic yeast strains, including *Pichia pastoris* (reclassified as *Komagataella phaffii*), *H. polymorpha*, and *C. boidinii* (Cregg *et al.*, 2000; Gellissen, 2000; Yurimoto, 2009). Another characteristic of yeast methanol metabolism is that methylotrophic growth is accompanied by massive development of a membrane-bound organelle, the peroxisome, in which AOD and other key methanol-assimilating enzymes are compartmentalized. Therefore, methylotrophic yeasts have also been used as model organisms to reveal the molecular machineries and mechanisms of peroxisomal protein import as well as peroxisome degradation by selective autophagy (pexophagy) (van der Klei *et al.*, 2006; Oku *et al.*, 2010).

Methanol is a volatile atmospheric carbon compound, and its annual emission is estimated to be 100 Tg (Galbally *et al.*, 2002). Methanol, which originates from methylesters in the plant cell wall constituent pectin, is emitted from plant leaves during plant cell elongation and division (Nemecek-Marshall *et al.*, 1995; Fall *et al.*, 1996). The phyllosphere, defined as the aerial portion of plants, has been recognized as a habitat for methylotrophic microorganisms. Phyllospheric methylotrophic bacteria were identified in the 1980s (Corpe *et al.*, 1982), and since then, the symbiotic relationship between plants and methylotrophic bacteria, *Methylobacterium* spp., which can promote plant growth, has been intensively investigated (Vorholt, 2012; Dourado *et al.*, 2015). Similarly, a number of methylotrophic yeast strains have been isolated from plant resources, e.g. forest soils, fallen leaves, and the skins of olives and grapes (Limtong *et al.*, 2008; Péter *et al.*, 2009). However, the ecology and physiology of methylotrophic yeasts in the phyllosphere have not been studied in detail.

In this chapter, after outlining methanol metabolism in methylotrophic yeasts, we describe how methylotrophic yeasts utilize carbon and nitrogen sources in the phyllosphere by regulating their cellular metabolism. Then the recently revealed molecular basis of methanol-inducible gene expression, including the methanol-sensing machinery is summarized.

Methanol metabolism in the methylotrophic yeasts

Fig. 11.1 summarizes the pathway of methanol metabolism, which so far is common to all isolated methylotrophic yeasts (Yurimoto *et al.*, 2002, 2011). Methanol metabolism begins with the oxidation of methanol to formaldehyde by AOD. AOD is a flavoprotein that contains FAD and uses O_2 as an electron acceptor to yield formaldehyde and H_2O_2, both of which are extremely toxic to cells. H_2O_2 is broken down by catalase (CTA) and peroxiredoxin (Pmp20) (Horiguchi *et al.*, 2001a,b). Formaldehyde, which is situated at the branch point between assimilation and dissimilation pathways, is a key central intermediate in methanol metabolism (Yurimoto *et al.*, 2005). In the assimilation pathway, formaldehyde (C1-compound) is fixed to xylulose 5-phosphate (Xu5P; C5 compound) by dihydroxyacetone synthase (DAS) to generate two C3 compounds, dihydroxyacetone (DHA) and glyceraldehyde 3-phosphate (GAP). DHA is phosphorylated by dihydroxyacetone kinase (DAK) to yield dihydroxyacetone phosphate (DHAP). GAP and DHAP are used for the synthesis of cell constituents and the regeneration of Xu5P after rearrangement reactions.

In the cells of methylotrophic yeasts, peroxisomes, the organelles specialized for methanol metabolism, massively proliferate during growth on methanol. The peroxisomal localization of AOD, CTA, Pmp20, and DAS has been described in detail. Recently, proteomic analysis of the peroxisomal fraction of methanol-grown *P. pastoris* cells revealed that further assimilatory metabolism catalysed by DAK and the sugar phosphate rearrangement reactions are also localized within peroxisomes (Russmayer *et al.*, 2015).

In the dissimilation pathway, formaldehyde is oxidized to CO_2 by the glutathione-dependent formaldehyde oxidation pathway. Formaldehyde reacts non-enzymatically with the reduced form of glutathione (GSH) to form *S*-hydroxymethyl

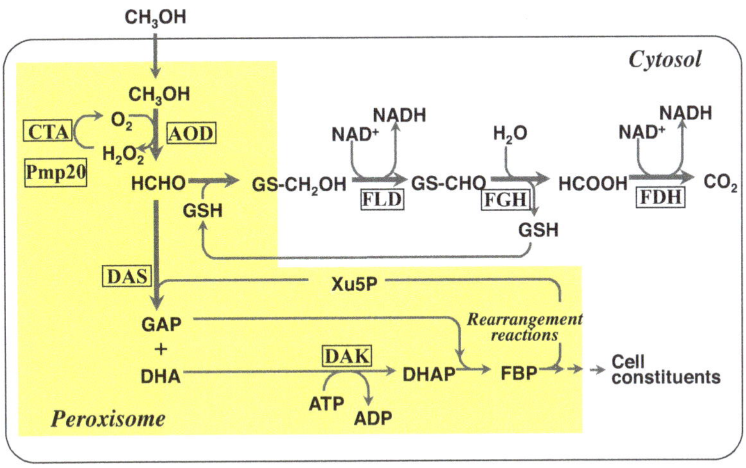

Figure 11.1 Methanol metabolism in methylotrophic yeasts. AOD, alcohol oxidase; CTA, catalase; DAK, dihydroxyacetone kinase; DAS, dihydroxyacetone synthase; FDH, formate dehydrogenase; FGH, *S*-formylglutathione hydrolase; FLD, formaldehyde dehydrogenase; Pmp20, peroxiredoxin (glutathione peroxidase). Abbreviations: DHA, dihydroxyacetone; DHAP, dihydroxyacetone phosphate; FBP, fructose 1,6-bisphosphate; GAP, glyceraldehyde 3-phosphate; GS-CH2OH, *S*-hydroxymethyl glutathione; GS-CHO, *S*-formylglutathione; GSH, reduced form of glutathione; Xu5P, xylulose 5-phosphate. Reactions occurring within peroxisomes are highlighted in yellow.

glutathione (*S*-HMG). Since GSH was shown to be present in peroxisomes, *S*-HMG can be formed within peroxisomes and then be exported to the cytosol (Horiguchi *et al.*, 2001b; Yurimoto *et al.*, 2003). *S*-HMG is oxidized to *S*-formylglutathione (*S*-FG) by NAD$^+$-linked and GSH-dependent formaldehyde dehydrogenase (FLD). *S*-FG is then hydrolysed to formate and GSH by *S*-formylglutathione hydrolase (FGH). Finally, formate is oxidized to CO_2 by NAD$^+$-linked formate dehydrogenase (FDH). Therefore, in methylotrophic yeasts, the formaldehyde oxidation pathway is physiologically responsible not only for formaldehyde detoxification but also for energy generation (Sakai *et al.*, 1997; Lee *et al.*, 2002; Yurimoto *et al.*, 2003).

Survival strategy in the phyllosphere: nutrient sources for methylotrophic yeasts

Plants emit various volatile organic compounds. Among them, 100 Tg of methanol per year is estimated to be emitted from plants (Galbally *et al.*, 2002; Laothawornkitkul *et al.*, 2009), and the main origin of methanol is thought to be the methylester residue of pectin, a major constituent of plant cell walls (Fall *et al.*, 1996). The above-ground part of plants, termed the phyllosphere, is one of the main habitats of methylotrophic microorganisms. Indeed, methylotrophic yeasts can often be isolated from plant material (Limtong *et al.*, 2008; Péter *et al.*, 2009). However, while the symbiotic relationship between plants and methylotrophic bacteria has been well characterized (Vorholt, 2012), it was not known whether methylotrophic yeasts can survive and proliferate in the phyllosphere.

Methanol emission from the phyllosphere was first reported by Nemecek-Marshall *et al.* (1995). In this report, methanol was detected in the atmospheric phase of a gas chamber in which a leaf of a growing plant was compartmentalized. Methanol was thought to be emitted from stomata because methanol emission was correlated with diurnal stomatal opening (Nemecek-Marshall *et al.*, 1995). However, it does not seem likely that phyllospheric microorganisms directly utilize gas-phase methanol. And until recently it was also unclear how much methanol is present in the phyllosphere and available to methylotrophic microorganisms.

In order to address these questions, we developed a cell-based methanol assay system using *C. boidinii* cells expressing the yellow fluorescent protein Venus tagged with peroxisome targeting signal 1 (Venus-PTS1) under the control of the methanol-inducible *DAS1* promoter, and established a method to quantify methanol concentration on the surface of plant leaves (Kawaguchi *et al.*, 2011). The *DAS1* promoter was chosen because it is the strongest methanol-inducible promoter and the level of methanol-independent expression (derepression) with the *DAS1* promoter is lower than that of the *AOD1* promoter (Yurimoto *et al.*, 2000). The methanol sensor cells were inoculated onto the leaves of *Arabidopsis thaliana*, and the fluorescence intensity of the cells was determined after a 4-h incubation. When cells were inoculated onto young leaves (grown 2–3 weeks after germination), the estimated methanol concentration on the leaves changed periodically by 0–0.2% (ca. 0–60 mM): higher in the dark period (25–60 mM) and lower in the light period (0–5 mM). This daily oscillation in methanol concentration was opposite from previous observations, which indicated that atmospheric methanol was higher in the light period during stomatal opening (Nemecek-Marshall *et al.*, 1995). Methanol, which accumulated in the spongy parenchyma of the leaf during stomatal closing in the dark period, diffuses to the surface of the leaf. In addition, quantitative transcript analyses revealed that the expression of methanol-inducible genes, such as *AOD1* and *DAS1*, was higher in the dark period and lower in the light period, corresponding to the methanol concentration in the phyllosphere (Kawaguchi *et al.*, 2011).

We confirmed that the methylotrophic yeasts *C. boidinii* and *P. pastoris* could proliferate on the leaf surface of growing *A. thaliana* (Kawaguchi *et al.*, 2011). *C. boidinii* cells expressing VENUS under the constitutive *ACT1* promoter were inoculated onto leaves and their growth was followed by fluorescence microscopy and quantitative PCR analysis for two weeks. *C. boidinii* cells proliferated approximately 3–4 times after 11 days of inoculation. We also examined phyllosphere proliferation of gene-disrupted mutant strains impaired in methanol metabolism, peroxisome biogenesis, and autophagy. Both *aod1Δ* and *das1Δ* strains were unable to proliferate on plant leaves, suggesting that *C. boidinii* proliferation in the phyllosphere was supported by methanol assimilation on the plant leaf surface. On the other hand, gene-disrupted mutant strains lacking formaldehyde dissimilation pathway enzymes (FLD and FDH) and peroxisomal anti-oxidant enzymes (CTA and Pmp20) could proliferate. These results suggest that methylotrophic yeasts in the phyllosphere acquire energy from some other compounds that do not repress the induction of methanol-assimilation genes and that the peroxisomal anti-oxidant system was not critical for survival in the phyllosphere. On young growing leaves, the number of peroxisomes in *C. boidinii* increased in the dark period, and gene disruption of Pex5, which is responsible for peroxisomal protein import, resulted in impaired proliferation in the phyllosphere, indicating that peroxisome assembly is required for *C. boidinii* proliferation in the phyllosphere. In contrast, the numbers of peroxisomes of *C. boidinii* on young leaves decreased in the light period in which the methanol concentration was low. These findings indicate that peroxisomes were degraded by autophagy in the light period. Indeed, the *C. boidinii atg1Δ* strain (defective in Atg1, a pivotal kinase for all autophagic pathways) and *atg30Δ* strain (defective in Atg30, a pexophagy receptor) were unable to proliferate on plant leaves. Thus, peroxisome biogenesis and degradation (pexophagy, a selective autophagy for peroxisome degradation) were dynamically regulated in the phyllosphere.

We also investigated nitrogen utilization of *C. boidinii* in the phyllosphere (Shiraishi *et al.*, 2015). *C. boidinii* can utilize several nitrogen sources including ammonium, nitrate, and methylamine. We tested whether the metabolism of nitrate or methylamine was required for growth and survival on plant leaves by analyses of gene expression and gene disruption of nitrate reductase (YNR) or amine oxidase (AMO). On young growing leaves of *A. thaliana*, the *YNR1* gene, not the *AMO* gene, was expressed, and its expression level fluctuated during the daily cycle. Since the *C. boidinii ynr1Δ* strain was unable to proliferate on young leaves, nitrate was concluded to be the major nitrogen source for *C. boidinii* on growing plant leaves.

Expression of the *AMO1* gene was not observed on young leaves, but the *AMO1* gene was significantly expressed in cells inoculated on wilting leaves. These results suggest that *C. boidinii* utilizes nitrate on young leaves, whereas the available nitrogen source changed and the yeast started to assimilate methylamine as a nitrogen source on aged leaves. Furthermore, *C. boidinii* YNR, which was essential for proliferation on young leaves, was transported to and degraded in the vacuole on aged leaves, by the cytoplasm-to-vacuole targeting (Cvt) pathway, one of the selective autophagic pathways.

In contrast to young leaves, the methanol concentration on wilting or dead leaves was estimated to be greater than 250 mM and did not show diurnal oscillation (Kawaguchi *et al.*, 2011). Huge developed peroxisomes were observed in *C. boidinii* inoculated onto these leaves. Since AMO is also a peroxisomal enzyme, peroxisomes of methylotrophic yeasts on the aged plant leaves may play important roles both in carbon and nitrogen metabolism. Peroxisomes are thought to be storage organelles for proteins that served as a source of amino acids in the natural environment because non-motile and asporogenous yeast cells on dead plants must survive until they obtain nutrients for further proliferation (Fig. 11.2).

Thus, methylotrophic yeasts proliferate and survive on plant leaf surfaces by adapting to environmental nutrient sources, which change not only during the day-night cycle but also during the life cycle of the plant. Such metabolic regulation was demonstrated not only by gene expression but also by autophagy (Fig. 11.2).

Transcription factors involved in regulation of methanol-inducible gene expression

When cells pre-grown on carbon sources other than methanol are transferred to methanol medium, cells must adapt to methanol medium by induction

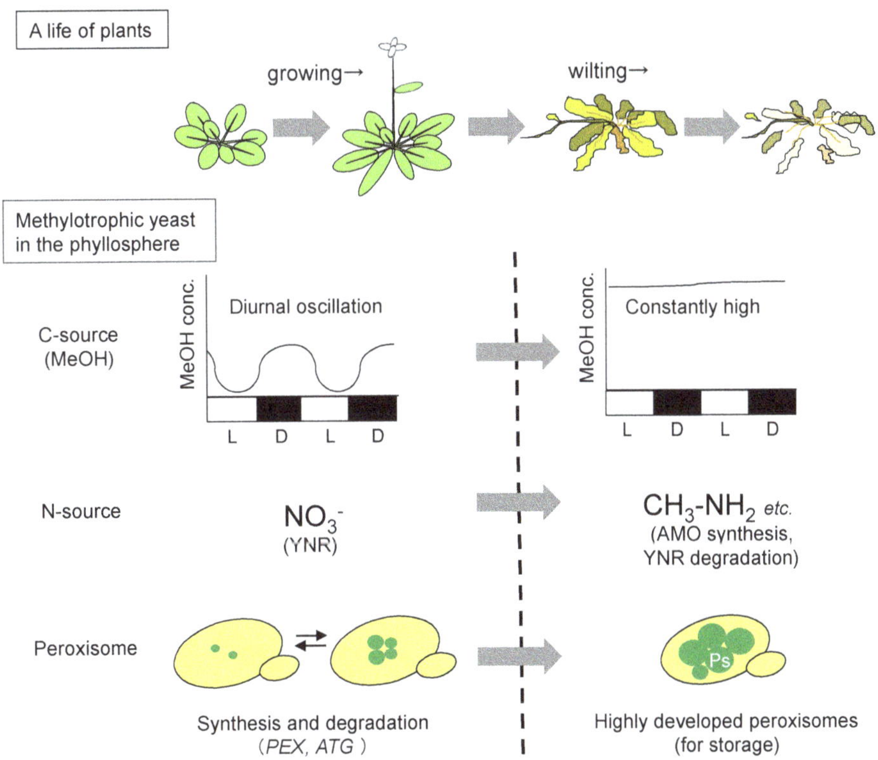

Figure 11.2 Lifestyle of methylotrophic yeasts on plant leaves. Methanol assimilatory enzymes and nitrate reductase (YNR) were required for proliferation of *C. boidinii* on young growing leaves. The methanol concentration (MeOH conc.) showed diurnal oscillation: high in the dark period (D) and low in the light period (L). Responding to methanol in the dark period, cells induced the methanol-inducible genes and peroxisomes (Ps). Responding to depletion of methanol in the light period, peroxisomes were degraded by autophagy. Not only synthesis of peroxisomes (conducted by *PEX* gene products), but also their autophagic degradation (conducted by *ATG* gene products) were necessary for *C. boidinii* to proliferate on young leaves.
On wilting or dead leaves, the MeOH conc. is generally high and peroxisomes are massively developed. Peroxisomes may function as storage organelles for proteins as a source of amino acids. While nitrate is the major nitrogen sources on young leaves, YNR is degraded by a type of selective autophagic pathway, the Cvt pathway, on wilting leaves. Under these conditions, AMO is induced to utilize methylamine as a nitrogen source.

of methanol metabolic genes and peroxisome proliferation. Since this metabolic shift includes drastic intracellular reorganization in addition to methanol-induced gene expression, it takes some time, ca. 2 h, before methanol consumption begins. During the metabolic shift, cells in transition were revealed to be a heterogenous population that exhibited differing levels of the methanol-induced gene expression based on single-cell analysis (Takeya *et al.*, 2018).

Based on the expression levels of methanol-inducible genes in media containing different carbon sources, there are three different regulatory states for methanol-inducible genes (Fig. 11.3) (Yurimoto *et al.*, 2000, 2011; Yurimoto, 2009). Glucose or ethanol completely represses methanol-inducible gene expression even in the presence of methanol [(i) repression]. Glycerol also represses methanol-inducible genes in *P. pastoris,* but not in *C. boidinii* and *H. polymorpha.* After depletion of the carbon source during metabolic shift (glucose in Fig. 11.3), repression is released and the genes are activated by (ii) derepression (independent of the presence of methanol). Finally, the presence of methanol induces maximum expression of methanol-inducible genes via (iii) methanol induction (Fig. 11.3) (Hartner *et al.*, 2006; Yurimoto *et al.*, 2011). So far, several transcription factors involved in regulation of methanol-inducible genes have been identified and characterized in *C. boidinii, H. polymorpha,* and *P. pastoris.* Gene disruption analyses revealed that each transcription factor is involved in one of the steps of gene regulation. Therefore, full activation of methanol-inducible genes is achieved by these three events in this order, (i) repression, (ii) derepression, and (iii) methanol induction, all of which are mediated by different transcription factors.

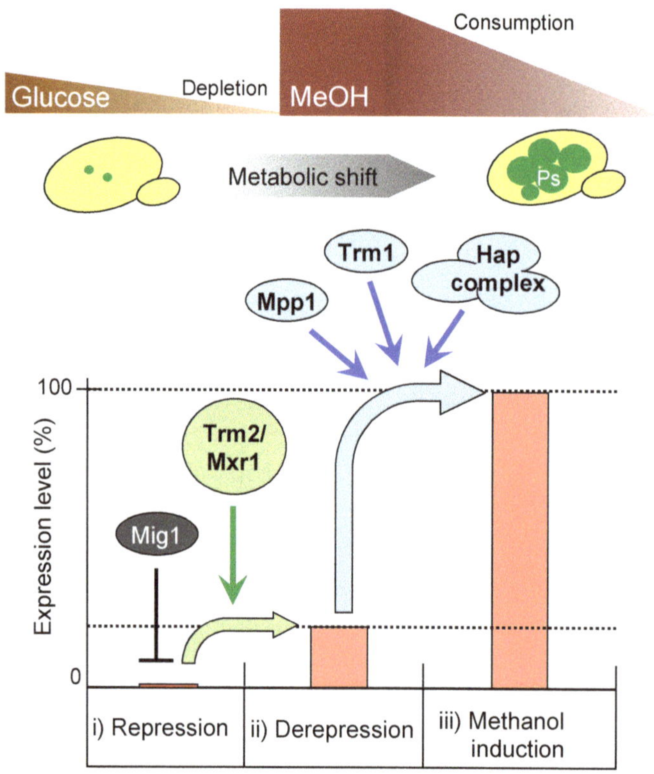

Figure 11.3 Regulation of methanol-inducible genes by repression and two-stepwise activation with transcription factors. Cells pre-grown on glucose adapt to the new carbon source methanol (MeOH) by induction of genes encoding metabolic enzymes and peroxisome (Ps) proliferation. Expression of methanol-inducible genes can be dissected into the following three regulatory events: (i) repression; (ii) derepression; and (iii) methanol induction. When glucose-grown cells are transferred to methanol medium, at first, (i) repression (regulated by Mig1) is released with consumption of glucose, followed by stepwise gene activation via (ii) derepression (by CbTrm2 and PpMxr1) and (iii) methanol-induction (by Mpp1, Trm1, and the Hap complex).

In *S. cerevisiae*, the key transcription factor of glucose repression is ScMig1, a zinc finger protein that binds to promoters of glucose-repressible genes (Klein *et al.*, 1998; Carlson, 1999; Schüller, 2003; Turcotte *et al.*, 2010). Homologues of ScMig1 in *H. polymorpha* and *C. boidinii* have been identified and characterized (Stasyk *et al.*, 2007; Zhai *et al.*, 2012). Analyses of the transcript level of the methanol-inducible *C. boidinii AOD1* gene in the *Cbmig1Δ* strain revealed that CbMig1 represses methanol-inducible genes in glucose-containing medium (Zhai *et al.*, 2012).

With respect to the transcription factors involved in (ii) derepression, homologues of the C_2H_2-type transcriptional factor ScAdr1, i.e. PpMxr1 and CbTrm2 were identified (Lin-Cereghino *et al.*, 2006; Sasano *et al.*, 2010). ScAdr1 functions in the activation of various genes involved in the metabolism of non-fermentable carbon sources such as glycerol, ethanol and oleate, as well as in peroxisome biogenesis (Simon *et al.*, 1991; Turcotte *et al.*, 2010). PpMxr1 and CbTrm2 share a number of functional similarities in the regulation of methanol-inducible gene expression, including binding to methanol-inducible promoters and subcellular localization (Lin-Cereghino *et al.*, 2006; Sasano *et al.*, 2010). However, growth of the gene-disrupted strains on non-fermentable carbon sources was different between *P. pastoris* and *C. boidinii*. The *Cbtrm2Δ* strain was unable to grow on methanol and oleate, and showed a reduced rate of growth on glycerol, but this strain did not exhibit growth defect on glucose and ethanol. On the other hand, the *Ppmxr1Δ* strain exhibited growth defects on methanol, ethanol, oleate, and glycerol, but no growth defect on glucose. CbTrm2 may have additional regulatory functions during ethanol metabolism since CbTrm2 is localized to the nucleus during growth on ethanol. PpMxr1 was shown to be inactivated by its phosphorylation in ethanol-grown cells (Parua *et al.*, 2012).

Several transcription factors unique to methanol induction are thought to be specific to and conserved only in methylotrophic yeasts. Three such transcription factors have been identified and characterized as responsible for (iii) methanol induction, i.e. (i) Mpp1 (HpMpp1 and PpMit1) (Leao-Helder *et al.*, 2003; Wang *et al.*, 2016), (ii) Trm1 (CbTrm1 and PpPrm1) (Sasano *et al.*, 2008; Sahu *et al.*, 2014), and (iii) the CbHap complex (Oda *et al.*, 2015; Oda *et al.*, 2016). Disruption of these genes resulted in severe growth defects on methanol but not on glucose or other non-fermentable carbon sources such as ethanol and glycerol. Mpp1 and Trm1 proteins belong to the $Zn(II)_2Cys_6$-type zinc cluster protein family. Mpp1 proteins were shown to be induced by methanol, whereas Trm1 proteins were constitutively expressed (Sasano *et al.*, 2008; van Zutphen *et al.*, 2010). Analysis of a double mutant lacking CbTrm1 and CbTrm2 and analyses of PpMit1 expression in gene disruptants of *PpMXR1* or *PpPRM1* revealed that Trm2 or Mxr1-dependent derepression occurred before methanol induction, which requires Trm1-dependent expression of Mpp1 (Fig. 11.3) (Sasano *et al.*, 2010; Wang *et al.*, 2016).

Recently, we revealed that the CbHap complex functions in (iii) methanol induction (Oda *et al.*, 2015, 2016). The Hap complex, a stable heterotrimer (Hap2/3/5), is highly conserved among all eukaryotes, from yeasts to humans and binds to a CCAAT consensus sequence (Buschlen *et al.*, 2003; McNabb *et al.*, 2005). In *S. cerevisiae*, the Hap complex together with another component, Hap4, activates genes involved in respiratory metabolism and mitochondrial biogenesis. Although the Hap complex functions in the metabolism of various non-fermentable carbon sources in methylotrophic yeasts, the CbHap complex was specifically responsible for methanol induction, since *Cbhap2Δ*, *Cbhap3Δ*, and *Cbhap5Δ* showed decreased transcript levels of methanol-inducible genes, resulting in a severe growth defect on methanol but not on glucose and other non-fermentable carbon sources (Oda *et al.*, 2015). Comparison of amino acid sequences of CbHap3 and ScHap3 revealed that CbHap3 contains a highly conserved region including a DNA binding motif in the N-terminal half of the protein. In addition. CbHap3 has a unique extended amino acids sequences at its C-terminus, which is conserved among methylotrophic yeasts but not present in ScHap3. Deletion of this C-terminal region of CbHap3 resulted in a decrease in the transcript level of methanol-induced genes but did not affect DNA-binding activity (Oda *et al.*, 2016). Furthermore, expression of CbHap3 in the *Schap3Δ* strain could complement the growth defect on glycerol, although expression of ScHap3 in the *Cbhap3Δ* strain could not complement the growth defect on methanol. Based on these results, it was concluded that the C-terminal extended region plays a unique role in (iii) methanol induction and the N-terminal region of CbHap3 has conserved roles, such as DNA binding and interaction with Hap2 and Hap5. We propose that, during evolution, Hap3 proteins of methylotrophic yeasts may have acquired the specific function of activating the methanol-inducible genes via their unique C-terminal region.

Methanol-sensing machinery

Multiple transcription factors have been found to function in the regulation of methanol-inducible genes in methylotrophic yeasts. However, how methylotrophic yeasts sense methanol, and how the methanol-sensing signal is transmitted to transcription factors, remained open questions for a long time. In nature, methylotrophic yeasts have to adapt to environments in which the methanol concentration is constantly changing, e.g. on the surface of plant leaves. The methanol concentration on young leaves changes in the range of 0–0.2% (ca. 0–60 mM) and that on wilting leaves reaches 1% (ca. 250 mM). Methylotrophic yeasts must therefore sense a wide range of methanol concentrations in the environment and regulate the expression of methanol-inducible genes in response. We found that the transcript levels of methanol-inducible genes increased in the presence of 0.001–0.1% methanol but decreased in the presence of more than 0.1% methanol (Ohsawa *et al.*, 2017). When the methanol concentration in the environment is high, excess AOD results in the accumulation of formaldehyde, which is toxic to cells. Therefore, cells avoid formaldehyde accumulation by regulating the expression levels of the methanol-inducible genes. On the other hand, a decrease in the transcript levels of methanol-inducible genes at higher methanol concentrations (> 0.1%) prevents efficient production of heterologous proteins under the control of methanol-inducible promoters. Therefore, revealing the molecular mechanism of methanol-sensing and the methanol-signalling pathway is important not only for understanding physiological functions of methylotrophic yeasts in nature but also for efficient production of heterologous proteins from methanol.

Recently, we found that the Wsc family proteins, which are plasma membrane spanning sensor proteins conserved from yeasts to mammalian cells, function as a methanol-sensing machinery in the methylotrophic yeast *P. pastoris* (Ohsawa *et al.*, 2017). Wsc family proteins activate the cell wall integrity (CWI) signalling pathway in response to cell surface stresses (heat, osmotic shock, high concentration of ethanol). *P. pastoris* has three putative genes encoding Wsc family proteins, and PpWsc1 and PpWsc3 were revealed to be involved in regulation of methanol-inducible gene expression. Growth of the *Ppwsc1Δ* strain on methanol was severely impaired and the transcript levels of methanol-inducible genes in the *Ppwsc1Δ* strain were significantly lower than those in the wild-type strain. There were no significant differences in both the growth or in the transcript levels between the *Ppwsc3Δ* strain and the wild-type strain. But the *Ppwsc1ΔPpwsc3Δ* strain exhibited a more severe growth defect than the *Ppwsc1Δ* strain on methanol medium and the transcript levels of methanol-inducible genes in the *Ppwsc1ΔPpwsc3Δ* strain grown on methanol were notably lower at concentrations above 0.005% methanol. Thus, PpWsc1 and PpWsc3 are responsible for the methanol-dose-dependent gene expression. PpWsc1 responded to lower methanol concentrations (0.01–0.05%) than PpWsc3 (0.1–0.5%). Therefore, PpWsc1 and PpWsc3 seem to cooperatively sense a wide range of methanol for regulation of methanol-inducible genes.

The *Ppwsc1Δ* strain, but not the *Ppwsc3Δ* strain, exhibited a high-temperature sensitive phenotype. Moreover, the expression of *PpWSC1* in the *Ppwsc1ΔPpwsc3Δ* strain could complement the high temperature-sensitive phenotype, but the expression of *PpWSC3* could not. Thus, PpWsc1 functions both in the high temperature stress response and in methanol-induced gene expression, while PpWsc3 functions specifically in methanol-induced gene expression. The *Ppwsc3Δ* strain grew normally on methanol at 28°C, but exhibited a severe growth defect on methanol at 37°C. During growth on methanol at high temperature, PpWsc1, which has to respond to both high-temperature stress and methanol, is likely to be more involved in the high-temperature stress response, and PpWsc3 may make a greater contribution to methanol sensing (Fig. 11.4).

To investigate whether PpWsc1 distinguishes two different environmental stimuli, i.e. high

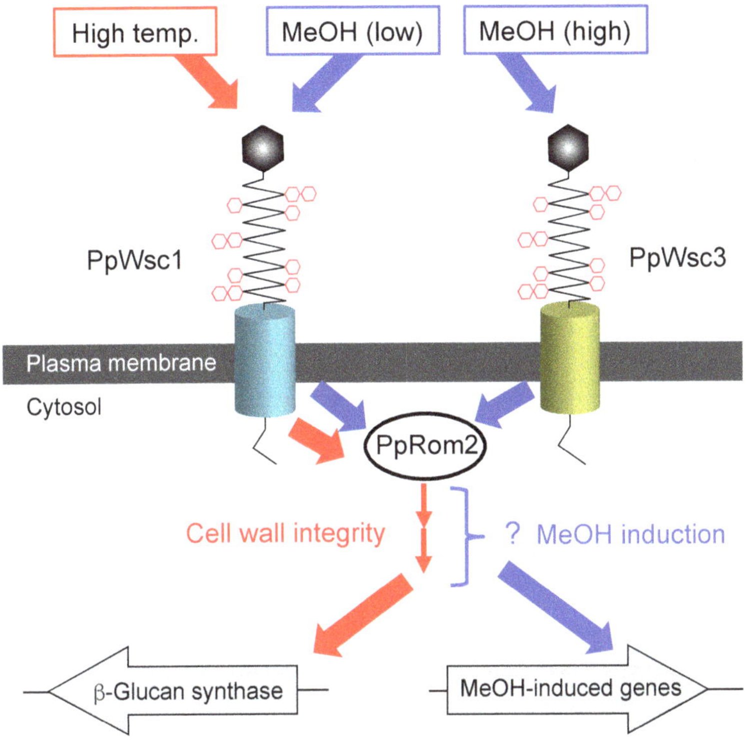

Figure 11.4 Proposed model of *P. pastoris* Wsc family protein function in methanol induction. PpWsc1 and PpWsc3 function in methanol sensing, for lower [MeOH (low)] and higher methanol concentrations [MeOH (high)], respectively. PpWsc1 also functions in the high temperature (High temp.) stress response, activating heat stress-inducible genes such as β-glucan synthase via the cell wall integrity (CWI) pathway. Both PpWsc1 and PpWsc3 interact with PpRom2 to activate methanol-inducible genes. We speculate that the methanol-sensing signal branches off from some point in the CWI pathway. The signal transduction pathway for the high temperature stress response is indicated by red arrows and that for methanol sensing is indicated by blue arrows.

temperature and methanol, we screened for site-directed mutants of PpWsc1 that were impaired in either methanol-induced gene expression or the high temperature stress response. A PpWsc1(Y53A) alanine substitution mutant was found to have decreased transcript levels of methanol-inducible genes compared with the wild-type PpWsc1, but did not show a high-temperature sensitive phenotype. On the other hand, a PpWsc1(Y53F) phenylalanine substitution mutant became sensitive to high temperature stress, but retained the normal transcript level of methanol-inducible genes. These results indicate that PpWsc1 senses both high temperature and methanol by different mechanisms. We hypothesize that PpWsc1 may adopt two different configurations in response to high temperature stress and methanol-sensing.

Rom2, interacting with Wsc family proteins, was known to be a downstream component in the CWI signalling pathway of *S. cerevisiae*. Interaction of PpWsc1 with PpRom2 was revealed to be responsible for both methanol induction and the CWI pathway (Fig. 11.4). Further investigation is necessary to elucidate the involvement of other components of the CWI pathway in the methanol induction pathway.

Development of a single-cell biosensor for detection of enzymatically produced methanol

Methanol is a promising alternative carbon source to replace coal and petroleum (Olah, 2005; Olah *et al.*, 2006), and efficient processes for methanol production from methane or renewable biomass are expected to be established. Methanol-producing enzymes, such as pectin methylesterase (PME) and methane monooxygenase, are candidates for enzymatic production of methanol. To exploit methanol-producing enzymes with high activity, a sensitive and high-throughput method to evaluate their activities at the single-cell level should be established. Methanol biosensors have been developed with microbial cells or immobilized enzymes; however, these biosensors are not appropriate for the detection of methanol that is endogenously produced by recombinant enzymes (Guilkbault *et al.*, 1983; Wen *et al.*, 2014). The detection limit of the *C. boidinii* methanol sensor, which is based on expression of the fluorescent protein gene under the control of the methanol-inducible *CbDAS1* promoter, was 0.25 mM (Kawaguchi *et al.*, 2011). However, a more sensitive sensor would be required to evaluate enzymatically produced levels of methanol by high-throughput analysis such as fluorescence-activated cell sorting (FACS).

Recently, we developed a *P. pastoris* methanol sensor and optimized the culture and methanol assay conditions by FACS (Takeya *et al.*, 2018). We successfully improved the detection limit to 2.5 μM, which is 100-fold lower than that of the fluorescence microscopy-based *C. boidinii* methanol sensor. Interestingly, *aox1Δ* cells, which are impaired in methanol metabolism, exhibited lower cellular fluorescence intensity than the wild-type cells, suggesting that the methanol sensor cells need to metabolize methanol for maximum production of the fluorescent protein. The preculture conditions of the sensor cells were also found to be critical for the sensitive detection of methanol. During preculture on glucose medium, methanol-inducible gene expression is completely repressed by glucose until mid-exponential phase, but glucose repression is released in late-exponential phase cells. Therefore, the physiological state of late-exponential phase cells is suitable for a sensitive response to low concentrations of methanol.

To substantiate the feasibility of this methanol biosensor for evaluating the activity of methanol-producing enzymes, the *Aspergillus niger* PME gene was heterologously expressed in the *P. pastoris* methanol sensor cells, and we revealed that the level of cellular enzyme activity correlated with the level of cellular fluorescence intensity. Thus, this methanol biosensor can be used for high-throughput screening of single cells harbouring an enzyme with high methanol-producing activity.

Conclusions

Methylotrophic yeasts have been used as hosts for heterologous protein production and model organisms to study peroxisome homeostasis for a long time, and understanding the transcription factors for methanol-inducible gene expression and peroxisome dynamics has been progressing. However, so far, we did not understand why methylotrophic yeasts have come to harbour such physiological traits as strong methanol-inducibility of gene expression and dynamic regulation of peroxisomal development and degradation, and how they live in nature. Regarding this point, we found that methylotrophic yeasts can proliferate on plant leaves where concentrations of methanol oscillate with the daily light–dark cycle (Kawaguchi *et al.*, 2011). Simultaneously, we established analytical methods that measure cell proliferation, gene expression, and intracellular organelle dynamics, and we elucidated metabolic and physiological functions required for the phyllospheric life of methylotrophic yeasts at the molecular level. We also constructed methanol sensor cells using methylotrophic yeasts expressing a fluorescent protein under the control of methanol-inducible gene promoters (Takeya *et al.*, 2018). These sensor cells were used to determine the methanol concentrations on plant leaves and the activity of methanol-producing enzymes.

In contrast to the transcription factors involved in methanol-inducible gene expression, the methanol-sensing machinery and associated signal transduction pathway have not been investigated in detail. Wsc family proteins, which are known to be cell-surface stress sensor proteins, are involved in methanol-sensing machinery in methylotrophic yeasts (Ohsawa *et al.*, 2017). PpWsc1 has two distinct functions, methanol sensing and response to high-temperature stress, and PpWsc3 specifically functions in methanol sensing. As in the case with the C-terminal conserved region of Hap3 in methylotrophic yeasts (Oda *et al.*, 2016), well-conserved Wsc family proteins have evolved to acquire methanol-sensing function in these yeasts.

Future trends

In this chapter, we described the recently obtained information about the physiological functions of methylotrophic yeasts in the phyllosphere. Our methylotrophic yeast-plant system, which includes newly established analytical methods, can combine molecular cellular biology with ecology, and can be used to explore the yeast life cycle at the molecular level in nature. Further studies on yeast methylotrophy in the phyllosphere will facilitate the characterization of new physiological functions and their importance in natural environments. A detailed understanding of the molecular basis of yeast methylotrophy, including gene activation, methanol-sensing machinery, and signal transduction pathways, will also lead to the development of new technology and applications, e.g. efficient heterologous gene expression and enzyme engineering by methanol-sensing technology.

References

Anthony, C., (1982). The Biochemistry of Methylotrophs. (Academic Press, London, UK).

Buschlen, S., Amillet, J.M., Guiard, B., Fournier, A., Marcireau, C., and Bolotin-Fukuhara, M. (2003). The *S. Cerevisiae* HAP complex, a key regulator of mitochondrial function, coordinates nuclear and mitochondrial gene expression. Comp. Funct. Genomics *4*, 37–46. https://doi.org/10.1002/cfg.254

Carlson, M. (1999). Glucose repression in yeast. Curr. Opin. Microbiol. 2, 202–207.

Corpe, W.A., and Basile, D.V. (1982). Methanol-utilizing bacteria associated with green plants. Dev. Ind. Microbiol. 23, 483–493.

Cregg, J.M., Cereghino, J.L., Shi, J., and Higgins, D.R. (2000). Recombinant protein expression in *Pichia pastoris*. Mol. Biotechnol. *16*, 23–52.

Dourado, M.N., Camargo Neves, A.A., Santos, D.S., and Araújo, W.L. (2015). Biotechnological and agronomic potential of endophytic pink-pigmented methylotrophic *Methylobacterium* spp. Biomed Res. Int. *2015*, 909016. https://doi.org/10.1155/2015/909016

Fall, R., and Benson, A.A. (1996). Leaf methanol-the simplest natural product from plants. Trends Plant. Sci. *1*, 296–301.

Galbally, I.E., and Kirstine, W. (2002). The production of methanol by flowering plants and the global cycle of methanol. J. Atmos. Chem. 43, 195–229.

Gellissen, G. (2000). Heterologous protein production in methylotrophic yeasts. Appl. Microbiol. Biotechnol. 54, 741–750.

Guilkbault, G.G., Danielsson, B., Mandenius, C.F., and Mosbach, K. (1983). Enzyme electrode and thermistor probes for determination of alcohols with alcohol oxidase. Anal. Chem. 55, 1582–1585.

Hartner, F.S., and Glieder, A. (2006). Regulation of methanol utilisation pathway genes in yeasts. Microb. Cell Fact. 5, 39.

Horiguchi, H., Yurimoto, H., Goh, T., Nakagawa, T., Kato, N., and Sakai, Y. (2001a). Peroxisomal catalase in the methylotrophic yeast *Candida boidinii*: transport efficiency and metabolic significance. J. Bacteriol. *183*, 6372–6383. https://doi.org/10.1128/JB.183.21.6372-6383.2001

Horiguchi, H., Yurimoto, H., Kato, N., and Sakai, Y. (2001b). Antioxidant system within yeast peroxisome. Biochemical and physiological characterization of CbPmp20 in the methylotrophic yeast *Candida boidinii*. J. Biol. Chem. 276, 14279–14288. https://doi.org/10.1074/jbc.M011661200

Kawaguchi, K., Yurimoto, H., Oku, M., and Sakai, Y. (2011). Yeast methylotrophy and autophagy in a methanol-oscillating environment on growing *Arabidopsis thaliana* leaves. PLOS ONE 6, e25257. https://doi.org/10.1371/journal.pone.0025257

Klein, C.J., Olsson, L., and Nielsen, J. (1998). Glucose control in *Saccharomyces cerevisiae*: the role of Mig1 in metabolic functions. Microbiology *144*, 13–24. https://doi.org/10.1099/00221287-144-1-13

Kolb, S. (2009). Aerobic methanol-oxidizing bacteria in soil. FEMS Microbiol. Lett. *300*, 1–10. https://doi.org/10.1111/j.1574-6968.2009.01681.x

Kurtzman, C.P. (2005). Description of *Komagataella phaffii* sp. nov. and the transfer of *Pichia pseudopastoris* to the methylotrophic yeast genus *Komagataella*. Int. J. Syst. Evol. Microbiol. 55, 973–976.

Laothawornkitkul, J., Taylor, J.E., Paul, N.D., and Hewitt, C.N. (2009). Biogenic volatile organic compounds in the Earth system. New Phytol. *183*, 27–51. https://doi.org/10.1111/j.1469-8137.2009.02859.x

Leao-Helder, A.N., Krikken, A.M., van der Klei, I.J., Kiel, J.A., and Veenhuis, M. (2003). Transcriptional down-regulation of peroxisome numbers affects selective peroxisome degradation in *Hansenula polymorpha*. J. Biol. Chem. *278*, 40749–40756. https://doi.org/10.1074/jbc.M304029200

Lee, B., Yurimoto, H., Sakai, Y., and Kato, N. (2002). Physiological role of the glutathione-dependent formaldehyde dehydrogenase in the methylotrophic yeast *Candida boidinii*. Microbiology *148*, 2697–2704.

Limtong, S., Srisuk, N., Yongmanitchai, W., Yurimoto, H., and Nakase, T. (2008). *Ogataea chonburiensis* sp. nov. and *Ogataea nakhonphanomensis* sp. nov., two thermotolerant, methylotrophic yeast species isolated in Thailand and the transfer of *Pichia siamensis* and *Pichia thermomethanolica* to the genus *Ogataea*. Int. J. Syst. Evol. Microbiol. 58, 302–307.

Lin-Cereghino, G.P., Godfrey, L., de la Cruz, B.J., Johnson, S., Khuongsathiene, S., Tolstorukov, I., Yan, M., Lin-Cereghino, J., Veenhuis, M., Subramani, S., *et al.* (2006). Mxr1p, a key regulator of the methanol utilization pathway and peroxisomal genes in *Pichia pastoris*. Mol. Cell. Biol. *26*, 883–897.

McNabb, D.S., and Pinto, I. (2005). Assembly of the Hap2p/Hap3p/Hap4p/Hap5p-DNA complex in *Saccharomyces* cerevisiae. Eukaryotic Cell *4*, 1829–1839.

Nemecek-Marshall, M., MacDonald, R.C., Franzen, J.J., Wojciechowski, C.L., and Fall, R. (1995). Methanol emission from leaves (enzymatic detection of gas-phase methanol and relation of methanol fluxes to stomatal conductance and leaf development). Plant Physiol. *108*, 1359–1368.

Oda, S., Yurimoto, H., Nitta, N., Sasano, Y., and Sakai, Y. (2015). Molecular characterization of hap complex components responsible for methanol-inducible gene expression in the methylotrophic yeast *Candida boidinii*. Eukaryotic Cell *14*, 278–285. https://doi.org/10.1128/EC.00285-14

Oda, S., Yurimoto, H., Nitta, N., and Sakai, Y. (2016). Unique C-terminal region of Hap3 is required for methanol-regulated gene expression in the methylotrophic yeast *Candida boidinii*. Microbiology *162*, 898–907. https://doi.org/10.1099/mic.0.000275

Ogata, K., Nishikawa, H., and Ohsugi, M. (1969). A yeast capable of utilizing methanol. Agric. Biol. Chem. *33*, 1519–1520.

Ohsawa, S., Yurimoto, H., and Sakai, Y. (2017). Novel function of Wsc proteins as a methanol-sensing machinery in the yeast Pichia pastoris. Mol. Microbiol. *104*, 349–363. https://doi.org/10.1111/mmi.13631

Oku, M., and Sakai, Y. (2010). Peroxisomes as dynamic organelles: autophagic degradation. FEBS J. *277*, 3289–3294. https://doi.org/10.1111/j.1742-4658.2010.07741.x

Olah, G.A. (2005). Beyond oil and gas: the methanol economy. Angew. Chem. Int. Ed. Engl. *44*, 2636–2639. https://doi.org/10.1002/anie.200462121

Olah, G.A., Goeppert, A., and Prakash, G.K.S., (2006). Beyond Oil and Gas: The Methanol Economy. (Wiley-VCH, Weinheim, Germany).

Parua, P.K., Ryan, P.M., Trang, K., and Young, E.T. (2012). *Pichia pastoris* 14-3-3 regulates transcriptional activity of the methanol inducible transcription factor Mxr1 by direct interaction. Mol. Microbiol. *85*, 282–298. https://doi.org/10.1111/j.1365-2958.2012.08112.x

Péter, G., Dlauchy, D., Tornai-Lehoczki, J., and Kurtzman, C.P. (2005). *Kuraishia molischiana* sp. nov., the teleomorph of *Candida molischiana*. Antonie Van Leeuwenhoek *88*, 241–247. https://doi.org/10.1007/s10482-005-7267-3

Péter, G., Tornai-Lehoczki, J., and Dlauchy, D. (2009). *Ogataea populialbae* sp. nov., a yeast species from white poplar. FEMS Yeast Res. *9*, 936–941. https://doi.org/10.1111/j.1567-1364.2009.00547.x

Rußmayer, H., Buchetics, M., Gruber, C., Valli, M., Grillitsch, K., Modarres, G., Guerrasio, R., Klavins, K., Neubauer, S., Drexler, H., *et al.* (2015). Systems-level organization of yeast methylotrophic lifestyle. BMC Biol. *13*, 80. https://doi.org/10.1186/s12915-015-0186-5

Sahu, U., Krishna Rao, K., and Rangarajan, P.N. (2014). Trm1p, a $Zn(II)_2Cys_6$-type transcription factor, is essential for the transcriptional activation of genes of methanol utilization pathway, in *Pichia pastoris*. Biochem. Biophys. Res. Commun. *451*, 158–164.

Sakai, Y., Murdanoto, A.P., Konishi, T., Iwamatsu, A., and Kato, N. (1997). Regulation of the formate dehydrogenase gene, *FDH1*, in the methylotrophic yeast *Candida boidinii* and growth characteristics of an *FDH1*-disrupted strain on methanol, methylamine, and choline. J. Bacteriol. *179*, 4480–4485.

Sasano, Y., Yurimoto, H., Yanaka, M., and Sakai, Y. (2008). Trm1p, a $Zn(II)_2Cys_6$-type transcription factor, is a master regulator of methanol-specific gene activation in the methylotrophic yeast *Candida boidinii*. Eukaryotic Cell 7, 527–536. https://doi.org/10.1128/EC.00403-07

Sasano, Y., Yurimoto, H., Kuriyama, M., and Sakai, Y. (2010). Trm2p-dependent derepression is essential for methanol-specific gene activation in the methylotrophic yeast *Candida boidinii*. FEMS Yeast Res. *10*, 535–544. https://doi.org/10.1111/j.1567-1364.2010.00640.x

Schüller, H.J. (2003). Transcriptional control of nonfermentative metabolism in the yeast *Saccharomyces cerevisiae*. Curr. Genet. *43*, 139–160. https://doi.org/10.1007/s00294-003-0381-8

Shiraishi, K., Oku, M., Kawaguchi, K., Uchida, D., Yurimoto, H., and Sakai, Y. (2015). Yeast nitrogen utilization in the phyllosphere during plant lifespan under regulation of autophagy. Sci. Rep. *5*, 9719. https://doi.org/10.1038/srep09719

Simon, M., Adam, G., Rapatz, W., Spevak, W., and Ruis, H. (1991). The *Saccharomyces cerevisiae ADR1* gene is a positive regulator of transcription of genes encoding peroxisomal proteins. Mol. Cell. Biol. *11*, 699–704.

Stasyk, O.G., van Zutphen, T., Ah Kang, H., Stasyk, O.V., Veenhuis, M., and Sibirny, A.A. (2007). The role of *Hansenula polymorpha MIG1* homologues in catabolite repression and pexophagy. FEMS Yeast Res. *7*, 1103–1113.

Suh, S.O., Blackwell, M., Kurtzman, C.P., and Lachance, M.A. (2006). Phylogenetics of *Saccharomycetales*, the ascomycete yeasts. Mycologia *98*, 1006–1017.

Takeya, T., Yurimoto, H., and Sakai, Y. (2018). A *Pichia pastoris* single-cell biosensor for detection of enzymatically produced methanol. Appl. Microbiol. Biotechnol. *102*, 7017–7027. https://doi.org/10.1007/s00253-018-9144-9

Tani, Y., (1984). Microbiology and biochemistry of methylotrophic yeasts. In Methylotrophs: Microbiology, Biochemistry, and Genetics, Hou, C.T., ed. (CRC Press Inc., Boca Raton), pp. 55–86.

Turcotte, B., Liang, X.B., Robert, F., and Soontorngun, N. (2010). Transcriptional regulation of nonfermentable carbon utilization in budding yeast. FEMS Yeast Res. *10*, 2–13. https://doi.org/10.1111/j.1567-1364.2009.00555.x

van der Klei, I.J., Yurimoto, H., Sakai, Y., and Veenhuis, M. (2006). The significance of peroxisomes in methanol metabolism in methylotrophic yeast. Biochim. Biophys. Acta *1763*, 1453–1462.

van Zutphen, T., Baerends, R.J., Susanna, K.A., de Jong, A., Kuipers, O.P., Veenhuis, M., and van der Klei, I.J. (2010). Adaptation of *Hansenula polymorpha* to methanol: a transcriptome analysis. BMC Genomics *11*, 1. https://doi.org/10.1186/1471-2164-11-1

Veenhuis, M., Van Dijken, J.P., and Harder, W. (1983). The significance of peroxisomes in the metabolism of one-carbon compounds in yeasts. Adv. Microb. Physiol. *24*, 1–82.

Vorholt, J.A. (2012). Microbial life in the phyllosphere. Nat. Rev. Microbiol. *10*, 828–840. https://doi.org/10.1038/nrmicro2910

Wang, X., Wang, Q., Wang, J., Bai, P., Shi, L., Shen, W., Zhou, M., Zhou, X., Zhang, Y., and Cai, M. (2016). Mit1 transcription factor mediates methanol signaling and regulates the alcohol oxidase 1 (*AOX1*) promoter in *Pichia pastoris*. J. Biol. Chem. *291*, 6245–6261. https://doi.org/10.1074/jbc.M115.692053

Wen, G., Wen, X., Shuang, S., and Choi, M.M.F. (2014). Whole-cell biosensor for determination of methanol. Sensors Actuators B Chem. *201*, 586–591.

Yurimoto, H. (2009). Molecular basis of methanol-inducible gene expression and its application in the methylotrophic yeast *Candida boidinii*. Biosci. Biotechnol. Biochem. *73*, 793–800.

Yurimoto, H., Komeda, T., Lim, C.R., Nakagawa, T., Kondo, K., Kato, N., and Sakai, Y. (2000). Regulation and evaluation of five methanol-inducible promoters in the methylotrophic yeast *Candida boidinii*. Biochim. Biophys. Acta *1493*, 56–63.

Yurimoto, H., Sakai, Y., and Kato, N., (2002). Methanol metabolism. In *Hansenula polymorpha* – Biology and Applications-. Gellissen, G., ed. (Wiley-VCH, Weinheim, Germany), pp. 61–75.

Yurimoto, H., Lee, B., Yano, T., Sakai, Y., and Kato, N. (2003). Physiological role of *S*-formylglutathione hydrolase in C1 metabolism of the methylotrophic yeast *Candida boidinii*. Microbiology *149*, 1971–1979.

Yurimoto, H., Kato, N., and Sakai, Y. (2005). Assimilation, dissimilation, and detoxification of formaldehyde, a central metabolic intermediate of methylotrophic metabolism. Chem. Rec. *5*, 367–375. https://doi.org/10.1002/tcr.20056

Yurimoto, H., Oku, M., and Sakai, Y. (2011). Yeast methylotrophy: metabolism, gene regulation and peroxisome homeostasis. Int. J. Microbiol. *2011*, 101298. https://doi.org/10.1155/2011/101298

Zhai, Z., Yurimoto, H., and Sakai, Y. (2012). Molecular characterization of *Candida boidinii MIG1* and its role in the regulation of methanol-inducible gene expression. Yeast *29*, 293–301. https://doi.org/10.1002/yea.2909

Specialized Metabolites from Methylotrophic Proteobacteria

12

Aaron W. Puri*

Department of Chemistry and the Henry Eyring Center for Cell and Genome Science, University of Utah, Salt Lake City, UT, USA.

*Correspondence: a.puri@utah.edu

https://doi.org/10.21775/9781912530045.12

Abstract

Biosynthesized small molecules known as specialized metabolites often have valuable applications in fields such as medicine and agriculture. Consequently, there is always a demand for novel specialized metabolites and an understanding of their bioactivity. Methylotrophs are an underexplored metabolic group of bacteria that have several growth features that make them enticing in terms of specialized metabolite discovery, characterization, and production from cheap feedstocks such as methanol and methane gas. This chapter will examine the predicted biosynthetic potential of these organisms and review some of the specialized metabolites they produce that have been characterized so far.

Introduction

Specialized metabolites, also known as secondary metabolites or natural products, form the basis of many bioactive compounds essential to modern medicine and agriculture (Demain and Sanchez, 2009; Cantrell *et al.*, 2012; Newman and Cragg, 2016). This makes intuitive sense because production of these compounds is genetically encoded in biosynthetic gene clusters (BGCs) and therefore the resulting structures have had millennia to evolve their roles. The great value of these molecules in manipulating biological systems means there is a constant demand for new sources of these compounds and strategies for determining their biological functions.

The explosion in bacterial genome sequences available in public databases as well as the availability of bioinformatics tools for analysing them has revealed that many bacterial species are potentially untapped sources for new molecules (Cimermancic *et al.*, 2014). This includes organisms beyond those traditionally relied upon for natural product discovery, and recent studies have shown that examining the biosynthetic potential of new species indeed reveals new classes of compounds (Pidot *et al.*, 2014; Pye *et al.*, 2017). This strategy is complementary to synthetic biology approaches focused on activating BGCs that are not normally expressed under laboratory conditions in strains traditionally used for natural product discovery, such as *Streptomyces* (Rutledge and Challis, 2015).

This chapter will focus on the methylotrophic Proteobacteria, which use reduced carbon compounds with no carbon–carbon bonds as their sole sources of carbon and energy. Specifically, the Proteobacteria are some of the most well-studied organisms within this metabolic group (Chistoserdova *et al.*, 2009). While several molecules have been isolated from methylotrophic Proteobacteria that have had an impact on the field of natural products (Kenney and Rosenzweig, 2018a; Khmelenina *et al.*, 2015), our understanding of the specialized metabolism of these organisms is still in its infancy.

Methylotrophic bacteria as an underexplored resource for specialized metabolite discovery

Methylotrophic bacteria are an attractive source for the discovery of new specialized metabolites for several reasons.

1. Obligate methylotrophs, including many species of methane-oxidizing bacteria, may have been overlooked during traditional 'grind and find' searches for new compounds that focused on the isolation of specific genera like *Streptomyces* using rich media.
2. The lack of complex carbon substrate in methylotroph growth medium can aid in the rapid analysis of new compounds produced by a culture without laborious prefractionation procedures, which may also help in high throughput screening approaches.
3. The chemically defined nature of the growth medium and simplicity of the carbon source also allows substrates with heavy labels such as $^{15}NO_3$ and $^{13}CH_4$ or $^{13}CH_3OH$ to be used for minimal cost, which can aid in structural elucidation using mass spectrometry and NMR.
4. If a molecule with attractive properties is identified, its methylotroph producer can be grown at scale using a feedstock such as methane or methanol, both of which have recently re-emerged as substrates of interest in industrial microbiology (Schrader *et al.*, 2009; Henard and Guarnieri, 2018).

It is therefore worthwhile to discover more molecules produced by this ecologically important group of bacteria.

Genome sequencing efforts supported by entities including the Joint Genome Institute (JGI) and Organization for Methanotroph Genome Analysis (OMeGA) have enabled a detailed examination of the predicted biosynthetic potential of many methylotrophic species (Fig. 12.1). Amongst the strains analysed here, more BGCs are generally predicted in methanotroph genomes compared with the genomes of non-methanotrophic methylotrophs. Overall the Alphaproteobacteria, including members of the *Methylobacterium* genus among the non-methanotrophs, appear to possess a large number of predicted BGCs. The *Methylobacterium* genus was also highlighted for its substantial biosynthetic potential in a past analysis of bacterial genomes (Cimermancic *et al.*, 2014). Many Alphaproteobacteria including methylotrophic species rely on the ethylmalonyl-CoA pathway to assimilate acetyl-CoA into central metabolism, and this pathway involves high flux through CoA-linked intermediates that are also often used in the synthesis of specialized metabolites such as polyketides (Alber, 2011). This makes the exploration of the specialized metabolism of these organisms particularly intriguing.

Bacteriocin and terpene BGCs were most commonly predicted within the categories classified by the antibiotic and secondary metabolite analysis shell (antiSMASH) (Blin *et al.*, 2017). However, the majority of predicted clusters were detected using ClusterFinder, which relies on more probabilistic (and less stringent) methods for BGC identification (Cimermancic *et al.*, 2014). This could suggest that methylotrophs produce different compounds than those made by traditionally studied organisms, which would be promising for the discovery of new classes of compounds. However, sequence homology-based comparisons to characterized BGCs and experimental verification will be essential to exploring the hypothesis that these gene clusters are producing novel products.

Intercellular chemical communication

Quorum sensing

Quorum sensing (QS) allows bacteria to control gene expression in a cell density-dependent manner. In Proteobacteria, a canonical QS signal is the acyl-homoserine lactone (acyl-HSL), which can vary in its acyl chain length, saturation, and oxidation. These signals are produced by LuxI-family acyl-HSL synthases, and detected by LuxR-family receptor/transcription factors. For reviews please see Papenfort and Bassler (2016) as well as Whiteley *et al.* (2017).

The first QS system to be characterized in a methylotroph was in the model strain *Methylobacterium extorquens* AM1 (Penalver *et al.*, 2006). Researchers found that AM1 possesses two LuxI-family synthases, termed MlaI and MsaI. While MsaI produces the short chain acyl-HSLs C_6- and C_8-HSL, MlaI produces the more unusual unsaturated long

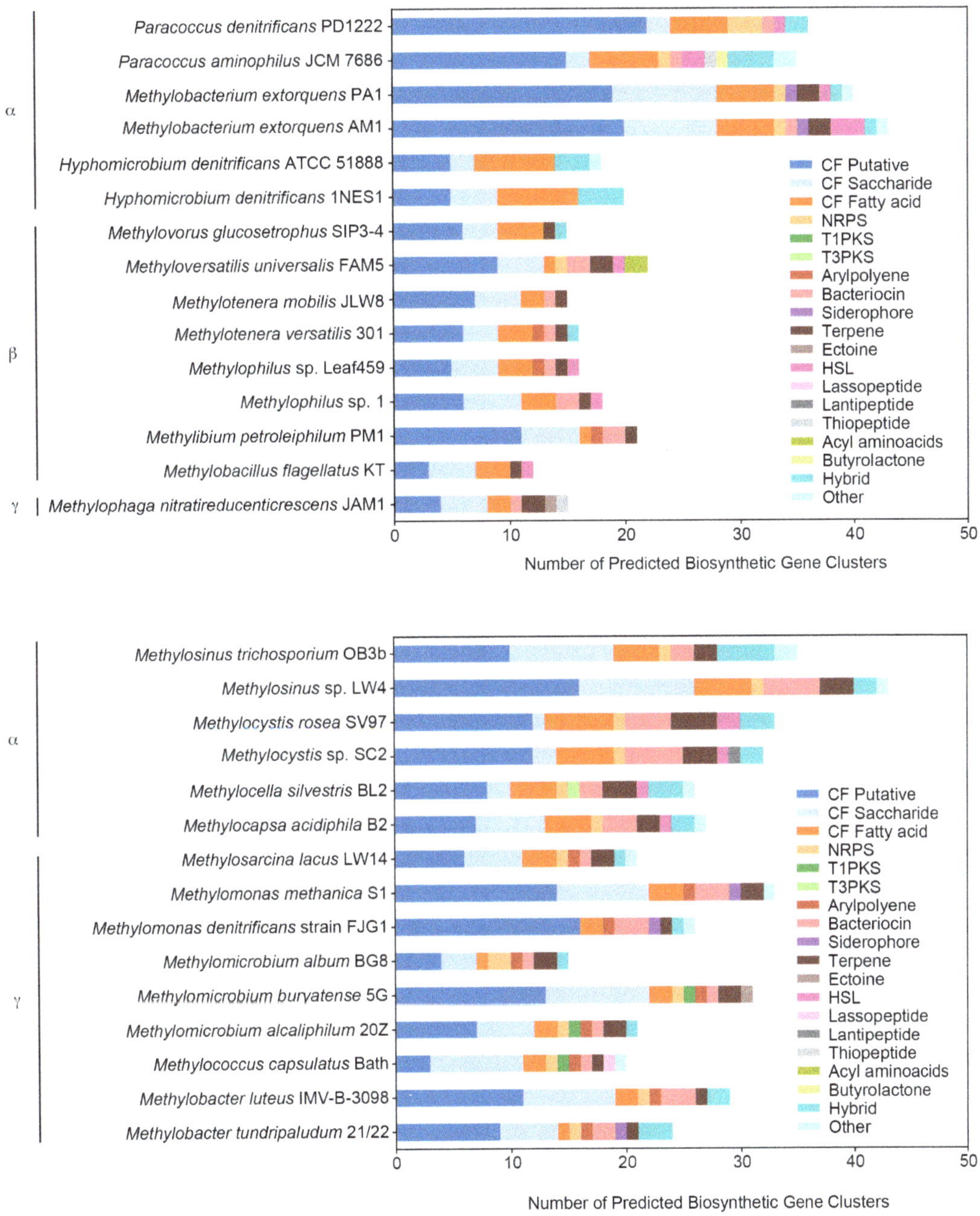

Figure 12.1 Predicted biosynthetic gene clusters in genomes of methylotrophic Proteobacteria. Predictions were made using antiSMASH 4.1.0 (Blin *et al*., 2017) with ClusterFinder (CF) (Cimermancic *et al*., 2014) enabled with default settings. HSL, homoserine lactone; NRPS, non-ribosomal peptide synthetase; PKS, polyketide synthase.

chain molecules 7*Z*-C_{14}-HSL and 2*E*,7*Z*-C_{14}-HSL only when AM1 is grown on methanol (Table 12.1). The short chain acyl-HSLs produced by *M. extorquens* AM1 increase the production of exopolysaccharide (Penalver *et al.*, 2006), and therefore could have a role in biofilm formation. In support of this, recently a highly homologous QS system in a *Methylobacterium populi* strain was found to regulate the structure and adherence of biofilms made by this organism (Morohoshi *et al.*, 2018).

Many *Methylobacterium* species have been found to produce acyl-HSLs as detected by bioassays (Poonguzhali *et al.*, 2007). In one study researchers found that the orange tree symbiont *Methylobacterium mesophilicum* SR 1.6/6 produces several signals, including novel variants (Table 12.1) (Pomini *et al.*, 2009). There is some evidence that increased concentrations of these molecules may regulate the transcription of metabolic genes in this bacterium (Dourado *et al.*, 2013), however the precise biological role of these QS signals will require further investigation.

A QS system was recently discovered and characterized in the methane-oxidizing Gammaproteobacterium *Methylobacter tundripaludum* 21/22, which was isolated from lake sediment (Table 12.1) (Puri *et al.*, 2017). This bacterium produces and responds to the signal 3-OH-C_{10}-HSL, which activates expression of a BGC co-located with the QS genes in *M. tundripaludum*. The same group of researchers subsequently identified the specialized metabolite product of this BGC, named tundrenone (see below) (Puri *et al.*, 2018). The fact that tundrenone is produced in a QS-dependent manner fits the paradigm that QS systems often activate the production of extracellular factors such as antibiotics (McGowan *et al.*, 1995; Duerkop *et al.*, 2009) and proteases (Pearson *et al.*, 1997), thereby allowing bacteria to affect their surrounding environment at high cell density.

It will be exciting to characterize more methylotroph QS systems in the future and determine their biological roles. It should be noted that methylotrophs occupy diverse ecological niches, including both acidic (Dedysh *et al.*, 2000) and alkaline (Khmelenina *et al.*, 1997) environments. The HSL ring is subject to base-catalysed hydrolysis (Schaefer *et al.*, 2000; Dong *et al.*, 2001), which may provide one explanation for why methylotrophs isolated from alkaline environments such as some *Methylomicrobium* species do not possess acyl-HSL-based quorum sensing systems in their genomes (Fig. 12.1).

Interspecies chemical communication

Specialized metabolites biosynthesized by methylotrophs have also been reported to be involved in interspecies interactions. The copper-binding small molecule methanobactin (see below) is produced by many methanotrophs, but its structure varies between species (Kenney and Rosenzweig, 2018a). It was reported that methanobactin produced by *Methylocystis* sp. str. SB2 may regulate the

Table 12.1 Acyl-homoserine lactone quorum sensing signals produced by methylotrophic Proteobacteria

Producing strain	Acyl-homoserine lactone signal	Reference
Methylobacterium extorquens AM1	C_6-HSL C_8-HSL 7Z-C_{14}-HSL 2E,7Z-C_{14}-HSL	Penalver *et al*. (2006)
Methylobacterium mesophilicum SR 1.6/6	C_{12}-HSL C_{13}-HSL C_{14}-HSL 2E-C_{12}-HSL 7Z-C_{14}-HSL 2E,7Z-C_{14}-HSL	Pomini *et al*. (2009)
Methylobacter tundripaludum 21/22	3-OH-C_{10}-HSL	Puri *et al*. (2017)
Methylobacterium populi P-1M	3-OH-C_{14}-HSL $C_{14:1}$-HSL[a]	Morohoshi *et al*. (2018)

Note: While the full stereochemistry has not been determined for all molecules, all biologically produced acyl-homoserine lactone (acyl-HSL) signals where stereochemistry has been determined possess an (*S*)-*N*- configuration.
[a]Stereochemistry of unsaturated bond not determined.

expression of genes related to methane oxidation in *Methylosinus trichosporium* OB3b via mechanisms beyond simply altering copper availability (Farhan Ul-Haque *et al.*, 2015). The ability of methanotrophs to interact with methanobactin produced by other species points to piracy between these organisms and suggests a fierce competition for copper in the environment (El Ghazouani *et al.*, 2012; Dassama *et al.*, 2016; DiSpirito *et al.*, 2016).

The presence of volatile specialized metabolites in interactions between methanotrophs and non-methylotrophic heterotrophs has also been surveyed (Veraart *et al.*, 2018). The methanotroph *Methylobacter luteus* was reported to produce the volatile bicyclic terpenoids cadinene and alpha-murolene in response to the presence of the heterotroph *Pseudomonas mandelli*. The role of these compounds in interactions between metabolically linked methanotrophs and non-methanotrophic species will require further investigation.

Examples of molecules produced by methylotrophs

Methanobactin

The particulate methane monooxygenase (pMMO) is found in almost all methane-oxidizing bacteria, and it requires copper to catalyse the oxidation of methane to methanol (Balasubramanian *et al.*, 2010). Consequently, methanotrophs have a great need for copper and use specialized mechanisms for its acquisition (Semrau *et al.*, 2010; Kenney and Rosenzweig, 2018b). Several methanotrophs that are members of the Alphaproteobacteria secrete a ribosomally produced and post-translationally modified peptide (RiPP) that tightly binds copper termed methanobactin (Table 12.2) (Kim *et al.*, 2004). Because of its relation to iron binding siderophores, methanobactin is known as a chalkophore. Methanobactin is perhaps the most well known of all characterized specialized metabolites produced by methylotrophs. For more comprehensive reviews of methanobactin please see DiSpirito *et al.* (2016) as well as Kenney and Rosenzweig (2018a).

Although the exact structure of methanobactin varies between species, all isolated versions coordinate copper using oxazolone rings or heterocycles with similar functionality and neighbouring enethiol/thioamide groups (Kenney and Rosenzweig, 2018a). The formation of the oxazole ring and thioamide in methanobactin was recently characterized, and involves the protein complex MbnBC, which contains domains of previously unrecognized function (Kenney *et al.*, 2018). Understanding the molecular mechanisms of methanobactin biosynthesis may ultimately enable the production of methanobactin analogues with varying properties.

Applications of methanobactin include its use as a chelation therapy for copper toxicity such as in Wilson disease (Lichtmannegger *et al.*, 2016). Methanobactin from *M. trichosporium* OB3b also binds other metals including gold with high affinity (Choi *et al.*, 2006), making it useful for biomining as well as nanoparticle preparations (DiSpirito *et al.*, 2016). This versatile compound also has growth inhibitory activity against a range of bacteria (DiSpirito *et al.*, 2007), which is a less discussed property of methanobactin but which may play a role in shaping the composition of methane-oxidizing bacterial communities in nature. Methanobactin production is activated by low copper availability, making it similar to other specialized metabolites that have tightly regulated production (Kim *et al.*, 2004; Kenney *et al.*, 2016; Gu and Semrau, 2017). Genome mining has also identified methanobactin BGCs in other non-methyltrophic species (Kenney and Rosenzweig, 2013), and it will be interesting to see what role methanobactins play in these bacteria.

Toblerols

Modular, non-iterative polyketide synthases (PKSs) follow an assembly line logic that often makes it possible to predict the structure of the metabolite product (Till and Race, 2016). The products of *trans*-acyltransferase (*trans*-AT) PKSs can be more difficult to predict, however detailed studies of the genes in these BGCs has resulted in a set of product prediction rules for these modular synthases as well (Nguyen *et al.*, 2008). Recently, researchers investigated an unusual *trans*-AT PKS gene cluster in *M. extorquens* AM1 that could not be classified using previously developed approaches (Ueoka *et al.*, 2017). The products of the cluster were found to be a family of novel molecules containing cyclopropanol moieties, and these compounds were subsequently named the toblerols (Table 12.2). The authors discovered that toblerol C suppresses the production of an additional factor

Table 12.2 Selected specialized metabolites produced by methylotrophic Proteobacteria

Compound name	Structure	Examples of producing strain(s)	Reference
4,4′-Diapolycopene-4,4′-dioic acid	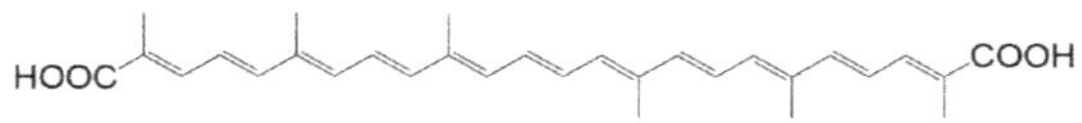	*Methylobacterium rhodium*, *Methylomonas* sp. str. 16a	Kleinig *et al.*, (1979), Tao *et al.* (2005)
Ectoine	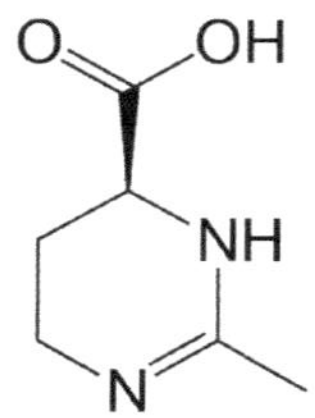	*Methylomicrobium alcaliphilum* 20Z	Reshetnikov *et al.* (2006)
Gibberellic acid (GA_3)	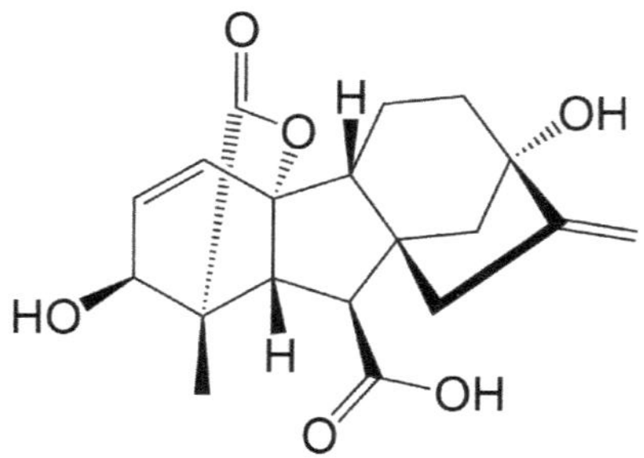	*Methylobacillus arboreus* Iva^T	Agafonova *et al.* (2018)
Indole-3-acetic acid (IAA)	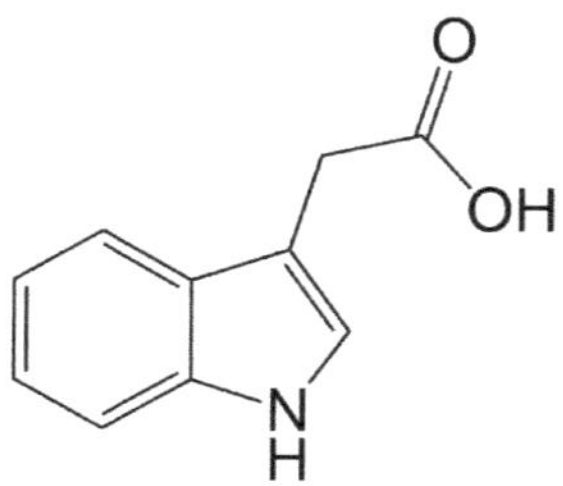	*Methylobacterium mesophilicum* DSM 1708	Ivanova *et al.* (2001)

Table 12.2 Continued

Compound name	Structure	Examples of producing strain(s)	Reference
Methanobactin (from OB3b)	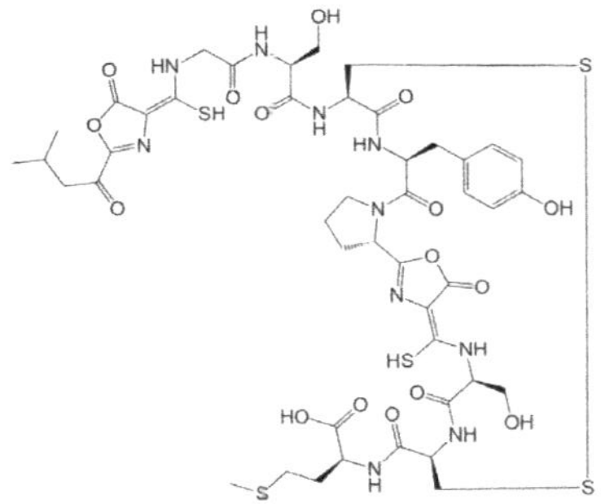	*Methylosinus trichosporium* OB3b (structural variants produced by other species)	Kim *et al.* (2004), Behling *et al.* (2008)
Toblerol C	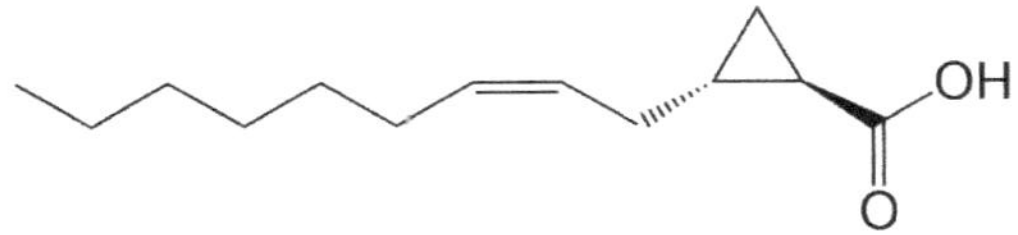	*Methylobacterium extorquens* AM1	Uekoa *et al*. (2017)
Tundrenone	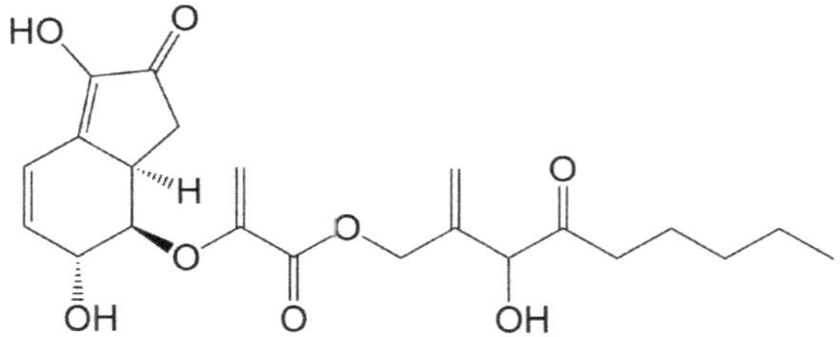	*Methylobacter tundripaludum* 21/22	Puri *et al.* (2018)
trans-Zeatin-riboside	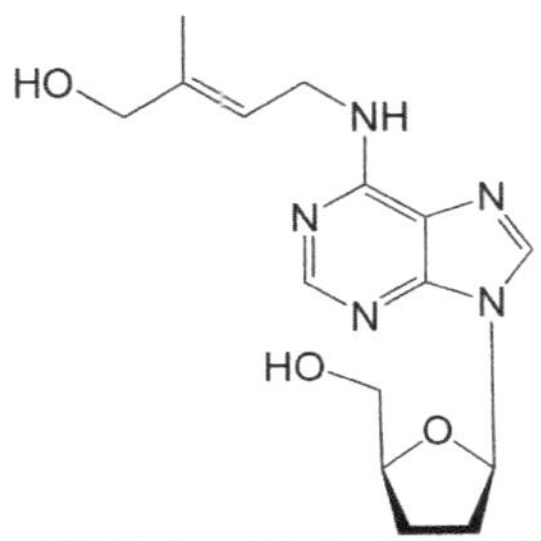	*Methybacterium mesophilicum* DSM 1708, *Methylovorus mays* VKM B-2221, *Methylobacterium extorquens* DSM 1337	Ivanova *et al*. (2000), Koenig *et al*. (2002)

that inhibits the growth of other *Methylobacterium* isolates. The actual factor that causes growth inhibition is currently under investigation, but the results so far point to an intriguing hierarchy of bioactive small molecule production and regulation in these bacteria. This study highlights an example where a methylotroph contains a non-canonical BGC that produces novel specialized metabolites, demonstrating the utility of methylotrophs as an underexplored source of these molecules.

Tundrenone

Tundrenone is a novel specialized metabolite produced by the methanotroph *M. tundripaludum* in a quorum sensing-dependent manner (Puri *et al.*, 2017, 2018). Tundrenone contains a tetrahydroindenone core predicted to be derived from the primary metabolite chorismate, as well as a lipid tail with unusual modifications (Table 12.2). These notable structural features suggest that an understanding of tundrenone's biosynthesis may uncover new enzymatic reactions that are useful for understanding and engineering the biosynthesis of other small molecules. The biological function of tundrenone is currently unknown, however the BGC responsible for tundrenone production is conserved in *M. tundripaludum* isolates from across the globe (Puri *et al.*, 2017), suggesting that this molecule may have an ecologically significant role for this organism.

Carotenoids

Many methylotrophs are pigmented, as the responsible carotenoids can provide antioxidant effects including protecting their hosts from UV damage (Birben *et al.*, 2012; Vorholt, 2012). This fact also has led to the commonly used descriptive grouping of pink pigmented facultative methylotrophs (PPFMs), which includes many species of the *Methylobacterium* genus that are found in association with plants (see below). *Methylobacterium rhodium* (formerly *Pseudomonas rhodos*) produces C_{30}-carotenoids including 4,4′-diapolycopene-4,4′-dioic acid diesters that contain conjugated sugars (Table 12.2) (Kleinig *et al.*, 1979). These unusual molecules are not known to be produced by many bacterial species, but they were also found to be produced by methane-oxidizing bacteria of the genus *Methylomonas* (Tao *et al.*, 2005). An oxidase in strain 16a termed CrtNb has been identified and characterized that functionalizes the termini of the precursor molecule 4,4′-diapolycopene with aldehyde moieties (Tao *et al.*, 2005). This enzyme may be useful for future modification strategies for these pigments to be employed in semisynthetic processes.

Production of the more traditional C_{40} carotenoids has also been studied in methylotrophs. *Methylobacterium* species are known to produce C_{40} carotenoids such as oscillaxanthin (Van Dien *et al.*, 2003; Konovalova *et al.*, 2007). Additionally, *Methylomonas* sp. str. 16a has been used as a chassis for the production of the C_{40} carotenoid astaxanthin (Ye *et al.*, 2007). An astaxanthin-producing methanotroph could be used in aquaculture as a single cell protein source that simultaneously enhances the pigmentation of the farmed animals.

Ectoine

Ectoine is a cyclic imino acid that functions as an osmoprotectant in many species of halotolerant and halophilic bacteria, including several methylotrophic species found in ecosystems with high salinity (Table 12.2) (Galinski *et al.*, 1985; Reshetnikov *et al.*, 2011). The biosynthesis of ectoine has been studied in the halotolerant methane-oxidizing bacterium *Methylomicrobium alcaliphilum* 20Z (Reshetnikov *et al.*, 2005, 2006), which was isolated from a soda lake in Russia (Khmelenina *et al.*, 1997; Kalyuzhnaya *et al.*, 2008). Ectoine is produced from aspartic acid by the products of the *ectABC* biosynthetic gene cluster, and its production markedly increases under osmotic stress (Reshetnikov *et al.*, 2011). As with many specialized metabolites, the biosynthesis of ectoine is transcriptionally regulated. The *ectABC* BGC also contains a co-located MarR-type transcriptional repressor, which is partially responsible for regulation of *ectABC* expression at different salt concentrations (Mustakhimov *et al.*, 2010). Production of ectoine is of commercial interest because this compatible solute is valuable for the preservation of biologics including those used in cosmetics and medicine, and currently ectoine is produced biosynthetically for commercial means (Graf *et al.*, 2008; Kunte *et al.*, 2014). Consequently there is interest in production of this specialized metabolite from methane gas, a cheap and abundant feedstock for industrial biotechnology.

Phytohormones

Plant stomata release methanol during growth as a by-product of pectin demethylation (Nemecek-Marshall *et al.*, 1995; Galbally and Kirstine, 2002), and this methanol is in turn catabolized by resident methylotrophs that occupy the phyllosphere (Vorholt, 2012). In addition to this metabolic link, the relationship between plants and methylotrophs is thought to be mutualistic, with methylotrophs providing phytohormones to their plant hosts (Holland *et al.*, 2002; Kutschera, 2007; Fedorov *et al.*, 2011).

In support of the close link between plants and methylotrophs, *Methylobacterium* species have been found to promote the growth of plants, and the molecular details of these interactions have been investigated. For example, strains of *M. mesophilicum* and *Methylovorus mays* were found to produce substances that promote the growth of *Amarantus caudatus* seedlings, including the cytokinin zeatin (Ivanova *et al.*, 2000). *trans*-Zeatin biosynthesis in *Methylobacterium* spp. was later found to involve tRNA precursors (Koenig *et al.*, 2002); however, removal of *trans*-zeatin alone from bacterial cultures had no effect on soybean seed germination stimulation.

Methylotrophs have also been reported to synthesize additional phytohormones, including auxins and gibberellins. Researchers found that *M. mesophilicum* and *Aminobacter aminovorans* strains produce the canonical auxin indole-3-acetic acid (IAA) (Ivanova *et al.*, 2001), which exerts major influences on plant growth and development (Duca *et al.*, 2014). Furthermore, a *Methylobacterium oryzae* strain (Siddikee *et al.*, 2010) as well as the obligate methylotroph *Methylobacillus arboreus* (Agafonova *et al.*, 2018) have been reported to produce bioactive gibberellic acid GA_3, and GA_3 from the latter strain stimulated the sprouting of lettuce seeds.

These molecules point to a complex chemical conversation between methylotrophs and their plant hosts, as plants are canonical producers of natural products.

Conclusions

The specialized metabolites that have been isolated and characterized to date from methylotrophic Proteobacteria have led to the discovery of novel structures, molecular functions, and enzymatic transformations. These efforts also underscore the continued utility of isolating bacterial strains and characterizing the compounds they produce naturally. While culturing these organisms may involve specialized knowledge or equipment, doing so can aid in the discovery of a metabolite's biological activity due to biological context. Isolating naturally produced compounds can also add some certainty that a molecule of interest is the actual product of the native pathway, which can be an issue for discovery efforts based on heterologous expression. The predicted biosynthetic potential of methylotrophic Proteobacteria, along with several practical advantages based on the growth conditions of these bacteria, make this an exciting time to investigate the specialized metabolism of these organisms.

Future directions

It has become increasingly clear that many bacterial BGCs are not expressed in axenic cultures grown under laboratory conditions (Milshteyn *et al.*, 2014; Rutledge and Challis, 2015). Consequently, biological context is necessary for both the production of many specialized metabolites and an understanding of their bioactivity. Methylotrophic bacteria occupy a diversity of niches, and one particularly intriguing future direction for the study of specialized metabolites produced by these organisms is in the context of microbe-microbe and microbe–host interactions. Examples include plant–microbe interactions, where recently a *Methylophilus* strain was found to produce factor(s) that inhibit the growth of several other phyllosphere constituents (Helfrich *et al.*, 2018). Methylotrophs are also symbionts of marine animals (Cavanaugh *et al.*, 1987; Petersen and Dubilier, 2009), and a sponge symbiont with a large biosynthetic repertoire was recently predicted to have the ability to catabolize methanol (Lackner *et al.*, 2017). Finally, the metabolic relationships between methane-oxidizing bacteria and the non-methanotrophic heterotrophs they support in the environment may provide context for the discovery of new molecules (Oshkin *et al.*, 2014; Krause *et al.*, 2017; Veraart *et al.*, 2018). It will be exciting to see the developments in each of these areas in the years to come.

Web resources

- Antibiotics and secondary metabolite analysis shell (antiSMASH): https://antismash.secondarymetabolites.org/
- MIBiG Repository: https://mibig.secondarymetabolites.org/repository.html
- JGI Integrated Microbial Genomes and Microbiomes: https://img.jgi.doe.gov/
- Carotenoids Database: http://carotenoiddb.jp/

Acknowledgements

A.W.P. was supported by the National Institute of General Medical Sciences of the National Institutes of Health under award number K99GM118762. The content is solely the responsibility of the author and does not necessarily represent the official views of the National Institutes of Health.

References

Agafonova, N.V., Kaparullina, E.N., Doronina, N.V., and Trotsenko, Y.A. (2018). Obligate methylotroph *Methylobacillus arboreus* IvaT synthesizes a plant hormone, gibberellic acid GA3. Microbiology *87*, 47–50 https://doi.org/10.1134/S0026261718010022.

Alber, B.E. (2011). Biotechnological potential of the ethylmalonyl-CoA pathway. Appl. Microbiol. Biotechnol. *89*, 17–25. https://doi.org/10.1007/s00253-010-2873-z

Balasubramanian, R., Smith, S.M., Rawat, S., Yatsunyk, L.A., Stemmler, T.L., and Rosenzweig, A.C. (2010). Oxidation of methane by a biological dicopper centre. Nature *465*, 115–119. https://doi.org/10.1038/nature08992

Birben, E., Sahiner, U.M., Sackesen, C., Erzurum, S., and Kalayci, O. (2012). Oxidative stress and antioxidant defense. World Allergy Organ. J. *5*, 9–19 https://doi.org/10.1097/WOX.0b013e3182439613

Blin, K., Wolf, T., Chevrette, M.G., Lu, X., Schwalen, C.J., Kautsar, S.A., Suarez Duran, H.G., de Los Santos, E.L.C., Kim, H.U., Nave, M., *et al.* (2017). antiSMASH 4.0-improvements in chemistry prediction and gene cluster boundary identification. Nucleic Acids Res. *45*, W36–W41. https://doi.org/10.1093/nar/gkx319

Behling, L.A., Hartsel, S.C., Lewis, D.E., DiSpirito, A.A., Choi, D.W., Masterson, L.R., Veglia, G., and Gallagher, W.H. (2008). NMR, mass spectrometry and chemical evidence reveal a different chemical structure for methanobactin that contains oxazolone rings. J. Am. Chem. Soc. *130*, 12604–12605. https://doi.org/10.1021/ja804747d

Cantrell, C.L., Dayan, F.E., and Duke, S.O. (2012). Natural products as sources for new pesticides. J. Nat. Prod. *75*, 1231–1242. https://doi.org/10.1021/np300024u

Cavanaugh, C.M., Levering, P.R., Maki, J.S., Mitchell, R., and Lidstrom, M.E. (1987). Symbiosis of methylotrophic bacteria and deep-sea mussels. Nature *325*, 346–348. https://doi.org/10.1038/325346a0

Chistoserdova, L., Kalyuzhnaya, M.G., and Lidstrom, M.E. (2009). The expanding world of methylotrophic metabolism. Annu. Rev. Microbiol. *63*, 477–499. https://doi.org/10.1146/annurev.micro.091208.073600

Choi, D.W., Do, Y.S., Zea, C.J., McEllistrem, M.T., Lee, S.W., Semrau, J.D., Pohl, N.L., Kisting, C.J., Scardino, L.L., Hartsel, S.C., *et al.* (2006). Spectral and thermodynamic properties of Ag(I), Au(III), Cd(II), Co(II), Fe(III), Hg(II), Mn(II), Ni(II), Pb(II), U(IV), and Zn(II) binding by methanobactin from *Methylosinus trichosporium* OB3b. J. Inorg. Biochem. *100*, 2150–2161.

Cimermancic, P., Medema, M.H., Claesen, J., Kurita, K., Wieland Brown, L.C., Mavrommatis, K., Pati, A., Godfrey, P.A., Koehrsen, M., Clardy, J., *et al.* (2014). Insights into secondary metabolism from a global analysis of prokaryotic biosynthetic gene clusters. Cell *158*, 412–421.

Dassama, L.M., Kenney, G.E., Ro, S.Y., Zielazinski, E.L., and Rosenzweig, A.C. (2016). Methanobactin transport machinery. Proc. Natl. Acad. Sci. U.S.A. *113*, 13027–13032.

Dedysh, S.N., Liesack, W., Khmelenina, V.N., Suzina, N.E., Trotsenko, Y.A., Semrau, J.D., Bares, A.M., Panikov, N.S., and Tiedje, J.M. (2000). *Methylocella palustris* gen. nov., sp. nov., a new methane-oxidizing acidophilic bacterium from peat bogs, representing a novel subtype of serine-pathway methanotrophs. Int. J. Syst. Evol. Microbiol. *50*, 955–969. https://doi.org/10.1099/00207713-50-3-955

Demain, A.L., and Sanchez, S. (2009). Microbial drug discovery: 80 years of progress. J. Antibiot. *62*, 5–16. https://doi.org/10.1038/ja.2008.16

DiSpirito, A., Zahn, J., Graham, D., Kim, H., Alterman, M., and Larive, C. (2007). Methanobactin: a copper binding compound having antibiotic and antioxidant activity isolated from methanotrophic bacteria. US7199099B2.

DiSpirito, A.A., Semrau, J.D., Murrell, J.C., Gallagher, W.H., Dennison, C., and Vuilleumier, S. (2016). Methanobactin and the Link between Copper and Bacterial Methane Oxidation. Microbiol. Mol. Biol. Rev. *80*, 387–409. https://doi.org/10.1128/MMBR.00058-15

Dong, Y.H., Wang, L.H., Xu, J.L., Zhang, H.B., Zhang, X.F., and Zhang, L.H. (2001). Quenching quorum-sensing-dependent bacterial infection by an *N*-acyl homoserine lactonase. Nature *411*, 813–817. https://doi.org/10.1038/35081101

Dourado, M.N., Bogas, A.C., Pomini, A.M., Andreote, F.D., Quecine, M.C., Marsaioli, A.J., and Araújo, W.L. (2013). *Methylobacterium*-plant interaction genes regulated by plant exudate and quorum sensing molecules. Braz. J. Microbiol. *44*, 1331–1339.

Duca, D., Lorv, J., Patten, C.L., Rose, D., and Glick, B.R. (2014). Indole-3-acetic acid in plant-microbe interactions. Antonie Van Leeuwenhoek *106*, 85–125. https://doi.org/10.1007/s10482-013-0095-y

Duerkop, B.A., Varga, J., Chandler, J.R., Peterson, S.B., Herman, J.P., Churchill, M.E., Parsek, M.R., Nierman, W.C., and Greenberg, E.P. (2009). Quorum-sensing control of antibiotic synthesis in *Burkholderia thailandensis*. J. Bacteriol. *191*, 3909–3918. https://doi.org/10.1128/JB.00200-09

El Ghazouani, A., Baslé, A., Gray, J., Graham, D.W., Firbank, S.J., and Dennison, C. (2012). Variations in methanobactin structure influences copper utilization by methane-oxidizing bacteria. Proc. Natl. Acad. Sci. U.S.A. *109*, 8400–8404. https://doi.org/10.1073/pnas.1112921109

Farhan Ul-Haque, M., Kalidass, B., Vorobev, A., Baral, B.S., DiSpirito, A.A., and Semrau, J.D. (2015). Methanobactin from *Methylocystis* sp. strain SB2 affects gene expression and methane monooxygenase activity in *Methylosinus trichosporium* OB3b. Appl. Environ. Microbiol. *81*, 2466–2473. https://doi.org/10.1128/AEM.03981-14

Fedorov, D.N., Doronina, N.V., and Trotsenko, I.u.A. (2011). Phytosymbiosis of aerobic methylobacteria: New facts and views. Mikrobiologiia *80*, 435–446.

Galbally, I.E., and Kirstine, W. (2002). The production of methanol by flowering plants and the global cycle of methanol. J. Atmospheric Chem. *43*, 195–229 https://doi.org/10.1023/A:1020684815474

Galinski, E.A., Pfeiffer, H.P., and Trüper, H.G. (1985). 1,4,5,6-Tetrahydro-2-methyl-4-pyrimidinecarboxylic acid. A novel cyclic amino acid from halophilic phototrophic bacteria of the genus *Ectothiorhodospira*. Eur. J. Biochem. *149*, 135–139.

Graf, R., Anzali, S., Buenger, J., Pfluecker, F., and Driller, H. (2008). The multifunctional role of ectoine as a natural cell protectant. Clin. Dermatol. *26*, 326–333. https://doi.org/10.1016/j.clindermatol.2008.01.002

Gu, W., and Semrau, J.D. (2017). Copper and cerium-regulated gene expression in *Methylosinus trichosporium* OB3b. Appl. Microbiol. Biotechnol. *101*, 8499–8516. https://doi.org/10.1007/s00253-017-8572-2

Helfrich, E.J.N., Vogel, C.M., Ueoka, R., Schäfer, M., Ryffel, F., Müller, D.B., Probst, S., Kreuzer, M., Piel, J., and Vorholt, J.A. (2018). Bipartite interactions, antibiotic production and biosynthetic potential of the *Arabidopsis* leaf microbiome. Nat. Microbiol. *3*, 909–919. https://doi.org/10.1038/s41564-018-0200-0

Henard, C.A., and Guarnieri, M.T. (2018). Metabolic engineering of methanotrophic bacteria for industrial biomanufacturing. In Methane Biocatalysis: Paving the Way to Sustainability, M.G. Kalyuzhnaya, and X.-H. Xing, eds. (Springer International Publishing, Cham), pp. 117–132 https://doi.org/10.1007/978-3-319-74866-5_8.

Holland, M., Long, R., and Polacco, J. (2002). *Methylobacterium* spp.: Phylloplane bacteria involved in cross-talk with the plant host? In Phyllosphere Microbiology, S. Lindow, E. Hect-Poinar, and V. Elliott, eds. (APS Press, St. Paul), pp. 125–135.

Ivanova, E.G., Doronina, N.V., Shepeliakovskaia, A.O., Laman, A.G., Brovko, F.A., and Trotsenko, I.u.A. (2000). Facultative and obligate aerobic methylobacteria synthesize cytokinins. Mikrobiologiia *69*, 764–769.

Ivanova, E.G., Doronina, N.V., and Trotsenko, I.u.A. (2001). Aerobic methylobacteria are capable of synthesizing auxins. Mikrobiologiia *70*, 452–458.

Kalyuzhnaya, M.G., Khmelenina, V., Eshinimaev, B., Sorokin, D., Fuse, H., Lidstrom, M., and Trotsenko, Y. (2008). Classification of halo(alkali)philic and halo(alkali) tolerant methanotrophs provisionally assigned to the genera *Methylomicrobium* and *Methylobacter* and emended description of the genus *Methylomicrobium*. Int. J. Syst. Evol. Microbiol. *58*, 591–596 https://doi.org/10.1099/ijs.0.65317-0

Kenney, G.E., and Rosenzweig, A.C. (2013). Genome mining for methanobactins. BMC Biol. *11*, 17. https://doi.org/10.1186/1741-7007-11-17

Kenney, G.E., and Rosenzweig, A.C. (2018a). Methanobactins: Maintaining copper homeostasis in methanotrophs and beyond. J. Biol. Chem. *293*, 4606–4615. https://doi.org/10.1074/jbc.TM117.000185

Kenney, G.E., and Rosenzweig, A.C. (2018b). Chalkophores. Annu. Rev. Biochem. *87*, 645–676. https://doi.org/10.1146/annurev-biochem-062917-012300

Kenney, G.E., Sadek, M., and Rosenzweig, A.C. (2016). Copper-responsive gene expression in the methanotroph *Methylosinus trichosporium* OB3b. Metallomics *8*, 931–940. https://doi.org/10.1039/c5mt00289c

Kenney, G.E., Dassama, L.M.K., Pandelia, M.E., Gizzi, A.S., Martinie, R.J., Gao, P., DeHart, C.J., Schachner, L.F., Skinner, O.S., Ro, S.Y., *et al.* (2018). The biosynthesis of methanobactin. Science *359*, 1411–1416. https://doi.org/10.1126/science.aap9437

Khmelenina, V.N., Kalyuzhnaya, M.G., Starostina, N.G., Suzina, N.E., and Trotsenko, Y.A. (1997). Isolation and characterization of halotolerant alkaliphilic methanotrophic bacteria from Tuva soda lakes. Curr. Microbiol. *35*, 257–261 https://doi.org/10.1007/s002849900249

Khmelenina, V.N., Rozova, N., But, C.Y., Mustakhimov, I.I., Reshetnikov, A.S., Beschastnyi, A.P., and Trotsenko, Y.A. (2015). Biosynthesis of secondary metabolites in methanotrophs: biochemical and genetic aspects (review). Prikl. Biokhim. Mikrobiol. *51*, 140–150.

Kim, H.J., Graham, D.W., DiSpirito, A.A., Alterman, M.A., Galeva, N., Larive, C.K., Asunskis, D., and Sherwood, P.M. (2004). Methanobactin, a copper-acquisition compound from methane-oxidizing bacteria. Science *305*, 1612–1615. https://doi.org/10.1126/science.1098322

Kleinig, H., Schmitt, R., Meister, W., Gerhard, E., and Thommen, H. (1979). New C30-carotenoic acid glucosyl esters from *Pseudomonas rhodos*. Z. Für Naturforschung C *34*, 181–185. https://doi.org/https://doi.org/10.1515/znc-1979-3-404

Koenig, R.L., Morris, R.O., and Polacco, J.C. (2002). tRNA is the source of low-level trans-zeatin production in *Methylobacterium* spp. J. Bacteriol. *184*, 1832–1842.

Konovalova, H.M., Shylin, S.O., and Rokytko, P.V. (2007). Characteristics of carotinoids of methylotrophic bacteria of *Methylobacterium* genus. Mikrobiol. Z. *69*, 35–41.

Krause, S.M., Johnson, T., Samadhi Karunaratne, Y., Fu, Y., Beck, D.A., Chistoserdova, L., and Lidstrom, M.E. (2017). Lanthanide-dependent cross-feeding of methane-derived carbon is linked by microbial community interactions. Proc. Natl. Acad. Sci. U.S.A. *114*, 358–363. https://doi.org/10.1073/pnas.1619871114

Kunte, H., Lentzen, G., and Galinksi, E. (2014). Industrial production of the cell protectant ectoine: protection mechanisms, processes, and products. Curr. Biotechnol. *3*, 10–25.

Kutschera, U. (2007). Plant-associated methylobacteria as co-evolved phytosymbionts: a hypothesis. Plant Signal. Behav. *2*, 74–78.

Lackner, G., Peters, E.E., Helfrich, E.J., and Piel, J. (2017). Insights into the lifestyle of uncultured bacterial natural product factories associated with marine sponges. Proc. Natl. Acad. Sci. U.S.A. *114*, E347–E356. https://doi.org/10.1073/pnas.1616234114

Lichtmannegger, J., Leitzinger, C., Wimmer, R., Schmitt, S., Schulz, S., Kabiri, Y., Eberhagen, C., Rieder, T., Janik, D., Neff, F., *et al.* (2016). Methanobactin reverses acute liver failure in a rat model of Wilson disease. J. Clin. Invest. *126*, 2721–2735. https://doi.org/10.1172/JCI85226

McGowan, S., Sebaihia, M., Jones, S., Yu, B., Bainton, N., Chan, P.F., Bycroft, B., Stewart, G.S., Williams, P., and Salmond, G.P. (1995). Carbapenem antibiotic production in *Erwinia carotovora* is regulated by CarR, a homologue of the LuxR transcriptional activator. Microbiology *141*, 541–550. https://doi.org/10.1099/13500872-141-3-541

Milshteyn, A., Schneider, J.S., and Brady, S.F. (2014). Mining the metabiome: identifying novel natural products from microbial communities. Chem. Biol. *21*, 1211–1223. https://doi.org/10.1016/j.chembiol.2014.08.006

Morohoshi, T., Xie, X., and Ikeda, T. (2018). *N*-Acylhomoserine lactone-mediated quorum sensing regulates biofilm structure in *Methylobacterium populi* P-1M, an isolate from a pink-pigmented household biofilm. Biosci. Biotechnol. Biochem. [Epub ahead of print]. https://doi.org/10.1080/09168451.2018.1518701

Mustakhimov, I.I., Reshetnikov, A.S., Glukhov, A.S., Khmelenina, V.N., Kalyuzhnaya, M.G., and Trotsenko, Y.A. (2010). Identification and characterization of EctR1, a new transcriptional regulator of the ectoine biosynthesis genes in the halotolerant methanotroph *Methylomicrobium alcaliphilum* 20Z. J. Bacteriol. *192*, 410–417. https://doi.org/10.1128/JB.00553-09

Nemecek-Marshall, M., MacDonald, R.C., Franzen, J.J., Wojciechowski, C.L., and Fall, R. (1995). Methanol Emission from Leaves (Enzymatic Detection of Gas-Phase Methanol and Relation of Methanol Fluxes to Stomatal Conductance and Leaf Development). Plant Physiol. *108*, 1359–1368.

Newman, D.J., and Cragg, G.M. (2016). Natural Products as Sources of New Drugs from 1981 to 2014. J. Nat. Prod. *79*, 629–661. https://doi.org/10.1021/acs.jnatprod.5b01055

Nguyen, T., Ishida, K., Jenke-Kodama, H., Dittmann, E., Gurgui, C., Hochmuth, T., Taudien, S., Platzer, M., Hertweck, C., and Piel, J. (2008). Exploiting the mosaic structure of *trans*-acyltransferase polyketide synthases for natural product discovery and pathway dissection. Nat. Biotechnol. *26*, 225–233. https://doi.org/10.1038/nbt1379

Nieto Penalver, C.G., Morin, D., Cantet, F., Saurel, O., Milon, A., and Vorholt, J.A. (2006). *Methylobacterium extorquens* AM1 produces a novel type of acyl-homoserine lactone with a double unsaturated side chain under methylotrophic growth conditions. FEBS Lett. *580*, 561–567.

Oshkin, I.Y., Beck, D.A., Lamb, A.E., Tchesnokova, V., Benuska, G., McTaggart, T.L., Kalyuzhnaya, M.G., Dedysh, S.N., Lidstrom, M.E., and Chistoserdova, L. (2014). Methane-fed microbial microcosms show differential community dynamics and pinpoint taxa involved in communal response. ISME J. *9*, 1119–1129. https://doi.org/10.1038/ismej.2014.203.

Papenfort, K., and Bassler, B.L. (2016). Quorum sensing signal-response systems in Gram-negative bacteria. Nat. Rev. Microbiol. *14*, 576–588. https://doi.org/10.1038/nrmicro.2016.89

Pearson, J.P., Pesci, E.C., and Iglewski, B.H. (1997). Roles of *Pseudomonas aeruginosa las* and *rhl* quorum-sensing systems in control of elastase and rhamnolipid biosynthesis genes. J. Bacteriol. *179*, 5756–5767.

Penalver, C.G., Cantet, F., Morin, D., Haras, D., and Vorholt, J.A. (2006). A plasmid-borne truncated *luxI* homolog controls quorum-sensing systems and extracellular carbohydrate production in *Methylobacterium extorquens* AM1. J. Bacteriol. *188*, 7321–7324.

Petersen, J.M., and Dubilier, N. (2009). Methanotrophic symbioses in marine invertebrates. Environ. Microbiol. Rep. *1*, 319–335. https://doi.org/10.1111/j.1758-2229.2009.00081.x

Pidot, S.J., Coyne, S., Kloss, F., and Hertweck, C. (2014). Antibiotics from neglected bacterial sources. Int. J. Med. Microbiol. *304*, 14–22. https://doi.org/10.1016/j.ijmm.2013.08.011

Pomini, A.M., Cruz, P.L., Gai, C., Araújo, W.L., and Marsaioli, A.J. (2009). Long-chain acyl-homoserine lactones from *Methylobacterium mesophilicum*: synthesis and absolute configuration. J. Nat. Prod. 72, 2125–2129. https://doi.org/10.1021/np900043j

Poonguzhali, S., Madhaiyan, M., and Sa, T. (2007). Production of acyl-homoserine lactone quorum-sensing signals is wide-spread in gram-negative *Methylobacterium*. J. Microbiol. Biotechnol. *17*, 226–233.

Puri, A.W., Schaefer, A.L., Fu, Y., Beck, D.A., Greenberg, E.P., and Lidstrom, M.E. (2017). Quorum sensing in a methane-oxidizing bacterium. J. Bacteriol. *199*, e00773-16.

Puri, A.W., Mevers, E., Ramadhar, T.R., Petras, D., Liu, D., Piel, J., Dorrestein, P.C., Greenberg, E.P., Lidstrom, M.E., and Clardy, J. (2018). Tundrenone: an atypical secondary metabolite from bacteria with highly restricted primary metabolism. J. Am. Chem. Soc. *140*, 2002–2006. https://doi.org/10.1021/jacs.7b12240

Pye, C.R., Bertin, M.J., Lokey, R.S., Gerwick, W.H., and Linington, R.G. (2017). Retrospective analysis of natural products provides insights for future discovery trends. Proc. Natl. Acad. Sci. U.S.A. *114*, 5601–5606. https://doi.org/10.1073/pnas.1614680114

Reshetnikov, A.S., Mustakhimov, I.I., Khmelenina, V.N., and Trotsenko, Y.A. (2005). Cloning, purification, and characterization of diaminobutyrate acetyltransferase from the halotolerant methanotroph *Methylomicrobium alcaliphilum* 20Z. Biochemistry Mosc. *70*, 878–883.

Reshetnikov, A.S., Khmelenina, V.N., and Trotsenko, Y.A. (2006). Characterization of the ectoine biosynthesis genes of haloalkalotolerant obligate methanotroph '*Methylomicrobium alcaliphilum* 20Z'. Arch. Microbiol. *184*, 286–297. https://doi.org/10.1007/s00203-005-0042-z

Reshetnikov, A.S., Khmelenina, V.N., Mustakhimov, I.I., Kalyuzhnaya, M., Lidstrom, M., and Trotsenko, Y.A. (2011). Diversity and phylogeny of the ectoine biosynthesis genes in aerobic, moderately halophilic methylotrophic bacteria. Extremophiles *15*, 653–663. https://doi.org/10.1007/s00792-011-0396-x

Rutledge, P.J., and Challis, G.L. (2015). Discovery of microbial natural products by activation of silent biosynthetic gene clusters. Nat. Rev. Microbiol. *13*, 509–523. https://doi.org/10.1038/nrmicro3496

Schaefer, A.L., Hanzelka, B.L., Parsek, M.R., and Greenberg, E.P. (2000). Detection, purification, and structural elucidation of the acylhomoserine lactone inducer of *Vibrio fischeri* luminescence and other related molecules. Meth. Enzymol. *305*, 288–301.

Schrader, J., Schilling, M., Holtmann, D., Sell, D., Filho, M.V., Marx, A., and Vorholt, J.A. (2009). Methanol-based industrial biotechnology: current status and future perspectives of methylotrophic bacteria. Trends Biotechnol. *27*, 107–115. https://doi.org/10.1016/j.tibtech.2008.10.009

Semrau, J.D., DiSpirito, A.A., and Yoon, S. (2010). Methanotrophs and copper. FEMS Microbiol. Rev. *34*, 496–531. https://doi.org/10.1111/j.1574-6976.2010.00212.x

Siddikee, M., Hamayun, M., Han, G., and Sa, T. (2010). Optimization of gibberellic acid production by *Methylobacterium oryzae* CBMB20. Korean Soc. Soil Sci. Fertil. *43*, 522–527.

Tao, L., Schenzle, A., Odom, J.M., and Cheng, Q. (2005). Novel carotenoid oxidase involved in biosynthesis of 4,4'-diapolycopene dialdehyde. Appl. Environ. Microbiol. *71*, 3294–3301.

Till, M., and Race, P.R. (2016). The assembly line enzymology of polyketide biosynthesis. Methods Mol. Biol. *1401*, 31–49. https://doi.org/10.1007/978-1-4939-3375-4_2

Ueoka, R., Bortfeld-Miller, M., Morinaka, B.I., Vorholt, J.A., and Piel, J. (2017). Toblerols, cyclopropanol-containing modulators of methylobacterial antibiosis generated by an unusual polyketide synthase. Angew. Chem. Int. Ed. 1–7 https://doi.org/10.1002/anie.201709056

Van Dien, S.J., Marx, C.J., O'Brien, B.N., and Lidstrom, M.E. (2003). Genetic characterization of the carotenoid biosynthetic pathway in *Methylobacterium extorquens* AM1 and isolation of a colorless mutant. Appl. Environ. Microbiol. *69*, 7563–7566.

Veraart, A.J., Garbeva, P., van Beersum, F., Ho, A., Hordijk, C.A., Meima-Franke, M., Zweers, A.J., and Bodelier, P.L.E. (2018). Living apart together-bacterial volatiles influence methanotrophic growth and activity. ISME J. *12*, 1163–1166. https://doi.org/10.1038/s41396-018-0055-7

Vorholt, J.A. (2012). Microbial life in the phyllosphere. Nat. Rev. Microbiol. *10*, 828–840. https://doi.org/10.1038/nrmicro2910

Whiteley, M., Diggle, S.P., and Greenberg, E.P. (2017). Progress in and promise of bacterial quorum sensing research. Nature *551*, 313–320. https://doi.org/10.1038/nature24624

Ye, R.W., Yao, H., Stead, K., Wang, T., Tao, L., Cheng, Q., Sharpe, P.L., Suh, W., Nagel, E., Arcilla, D., *et al.* (2007). Construction of the astaxanthin biosynthetic pathway in a methanotrophic bacterium *Methylomonas* sp. strain 16a. J. Ind. Microbiol. Biotechnol. *34*, 289–299. https://doi.org/10.1007/s10295-006-0197-x

Bioconversion of Methanol into Value-added Chemicals in Native and Synthetic Methylotrophs

13

Min Zhang[1], Xiao-jie Yuan[1], Cong Zhang[1], Li-ping Zhu[1], Xu-hua Mo[1], Wen-jing Chen[1] and Song Yang[1,2*]

[1]School of Life Science, Qingdao Agricultural University, Shandong Province Key Laboratory of Applied Mycology, and Qingdao International Center on Microbes Utilizing Biogas, Qingdao, China.
[2]Key Laboratory of Systems Bioengineering, Ministry of Education, Tianjin University, Tianjin, China.

*Correspondence: yangsong1209@163.com

https://doi.org/10.21775/9781912530045.13

Abstract

Methanol, commercially generated from methane, is a renewable chemical feedstock that is highly soluble, relatively inexpensive, and easy to handle. The concept of native methylotrophic bacteria serving as whole cell catalysts for production of chemicals and materials using methanol as a feedstock is highly attractive. In recent years, the available omics data for methylotrophic bacteria, especially for *Methylobacterium extorquens*, the best-characterized model methylotroph, have provided a solid platform for rational engineering of methylotrophic bacteria for industrial production. In addition, there is a strong interest in converting the more traditional heterotrophic production platforms towards the use of single carbon substrates, including methanol, through metabolic engineering. In this chapter, we review the recent progress towards achieving the desired growth and production yields from methanol, by genetically engineered native methylotrophic strains and by the engineered synthetic methylotrophs.

Introduction

One goal of metabolic engineering is to develop renewable and sustainable alternatives in production of value-added chemicals. In particular, methanol is an important one-carbon (C1) feedstock that can be generated from either synthesis gas (a mixture of CO and H_2) or from biogas, assuming that large quantities of this feedstock could be produced at relatively low market price (Clomburg *et al.*, 2017). Methylotrophic bacteria are a diverse group of microbes that can use reduced C1 compounds such as methanol and methane as sole sources of both energy and carbon (Chistoserdova *et al.*, 2009; Chistoserdova, 2018). Some methylotrophic bacteria, such as *Methylobacterium extorquens*, *Methylococcus capsulatus*, and *Methylomicrobium buryatense*, show robust growth under laboratory conditions (Bird *et al.*, 1971; Guo and Lidstrom, 2006; de la Torre *et al.*, 2015). The availability of whole genome sequences, combined with biochemical studies (Peyraud *et al.*, 2011; Kalyuzhnaya *et al.*, 2013), have provided important insights into their metabolism that are crucial for enabling engineering these types of microorganism for bulk chemical production. Meantime, as an alternative approach, more traditional model microorganisms such has *Escherichia coli*, *Corynebacterium glutamicum* and *Saccharomyces cerevisiae*, have been genetically modified towards acquiring the ability to utilize methanol for growth, with a potential for converting methanol into value-added compounds (Whitaker *et al.*, 2015; Bennett *et al.*, 2018b). In this chapter, we review the current status of the fields of engineering native as well as synthetic methylotrophs and provide examples of successfully engineered strains with a potential in producing commercial value-added chemicals.

Metabolic engineering of native methylotrophs for value-added chemicals production

M. extorquens AM1, a pink-pigmented facultative methylotrophic α-proteobacterium capable of growth on both C1 and multi-carbon compounds, has served for decades as the model organism for studying C1 metabolism (Chistoserdova *et al.*, 2003), including its biotechnological applications (Ochsner *et al.*, 2015). Since the 1960s, it has been reported that *M. extorquens* was successfully employed as a producer of certain products such as amino acids (Sirirote *et al.*, 1988). Based on the current understanding, *M. extorquens* employs three interlocked metabolic cycles for carbon assimilation: the serine cycle, the ethylmalonyl-CoA (EMC) pathway, and the poly-3-hydroxybutyrate (PHB) cycle (Fig. 13.1) (Anthony, 2011; Cui *et al.*, 2016). The main function of the EMC pathway is to regenerate glyoxylate from acetyl-CoA for reincorporating it into the serine cycle during C1 assimilation (Erb *et al.*, 2007; Peyraud *et al.*, 2009). It has been shown that significant metabolic flux goes through the serine cycle and the EMC pathway when cells are grown on methanol, generating a stable supply of acetyl-CoA, which could serve as a precursor for the production of value-added chemicals in engineered *M. extorquens* AM1 strains (Fig. 13.1) (Peyraud *et al.*, 2011; Fu *et al.*, 2016).

Production of mevalonate and terpenoids

Mevalonate and its derivatives terpenoids are promising bulk chemicals. Terpenoids are used as flavours, fragrances, pharmaceuticals, and biofuels, representing the largest and the most structurally diverse family of chemicals in nature (Zhang *et al.*, 2017). To synthesize considerable amounts of terpenoids, a commonly used strategy in bacteria is to introduce an exogenous mevalonate (MEV) pathway instead of utilizing the native methylerythritol phosphate (MEP) pathway for circumventing the regulatory control of the MEP pathway (Zhao *et al.*, 2013). Thus, pathway engineering for mevalonate production is a feasible first step as a platform for diverse terpenoid synthesis. Although *E. coli* is the primary industrial microbe for the production of mevalonate (Nagai *et al.*, 2018), other engineered hosts have the potential to generate it from alternative carbon feedstocks via synthetic metabolic pathways. Zhu and colleagues

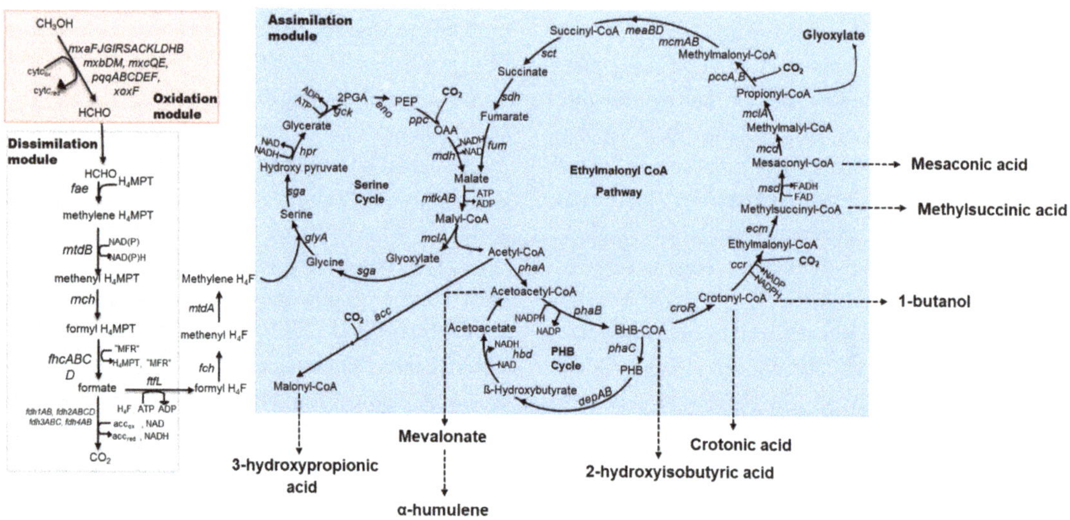

Figure 13.1 Schematic of methanol metabolism in *M. extorquens* and target precursors for synthesis of value added compounds. Pink, grey and blue backgrounds represent the primary oxidation module, the dissimilation module, and the assimilation modules, respectively.

engineered a mevalonate pathway by using acetyl-CoA as the initial precursor in *M. extorquens* AM1, to produce mevalonate from methanol (Zhu *et al.*, 2016). A natural operon (MVH) harbouring the *mvaS* and *mvaE* genes from *Enterococcus faecalis,* as well as an artificial operon (MVH) harbouring the *hmgcs1* gene from *Blattella germanica* and the *tchmgr* gene from *Trypanosoma cruzi* were constructed in a plasmid pCM110. Expression of these two operons resulted in the titres of mevalonate of 56 mg/l and 66 mg/l, respectively, in flask cultivations. Further introduction of the *phaA* gene from *Ralstonia eutropha* into the MVH operon, with the goal of improving the conversion from acetyl-CoA into acetoacetyl-CoA increased the mevalonate titre to 180 mg/l, 3.2-fold higher than that of the natural MVE operon. Further modification of the expression level of the *phaA* gene by regulating the strength of the ribosomal binding site (RBS) resulted in an additional increase in mevalonate production to 215 mg/l (Zhu *et al.*, 2016). Moreover, it has been demonstrated that, under these conditions, the supply of acetyl-CoA was limited in engineered *M. extorquens* AM1. To overcome this limitation, a strategy termed mevalonate sensor-assisted transcriptional regulator engineering (SATRE) was employed, to control metabolic flux redistribution toward increasing acetyl-CoA flux from methanol for mevalonate production (Fig. 13.2) (Liang *et al.*, 2017a). A 60% increase in mevalonate concentration was further achieved following sensor-based high-throughput screening of a QscR transcriptional regulator library. Finally, a fed-batch fermentation in a 5 l bioreactor yielded a mevalonate concentration of 2.67 g/l, which was equivalent to an overall yield of 0.055 mol acetyl-CoA/mol methanol. In addition, Cui and colleagues utilized atmospheric and room temperature plasma (ARTP) mutagenesis, in combination with adaptive laboratory evolution (ALE), generating a mutant with high methanol tolerance whose cell density was 7.1-fold higher than that of the parent strain in 1.25 M (5%, v/v) methanol (Cui *et al.*, 2018). Accordingly, the mevalonate productivity of the mutant strain carrying a mevalonate synthesis pathway was 65% higher than that of the wild-type strain in methanol fed-batch fermentation.

In a separate study, Sonntag and colleagues introduced a mevalonate pathway from *Myxococcus xanthus,* in combination with RBS optimization of α-humulene and farnesyl pyrophosphate synthases, resulting in the increase of titres of α-humulene from 18 mg/l to 54 mg/l in *M. extorquens* AM1 (Sonntag *et al.*, 2015a). Furthermore, the recombinant *M. extorquens* AM1 was grown in methanol-limited fed-batch conditions, with the addition of dodecane to avoid the accumulation of α-humulene in the medium. An average OD_{600} value of 80–90 were achieved, and final product concentrations were up to 1.65 g/l. This is the highest titre of *de*

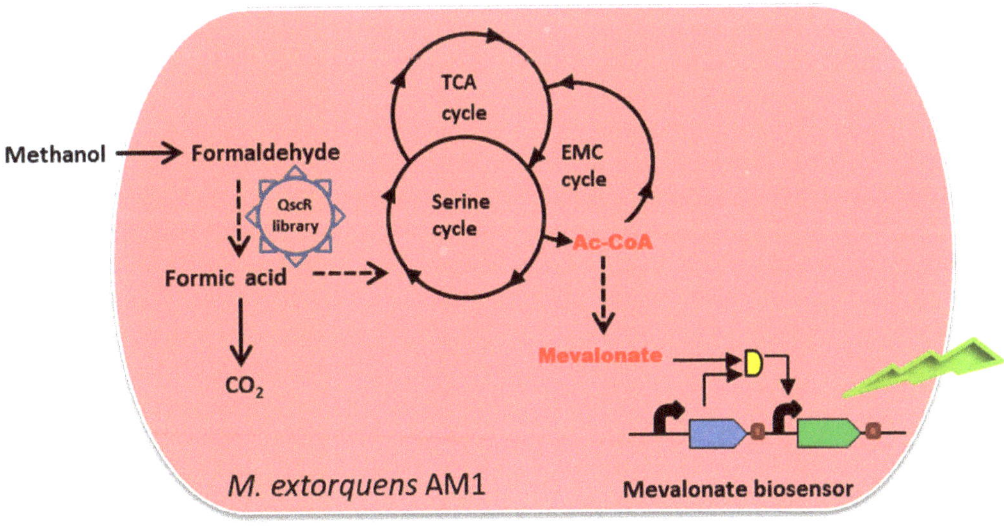

Figure 13.2 Biosensor-assisted transcriptional regulator engineering approach to increase mevalonate production in *M. extorquens* AM1. QscR is the regulator of the serine cycle for methanol assimilation (Liang *et al.*, 2017).

novo synthesized α-humulene reported to date, which corresponds to 12% of the maximum theoretical yield. The product titres of mevalonate and α-humulene described in these studies are already in a competitive range compared to the industrial platform strains (Harada *et al.*, 2009), demonstrating the potential of *M. extorquens* as a future platform for the production of high-value terpenoids.

Production of carboxylic acids

3-Hydroxypropionic acid (3-HP) is an important platform chemical, serving as a precursor for a wide range of industrial applications, such as the production of acrylic acid and 1,3-propanediol. In particular, acrylic acid has an annual market exceeding $10 billion. 3-HP can be further polymerized into poly-3-hydroxypropionate, which is a biodegradable and biocompatible thermoplastic polymer. Recently, Yang and colleagues constructed a malonyl-CoA pathway in *M. extorquens* AM1 that reduced malonyl-CoA to synthesize 3-HP through a bifunctional reductase encoded by the *mcr* gene from *Chloroflexus aurantiacus* (Yang *et al.*, 2017). The engineered strain demonstrated 3-HP production from methanol with initial titre of 6.8 mg/l in shake flask cultivation, which was then improved to 91.0 mg/l by increasing the strength of the promoter and the *mcr* gene copy number, and by decreasing the concentration of sodium ions in the culture medium. Notably, when the cells reached the stationary growth phase, 3-HP content was rapidly reduced to lower than 10.0 mg/l within 10 h. Furthermore, based on the combined results from *in vitro* enzyme assays, ^{13}C-labeling analysis, metabolomics and gene knockouts, major routes were identified responsible for the degradation of the 3-HP during this phase of growth. This work provides a solid base for future engineering efforts towards bioconversion of methanol into the economically important product 3-HP.

Recently, Rohde and colleagues introduced (R)-3-hydroxybutyryl-CoA-specific, coenzyme B12-dependent mutases from *Bacillus massiliosenegalensis* JC6 into *M. extorquens* AM1, to convert the intermediate of 3-hydroxybutyrl-CoA towards production of another carboxylic acid, 2-hydroxyisobutyric acid, a building block with a high potential for poly-methylmethacrylate production (Rohde *et al.*, 2017). 2-hydroxyisobutyric acid titres of 2.1 g/l were further achieved in fed-batch bioreactor without affecting the biomass yield, with the yield of 0.11 g/g from methanol. In addition, Sonntag and colleagues overexpressed *yciA* encoding thioesterases, from *E. coli* in *M. extorquens* AM1, an engineering step designed to generate two dicarboxylic acids from their CoA derivatives involved in the EMC pathway, mesaconic acid (with 70 mg/l yield), and methylsuccinic acid (with 60 mg/l yield), respectively (Sonntag *et al.*, 2014). Further increase in accumulation of mesaconyl-CoA (16-fold) and methylsuccinyl-CoA (6-fold) was achieved by adjusting the concentration of cobalt, a growth limiting factor affecting the activities of ethylmalonyl-CoA and methylmalonyl-CoA mutases, in the culture medium, with a combined titre of the two dicarboxylic acids of 0.65 g/l (Sonntag *et al.*, 2015b). This titre represented a yield of 0.17 g/g methanol. Not only these important achievements shed light on the potential in engineering *M. extorquens* AM1 as an alternative microbial factory for producing common carboxylic acids and uncommon dicarboxylic acids from methanol, but they also illustrate the complex metabolic regulation of methylotrophy in this organism.

Production of 1-butanol

1-Butanol has been proposed to be a better alternative to ethanol as a replacement for gasoline because of its many advantages. For instance, 1-butanol is less soluble in water than ethanol, reducing the possibility of contamination in groundwater, and this trait also makes it feasible to use liquid-liquid extraction for 1-butanol recovery in industrial fermentation. Additionally, 1-butanol is compatible with the current petroleum infrastructure and can be blended with gasoline at high ratios. Although clostridia are the primary industrial microbes for the fermentative production of 1-butanol, clostridia-based industrial fermentation has some limitations, such as formation of by-products, spore formation, and low cell density (Lee *et al.*, 2008). Recently, Hu and colleagues engineered a modified CoA-dependent pathway for 1-butanol production from the precursor crotonyl-CoA, by introducing the *ter* gene from *Treponema denticola* and the *adhE2* from *Clostridium acetobutylicum* into *M. extorquens* AM1 (Hu and Lidstrom, 2014). Although tests for 1-butanol production by cells grown on methanol were negative, the engineered strain was able to generate 1-butanol at 15.2 mg/l when grown on ethylamine while overexpressing the *croR* gene as a means of increasing the supply of crotonyl-CoA (Hu and Lidstrom, 2014). Adaptive laboratory evolution (ALE) was further employed, resulting in isolation of a mutant with 1-butanol tolerance of up to 0.5%, after 200 generations. The adapted strain was able to produce 25.5 mg/l 1-butanol, representing an 87% increase in production compared with the original strain (Hu *et al.*, 2016). Whole genome sequencing revealed a SNP mutation in the adapted strain, in *kefB*, a gene encoding a potassium/proton antiporter. Analysis of a knockout mutant in this gene confirmed that this mutation was indeed responsible for the increased 1-butanol tolerance. Global metabolomics analysis further uncovered that the pools of multiple key metabolites, including fatty acids, amino acids, disaccharides and carotenoids were either increased or decreased in the adapted strain in response to 1-butanol stress. Overall, this study uncovered an array of cellular responses by *M. extorquens* AM1 to solvent stresses, thus providing a genetic blueprint for further rational design of derivatives of this strain for increased 1-butanol production.

Novel insights into methylotrophy in Methylobacterium

Although natural methylotrophs represent promising candidates as new microbial cell factories, further efforts are required to better understand the regulation of methylotrophs in these species and to further develop genetic tools essential for improving the product yields. Recently, Ochsner and colleagues carried out experiments involving hyper-saturating transposon mutagenesis, followed up by high-throughput sequencing of mutant populations (Ochsner *et al.*, 2017). This study uncovered 95 genes that have not been previously known to play a role in methylotrophy (Bringel and Vuilleumier, 2017; Ochsner *et al.*, 2017). Notably, this study uncovered that ribulose-1, 5-bisphosphate (RuBP) functions as a master regulator metabolite in methylotrophic metabolism, by uncovering that *prk*, a gene encoding phosphoribulokinase (Prk), responsible for the production of RuBP was essential for growth on methanol. Prk was ultimately demonstrated to be a co-inducer of QscR, a major regulator of the serine cycle.

In another study, Schada von Borzyskowski and colleagues explored the possibility of converting the heterotrophic *M. extorquens* AM1 into an autotrophic derivative (Schada von Borzyskowski *et al.*, 2018). They constructed a chassis that utilized methanol solely for energy acquisition while forming the biomass entirely from CO_2. First, they engineered a *M. extorquens* AM1 derivative with deletion in *ftfl*, a gene encoding formate tetrahydrofolate ligase, causing the inability to grow on methanol. They then heterologously expressed Rubisco from *Rhodospirillum rubrum*, in order to enable CO_2 fixation. As a result, *M. extorquens* AM1Δ*ftfL* expressing the newly built Calbin-Benson-Bassham (CBB) cycle could grow in methanol minimal medium with 5% CO_2, although this strain did not sustain long-lasting growth under this condition. This work is the first example of decoupling energy metabolism and carbon assimilation in *M. extorquens* AM1 and establishing synthetic autotrophy with a heterologously derived CBB cycle. This work also demonstrated the importance of the continuous operation of carbon assimilatory cycles in balancing carbon fluxes between biomass building and draining metabolites from central metabolism into the designed biosynthetic pathways. Taken together, these new contributions offer opportunities for elucidating genotype to phenotype relationships in C_1 assimilation and providing novel insights into its regulation, which will be key in engineering native methylotrophs for biotechnological application in the future.

Constructing synthetic methylotrophs for direct methanol conversion

In recent years, there has been increasing interest in engineering non-methylotrophic bacteria toward the use of C1 compounds, through metabolic engineering (Bennett *et al.*, 2018b). This strategy not only increases our fundamental understanding of methylotrophy, but also provides an opportunity to use established industrial bacteria as platforms for converting methanol into value-added products. The efforts to engineer methanol utilization in traditional platform organisms, both prokaryotic (*E. coli, C. glutamicum*; Müller *et al.*, 2015b; Witthoff *et al.*, 2015) and eukaryotic *S. cerevisiae* (Dai *et al.*, 2017) have been recently reported.

Overview of assimilatory C_1 metabolism

Generally, there are two types of methanol dehydrogenase. One is PQQ-dependent Mdh in Gram-negative methylotrophs such as *M. extorquens*, another is NAD-dependent Mdh in Gram-positive *Bacillus methanolicus* (Müller *et al.*, 2015a). At least 14 gene products are necessary for PQQ-dependent Mdh function (Lidstrom, 2006), but NAD-dependent Mdh requires one gene for converting methanol to formaldehyde, and its activity can be greatly enhanced by an activator protein ACT (Anne *et al.*, 2013). Moreover, the PQQ cofactor that is required for PQQ-Mdh is not synthesized by the industrial strains such as *E. coli* or *C. glutamicum*. Therefore, NAD-dependent Mdh has become an obvious choice for implementation of methanol to formaldehyde conversion in *E. coli* and *C. glutamicum* (Fig. 13.3).

For assimilation of formaldehyde into central metabolism, two major pathways in natural methylotrophs have been reported, the ribulose monophosphate (RuMP) and the serine cycles. The RuMP cycle is present in several methylotrophs such as *Methylococcus capsulatus* Bath and *Methylobacillus flagellatus* KT (Strom *et al.*, 1974; Chistoserdova *et al.*, 2000). Formaldehyde is condensed with RuMP by hexulose-6-phosphate synthase (HPS), to generate hexulose-6-phosphate. Hexulose-6-phosphate is further isomerized into fructose 6-phosphate by phosphohexuloisomerase (PHI). Fructose 6-phosphate can be cleaved into C_3 intermediates either via the glycolysis pathway or via the Entner–Doudoroff pathway reactions. The RuMP pathway is quite an efficient route for formaldehyde assimilation in terms of ATP consumption and biomass yield. Per one molecule of pyruvate generated, 1 molecule of ATP and 3 molecules of NAD(P)H are produced.

The serine cycle operates in methylotrophic bacteria such as *M. extorquens* AM1 (Chistoserdova and Lidstrom, 1994). In this cycle, formaldehyde is first oxidized to formate, which is subsequently converted into methylene tetrahydrofolate (THF) via the THF-linked C_1 transfer pathway. Methylene-THF is then condensed with glycine to produce serine, which is subsequently transformed to form C_3 and C_4 intermediates for biomass synthesis. In contrast to the RuMP cycle, the serine cycle leads to the consumption of four molecules of NAD(P)H and three molecules of ATP to produce one molecule of pyruvate.

Engineering bacteria for methanol conversion

In silico modelling predicted that expression of a NAD-linked Mdh and the two first enzymes of the RuMP cycle, Hps and Phi, into a non-methylotroph such as *E. coli* should be sufficient to allow biomass formation from methanol (Müller *et al.*, 2015b). In order to test this hypothesis, Müller and colleagues heterologously expressed respective genes in *E. coli* (Fig. 13.3) (Müller *et al.*, 2015b). The catalytic activities of several candidate enzymes from different microorganisms such as *B. methanolicus* MGA3, *Lysinibacillus fusiformis*, *M. flagellatus* KT, *M. capsulatus* Bath were compared in cell lysates, and subsequently *in vivo* by measuring formaldehyde formation and assimilation. The three genes originating from *B. methanolicus* MGA3 were found to be the most effective in, respectively, methanol oxidation and formaldehyde conversion by *E. coli*. To further demonstrate how the metabolism of *E. coli* was affected by the introduction of the C1 modules, dynamic labelling experiments using ^{13}C-labelled methanol were performed in an *E. coli* strain lacking the *frmA* gene that is responsible for formaldehyde detoxification in native *E. coli*. The incorporation of ^{13}C into hexose 6-phosphates was measured to be within approximately 40%. Other metabolites of the RuMP cycle, such as ribose 5-phosphate and ribulose 5-phosphate, as well as intermediates of the glycolysis pathway, such as dihydroxyacetone phosphate and phosphoenolpyruvate were detected as well, demonstrating that at least several turnovers have occurred through the synthetic RuMP cycle. However, no methanol-dependent growth of *E. coli* was achieved. Meantime, a similar engineering scheme has also been applied to *C. glutamicum*, demonstrating utilization of methanol as a co-carbon source in a glucose-based medium in (Lessmeier *et al.*, 2015; Witthoff *et al.*, 2015).

Since Mdh from *B. methanolicus* has low affinity toward methanol (Km of 200 mM), Whitaker and colleagues employed and alternative enzyme from *B. stearothermophilus*, possessing a Km of 20 mM and V_{max} of 2.1 U/mg for methanol (Price *et al.*, 2016; Whitaker *et al.*, 2017). They engineered an E. coli strain expressing *B. stearothermophilus* Mdh

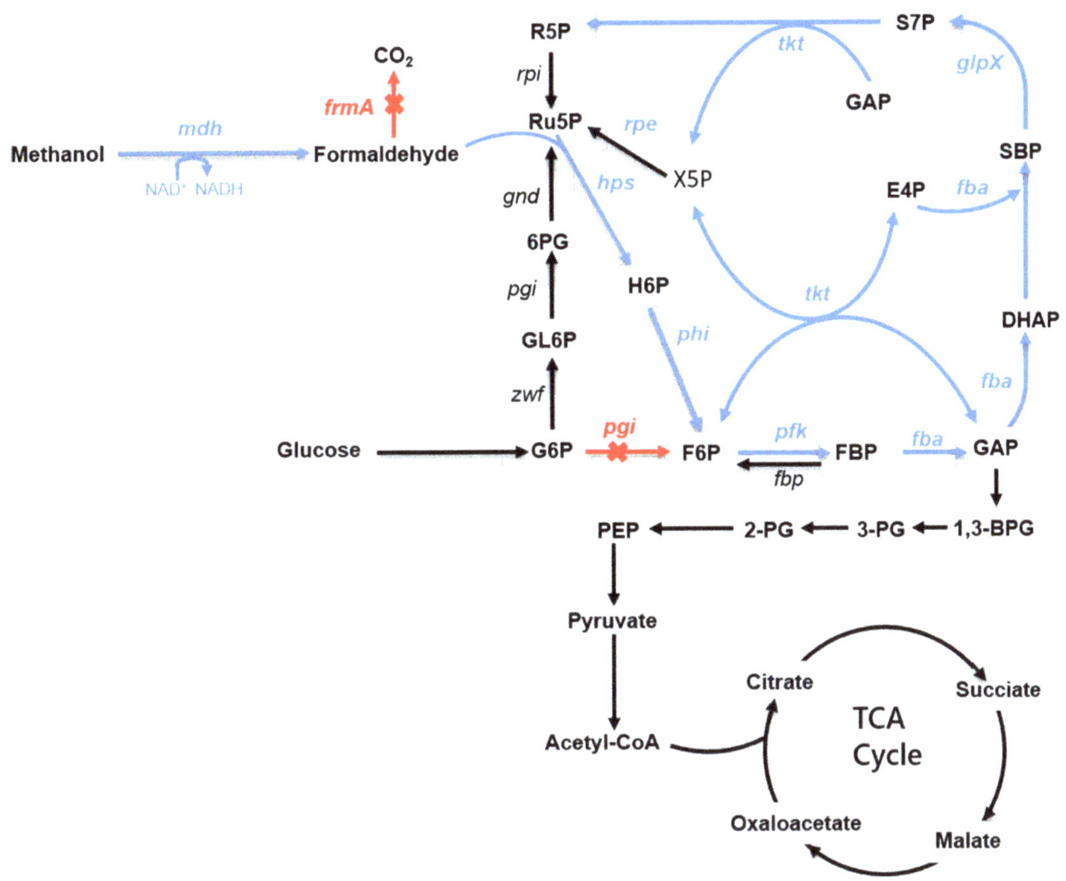

Figure 13.3 Methanol metabolism in synthetic methylotrophs. Blue solid lines represent reactions mediated by the heterologously expression genes. Red solid lines represent gene knockouts. Enzymes (genes): methanol dehydrogenase (*mdh*), hexulose phosphate synthase (*hps*), phosphohexuloisomerase (*phi*), phosphoglucoisomerase (*pgi*), formaldehyde dehydrogenase (*frmA*), phosphofructokinase (*pfk*), fructose-bisphosphate aldolase (*fba*), transketolase (*tkt*), ribulose phosphate epimerase (*rpe*), sedoheptulose bisphosphatase (*glpX*), fructose bisphosphatase (*fbp*), ribose phosphate isomerase (*rpi*), glucose-6-phosphate dehydrogenase (*zwf*). Metabolites: glucose 6-phosphate (G6P), 6-phosphogluconolactone (GL6P), 6 phosphogluconate (6PG), ribulose 5-phosphate (Ru5P), ribose 5-phosphate (R5P), hexulose 6-phosphate (H6P), fructose 6-phosphate (F6P), fructose-1, 6-bisphosphate (FBP), dihydroxyacetone phosphate (DHAP), glyceraldehyde 3-phosphate (GAP), xylulose 5-phosphate (X5P), erythrose 4-phosphate (E4P), sedoheptulose bisphosphate (SBP), sedoheptulose 7-phosphate (S7P), 1,3-bisphosphoglycerate (1,3-BPG), 3-phosphoglycerate (3-PG), 2-phosphoglycerate (2-PG), phosphoenolpyruvate (PEP)

and *B. methanlicus* Hps and Phi, and found that small amounts of supplements such as yeast extract enabled initiation of growth of *E. coli* on methanol. When the engineered *E. coli* strain was grown in M9 minimal medium with 1 g/l yeast extract and 60 mM methanol, a cell biomass increase by 30% was observed, corresponding to biomass yield of 0.289 gCDW/g methanol. This yield is close to biomass yields reported for native methylotrophs. ^{13}C labelling analysis demonstrated that not only the RuMP cycle intermediates were labelled, but also the metabolites in the TCA cycle and related amino acids, supporting biomass generation from methanol. Moreover, in this study, a 65% improvement in flavonoid production was observed in the presence of 60 mM methanol. Additional engineering effort included heterologous expression of the non-oxidative pentose phosphate pathway (PPP) consisting of five enzymes (ribulose phosphate epimerase (*rpe*), fructose-bisphosphate aldolase (*fba*), sedoheptulose bisphosphatase (*glpX*), phosphofructokinase (*pfk*) and transketolase (*tkt*)

(Bennet *et al.*, 2018a). The introduction of the five genes, all originating from *B. methanolius,* resulted in increased flux through the PPP pathway, further enhancing methanol assimilation by the engineered *E. coli* (Fig. 13.3) (Bennett *et al.*, 2018a). In addition, Bennett and colleagues deleted the phosphoglucoisomerase gene (*pgi*), further increasing carbon flux through the oxidative PPP, for increased supply of Ru5P. This change led to a significant improvement of methanol assimilation and to increased acetone titres during co-culturing on methanol and glucose.

In order to further increase methanol consumption, Meyer and colleagues used genetic constraints to develop an approach to link methanol assimilation to growth supported by a multi-carbon source (Meyer *et al.*, 2018). Via *in silico* analysis of a metabolic scenario including the presence of the *mdh-hps-phi* operon, they predicted that the deletion of *edd,* encoding phosphogluconate dehydratase, and *rpiAB,* encoding ribose-5-phosphate isomerase, would resulted in failure to grow on gluconate, but this variant would growth in the presence of methanol. ALE was then employed to improve co-consumption of methanol and gluconate. After 35 generations, the adapted strain showed 3-fold increase in final $OD_{600,}$ compared with the initially engineered strain. Furthermore, deletion of the NAD-dependent malate dehydrogenase (*maldh*), which is involved in the TCA cycle, was predicted to increase the ratio of NAD+/NADH, potentially increasing methanol oxidation and methanol dependency. After additional rounds of ALE of the Δ*edd* Δ*rpiAB* Δ*maldh* variant of *E. coli,* growth rate was further improved, from 0.017/h to 0.081/h when during co-cultivated on methanol and gluconate. This evolved strain had a methanol consumption rate of 13 mmol/gCDW/h, which is comparable with the reported rate for *M. extorquens* AM1. This work provided an important basis for further optimizing synthetic methylotrophic organisms, in the future.

Engineering a formaldehyde-inducible promoter for improved growth on methanol

In synthetic methylotrophs, methanol is converted to the toxic intermediate formaldehyde by Mdh, catalyzing the initial reaction of methanol conversion. *E. coli* has a native formaldehyde-inducible promoter, *Pfrm,* which negatively regulates the formaldehyde detoxification operon *frmRAB* (Fig. 13.4) (Danilov and Ismailov, 1989). Recently, Rohlhill and Sandoval constructed a library of mutant variants of *Pfrm* and used the combination of fluorescence-activated cell sorting and high-throughput sequencing (sort-seq) to screen formaldehyde-responsive promoters (Rohlhill and Sandoval, 2017). Two 19 base pair repressor binding sites, one upstream of the −35 RNA polymerase (RNAP) binding site and one overlapping with the −10 site were identified and subsequently engineered, resulting in 13-fold higher expression compared with the native promoter. They also found that the mutant *Pfrm*, placed upstream of the methanol assimilation module (i.e. the *mdh-hps-phi* operon), resulted in increased biomass from methanol, likely due to the improved balance in formaldehyde production/consumption. Further, Woolston and colleagues rationally designed the sequence between −10 and −35 region of the *Pfrm,* and found that the engineered promoter possessed increased sensitivity and wider dynamic range of sensing formaldehyde, at levels as low as 1 μM. This synthetic biosensor was also employed for optimizing the expression levels of *mdh* and *hps/phi,* to maximize the methanol assimilation flux (Woolston *et al.*, 2018).

Preliminary test for direct methanol utilization in yeast

Besides prokaryotic methylotrophs, *S. cerevisiae* also presents a great potential as a microbial chassis for methanol conversion, due to its robust tolerance to high concentrations of methanol. Dai and colleagues integrated a methanol oxidation pathway including alcohol oxidases (AOX), catalases (CAT), dihydroxyacetone synthases (DAS1/2), and dihydroxyacetone kinase (DAK), originating from the methylotrophic yeast *Pichia pastoris,* into the chromosome of *S. cerevisiae* (Dai *et al.*, 2017). These methylotrophic enzymes with targeting signals from *P. pastoris* were found to be functionally expressed and located in the peroxisomes in the recombinant *S. cerevisiae.* The engineered *S. cerevisiae* showed 1.04 g/l consumption of methanol and 3.13% increase in cell growth, accompanied with 0.26 g/l of pyruvate production in the fermentation broth when using methanol as the sole carbon source. In another study, *P. pastoris* was engineered to produce lovastatin and its precursor monacolin

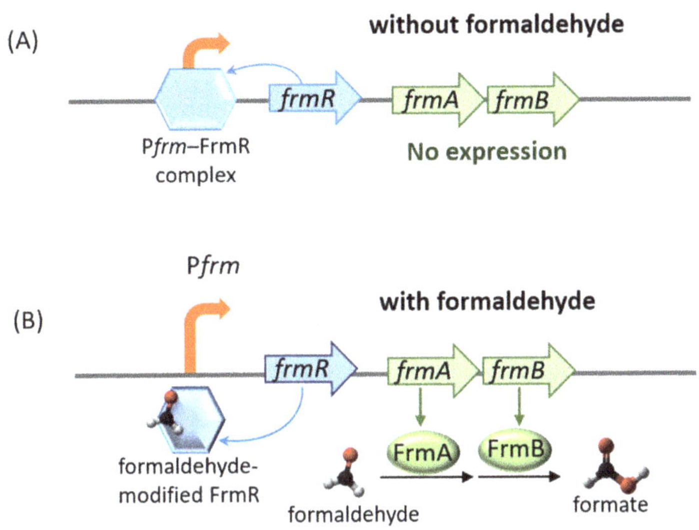

Figure 13.4 Regulation of the *fmr* operon. (A) FrmR binds to the promoter (P*frm*) upstream of the *frmRAB* gene cluster, repressing transcription. (B) FrmR is conformationally modified by binding formaldehyde, permitting transcription.

J from methanol (Liu *et al.*, 2018a,b). A co-culture of *P. pastoris* was optimized for bioreactor fermentations with the yields of 593.9 mg/l monacolin J and 250.8 mg/l lovastatin. Although much further work is required to improve methanol utilization and concomitant production of value-added chemicals, these preliminary studies demonstrate a potential in engineering non-methylotrophic eukaryotic microbes for methanol utilization and scaling their use toward industrial biotechnological processes in the future.

Conclusions

In the recent years, native methylotrophs such as *M. extorquens* AM1 have gained interest highlighting the important role of methanol as a future alternative carbon feedstock. With understanding of native metabolic pathways and further development of metabolic engineering tools in *M. extorquens* AM1, it may be feasible to rationally manipulate this microorganism for producing bio-based chemicals and biofuels (Lieven *et al.*, 2018). However, further strain optimization is required to make this system industrially relevant. Alternatively, engineering synthetic methylotrophic bacteria and yeasts provide attractive alternative platforms for methanol conversion. Examples of methanol assimilation by engineered synthetic methylotrophs demonstrate the potential for using traditional industrial microorganism as chassis for production of value-added chemicals from methanol.

Future trends

Two key roadblocks were identified while attempting to engineer native methylotrophs for methanol conversion. First is the insufficient knowledge on the regulatory mechanisms. For example, engineering even one of the best-studied methylotrophs, *M. extorquens,* for biotechnological applications requires more knowledge on both the essential genes and their associated regulatory networks. The integrated multiple 'omics' approaches and the genome editing tools that are becoming available for a multitude of organisms, will aid in elucidating the essential sets of proteins-metabolites that control the regulatory networks (Jiang *et al.*, 2013; Diether and Sauer, 2017). Recent discovery of the phosphoribulokinase gene playing an unexpected but essential regulatory role provides one example (Ochsner *et al.*, 2017). The second roadblock is the lack of well-developed genetic tools and genetic elements. Schada von Borzyskowski and collaborators demonstrated that extensive sets of plasmids with different properties are required to allow for efficient construction of synthetic operons with different promoter strengths in *M. extorquens* AM1 (Schada von Borzyskowski *et al.*, 2015). Liang and colleagues developed biosensor-assisted transcriptional regulator engineered to control metabolic flux re-distribution between C1 assimilation and chemicals biosynthesis (Liang *et al.*, 2017a). Future work is necessary to explore regulatory mechanism that may not be directly involved in methanol metabolism, but could potentially improve methanol assimilation. It is also necessary to develop advanced engineering tools, such as promoters and RBS's of different strengths to allow for fine tuning gene expression, as well as the convenient Cre/loxP systems for integrating entire heterologous pathways into the genomes of *M. extorquens* (Liang *et al.*, 2017b) and other potential platforms.

For synthetic methylotrophs, a persisting challenge remains the lack of growth on methanol as the sole carbon and energy source. One limitation is the efficiency of methanol oxidation catalyzed by NAD-dependent Mdh. So far, the most studied is the Mdh from *B. methanolicus*, which exhibits low enzyme specificity toward methanol and dependency on an activator protein (Price *et al.*, 2016). Recently, Wu and collaborators characterized a NAD-dependent alcohol dehydrogenase (Mdh2) from *Cupriavidus necator* that does not require an activator (Wu *et al.*, 2016). Further, they engineered a variant of Mdh2 by using directed molecular evolution that showed a 6-fold increase in $Kcat/Km$ for methanol and 10-fold decrease in $Kcat/Km$ for 1-butanol than that of Mdh from *B. methanolicus*. Another limitation is the insufficient flux toward regeneration of Ru5P. Bennett and colleagues demonstrated significant progress towards solving this problem, but further progress is needed (Bennett *et al.*, 2018a). Recently, Cui and colleagues used ARTP mutagenesis to identify a mutant with obvious improvement of methanol tolerance and consumption (Cui *et al.*, 2018). This state-of-the-art technique could also be adapted to evolve current synthetic methylotrophs and screen mutants for increased methanol conversion in the future.

References

Anne, K., Heggeset, T.M.B., Müller, J.E.N., Kupper, C.E., Olha, S., Vorholt, J.A., Ellingsen, T.E., and Trygve, B. (2013). Methylotrophic *Bacillus methanolicus* encodes two chromosomal and one plasmid Born NAD+-dependent methanol dehydrogenase paralogs with different catalytic and biochemical properties. PLOS ONE *8*, e59188.

Anthony, C. (2011). How half a century of research was required to understand bacterial growth on C1 and C2 compounds; the story of the serine cycle and the ethylmalonyl-CoA pathway. Sci. Prog. *94*, 109–137.

Bennett, R.K., Gonzalez, J.E., Whitaker, W.B., Antoniewicz, M.R., and Papoutsakis, E.T. (2018a). Expression of heterologous non-oxidative pentose phosphate pathway from *Bacillus methanolicus* and phosphoglucose isomerase deletion improves methanol assimilation and metabolite production by a synthetic *Escherichia coli* methylotroph. Metab. Eng. *45*, 75–85.

Bennett, R.K., Steinberg, L.M., Chen, W., and Papoutsakis, E.T. (2018b). Engineering the bioconversion of methane and methanol to fuels and chemicals in native and synthetic methylotrophs. Curr. Opin. Biotechnol. *50*, 81–93.

Bird, C.W., Lynch, J.M., Pirt, F.J., and Reid, W.W. (1971). Steroids and squalene in *Methylococcus capsulatus* grown on methane. Nature *230*, 473–474.

Bringel, F., and Vuilleumier, S. (2017). Metabolic regulation: a master role for ribulose-1,5-bisphosphate in one-carbon assimilation. Curr. Biol. *27*, R1127–R1129.

Chistoserdova, L. (2018). Applications of methylotrophs: can single carbon be harnessed for biotechnology? Curr. Opin. Biotechnol. *50*, 189–194.

Chistoserdova, L.V., and Lidstrom, M.E. (1994). Genetics of the serine cycle in *Methylobacterium extorquens* AM1: identification of sgaA and mtdA and sequences of sgaA, hprA, and mtdA. J. Bacteriol. *176*, 1957–1968.

Chistoserdova, L., Gomelsky, L., Vorholt, J.A., Gomelsky, M., Tsygankov, Y.D., and Lidstrom, M.E. (2000). Analysis of two formaldehyde oxidation pathways in *Methylobacillus flagellatus* KT, a ribulose monophosphate cycle methylotroph. Microbiology *146*, 233–238. https://doi.org/10.1099/00221287-146-1-233

Chistoserdova, L., Chen, S.W., Lapidus, A., and Lidstrom, M.E. (2003). Methylotrophy in *Methylobacterium extorquens* AM1 from a genomic point of view. J. Bacteriol. *185*, 2980–2987.

Chistoserdova, L., Kalyuzhnaya, M.G., and Lidstrom, M.E. (2009). The expanding world of methylotrophic metabolism. Annu. Rev. Microbiol. *63*, 477–499. https://doi.org/10.1146/annurev.micro.091208.073600

Clomburg, J.M., Crumbley, A.M., and Gonzalez, R. (2017). Industrial biomanufacturing: The future of chemical production. Science *355*, aag0804.

Cui, J., Good, N.M., Hu, B., Yang, J., Wang, Q., Sadilek, M., and Yang, S. (2016). Metabolomics revealed an association of metabolite changes and defective growth in methylobacterium extorquens AM1 overexpressing ecm during growth on methanol. PLOS ONE *11*, e0154043. https://doi.org/10.1371/journal.pone.0154043

Cui, L.Y., Wang, S.S., Guan, C.G., Liang, W.F., Xue, Z.L., Zhang, C., and Xing, X.H. (2018). Breeding of methanol-tolerant *Methylobacterium extorquens* AM1 by atmospheric and room temperature plasma mutagenesis combined with adaptive laboratory evolution. Biotechnol. J. *13*, e1700679. https://doi.org/10.1002/biot.201700679

Dai, Z., Gu, H., Zhang, S., Xin, F., Zhang, W., Dong, W., Ma, J., Jia, H., and Jiang, M. (2017). Metabolic construction strategies for direct methanol utilization in *Saccharomyces cerevisiae*. Bioresour. Technol. *245*, 1407–1412.

Danilov, V.S., and Ismailov, A.D. (1989). Bacterial luciferase as a biosensor of biologically active compounds. Biotechnology *11*, 39–78.

de la Torre, A., Metivier, A., Chu, F., Laurens, L.M., Beck, D.A., Pienkos, P.T., Lidstrom, M.E., and Kalyuzhnaya, M.G. (2015). Genome-scale metabolic reconstructions and theoretical investigation of methane conversion in Methylomicrobium buryatense strain 5G(B1). Microb. Cell Fact. *14*, 188. https://doi.org/10.1186/s12934-015-0377-3

Diether, M., and Sauer, U. (2017). Towards detecting regulatory protein-metabolite interactions. Curr. Opin. Microbiol. *39*, 16–23.

Erb, T.J., Berg, I.A., Brecht, V., Müller, M., Fuchs, G., and Alber, B.E. (2007). Synthesis of C5-dicarboxylic acids from C2-units involving crotonyl-CoA carboxylase/reductase: the ethylmalonyl-CoA pathway. Proc. Natl. Acad. Sci. U.S.A. *104*, 10631–10636.

Fu, Y., Beck, D.A., and Lidstrom, M.E. (2016). Difference in C3-C4 metabolism underlies tradeoff between growth rate and biomass yield in *Methylobacterium extorquens* AM1. BMC Microbiol. *16*, 156. https://doi.org/10.1186/s12866-016-0778-4

Guo, X., and Lidstrom, M.E. (2006). Physiological analysis of *Methylobacterium extorquens* AM1 grown in continuous and batch cultures. Arch. Microbiol. *186*, 139–149. https://doi.org/10.1007/s00203-006-0131-7

Harada, H., Yu, F., Okamoto, S., Kuzuyama, T., Utsumi, R., and Misawa, N. (2009). Efficient synthesis of functional isoprenoids from acetoacetate through metabolic pathway-engineered *Escherichia coli*. Appl. Microbiol. Biotechnol. *81*, 915–925. https://doi.org/10.1007/s00253-008-1724-7

Hu, B., and Lidstrom, M.E. (2014). Metabolic engineering of *Methylobacterium extorquens* AM1 for 1-butanol production. Biotechnol. Biofuels *7*, 156. https://doi.org/10.1186/s13068-014-0156-0

Hu, B., Yang, Y.M., Beck, D.A., Wang, Q.W., Chen, W.J., Yang, J., Lidstrom, M.E., and Yang, S. (2016). Comprehensive molecular characterization of *Methylobacterium extorquens* AM1 adapted for 1-butanol tolerance. Biotechnol. Biofuels *9*, 84. https://doi.org/10.1186/s13068-016-0497-y

Jiang, W., Bikard, D., Cox, D., Zhang, F., and Marraffini, L.A. (2013). RNA-guided editing of bacterial genomes using CRISPR-Cas systems. Nat. Biotechnol. *31*, 233–239. https://doi.org/10.1038/nbt.2508

Kalyuzhnaya, M.G., Yang, S., Rozova, O.N., Smalley, N.E., Clubb, J., Lamb, A., Gowda, G.A., Raftery, D., Fu, Y., Bringel, F., *et al.* (2013). Highly efficient methane biocatalysis revealed in a methanotrophic bacterium. Nat. Commun. *4*, 2785. https://doi.org/10.1038/ncomms3785

Lee, S.Y., Park, J.H., Jang, S.H., Nielsen, L.K., Kim, J., and Jung, K.S. (2008). Fermentative butanol production by Clostridia. Biotechnol. Bioeng. *101*, 209–228. https://doi.org/10.1002/bit.22003

Leßmeier, L., Pfeifenschneider, J., Carnicer, M., Heux, S., Portais, J.C., and Wendisch, V.F. (2015). Production of carbon-13-labeled cadaverine by engineered *Corynebacterium glutamicum* using carbon-13-labeled methanol as co-substrate. Appl. Microbiol. Biotechnol. *99*, 10163–10176. https://doi.org/10.1007/s00253-015-6906-5

Liang, W.F., Cui, L.Y., Cui, J.Y., Yu, K.W., Yang, S., Wang, T.M., Guan, C.G., Zhang, C., and Xing, X.H. (2017a). Biosensor-assisted transcriptional regulator engineering for *Methylobacterium extorquens* AM1 to improve mevalonate synthesis by increasing the acetyl-CoA supply. Metab. Eng. *39*, 159–168.

Liang, W.F., Sun, M.Y., Cui, L.Y., Zhang, C., and Xing, X.H. (2017b). Cre/loxP-mediated multicopy integration of the mevalonate operon into the genome of *Methylobacterium extorquens* AM1. Appl. Biochem. Biotechnol. *2*, 1–13.

Lidstrom, M.E. (2006). Aerobic Methylotrophic Prokaryotes. In The Prokaryotes, Vol. 2: Ecophysiology and Biochemistry, Dworkin, M., Falkow, S., Rosenberg, E., Schleifer, K.-H., and Stackebrandt, E., eds. (Springer-Verlag, New York, NY), pp. 618–634.

Lieven, C., Herrgård, M.J., and Sonnenschein, N. (2018). Microbial Methylotrophic metabolism: recent metabolic modeling efforts and their applications in industrial biotechnology. Biotechnol. J. *13*, e1800011. https://doi.org/10.1002/biot.201800011

Liu, Y., Bai, C., Xu, Q., Yu, J., Zhou, X., Zhang, Y., and Cai, M. (2018a). Improved methanol-derived lovastatin production through enhancement of the biosynthetic pathway and intracellular lovastatin efflux in methylotrophic yeast. Bioresources & Bioprocessing *5*, 22.

Liu, Y., Tu, X., Xu, Q., Bai, C., Kong, C., Liu, Q., Yu, J., Peng, Q., Zhou, X., Zhang, Y., *et al.* (2018b). Engineered monoculture and co-culture of methylotrophic yeast for de novo production of monacolin J and lovastatin from methanol. Metab. Eng. *45*, 189–199.

Meyer, F., Keller, P., Hartl, J., Gröninger, O.G., Kiefer, P., and Vorholt, J.A. (2018). Methanol-essential growth of *Escherichia coli*. Nat. Commun. *9*, 1508. https://doi.org/10.1038/s41467-018-03937-y

Müller, J.E., Meyer, F., Litsanov, B., Kiefer, P., and Vorholt, J.A. (2015a). Core pathways operating during methylotrophy of *Bacillus methanolicus* MGA3 and induction of a bacillithiol-dependent detoxification pathway upon formaldehyde stress. Mol. Microbiol. *98*, 1089–1100. https://doi.org/10.1111/mmi.13200

Müller, J.E.N., Meyer, F., Litsanov, B., Kiefer, P., Potthoff, E., Heux, S., Quax, W.J., Wendisch, V.F., Brautaset, T., Portais, J.C., *et al.* (2015b). Engineering *Escherichia coli* for methanol conversion. Metab. Eng. *28*, 190–201.

Nagai, H., Masuda, A., Toya, Y., Matsuda, F., and Shimizu, H. (2018). Metabolic engineering of mevalonate-producing Escherichia coli strains based on thermodynamic analysis. Metab. Eng. *47*, 1–9.

Ochsner, A.M., Sonntag, F., Buchhaupt, M., Schrader, J., and Vorholt, J.A. (2015). *Methylobacterium extorquens*: methylotrophy and biotechnological applications. Appl. Microbiol. Biotechnol. *99*, 517–534. https://doi.org/10.1007/s00253-014-6240-3

Ochsner, A.M., Christen, M., Hemmerle, L., Peyraud, R., Christen, B., and Vorholt, J.A. (2017). Transposon sequencing uncovers an essential regulatory function of phosphoribulokinase for methylotrophy. Curr. Biol. *27*, 2579–2588.e6.

Peyraud, R., Kiefer, P., Christen, P., Massou, S., Portais, J.C., and Vorholt, J.A. (2009). Demonstration of the ethylmalonyl-CoA pathway by using 13C metabolomics. Proc. Natl. Acad. Sci. U.S.A. *106*, 4846–4851. https://doi.org/10.1073/pnas.0810932106

Peyraud, R., Schneider, K., Kiefer, P., Massou, S., Vorholt, J.A., and Portais, J.C. (2011). Genome-scale reconstruction and system level investigation of the metabolic network of *Methylobacterium extorquens* AM1. BMC Syst. Biol. *5*, 189. https://doi.org/10.1186/1752-0509-5-189

Price, J.V., Chen, L., Whitaker, W.B., Papoutsakis, E., and Chen, W. (2016). Scaffoldless engineered enzyme assembly for enhanced methanol utilization. Proc. Natl. Acad. Sci. U.S.A. *113*, 12691–12696. https://doi.org/10.1073/pnas.1601797113

Rohde, M.T., Tischer, S., Harms, H., and Rohwerder, T. (2017). Production of 2-hydroxyisobutyric acid from methanol by *Methylobacterium extorquens* AM1 Expressing (R)-3-Hydroxybutyryl Coenzyme A-Isomerizing Enzymes. Appl. Environ. Microbiol. *83*, e02622–16.

Rohlhill, J., Sandoval, N.R., and Papoutsakis, E.T. (2017). Sort-seq approach to engineering a formaldehyde-inducible promoter for dynamically regulated *Escherichia coli* growth on methanol. ACS Synth. Biol. *6*, 1584–1595. https://doi.org/10.1021/acssynbio.7b00114

Schada von Borzyskowski, L., Remus-Emsermann, M., Weishaupt, R., Vorholt, J.A., and Erb, T.J. (2015). A set of versatile brick vectors and promoters for the assembly, expression, and integration of synthetic operons in *Methylobacterium extorquens* AM1 and other alphaproteobacteria. ACS Synth. Biol. *4*, 430–443. https://doi.org/10.1021/sb500221v

Schada von Borzyskowski, L., Carrillo, M., Leupold, S., Glatter, T., Kiefer, P., Weishaupt, R., Heinemann, M., and Erb, T.J. (2018). An engineered Calvin-Benson-Bassham cycle for carbon dioxide fixation in *Methylobacterium extorquens* AM1. Metab. Eng. *47*, 423–433.

Sirirote, P., Tsuneo, Y., and Shoichi, S. (1988). l-serine production from methanol and glycine with an immobilized methylotroph. J. Ferment. Technol. *66*, 291–297.

Sonntag, F., Buchhaupt, M., and Schrader, J. (2014). Thioesterases for ethylmalonyl-CoA pathway derived dicarboxylic acid production in *Methylobacterium extorquens* AM1. Appl. Microbiol. Biotechnol. *98*, 4533–4544. https://doi.org/10.1007/s00253-013-5456-y

Sonntag, F., Kroner, C., Lubuta, P., Peyraud, R., Horst, A., Buchhaupt, M., and Schrader, J. (2015a). Engineering *Methylobacterium extorquens* for de novo synthesis of the sesquiterpenoid α-humulene from methanol. Metab. Eng. *32*, 82–94.

Sonntag, F., Müller, J.E., Kiefer, P., Vorholt, J.A., Schrader, J., and Buchhaupt, M. (2015b). High-level production of ethylmalonyl-CoA pathway-derived dicarboxylic acids by *Methylobacterium extorquens* under cobalt-deficient conditions and by polyhydroxybutyrate negative strains. Appl. Microbiol. Biotechnol. *99*, 3407–3419. https://doi.org/10.1007/s00253-015-6418-3

Strom, T., Ferenci, T., and Quayle, J.R. (1974). The carbon assimilation pathways of Methylococcus capsulatus, Pseudomonas methanica and Methylosinus trichosporium (OB3B) during growth on methane. Biochem. J. *144*, 465–476.

Whitaker, W.B., Sandoval, N.R., Bennett, R.K., Fast, A.G., and Papoutsakis, E.T. (2015). Synthetic methylotrophy: engineering the production of biofuels and chemicals based on the biology of aerobic methanol utilization. Curr. Opin. Biotechnol. *33*, 165–175. https://doi.org/10.1016/j.copbio.2015.01.007

Whitaker, W.B., Jones, J.A., Bennett, R.K., Gonzalez, J.E., Vernacchio, V.R., Collins, S.M., Palmer, M.A., Schmidt, S., Antoniewicz, M.R., Koffas, M.A., *et al.* (2017). Engineering the biological conversion of methanol to specialty chemicals in Escherichia coli. Metab. Eng. *39*, 49–59.

Witthoff, S., Schmitz, K., Niedenführ, S., Nöh, K., Noack, S., Bott, M., and Marienhagen, J. (2015). Metabolic engineering of *Corynebacterium glutamicum* for methanol metabolism. Appl. Environ. Microbiol. *81*, 2215–2225. https://doi.org/10.1128/AEM.03110-14

Woolston, B.M., Roth, T., Kohale, I., Liu, D.R., and Stephanopoulos, G. (2018). Development of a formaldehyde biosensor with application to synthetic methylotrophy. Biotechnol. Bioeng. *115*, 206–215. https://doi.org/10.1002/bit.26455

Wu, T.Y., Chen, C.T., Liu, J.T., Bogorad, I.W., Damoiseaux, R., and Liao, J.C. (2016). Characterization and evolution of an activator-independent methanol dehydrogenase from Cupriavidus necator N-1. Appl. Microbiol. Biotechnol. *100*, 4969–4983. https://doi.org/10.1007/s00253-016-7320-3

Yang, Y.M., Chen, W.J., Yang, J., Zhou, Y.M., Hu, B., Zhang, M., Zhu, L.P., Wang, G.Y., and Yang, S. (2017). Production of 3-hydroxypropionic acid in engineered *Methylobacterium extorquens* AM1 and its reassimilation through a reductive route. Microb. Cell Fact. *16*, 179. https://doi.org/10.1186/s12934-017-0798-2

Zhang, L., Xiao, W.H., Wang, Y., Yao, M.D., Jiang, G.Z., Zeng, B.X., Zhang, R.S., and Yuan, Y.J. (2017). Chassis and key enzymes engineering for monoterpenes production. Biotechnol. Adv. *35*, 1022–1031.

Zhao, L., Chang, W.C., Xiao, Y., Liu, H.W., and Liu, P. (2013). Methylerythritol phosphate pathway of isoprenoid biosynthesis. Annu. Rev. Biochem. *82*, 497–530. https://doi.org/10.1146/annurev-biochem-052010-100934

Zhu, W.L., Cui, J.Y., Cui, L.Y., Liang, W.F., Yang, S., Zhang, C., and Xing, X.H. (2016). Bioconversion of methanol to value-added mevalonate by engineered *Methylobacterium extorquens* AM1 containing an optimized mevalonate pathway. Appl. Microbiol. Biotechnol. *100*, 2171–2182. https://doi.org/10.1007/s00253-015-7078-z

Synthetic Methanol and Formate Assimilation Via Modular Engineering and Selection Strategies

14

Nico J. Claassens, Hai He and Arren Bar-Even*

Max Planck Institute of Molecular Plant Physiology, Potsdam-Golm, Germany.

*Correspondence: bar-even@mpimp-golm.mpg.de

https://doi.org/10.21775/9781912530045.14

Abstract

One-carbon (C1) feedstocks can provide a vital link between cheap and sustainable abiotic resources and microbial bioproduction. Soluble C1 substrates – methanol and formate – could prove to be more suitable than gaseous feedstocks as they avoid mass transfer barriers. However, microorganisms that naturally assimilate methanol and formate are limited by a narrow product spectrum and a restricted genetic toolbox. Engineering biotechnological organisms to assimilate these soluble C1 substrates has therefore become an attractive goal. Here, we discuss the use of a step-wise, modular engineering approach for the implementation of C1 assimilation pathways. In this strategy, pathways are divided into metabolic modules, the activities of which are selected for in dedicated gene-deletion strains whose growth directly depends on module activity. This provides an easy way to identify and resolve metabolic barriers hampering pathway performance. Optimization of gene expression levels and adaptive laboratory evolution can be used to establish the desired activity if direct selection fails. We exemplify this approach using several pathways, focusing especially on the ribulose monophosphate cycle for methanol assimilation and the reductive glycine pathway for formate assimilation. We argue that such modular engineering and selection strategies will prove essential for rewiring microbial metabolism towards new growth phenotypes and sustainable bioproduction.

Introduction

One-carbon (C1) compounds could prove to be a crucial link between the abiotic and the biotic worlds. These feedstocks can be obtained from low-cost and abundant sources, such as syngas and natural gas (Dürre and Eikmanns, 2015; Clomburg *et al.*, 2017), and can be produced directly from CO_2 using energy sources such as sunlight and renewable electricity (Kumar *et al.*, 2012; Martín *et al.*, 2015; Claassens *et al.*, 2018; Jouny *et al.*, 2018). Multiple microorganisms can be cultivated on C1 compounds as sole carbon and energy sources, thus opening new avenues for sustainable bioproduction.

However, the use of microorganisms that can naturally grow on C1 substrates is limited by multiple factors, including a narrow product spectrum, low yields, titres, and productivities, a restricted genetic toolbox for engineering, and gaps in our understanding of their cellular physiology and metabolism (Whitaker *et al.*, 2015; Clomburg *et al.*, 2017). To overcome these difficulties, recent metabolic engineering efforts are aiming to introduce C1 assimilation pathways into model biotechnological microorganisms that are easier to engineer and that can be better optimized for industrially relevant conditions. These efforts use either natural pathways that are known to sustain high yields, or, more boldly, synthetic pathways with low ATP cost that could theoretically support increased yields (Bar-Even *et al.*, 2013; Siegel *et al.*, 2015; Bar-Even,

2016;). Some of these synthetic pathways can be established by combining naturally existing enzymes, while others include novel enzyme activities that can be realized by protein engineering (Erb *et al.*, 2017). In fact, engineered enzymes have already been demonstrated *in vitro* to support formate assimilation (Siegel *et al.*, 2015) and carbon fixation (Schwander *et al.*, 2016).

In this review, we discuss metabolic engineering studies aiming to introduce pathways for the assimilation of the soluble C1 compounds methanol and formate, the utilization of which bypasses the challenges associated with mass transfer of gaseous C1 substrates, such as methane and carbon monoxide (Henstra *et al.*, 2007; Fei *et al.*, 2014). We specifically focus on modular and selection-based engineering strategies in which the activity of pathway segments is coupled to cellular growth. We show that this step-wise approach is vital for the realization of synthetic C1 assimilation.

Modularity and selection as metabolic engineering strategies

Engineering synthetic C1 metabolism requires the overexpression of pathway enzymes, especially those that are missing in the host or that are natively expressed at insufficient levels. However, simple overexpression is unlikely to be sufficient for realizing the activity of the entire pathway. This is mainly because of the overlap between the introduced pathway and the host central metabolism, resulting in disrupted fluxes through both systems. To better identify and resolve problematic metabolic interactions, it is helpful to divide the synthetic pathway into smaller metabolic modules, i.e. sub-pathways consisting of several reactions (Fig. 14.1). The *in vivo* implementation of these modules can be considerably easier than the full pathway and provide vital information on the metabolic context that enables or constrains the newly introduced

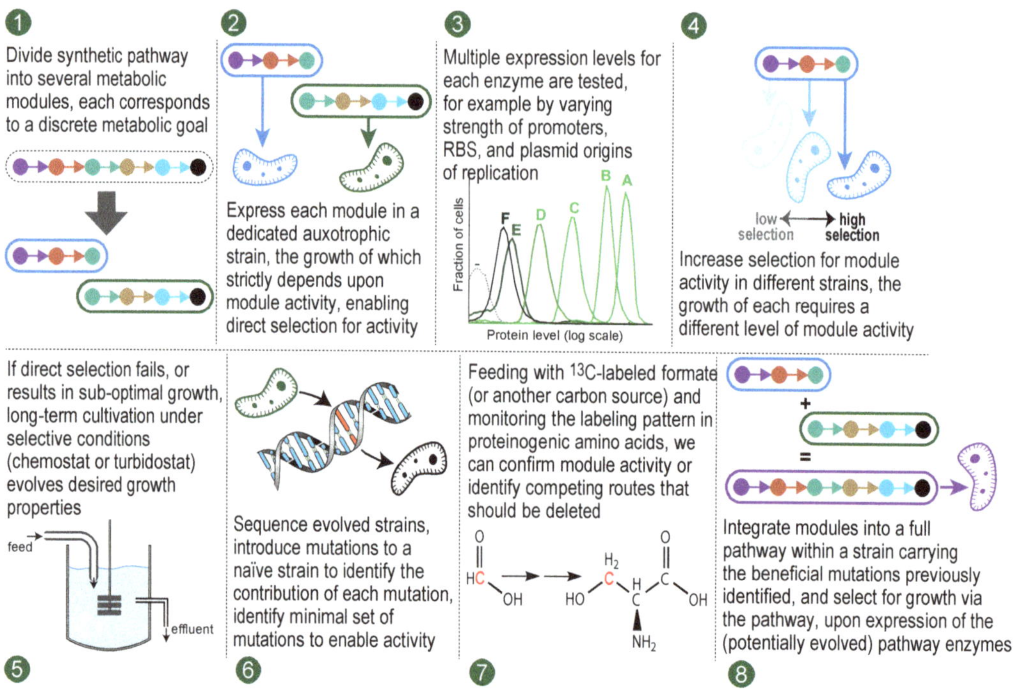

Figure 14.1 A schematic representation of the modular engineering and selection approach outlined in this paper.

activities. To probe the implementation of metabolic modules, it is useful to couple their activity with the growth of the host.

Coupling module activity with growth usually requires modifying the metabolic network of the host by performing strategic gene deletions. These are made to generate a strain auxotrophic for certain essential metabolic intermediates – for example, an amino acid – which can be exclusively synthesized via the synthetic module. As a result, cellular growth becomes dependent on the activity of the module. A range of selection strains can be designed with increasing selection pressure for pathway activity: a 'minimal' selection is sustained if the module provides a single required metabolite, higher selection pressure is obtained when module activity is responsible for the biosynthesis of multiple building blocks, and very high selection pressure is imposed when the biosynthesis of all or most biomass is dependent on the module.

The design of modules and selection strains can be assisted by computational tools based on Flux Balance Analysis, for example OptKnock or FlexFlux (Burgard *et al.*, 2003; Marmiesse *et al.*, 2015; Meyer *et al.*, 2018). Yet, in most cases, manual design based on biochemical and metabolic knowledge suffices. Specifically, when dividing a pathway into metabolic modules, several factors should be taken into consideration (Wenk *et al.*, 2018): (1) the module should be linked to a clear growth phenotype within an appropriate selection strain, resulting in growth versus no growth readout; (2) the number of enzymes in a module should be limited, to enable easy expression and optimization, and to allow straightforward interpretation of growth phenotypes; (3) modules should together cover the whole pathway and could overlap with one another, such that enzymes occurring in multiple modules can be tested in different metabolic contexts; (4) ideally, modules should be easy to combine, that is, dedicated selection strains – whose growth is dependent on the activity of several consecutive modules – should be easy to construct.

When direct selection for module activity fails or results in poor growth, further optimization is required (Fig. 14.1). This can be achieved by modulating the expression levels of the pathway enzymes (Zelcbuch *et al.*, 2013; Wenk *et al.*, 2018), and potentially also of related host enzymes, e.g. deletion or down-regulation of enzymes that divert metabolic intermediates from the pathway. In addition, different enzyme variants or codon optimization of the relevant genes can support increased expression and activity. Another method, which does not rely on genetic tools, is the addition of small molecules that specifically inhibit interfering enzymes, as demonstrated for the glycolytic glyceraldehyde 3-phosphate dehydrogenase in the engineering of methanol assimilation in *Escherichia coli* (Woolston *et al.*, 2018a).

If these approaches fail to establish module activity, adaptive laboratory evolution (ALE) can be performed to increase module functionality and establish module-dependent growth (Portnoy *et al.*, 2011). For this process, the overexpressed genes should preferably be integrated into the genome rather than carried on a plasmid as to increase the chance of beneficial mutations to be fixed in the population. Different types of ALE can be applied; a prominent approach being continuous cultivation on a selective medium, with limiting amounts of the compounds for which the cells are auxotrophic. This method was applied for the successful engineering of the CO_2-fixing Calvin cycle in *E. coli* (Antonovsky *et al.*, 2016), where the concentration of the limiting substrate xylose was gradually decreased as growth improved. Another approach involves swapping between permissive and selection media, where the former contains the auxotrophy-related compounds and the latter does not (Marlière *et al.*, 2011). The relative dosages of the different media are coupled to the growth of the population, such that increased cell density leads to addition of selection medium, and a decrease in cell density results in more permissive medium. Such cultivation regime adapts the population towards using the selection medium until the permissive medium is no longer required (Marlière *et al.*, 2011; Bouzon *et al.*, 2017; Döring *et al.*, 2018). Following successful evolution of module activity, the evolved genomes can be sequenced to identify the accumulated mutations. The specific effect of different mutations can be interpreted and further studied by reintroducing them into a non-evolved selection strain.

Synthetic methanol assimilation

Methanol can be produced from diverse fossil and renewable sources and has been proposed as a promising feedstock for industrial applications and microbial growth (Olah *et al.*, 2009; Olah, 2013; Zhang *et al.*, 2017). Hence, biological conversion of methanol to products has received considerable attention (Schrader *et al.*, 2009; Whitaker *et al.*, 2015; Pfeifenschneider *et al.*, 2017; Bennett *et al.*, 2018a). As genetic toolboxes for the engineering of most promising natural methylotrophs (e.g. *Bacillus methanolicus*) are still underdeveloped, engineering model biotechnological microbes for growth on methanol has become an attractive target.

Three native pathways are known to support growth on methanol: the ribulose monophosphate (RuMP) cycle, the xylulose monophosphate cycle, and the serine pathway (Kato *et al.*, 2006; Chistoserdova *et al.*, 2009). Among these, the engineering of the RuMP cycle has received most attention, as this route supports the highest yield (Bar-Even *et al.*, 2013). The heterologous establishment of the serine pathway has not yet been reported and only a single study aimed at engineering the xylulose monophosphate pathway in *Saccharomyces cerevisiae* (Dai *et al.*, 2017).

A synthetic methanol assimilation pathway has also been proposed, where three formaldehyde molecules are condensed to dihydroxyacetone by a rationally engineered formolase enzyme. This pathway was recently introduced in *E. coli*, but did not lead to substantial methanol assimilation, probably due to the poor kinetics of the formolase enzyme (Wang *et al.*, 2017). Another proposed synthetic methanol condensation cycle, which was demonstrated *in vitro*, consists of the RuMP cycle combined with non-oxidative glycolysis (Bogorad *et al.*, 2014). In this pathway, fructose-6-phosphate (F6P) is cleaved by phosphoketolases to produce acetyl-CoA, bypassing pyruvate decarboxylation and preventing carbon loss. Finally, an alternative, synthetic structure of the serine pathway was proposed, in which serine is deaminated to pyruvate, and glycine is recycled via threonine biosynthesis and degradation, further generating acetyl-CoA as a biomass precursor (Bar-Even, 2016). A recent study describes the successful selection for key metabolic modules of this synthetic route, that is, formate assimilation into serine via the tetrahydrofolate (THF) system and glycine production solely from threonine cleavage (Yishai *et al.*, 2017).

As the RuMP cycle (Fig. 14.2a) attracts the most research attention so far, we will focus on attempts for its engineering in biotechnological hosts. Non-methylotrophic hosts typically lack only three enzymes of the pathway: methanol dehydrogenase, hexulose-6-phosphate synthase and 6-phospho-3-hexulosiomerase. Several studies have overexpressed these enzymes in *E. coli* (Müller *et al.*, 2015; Price *et al.*, 2016; Whitaker *et al.*, 2017; Bennett *et al.*, 2018b; Gonzalez *et al.*, 2018) and *Corynebacterium glutamicum* (Leßmeier *et al.*, 2015; Witthoff *et al.*, 2015) and have demonstrated methanol assimilation, albeit at low rates and assimilation efficiencies. These efforts, however, did not apply a selection strategy, that is, cellular growth was not dependent on methanol assimilation.

The RuMP cycle can be divided into three main modules: methanol oxidation to formaldehyde, formaldehyde assimilation into central metabolism, and regeneration of the acceptor metabolite ribulose 5-phosphate (Ru5P) (Fig. 14.2b). In the assimilation module, formaldehyde is condensed with Ru5P to generate F6P, which can be metabolized to all biomass building blocks. The assimilation and regeneration modules can be supported by several alternative metabolic structures, the most efficient one uses glycolysis and the non-oxidative pentose phosphate pathway (Quayle and Ferenci, 1978; Zhang *et al.*, 2017).

Several recent studies have attempted direct selection for the activities of the methanol oxidation module and the formaldehyde assimilation module in *E. coli* (Chen *et al.*, 2018; Meyer *et al.*, 2018) and *C. glutamicum* (Tuyishime *et al.*, 2018). These studies generated a selection strain in which ribose 5-phosphate isomerase is deleted (*ΔrpiAB*). This knockout blocked growth on xylose or gluconate (the latter with the additional deletion of *edd*, encoding phosphogluconate dehydratase). The methanol oxidation and formaldehyde assimilation modules were expected to rescue growth on xylose or gluconate by enabling their assimilation via a 'RuMP shunt', a linear route converting Ru5P to F6P via condensation with formaldehyde (Fig. 14.2c and d). Yet, none of these studies was able to demonstrate the required activity upon direct

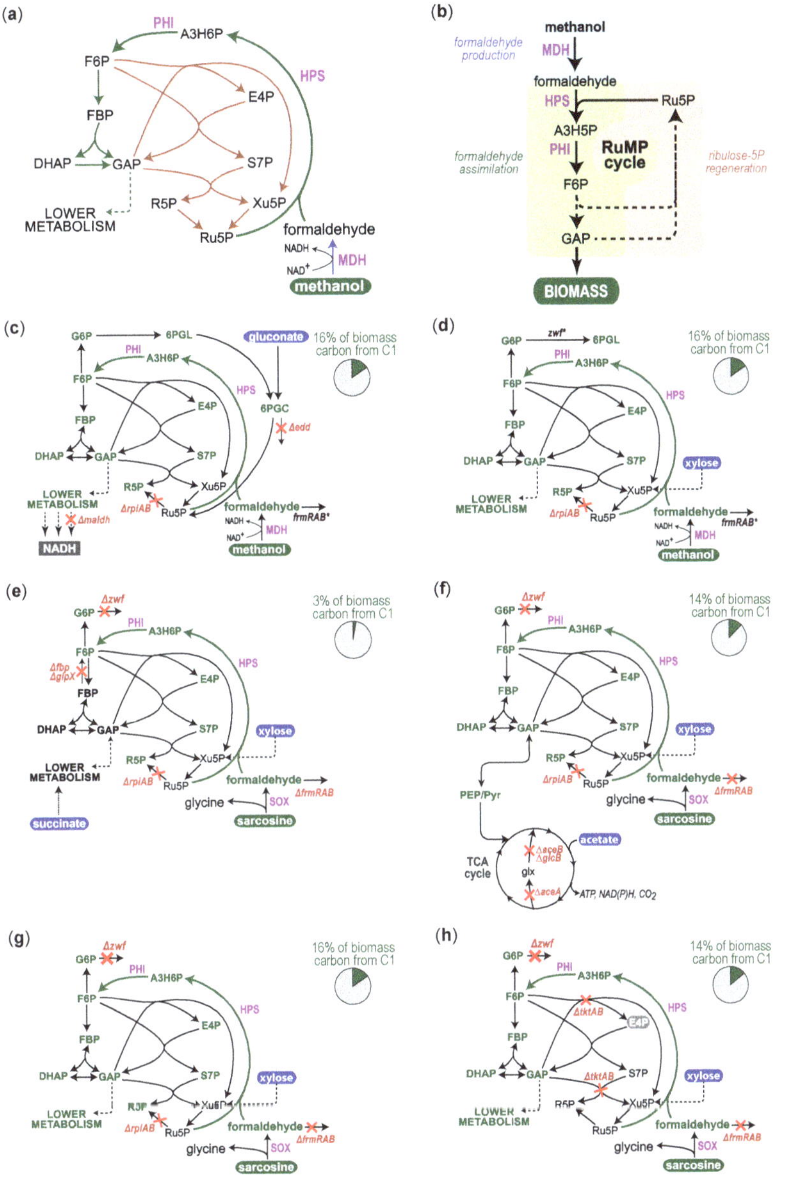

Figure 14.2 Engineering the RuMP shunt in *E. coli*. (a) Metabolic structure of a variant of the RuMP cycle. (b) Subdivision of the RuMP cycle into metabolic modules; (c, d) Selection schemes for methanol-dependent growth of *E. coli* via the RuMP shunt as described by Meyer *et al.* (2018) (c) and Chen *et al.* (2018) (d). (e–h) Selection schemes for growth via the RuMP shunt at different selection strengths as described in He *et al.* (2018). Pie charts indicate the minimum fraction of carbons in biomass that are derived from C1 (formaldehyde or methanol) in the different selection schemes, as calculated using (Neidhardt *et al.*, 1990). Gene deletions are shown in red; overexpressed enzymes in purple; enzymes that mutated during ALE in brown; metabolites that are dependent on the RuMP shunt in green; and substrates that have to be co-fed in blue. Enzyme abbreviations: aceA, isocitrate lyase; aceB, malate synthase B; edd, phosphogluconate dehydratase; fbp; fructose-1,6-bisphosphatase; frmRAB, glutathione-dependent formaldehyde detoxification system; glcB, malate synthase G; glpX, fructose-1,6-bisphosphatase 2; HPS, 3-hexulose-6-phosphate synthase; maldh, malate dehydrogenase; MDH, methanol dehydrogenase; PHI, 6-phospho-3-hexuloisomerase; rpiAB, ribose 5-phosphate isomerase; SOX, sarcosine oxidase; tktAB, transketolase A and B; zwf, glucose 6-phosphate dehydrogenase. Metabolite abbreviations: A3H6P, arabino 3-hexulose 6-phosphate; DHAP, dihydroxyacetone phosphate; E4P, erythrose 4-phosphate; F6P, fructose 6-phosphate; FBP, fructose bisphosphate; GAP, glyceraldehyde 3-phosphate; 6PGC, 6-phospho-gluconate; 6PGL, 6-phospho-glucono-lactone; R5P, ribose 5-phosphate; Ru5P, ribulose 5-phosphate; S7P, sedoheptulose 7-phosphate; Xu5P, xylulose 5-phosphate.

selection, and ALE was necessary to establish methanol-assimilation-dependent growth.

The strict requirement for ALE to achieve RuMP shunt-dependent growth can be attributed to the unfavorability of methanol oxidation. Supporting this notion, all of these studies found mutations linked to a decreased NADH/NAD ratio, e.g. by interrupting or inhibiting the TCA cycle (Chen *et al.*, 2018; Meyer *et al.*, 2018; Tuyishime *et al.*, 2018). This likely relates to the fact that methanol oxidation is thermodynamically limited by a high NADH/NAD ratio. Using deuterated methanol (CD_3OD), it was shown that methanol dehydrogenase also kinetically limits the activity of the RuMP shunt in *E. coli* (Woolston *et al.*, 2018a). Methanol oxidation could be improved by identifying or engineering kinetically superior variants, which could be directly screened using a formaldehyde biosensor (Woolston *et al.*, 2018b) or selected for in appropriate selection strains.

To overcome the barriers associated with methanol oxidation, we decided to separate the formaldehyde production module from that of formaldehyde assimilation, such that we could test the latter in more detail. Towards this aim, we replaced methanol dehydrogenase with the kinetically efficient and thermodynamically favourable sarcosine oxidase that metabolizes sarcosine to formaldehyde and glycine (He *et al.*, 2018). We further constructed several selection strains, imposing different levels of selective pressure on formaldehyde assimilation via the RuMP shunt. These included (i) *ΔfrmRAB ΔrpiAB Δfbp ΔglpX Δzwf* strain in which the synthetic shunt provides only essential sugar phosphates while succinate serves as the main growth substrate (Fig. 14.2e); (ii) *ΔfrmRAB ΔrpiAB ΔaceBA ΔglcB Δzwf* strain in which the RuMP shunt provides almost all cellular building blocks while acetate oxidation provides reducing power and energy (Fig. 14.2f); (iii) *ΔfrmRAB ΔrpiAB Δzwf* strain, where the shunt is responsible for satisfying all cellular carbon and energy needs (Fig. 14.2g); and (iv) *ΔfrmRAB ΔtktAB Δzwf* strain in which almost all cellular carbon and energy needs are supported by the RuMP shunt (Fig. 14.2h).

We were able to directly select for growth of all of these strains via the RuMP shunt without the need for ALE (He *et al.*, 2018), most probably because our selection scheme bypasses the challenge of methanol oxidation. However, the growth we obtained with the *ΔfrmRAB ΔrpiAB Δzwf* strain was considerably poorer than that observed using the *ΔfrmRAB ΔtktAB Δzwf* strain. This serves as a clear demonstration of the importance of the metabolic context for establishing activity of a newly introduced pathway. In this specific case, the poor growth associated with the *ΔfrmRAB ΔrpiAB Δzwf* strain probably stems from costly metabolism of F6P 'back' to xylulose-5-phosphate and Ru5P, and potentially from the inhibition of essential transketolase reactions due to the accumulation of the coproduct xylulose-5-phosphate (He *et al.*, 2018).

The major challenge of establishing a fully functional RuMP cycle lies in the regeneration module. To make things even more difficult, the establishment of autocatalytic cycles, such as the RuMP cycle, requires the kinetic parameters of enzymes to be carefully balanced as to avoid draining the pathway intermediates (Barenholz *et al.*, 2017). A recent study suggested an interesting way to tackle this challenge, by overexpressing the irreversible SBPase to force flux towards the regeneration of Ru5P (Woolston *et al.*, 2018a). Another approach is to overexpress the enzymes of the pentose phosphate pathway from an organism that supports more efficient regeneration of Ru5P (Bennett *et al.*, 2018b). Still, as was previously shown with ALE of *E. coli* to achieve a functional Calvin Cycle, down-regulation of branching enzymes, e.g. ribose-phosphate diphosphokinase, might be necessary to establish a sustainable cyclic flux (Antonovsky *et al.*, 2017; Herz *et al.*, 2017).

Synthetic formate assimilation

Formate is a promising microbial feedstock that can be efficiently produced from CO_2 by electrochemical and photochemical processes (Kopljar *et al.*, 2016; Zhou *et al.*, 2016; Yang *et al.*, 2017). Formate can also be obtained from partial oxidation of biomass or natural gas and by hydrogenation of CO_2 (Shen *et al.*, 2015; Wang *et al.*, 2015).

Microbial growth on formate as a carbon and energy source is reported for diverse groups of microorganisms. Acetogenic and methanogenic microbes can grow on formate using the reductive acetyl-CoA pathway (Kerby and Zeikus, 1987), representing the most efficient way to convert this feedstock into a product (Bar-Even *et al.*, 2013).

However, this anaerobic pathway can support only a narrow product spectrum (Bertsch and Müller, 2015). Aerobic growth on formate is sustained by the Calvin cycle and the serine pathway; however, the high ATP costs of these pathways reduce the potential biomass and product yields (Bar-Even *et al.*, 2013). To overcome this problem, several synthetic pathways have been designed to support high yields under aerobic conditions (Bar-Even *et al.*, 2013; Siegel *et al.*, 2015; Bar-Even, 2016).

In the synthetic formolase pathway, formate is first reduced to formaldehyde by promiscuous activity of two enzymes: acetyl-CoA synthase and acetylaldehyde dehydrogenase (Siegel *et al.*, 2015). Then, three formaldehyde molecules are condensed into dihydroxyacetone by the formolase enzyme described before. The activity of the pathway was demonstrated *in vitro*, but its *in vivo* functionality was very poor due to the low activities of formaldehyde dehydrogenase and the formolase enzyme (Siegel *et al.*, 2015). To boost the activities of these limiting enzymes, a modular selection strategy can be used. For example, a module responsible for formate reduction to formaldehyde can be tested in a formaldehyde-dependent strain – for example, the RuMP shunt-dependent strains described above – and ALE can be used to increase this activity. Similarly, activity of the formolase enzyme might be initially tested and optimized in a strain which produces formaldehyde via sarcosine oxidation and assimilate dihydroxyacetone to provide only a fraction of cellular building blocks. For example, by deleting phosphoglycerate mutase, central metabolism can be divided into upper and lower segments, where succinate provides carbon and energy for lower metabolism and dihydroxyacetone phosphorylation provides carbon only for upper metabolism (Zelcbuch *et al.*, 2015). This would impose moderate selection for the activity of the formolase enzyme and would thus be a more reasonable initial target.

A more feasible approach to establish formate assimilation would be to focus on existing enzymatic activities that could be combined to realize a new pathway. This is exactly the case of the synthetic reductive glycine pathway (Fig. 14.3a) (Bar-Even *et al.*, 2013). All the reactions of this pathway are catalysed by known and ubiquitous enzymes. It is even possible that the pathway operates endogenously in some microbes (Figueroa *et al.*, 2017). Its limited overlap with central metabolism and its very high ATP-efficiency further make the reductive glycine pathway especially promising to support aerobic formate assimilation.

To facilitate the implementation of the reductive glycine pathway in *E. coli,* we divided it into four modules (Fig. 14.3a): (i) a C1 activation module in which formate is condensed with the THF system and reduced to 5,10-methylene-THF; (ii) a glycine biosynthesis module that condenses the C1 moiety from 5,10-methylene-THF with CO_2 and ammonia to generate the C2 metabolite glycine; (iii) a serine biosynthesis module that condenses glycine with another C1 moiety to give the C3 metabolite serine; and (iv) a serine assimilation module that deaminates this amino acid to produce pyruvate as a biomass precursor.

We constructed several selection strains to demonstrate the activities of the pathway modules in *E. coli* (Yishai *et al.*, 2017, 2018). First, we constructed a C1-auxotroph strain (Δ*glyA* Δ*gcvTHP*) and showed that overexpression of formate THF ligase (FTL) enabled formate to serve as sole source of all C1-dependent building blocks, including purines, thymidine, and methionine (Fig. 14.3b). This confirmed the efficient activity of the C1-activation module. Next, we selected for the combined activity of the C1-activation module and the serine biosynthesis module. Towards this aim, we constructed a strain deleted in 3-phosphoglycerate dehydrogenase (Δ*serA*, the first enzyme of serine biosynthesis) and the glycine cleavage system (Δ*gcvTHP*). Only upon overexpression of FTL and the native bifunctional 5,10-methenyl-THF cyclohydrolase/5,10-methylene-THF dehydrogenase (FolD), glycine and formate could serve as C1 and serine sources to support cell growth (Fig. 14.3c). This demonstrates that testing a module – in this case the C1 activation module – in different metabolic contexts is important to uncover hidden bottlenecks: while the endogenous activity of FolD sufficed for the first selection strain, the higher activity required for the growth of the second selection strain necessitated dedicated overexpression of the enzyme (Yishai *et al.*, 2017).

In a follow-up study, we selected for the activities of the C1-assimilation module and the glycine biosynthesis module (Yishai *et al.*, 2018). We constructed a strain auxotrophic to C1 and glycine (Δ*glyA* Δ*ltaE* Δ*kbl* Δ*aceA*). We could establish

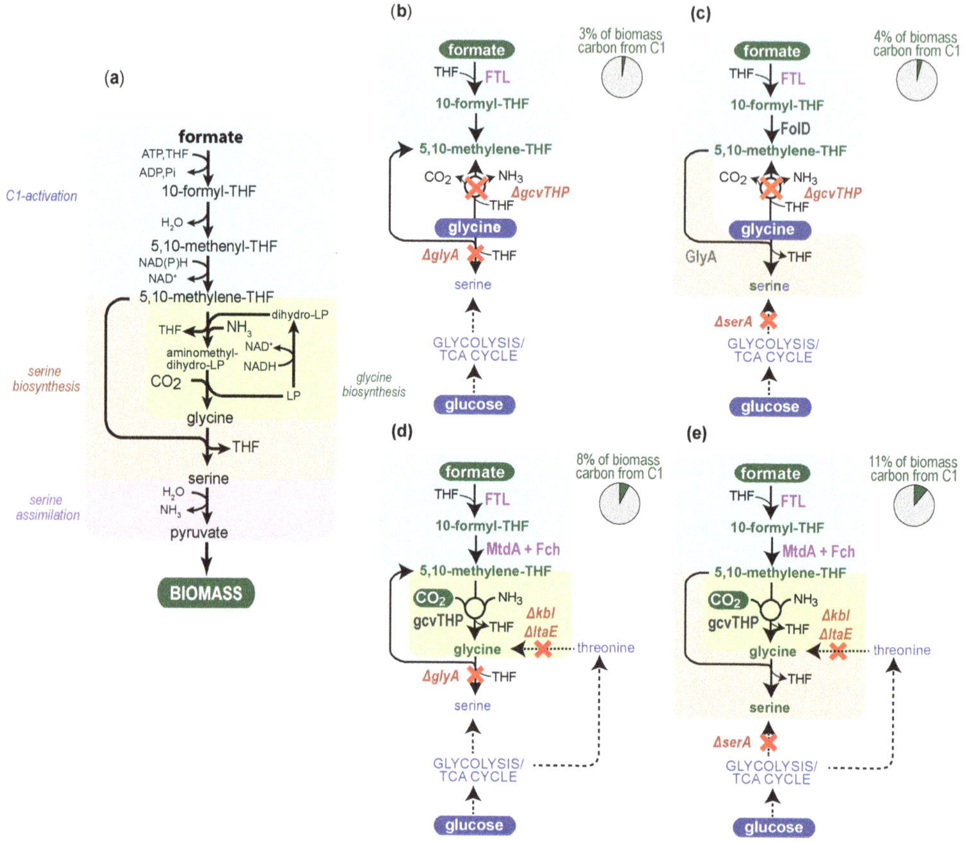

Figure 14.3 Engineering the reductive glycine pathway in *E. coli*. (a) Metabolic scheme of the reductive glycine pathway and its subdivision into modules; (b–e) Selection schemes for the activity of different modules of the reductive glycine pathway, as demonstrated in *E. coli* (Yishai *et al*., 2017, 2018). Pie charts indicate the minimum fraction of carbons in biomass that are derived from formate and CO_2 in the different selection schemes, as calculated using (Neidhardt *et al*., 1990). Gene deletions are shown in red; overexpressed foreign enzymes in purple; overexpressed native enzymes in grey; metabolites that are produced from formate and CO_2 in green; and substrates that have to be co-fed in blue. Abbreviations: Fch, 5,10-methenyl-THF cyclohydrolase; FolD, bifunctional 5,10-methenyl-THF cyclohydrolase/5,10-methylene-THF dehydrogenase; FTL, formate-THF ligase; gcvTHP(L), glycine cleavage system subunits T, H and P (and lipoamide dehydrogenase subunit); glyA, serine hydroxymethyltransferase; ltaE, threonine aldolase; kbl, 2-amino-3-ketobutyrate CoA ligase; LP, lipoamide-protein; MtdA, 5,10-methylene-THF dehydrogenase; serA, 3-phosphoglycerate dehydrogenase; THF, tetrahydrofolate.

formate conversion to C1-activated-THF and glycine upon overexpression of the enzymes of the native glycine cleavage system as well as FTL, 5,10-methenyl-THF cyclohydrolase, and 5,10-methylene-THF dehydrogenase from *Methylobacterium extorquens* (Fig. 14.3d). The latter two enzymes were necessary to replace the native FolD, whose activity was too low to support the required flux even when overexpressed, probably since it is inhibited by the key intermediate 10-formyl-THF (Yishai *et al.*, 2018). Next, to select for the activity of three of the pathway modules, we constructed a C1-glycine-serine auxotrophic strain (Δ*serA* Δ*ltaE* Δ*kbl* Δ*aceA*) (Fig. 14.3e). Overexpression of enzymes mentioned above resulted in C1, glycine and serine production solely from formate and CO_2. Overall, ≈ 10% of carbons in biomass were provided by the pathway and the fast growth obtained (doubling rate of ≈ 1.7 hours) indicates high activity of all pathway components.

Another recent study also focused on modular engineering of the reductive glycine pathway in *E. coli* (Tashiro *et al.*, 2018). Similar to the approach described above, this study demonstrated that the overexpression of three foreign genes from *Clostridium ljungdahlii* can support C1 and serine biosynthesis from formate and glycine (Fig. 14.3c). However, for demonstration of glycine and serine biosynthesis via the reductive activity of the glycine cleavage system, leaky auxotrophic strains were employed in which only *serA* or *glyA* were deleted, while threonine cleavage to glycine (via Kbl or LtaE) was left untouched. Consequently, as is supported by the labelling data, in these strains, the majority of glycine and serine were most probably derived from threonine cleavage rather than formate assimilation.

In a parallel effort to the rational engineering approach, we demonstrated the establishment of the three modules of the reductive glycine pathway using ALE (Döring *et al.*, 2018). This work used the same tight selection strategies as above but introduced only a single foreign enzyme: *Clostridial* FTL that was integrated into the genome. The cells were cultivated continuously and provided with two types of alternating media: a selection medium (which does not contain the auxotrophy-relieving compounds) that was added when the culture turbidity surpassed a threshold, and a permissive medium (containing the compounds for which the strains are auxotrophic) that was supplied upon decrease of the culture turbidity below this threshold. This procedure was used to select for growth on the stressing medium and, using sequential rounds of evolution, established metabolism of formate to C1-activated-THF, glycine, and, finally, serine. Genome sequencing of these strains revealed several mutations, including in the coding region of the key limiting enzyme FolD. Several genes were duplicated in the genome, including FTL and FolD, probably to increase their expression levels. This study confirms that the same metabolic goals can be achieved either via a rational engineering approach or via long term evolution.

To complete the establishment of the reductive glycine pathway, the next challenge is to integrate the first three modules – converting formate to serine – with the last one, that is, serine assimilation into central metabolism. This would require an order of magnitude increase in the flux via the pathway and would most probably require ALE to adapt the cellular physiology for the novel growth mode.

Conclusions

Recent efforts using modular and selection-based engineering approach have enabled C1-routes to support the biosynthesis of a substantial fraction of the host's biomass. These studies demonstrate a general approach of iterative design, build, test and learn cycles. First, pathway modules and selection strains are designed and constructed. Testing module activity using these selection strains provides insights into the metabolic constraints that limit flux, from which better designs and improved activity can emerge. Selection strains are therefore an essential tool for these cycles, as they provide simple performance readout that facilitates optimization of module and pathway activity.

As mentioned above for the RuMP cycle, a major challenge for establishing C1 pathways relates to the regulation of flux that is diverted away from the pathway towards other biosynthesis routes. This is especially true for autocatalytic cycles in which the product is also an intermediate of the pathway. To realize stable activity, proper balancing is required between the rates of pathway enzymes and those of the branching enzymes. As previously demonstrated, for autocatalytic cycles, this likely requires ensuring low affinities of the branching enzymes towards the pathway metabolites (Barenholz *et al.*, 2017). Indeed, the use of ALE for the establishment of a functional Calvin cycle in *E. coli* resulted in lower activity of branching reactions (Antonovsky *et al.*, 2016). Fortunately, branching reactions are easy to identify such that a rational design approach might be able to achieve the same goals as ALE, e.g. genetic or protein engineering aiming at lower expression levels or affinities of branching enzymes could directly establish the desired growth phenotype.

Most efforts to establish synthetic C1 metabolism have so far focused on the RuMP cycle, the reductive glycine pathway, and the formolase pathway. Yet, given the diversity of metabolic solutions for C1 assimilation, more pathways are likely to be designed and tested, and the modular selection approach described here would be vital for their implementation. Currently, *E. coli* serves as the prime host for engineering synthetic C1 assimilation, mainly due to the highly developed genetic toolbox for its engineering and our extensive knowledge of its physiology and metabolism. However, other microbial hosts might prove to be more suitable. Specifically, some microbes can produce specific chemicals better than *E. coli* and others display higher tolerance towards substrates or products. For example *Corynebacterium glutamicum* is arguably the best host to produce amino acids (Wendisch *et al.*, 2006), and *Saccharomyces cerevisiae* can tolerate high concentrations of the formate feedstock and alcohol products (Overkamp *et al.*, 2002; Mohd Azhar *et al.*, 2017). Also, some metabolic modules can directly integrate with C1 metabolism, making microbes that naturally harbour these better hosts. A primary example is the metal-dependent formate dehydrogenase, which can support highly efficient utilization of formate as cellular energy source but is difficult to introduce into a new host (Maia *et al.*, 2017).

In the next few years, we will likely witness the completion of synthetic C1 metabolism in model microbes. While we cannot predict which pathway will prove to be the most beneficial in the long term, it is clear that efforts such as those described in this review are bringing efficient bioproduction from C feedstocks closer.

Acknowledgements

The authors thank Charles Cotton for his feedback and critical reading of this manuscript. This work was funded by the Max Planck Society and European Commission Horizon 2020 grant eForFuel. Nico Claassens is supported by The Netherlands Organisation for Scientific Research (NWO) through a Rubicon Grant (Project 019.163LW.035). Hai He is funded by the China Scholarship Council.

References

Antonovsky, N., Gleizer, S., Noor, E., Zohar, Y., Herz, E., Barenholz, U., Zelcbuch, L., Amram, S., Wides, A., Tepper, N., *et al.* (2016). Sugar synthesis from CO_2 in *Escherichia coli*. Cell *166*, 115–125. https://doi.org/10.1016/j.cell.2016.05.064

Antonovsky, N., Gleizer, S., and Milo, R. (2017). Engineering carbon fixation in *E. coli*: from heterologous RuBisCO expression to the Calvin-Benson-Bassham cycle. Curr. Opin. Biotechnol. 47, 83–91.

Bar-Even, A. (2016). Formate assimilation: The metabolic architecture of natural and synthetic pathways. Biochemistry 55, 3851–3863. https://doi.org/10.1021/acs.biochem.6b00495

Bar-Even, A., Noor, E., Flamholz, A., and Milo, R. (2013). Design and analysis of metabolic pathways supporting formatotrophic growth for electricity-dependent cultivation of microbes. Biochim. Biophys. Acta *1827*, 1039–1047. https://doi.org/10.1016/j.bbabio.2012.10.013

Barenholz, U., Davidi, D., Reznik, E., Bar-On, Y., Antonovsky, N., Noor, E., and Milo, R. (2017). Design principles of autocatalytic cycles constrain enzyme kinetics and force low substrate saturation at flux branch points. Elife *6*, e20667. https://doi.org/10.7554/eLife.20667

Bennett, R.K., Steinberg, L.M., Chen, W., and Papoutsakis, E.T. (2018a). Engineering the bioconversion of methane and methanol to fuels and chemicals in native and synthetic methylotrophs. Curr. Opin. Biotechnol. *50*, 81–93.

Bennett, R.K., Gonzalez, J.E., Whitaker, W.B., Antoniewicz, M.R., and Papoutsakis, E.T. (2018b). Expression of heterologous non-oxidative pentose phosphate pathway from *Bacillus methanolicus* and phosphoglucose isomerase deletion improves methanol assimilation and metabolite production by a synthetic *Escherichia coli* methylotroph. Metab. Eng. *45*, 75–85.

Bertsch, J., and Müller, V. (2015). Bioenergetic constraints for conversion of syngas to biofuels in acetogenic bacteria. Biotechnol. Biofuels *8*, 210. https://doi.org/10.1186/s13068-015-0393-x

Bogorad, I.W., Chen, C.T., Theisen, M.K., Wu, T.Y., Schlenz, A.R., Lam, A.T., and Liao, J.C. (2014). Building carbon-carbon bonds using a biocatalytic methanol condensation cycle. Proc. Natl. Acad. Sci. U.S.A. *111*, 15928–15933. https://doi.org/10.1073/pnas.1413470111

Bouzon, M., Perret, A., Loreau, O., Delmas, V., Perchat, N., Weissenbach, J., Taran, F., and Marlière, P. (2017). A Synthetic Alternative to Canonical One-Carbon Metabolism. ACS Synth. Biol. *6*, 1520–1533. https://doi.org/10.1021/acssynbio.7b00029

Burgard, A.P., Pharkya, P., and Maranas, C.D. (2003). Optknock: a bilevel programming framework for identifying gene knockout strategies for microbial strain optimization. Biotechnol. Bioeng. *84*, 647–657. https://doi.org/10.1002/bit.10803

Chen, C.T., Chen, F.Y., Bogorad, I.W., Wu, T.Y., Zhang, R., Lee, A.S., and Liao, J.C. (2018). Synthetic methanol auxotrophy of *Escherichia coli* for methanol-dependent growth and production. Metab. Eng. *49*, 257–266.

Chistoserdova, L., Kalyuzhnaya, M.G., and Lidstrom, M.E. (2009). The expanding world of methylotrophic metabolism. Annu. Rev. Microbiol. *63*, 477–499. https://doi.org/10.1146/annurev.micro.091208.073600

Claassens, N.J., Sánchez-Andrea, I., Sousa, D.Z., and Bar-Even, A. (2018). Towards sustainable feedstocks: A guide to electron donors for microbial carbon fixation. Curr. Opin. Biotechnol. *50*, 195–205.

Clomburg, J.M., Crumbley, A.M., and Gonzalez, R. (2017). Industrial biomanufacturing: The future of chemical production. Science *355*, aag0804.

Dai, Z., Gu, H., Zhang, S., Xin, F., Zhang, W., Dong, W., Ma, J., Jia, H., and Jiang, M. (2017). Metabolic construction strategies for direct methanol utilization in *Saccharomyces cerevisiae*. Bioresour. Technol. *245*, 1407–1412. https://doi.org/10.1016/j.biortech.2017.05.100

Döring, V., Darii, E., Yishai, O., Bar-Even, A., and Bouzon, M. (2018). Implementation of a reductive route of one-carbon assimilation in *Escherichia coli* through directed evolution. ACS Synth. Biol. 7, 2029–2036. https://doi.org/10.1021/acssynbio.8b00167

Dürre, P., and Eikmanns, B.J. (2015). C1-carbon sources for chemical and fuel production by microbial gas fermentation. Curr. Opin. Biotechnol. *35*, 63–72. https://doi.org/10.1016/j.copbio.2015.03.008

Erb, T.J., Jones, P.R., and Bar-Even, A. (2017). Synthetic metabolism: metabolic engineering meets enzyme design. Curr. Opin. Chem. Biol. *37*, 56–62.

Fei, Q., Guarnieri, M.T., Tao, L., Laurens, L.M., Dowe, N., and Pienkos, P.T. (2014). Bioconversion of natural gas to liquid fuel: opportunities and challenges. Biotechnol. Adv. *32*, 596–614. https://doi.org/10.1016/j.biotechadv.2014.03.011

Figueroa, I.A., Barnum, T.P., Somasekhar, P.Y., Carlström, C.I., Engelbrektson, A.L., and Coates, J.D. (2018). Metagenomics-guided analysis of microbial chemolithoautotrophic phosphite oxidation yields evidence of a seventh natural CO_2 fixation pathway. Proc. Natl. Acad. Sci. U.S.A. *115*, E92–E101. https://doi.org/10.1073/pnas.1715549114

Gonzalez, J.E., Bennett, R.K., Papoutsakis, E.T., and Antoniewicz, M.R. (2018). Methanol assimilation in *Escherichia coli* is improved by co-utilization of threonine and deletion of leucine-responsive regulatory protein. Metab. Eng. *45*, 67–74.

He, H., Edlich-Muth, C., Lindner, S.N., and Bar-Even, A. (2018). Ribulose monophosphate shunt provides nearly all biomass and energy required for growth of *E. coli*. ACS Synth. Biol. *7*, 1601–1611. https://doi.org/10.1021/acssynbio.8b00093

Henstra, A.M., Sipma, J., Rinzema, A., and Stams, A.J. (2007). Microbiology of synthesis gas fermentation for biofuel production. Curr. Opin. Biotechnol. *18*, 200–206.

Herz, E., Antonovsky, N., Bar-On, Y., Davidi, D., Gleizer, S., Prywes, N., Noda-Garcia, L., Lyn Frisch, K., Zohar, Y., Wernick, D.G., *et al.* (2017). The genetic basis for the adaptation of E. coli to sugar synthesis from CO_2. Nat. Commun. *8*, 1705. https://doi.org/10.1038/s41467-017-01835-3

Jouny, M., Luc, W.W., and Jiao, F. (2018). A general techno-economic analysis of CO_2 electrolysis systems. Ind. Eng. Chem. Res. *57*, 2165–2177. https://doi.org/10.1021/acs.iecr.7b03514

Kato, N., Yurimoto, H., and Thauer, R.K. (2006). The physiological role of the ribulose monophosphate pathway in bacteria and archaea. Biosci. Biotechnol. Biochem. *70*, 10–21.

Kerby, R., and Zeikus, J.G. (1987). Anaerobic catabolism of formate to acetate and CO_2 by *Butyribacterium methylotrophicum*. J. Bacteriol. *169*, 2063–2068.

Kopljar, D., Wagner, N., and Klemm, E. (2016). Transferring electrochemical CO_2 reduction from semi-batch into continuous operation mode using gas diffusion electrodes. Chem. Eng. Technol. *39*, 2042–2050. https://doi.org/10.1002/ceat.201600198

Kumar, B., Llorente, M., Froehlich, J., Dang, T., Sathrum, A., and Kubiak, C.P. (2012). Photochemical and photoelectrochemical reduction of CO2. Annu. Rev. Phys. Chem. *63*, 541–569. https://doi.org/10.1146/annurev-physchem-032511-143759

Leßmeier, L., Pfeifenschneider, J., Carnicer, M., Heux, S., Portais, J.C., and Wendisch, V.F. (2015). Production of carbon-13-labeled cadaverine by engineered Corynebacterium glutamicum using carbon-13-labeled methanol as co-substrate. Appl. Microbiol. Biotechnol. *99*, 10163–10176. https://doi.org/10.1007/s00253-015-6906-5

Maia, L.B., Moura, I., and Moura, J.J.G. (2017). Molybdenum and tungsten-containing formate dehydrogenases: Aiming to inspire a catalyst for carbon dioxide utilization. Inorganica Chim. Acta *455*, 350–363. https://doi.org/10.1016/j.ica.2016.07.010

Marlière, P., Patrouix, J., Döring, V., Herdewijn, P., Tricot, S., Cruveiller, S., Bouzon, M., and Mutzel, R. (2011). Chemical evolution of a bacterium's genome. Angew. Chem. Int. Ed. Engl. *50*, 7109–7114. https://doi.org/10.1002/anie.201100535

Marmiesse, L., Peyraud, R., and Cottret, L. (2015). FlexFlux: combining metabolic flux and regulatory network analyses. BMC Syst. Biol. *9*, 93. https://doi.org/10.1186/s12918-015-0238-z

Martín, A.J., Larrazábal, G.O., and Pérez-Ramírez, J. (2015). Towards sustainable fuels and chemicals through the electrochemical reduction of CO2: lessons from water electrolysis. Green Chem. *17*, 5114–5130. https://doi.org/10.1039/C5GC01893E

Meyer, F., Keller, P., Hartl, J., Gröninger, O.G., Kiefer, P., and Vorholt, J.A. (2018). Methanol-essential growth of *Escherichia coli*. Nat. Commun. *9*, 1508. https://doi.org/10.1038/s41467-018-03937-y

Mohd Azhar, S.H., Abdulla, R., Jambo, S.A., Marbawi, H., Gansau, J.A., Mohd Faik, A.A., and Rodrigues, K.F. (2017). Yeasts in sustainable bioethanol production: A review. Biochem. Biophys. Rep. *10*, 52–61. https://doi.org/10.1016/j.bbrep.2017.03.003

Müller, J.E.N., Meyer, F., Litsanov, B., Kiefer, P., Potthoff, E., Heux, S., Quax, W.J., Wendisch, V.F., Brautaset, T., Portais, J.C., *et al.* (2015). Engineering *Escherichia coli* for methanol conversion. Metab. Eng. *28*, 190–201.

Neidhardt, F.C., Ingraham, J.L., and Schaechter, M. (1990). Building blocks needed to produce 1g of *E. coli* protoplasm. In Physiology of the Bacterial Cell: A Molecular Approach, Neidhardt, F.C., Ingraham, J.L., and Schaechter, M., eds. (Sinaver Associates, Sunderland, MA), pp. 134–143.

Olah, G.A. (2013). Towards oil independence through renewable methanol chemistry. Angew. Chem. Int. Ed. Engl. *52*, 104–107. https://doi.org/10.1002/anie.201204995

Olah, G.A., Goeppert, A., and Prakash, G.K.S. (2009). Beyond Oil and Gas: The Methanol Economy, Second Edition. (Wiley VCH, Weinheim), pp. 1–334 https://doi.org/10.1002/9783527627806.

Overkamp, K.M., Kötter, P., van der Hoek, R., Schoondermark-Stolk, S., Luttik, M.A., van Dijken, J.P., and Pronk, J.T. (2002). Functional analysis of structural genes for NAD(+)-dependent formate dehydrogenase in *Saccharomyces cerevisiae*. Yeast *19*, 509–520.

Pfeifenschneider, J., Brautaset, T., and Wendisch, V.F. (2017). Methanol as carbon substrate in the bio-economy: Metabolic engineering of aerobic methylotrophic bacteria for production of value-added chemicals. Biofuels Bioprod. Biorefining *11*, 719–731. https://doi.org/10.1002/bbb.

Portnoy, V.A., Bezdan, D., and Zengler, K. (2011). Adaptive laboratory evolution – harnessing the power of biology for metabolic engineering. Curr. Opin. Biotechnol. 22, 590–594. https://doi.org/10.1016/j.copbio.2011.03.007

Price, J.V., Chen, L., Whitaker, W.B., Papoutsakis, E., and Chen, W. (2016). Scaffoldless engineered enzyme assembly for enhanced methanol utilization. Proc. Natl. Acad. Sci. U.S.A. *113*, 12691–12696. https://doi.org/10.1073/pnas.1601797113

Quayle, J.R., and Ferenci, T. (1978). Evolutionary aspects of autotrophy. Microbiol. Rev. *42*, 251–273.

Schrader, J., Schilling, M., Holtmann, D., Sell, D., Filho, M.V., Marx, A., and Vorholt, J.A. (2009). Methanol-based industrial biotechnology: current status and future perspectives of methylotrophic bacteria. Trends Biotechnol. 27, 107–115. https://doi.org/10.1016/j.tibtech.2008.10.009

Schwander, T., Schada von Borzyskowski, L., Burgener, S., Cortina, N.S., and Erb, T.J. (2016). A synthetic pathway for the fixation of carbon dioxide in vitro. Science *354*, 900–904.

Shen, Y., Jarboe, L., Brown, R., and Wen, Z. (2015). A thermochemical-biochemical hybrid processing of lignocellulosic biomass for producing fuels and chemicals. Biotechnol. Adv. 33, 1799–1813. https://doi.org/10.1016/j.biotechadv.2015.10.006

Siegel, J.B., Smith, A.L., Poust, S., Wargacki, A.J., Bar-Even, A., Louw, C., Shen, B.W., Eiben, C.B., Tran, H.M., Noor, E., *et al.* (2015). Computational protein design enables a novel one-carbon assimilation pathway. Proc. Natl. Acad. Sci. U.S.A. *112*, 3704–3709. https://doi.org/10.1073/pnas.1500545112

Tashiro, Y., Hirano, S., Matson, M.M., Atsumi, S., and Kondo, A. (2018). Electrical-biological hybrid system for CO2 reduction. Metab. Eng. *47*, 211–218.

Tuyishime, P., Wang, Y., Fan, L., Zhang, Q., Li, Q., Zheng, P., Sun, J., and Ma, Y. (2018). Engineering Corynebacterium glutamicum for methanol-dependent growth and glutamate production. Metab. Eng. *49*, 220–231.

Wang, W.H., Himeda, Y., Muckerman, J.T., Manbeck, G.F., and Fujita, E. (2015). CO_2 hydrogenation to formate and methanol as an alternative to photo- and electrochemical CO_2 reduction. Chem. Rev. *115*, 12936–12973. https://doi.org/10.1021/acs.chemrev.5b00197

Wang, X., Wang, Y., Liu, J., Li, Q., Zhang, Z., Zheng, P., Lu, F., and Sun, J. (2017). Biological conversion of methanol by evolved *Escherichia coli* carrying a linear methanol assimilation pathway. Bioresour. Bioprocess. *4*, 4–9. https://doi.org/10.1186/s40643-017-0172-6.

Wendisch, V.F., Bott, M., and Eikmanns, B.J. (2006). Metabolic engineering of *Escherichia coli* and *Corynebacterium glutamicum* for biotechnological production of organic acids and amino acids. Curr. Opin. Microbiol. 9, 268–274.

Wenk, S., Yishai, O., Lindner, S.N., and Bar-Even, A. (2018). An engineering approach for rewiring microbial metabolism. Meth. Enzymol. *608*, 329–367.

Whitaker, W.B., Sandoval, N.R., Bennett, R.K., Fast, A.G., and Papoutsakis, E.T. (2015). Synthetic methylotrophy: engineering the production of biofuels and chemicals based on the biology of aerobic methanol utilization. Curr. Opin. Biotechnol. *33*, 165–175. https://doi.org/10.1016/j.copbio.2015.01.007

Whitaker, W.B., Jones, J.A., Bennett, R.K., Gonzalez, J.E., Vernacchio, V.R., Collins, S.M., Palmer, M.A., Schmidt, S., Antoniewicz, M.R., Koffas, M.A., *et al.* (2017). Engineering the biological conversion of methanol to specialty chemicals in *Escherichia coli*. Metab. Eng. *39*, 49–59.

Witthoff, S., Schmitz, K., Niedenführ, S., Nöh, K., Noack, S., Bott, M., and Marienhagen, J. (2015). Metabolic engineering of Corynebacterium glutamicum for methanol metabolism. Appl. Environ. Microbiol. *81*, 2215–2225. https://doi.org/10.1128/AEM.03110-14

Woolston, B.M., King, J.R., Reiter, M., Van Hove, B., and Stephanopoulos, G. (2018a). Improving formaldehyde consumption drives methanol assimilation in engineered E. coli. Nat. Commun. *9*, 2387. https://doi.org/10.1038/s41467-018-04795-4

Woolston, B.M., Roth, T., Kohale, I., Liu, D.R., and Stephanopoulos, G. (2018b). Development of a formaldehyde biosensor with application to synthetic methylotrophy. Biotechnol. Bioeng. *115*, 206–215. https://doi.org/10.1002/bit.26455

Yang, H., Kaczur, J.J., Sajjad, S.D., and Masel, R.I. (2017). CO2 Conversion to formic acid in a three compartment cell with Sustainion™ membranes. ECS Trans. *77*, 1425–1431. https://doi.org/10.1149/07711.1425ecst.

Yishai, O., Goldbach, L., Tenenboim, H., Lindner, S.N., and Bar-Even, A. (2017). Engineered assimilation of exogenous and endogenous formate in *Escherichia coli*. ACS Synth. Biol. *6*, 1722–1731. https://doi.org/10.1021/acssynbio.7b00086

Yishai, O., Bouzon, M., Döring, V., and Bar-Even, A. (2018). In vivo assimilation of one-carbon via a synthetic reductive glycine pathway in *Escherichia coli*. ACS Synth. Biol. 7, 2023–2028. https://doi.org/10.1021/acssynbio.8b00131

Zelcbuch, L., Antonovsky, N., Bar-Even, A., Levin-Karp, A., Barenholz, U., Dayagi, M., Liebermeister, W., Flamholz, A., Noor, E., Amram, S., *et al.* (2013). Spanning high-dimensional expression space using ribosome-binding site combinatorics. Nucleic Acids Res. *41*, e98. https://doi.org/10.1093/nar/gkt151

Zelcbuch, L., Razo-Mejia, M., Herz, E., Yahav, S., Antonovsky, N., Kroytoro, H., Milo, R., and Bar-Even, A. (2015). An in vivo metabolic approach for deciphering the product specificity of glycerate kinase proves that both *E. coli*'s glycerate kinases generate 2-phosphoglycerate. PLOS ONE *10*, e0122957. https://doi.org/10.1371/journal.pone.0122957

Zhang, W., Zhang, T., Wu, S., Wu, M., Xin, F., Dong, W., Ma, J., Zhang, M., and Jiang, M. (2017). Guidance for engineering of synthetic methylotrophy based on methanol metabolism in methylotrophy. RSC Adv. 7, 4083–4091. https://doi.org/10.1039/C6RA27038G

Zhou, X., Liu, R., Sun, K., Chen, Y., Verlage, E., Francis, S.A., Lewis, N.S., and Xiang, C. (2016). Solar-driven reduction of 1 atm of CO_2 to formate at 10% energy-conversion efficiency by use of a TiO2-protected III-V tandem photoanode in conjunction with a bipolar membrane and a Pd/C cathode. ACS Energy Lett. *1*, 764–770. https://doi.org/10.1021/acsenergylett.6b00317

Experimental Evolution of *Methylobacterium*: 15 Years of Planned Experiments and Surprise Findings

15

Christopher J. Marx*

Department of Biological Sciences, Institute for Bioinformatics and Evolutionary Studies, Center for Modeling Complex Interactions, University of Idaho, Moscow, ID, USA.

*Correspondence: cmarx@uidaho.edu

https://doi.org/10.21775/9781912530045.15

Abstract

Experimental evolution has become an increasingly common approach for studying evolutionary phenomena, as well as uncovering physiological connections in a manner complementary to traditional genetics. Here I describe the development of *Methylobacterium* as a model system for using experimental evolution to study questions at the intersection of metabolism and evolution. Each experiment was initiated to address a particular question inspired by patterns in natural methylotrophs, such as trade-offs between single-carbon and multi-carbon growth, or the challenges involved in incorporating novel metabolic pathways or genes with poor codon usage that are acquired via horizontal gene transfer. What I could not have appreciated initially, however, was just how many fortuitous, surprise findings would emerge. These have ranged from the repeatability of evolution, complex dynamics within populations, epistasis between beneficial mutations, and even the ability to use simple mathematical models to generate testable, quantitative hypotheses about the fitness landscape.

Introduction

Experimental evolution of populations in the laboratory allows a researcher to simultaneously address evolutionary and physiological questions. The great advantage from an evolutionary perspective is that – in a typical experimental design – replicate populations initially have no within- or between-population genetic variation, and the selective conditions are under the control of the experimenter. This allows for a 'reductionist' approach to evolutionary questions, whereby the influence of one or a few individual factors upon the outcome can be ascertained (reviewed in Lenski, 2017). From a physiological perspective, experimental evolution is simply a patient version of a genetic selection experiment (reviewed in Marx, 2011). Rather than requiring a discrete change in phenotype to be immediately apparent upon plating, the continued transfers of the experiment permit mutations of 'modest' effect – such as a 10% increase in growth rate – to occur, escape drift, and rise towards fixation. Furthermore, the advent of high-throughput sequencing has revolutionized the ability to address both the evolutionary and physiological questions (reviewed in Bruger and Marx, 2018).

The kind invitation I received to write this chapter was a request to specifically describe the work in my laboratory where we have repeatedly used experimental evolution with *Methylobacterium extorquens* to address both evolutionary and physiological questions. Given that charge, I will shamelessly describe themes arising from our own work, but my primary goal is to highlight two broader messages. The first message is that experimental evolution can be utilized as a complementary approach to address some of the same questions that emerge from studying natural populations. Second, the unexpected, surprise results from experimental evolution can be as impactful as answering the questions that were part of the intended experimental design. For this reason, I have divided the topics along these lines – the planned questions and the answers we uncovered (section 'The planned: experimental evolution to address questions inspired by natural methylotrophs'), and then the several categories of surprise findings (section 'Surprises: unexpected lessons emerging from evolution experiments') – but have tried to illuminate the connections between these as best as I can.

The planned: experimental evolution to address questions inspired by natural methylotrophs

Thus far, my laboratory has published experimental evolution stories using *M. extorquens* that were designed to address three major topics: evolution of metabolic trade-offs, adaptation to utilize a foreign metabolic pathway, and the use of experimental evolution to illuminate how codon usage can evolve.

Trade-offs between C_1 and multi-C metabolism

My initial foray into experimental evolution was to explore metabolic trade-offs between C_1 and multi-C metabolism. Many methylotrophs have a fairly limited breadth of multi-C use. In particular, although *Methylobacterium* are a member of Bradyrhizobiales, which generally have a wide substrate breadth, the major model systems of *M. extorquens*, strains AM1 and PA1 (Peel and Quayle, 1961; Knief *et al.*, 2010; Nayak and Marx, 2014a), have not been reported to use many common multi-C substrates, such as sugars. Having studied the function of C_1 pathways as a grad student in Mary Lidstrom's lab, when I applied for an NSF postdoctoral fellowship to work with Rich Lenski, I focused half of my proposal upon exploring whether C_1 versus multi-C trade-offs would emerge over the short term in the laboratory, if they would be consistent and symmetric between substrates, and whether they would be driven by selection or arise neutrally due to a lack of selection.

Before moving to our results, let me explain how this experimental evolution – and the other ones mentioned below – were actually carried out? For this experiment (Lee *et al.*, 2009), I set up eight populations to evolve on methanol (15 mM) in minimal medium, and eight to evolve on succinate (3.5 mM). These somewhat lower concentrations than usual were chosen to prevent large changes in medium pH (which would exert a second selective pressure). Critically, each population was initiated from a single colony of the given ancestral strain, thus ensuring that if any identical mutations emerged between replicates, these could be interpreted as having risen independently, rather than from shared ancestry, as in a Luria-Delbrück experiment (Luria and Delbrück, 1943). Culturing took place in 9.6 ml of liquid medium in 50 ml flasks, shaken, at 30 °C. A volume of 150 µl was transferred to fresh medium every two days, which permitted a $2^6 = 64$-fold net increase in cells, and thus six generations per transfer. These particular populations were carried forward for 1500 generations in this manner. There are a couple differences in experimental design I will note later, but otherwise, this style of batch transfer experiments were the norm throughout what I present here. Unsurprisingly, given where I did my postdoc, this style of experimental evolution is analogous to that used during the very well-studied Lenski long-term experimental evolution populations of *Escherichia coli* (Lenski *et al.*, 1991; Lenski, 2017).

The second generic aspect to convey is how fitness is assayed, as natural selection depends upon the fitness (i.e. relative net reproductive success) of genotypes. The classical method for assaying fitness is a pairwise competition between the strain of interest and a common competitor, generally the ancestor of the populations. These strains must be distinguishable in order to enumerate each of them in a mixture, and we did possess the ability to use pigmentation and count colonies (Van Dien *et al.*, 2003). Given the time intensity of plating and counting, however, as well as the large measurement noise inherent in counting relatively small numbers of colonies on plates, for our first paper on experimental evolution we invested time in developing a flow cytometry-based method (Lee *et al.*, 2009). This allowed us to use a fluorescently labelled ancestor as a common competitor against a non-fluorescent evolved isolate (or a genetically constructed strain). If possible, even better resolution can be obtained by using two different fluorescent proteins, one in each competitor, but this was not yet established in our early work. We found that either Venus or mCherry expressed at a modest level from a chromosomal locus was effectively neutral (fitnesses between 0.995 and 1.005) and could be measured accurately (95% confidence intervals of less than 0.3% of the mean).

Finally, it is critical to realize that fitness is not synonymous with maximal growth rate, but in this sort of evolution regime the two are often very highly correlated. Under batch culture conditions in well-mixed medium, exponential growth rate is the primary fitness component, with lag or survival in stationary phase generally contributing to a far lesser extent (Vasi *et al.*, 1994). Note that final yield (in terms of biomass converted or cfu/ml) is not a fitness component during batch growth (Vasi *et al.*, 1994). The Malthusian fitness parameter, W, is calculated from the change in the ratio of competitors from the initial to final timepoint, as well as the number of net generations. As such, W is a relative, 'per generation' fitness value averaged over the entire growth cycle. For example, a fitness of 1.2 would indicate the strain in question is 20% more fit than the one it is being compared with, such as its ancestor. If the two strains only differ in growth rate, this would also equate to a 20% difference in growth rates. Indeed, in many experiments we have observed a tremendously tight correlation between fitness measured in a competition and growth rate measured individually (e.g. R^2 of 0.98; Agashe *et al.*, 2013). If, however, there are interactions between strains beyond competition, such as toxins or cross-fed metabolites, this makes fitness both non-transitive across multiple strains, and frequency-dependent between strains (i.e. different values depending upon the ratios of competitors). Furthermore, the fitness can even depend upon the absolute densities of competitors rather than their ratio, as discussed below (section 'Swapping formaldehyde oxidation pathways'). There was very little of these sort of complex non-transitive fitness effects for most of the work I describe here on *M. extorquens* alone, but it can be quite common in other scenarios. For example, non-transitive fitness interactions between genotypes were the norm in as a series of experiments we performed with two- and three-species consortia that exchanged metabolites as public goods in spatially structured environments (Harcombe *et al.*, 2014, 2016; Douglas *et al.*, 2016, 2017).

Having covered how the experimental evolution was performed, what did we uncover regarding trade-offs in performance on C_1 versus multi-C substrates? This work, led by a former graduate student, Ming-Chun 'Miki' Lee, generated two primary findings (Lee *et al.*, 2009). First, the trade-offs were asymmetric: succinate-evolved became worse in methanol, but the methanol-evolved became better in succinate. Correlated improvements in substrates other than the one experienced in the selective environment are in fact quite common in such experiments, even over quite long timescales (e.g. Leiby and Marx, 2014). There are many dimensions to the selective environment (temperature, media, gas exchange, timescale of transferring) that are identical even when testing different substrates. In this regard, the methanol-evolved populations simply improved to the overall laboratory conditions. The second primary finding was that, although the succinate-evolved populations on *average* decreased in performance on methanol, this obscured a bimodal evolutionary response. Approximately half of the populations actually improved in growth on methanol, whereas the other half entirely lost the ability to grow on methanol at all!

Was the loss of ability to grow on methanol a side-product of a mutation that was beneficial during succinate growth (and was thus driven by selection, a phenomenon known as antagonistic pleiotropy), or did it occur due to lack of purifying selection to maintain methylotrophy in a succinate-only world (and thus occurred neutrally, a phenomenon known as mutation accumulation). Subsequent genome sequencing of one isolate revealed a loss-of-function mutation in formate-tetrahydrofolate ligase (encoded by *ftfL*; Marx *et al.*, 2003a,b). Miki Lee discovered that all of the methanol-minus isolates – across independent populations – had non-identical mutations that caused loss-of-function of *ftfL*. So were these *ftfL* null alleles beneficial during succinate growth? A former postdoc, Sean Carroll, picked up this work and found that there was not a simple answer. A loss-of-function *ftfL* allele was beneficial in the strains in which it emerged during succinate growth, but the same allele was neutral, or even deleterious, in strains that had retained the ability to grow on methanol. This dependence of a mutation's phenotypic effect upon other alleles – i.e. epistasis – indicated that one (or more) of the other beneficial alleles that had arisen during succinate adaptation either opened the door to selection-driven loss of methylotrophy, or would prevent it, at least from this mechanism. We thus had a story that began due to a desire to explore trade-offs, and ended up uncovering a key role for epistasis in the modest level of evolutionary repeatability, themes covered in more detail in the sections 'Epistatic interactions between beneficial mutations' and 'Repeatability in adaptation', respectively.

Adaptation requiring use of an introduced enzyme or metabolic pathway

The second half of my postdoctoral fellowship proposal focused upon the concept of using an 'engineered horizontal gene transfer' to study how bacteria adapt to use foreign metabolic pathways. This was written in 2002, when the appreciation of horizontal gene transfer from phylogenetic and genomic analyses was sky-rocketing, but we had very little understanding of the evolutionary steps that proceed immediately after gene acquisition in the cases that turn out to be successful (which I co-authored a review about; Michener and Marx, 2015). Beyond my initial project involving formaldehyde oxidation, three other systems involving primary oxidation enzymes for C_1 compounds were later established by members of the laboratory, and I will discuss each of these in turn.

Swapping formaldehyde oxidation pathways

The first example of adaptation with a foreign pathway involved swapping unrelated formaldehyde oxidation pathways. During my graduate work, I introduced a pathway with known function to try to decipher the roles of two apparently redundant pathways with unclear functions (Marx *et al.*, 2003c). The glutathione (GSH)-dependent formaldehyde oxidation pathway from a distantly related methylotroph, *Paracoccus denitrificans*, is known to function irreversibly in the dissimilatory direction (Ras *et al.*, 1995; Harms *et al.*, 1996). *M. extorquens*, on the other hand, has two pathways that were thought to function in formaldehyde oxidation: the tetrahydromethanopterin (H_4MPT)-dependent C_1 transfer pathway and the tetrahydrofolate (H_4F)-dependent pathway. Mutants with lesions in either pathway failed to grow on methanol, despite that fact that either pathway was thought at the time to be sufficient for formaldehyde oxidation (Chistoserdova and Lidstrom, 1994; Chistoserdova *et al.*, 1998; Marx *et al.*, 2003b,c; Marx and Lidstrom, 2004). When I introduced the foreign GSH-dependent pathway on an expression plasmid into mutants defective for one of these two pathways, I found that it could complement the H_4MPT-dependent pathway mutants, but not the H_4F-dependent pathway ones (Marx *et al.*, 2003c). This, combined with analytical chemistry experiments to follow radio-isotopes or stable isotopes through the cell led to the realization that only the H_4MPT-dependent pathway functions oxidatively *in vivo;* the H_4F-dependent pathway functions in the reverse direction to feed the serine cycle, and thus the flow of carbon splits to assimilation versus dissimilation at the level of formate, not formaldehyde (Marx *et al.*, 2003b, 2005; Crowther *et al.*, 2008). The replacement of the H_4MPT-dependent pathway with the GSH-dependent one meant that an unrelated set of enzymes could handle the central flow of carbon in the cell, yet it did so substantially slower (3-fold) than wild-type, suggesting there may be clear room for evolutionary improvement.

What I proposed in my fellowship application was that the slowed growth of the GSH-dependent *M. extorquens* strains presented a remarkable opportunity to study adaptation subsequent to acquiring a new pathway – as often occurs via horizontal gene transfer – because it focuses selection upon a particular module of the cell. Owing to this, I hypothesized that the biggest benefit mutations – which would be the most likely to escape drift and rise in frequency in the populations – would be constrained to occur within the introduced pathway, or functions that were directly connected to it.

As in the above experiment, I set up eight populations that were passaged through 64-fold dilutions and evolved these for a period of time. Owing to their very slow initial growth ($t_D = 11$ h), they were passaged every four days for the first 300 generations, before bumping all populations up to the 2-day schedule I was using with my trade-off work. The majority of our work was on isolates taken from 600 generations, although some of the dynamics were followed for 900 generations (see 'Complex allele dynamics during adaptation').

Could *M. extorquens* substantially improve over this timescale in terms of its ability to utilize such a different pathway for formaldehyde oxidation? I had not really even considered this as a 'hypothesis' for the project when I started it, but looking back I had implicitly assumed they would improve, and likely due to multiple mutations, rather than just a single quick-fix. My graduate student, Hsin-Hung 'David' Chou spearheaded the majority of this project. He found tremendous improvements in fitness: over 600 generations the average improvement was to nearly double fitness (Chou *et al.*, 2011). And unlike wild-type having evolved on methanol, improvement of this strain was specific to methanol, and not generic adaptation to lab conditions. Similar gains were noted for methylamine growth, which also requires formaldehyde oxidation, but not for formate growth, which is downstream of that step (Carroll and Marx, 2013). Adaptation was also quite rapid, with a full 50% of the observed improvement occurring in just the first 150 generations. Some of the populations improved much more than others, however, and the shape of fitness increases over time suggested that some may have gotten stuck on a lower trajectory of improvement than others (see more about epistasis below; Lee and Marx, 2013).

Was adaptation largely due to mutations in either the GSH-dependent pathway or one physiological process removed from it? As it turned out, yes! David collaborated with another former graduate student, Nigel Delaney, to use Illumina to generate the first re-sequenced genome from an evolved isolate from our lab. This strain, CM1145, will come up several times below. We identified a total of nine mutations that had occurred over 600 generations in CM1145 (Chou *et al.*, 2011). Three of these fit our expectations, and using allelic exchange (Marx, 2008) to generate various mutational combinations, these collectively accounted for ≈85% of the total improvement. These three loci were: *fghA*, which encodes the second of the two GSH-dependent pathway enzymes; *gshA*, which encodes the first step of GSH biosynthesis; and *pntAB*, which encodes transhydrogenase (the GSH-dependent pathway cannot make NADPH directly). This makes quite clear the point that use of a novel pathway can require coordinated mutational changes in both the new genes and in the rest of the recipient genome (Michener and Marx, 2015). Much more emerged from this experiment, but these surprise findings about repeatability (section 'Repeatability during adaptation to use of an introduced enzyme or metabolic pathway'), evolutionary dynamics (section 'Complex allele dynamics during adaptation'), epistasis (section 'Epistatic interactions between beneficial mutations'), and modelling the fitness landscape (see 'Simple mathematical models to understand epistasis between beneficial mutations') are discussed below.

As a side note, the eight populations mentioned above were not actually the first ones initiated from the GSH-dependent *M. extorquens* strain, but rather were a re-start to correct an initial 'failure' that turned out to have its own interesting story (Marx, 2012). During my postdoc, the first eight populations I started behaved quite strangely. Despite having tested what dilution scheme would work with this slow strain, when I started these populations, six of the eight crashed to final cell densities 60- to 400-fold lower than their cohorts. Five of these recovered soon after, but one population persisted with final densities 10-fold lower than expected for over 100 generations. It turns out that specifically for this strain growing on methanol, there was an absolute density threshold for growth. The one population that remained at low density rapidly adapted to this scenario: by generation 84 it had a fitness more than 3-fold greater than the ancestor if competed at a starting density 16-fold lower than normal, but had a fitness less than half the ancestor if started at the typical initial density of $\approx 1 \times 10^6$/ml. By the time this population recovered to a typical final density by generation 180, its advantage at low density had waned. These data suggest that adaptation to low densities may come with trade-offs to growth at high densities. This whole mistake occurred because I went straight from single colonies – to eliminate the possibility of shared mutations – to growth in methanol media. Owing to the very slow growth on methanol, four days was not sufficient for the tiny inoculum of a single colony to reach full density, and thus the cultures remained too dilute after the next 1/64 dilution. I avoided this for the second set of populations with a simple change in protocol: for the first transfer cycle, each colony was inoculated into a mixture of succinate and methanol, and then all flasks thereafter grew on methanol alone. This allowed the populations to rapidly use succinate to pop up from a single colony to half the typical final density in the flask (for this strain, $\approx 1 \times 10^6$/ml), and then during the remaining time before the first transfer they could switch metabolism to utilization of methanol.

Use of the novel, toxic C_1 compound dichloromethane

The second experimental example of this sort involved introduction of a metabolic capacity not present in *M. extorquens* AM1. A couple *M. extorquens* strains have the ability to utilize dichloromethane (DCM), an industrial solvent, as a growth substrate. In strains such as *M. extorquens* DM4, DCM is converted – via a genotoxic intermediate – to one formaldehyde and two hydrochloric acids (Muller *et al.*, 2011). What a way to live! A former postdoc, Josh Michener, developed a collaboration with two experts on DCM growth, Stéphane Vuilleumier and Françoise Bringel from Université de Strasbourg, to explore how easily this novel trait could be moved into other *Methylobacterium* strains. In principle, DCM use only needs a single enzyme, DCM dehalogenase (encoded by *dcmA*), to generate formaldehyde, and thus feed into metabolism in the same manner as methanol would be utilized. Expressing *dcmA* in various *Methylobacterium* strains within and beyond the *M. extorquens* clade, he found that strains were quite variable in their ability to use DCM, and it did not correlate with phylogenetic relatedness (Michener *et al.*, 2014a). Inspired by this finding, Josh initiated replicate populations for each of five initial strains bearing the same *dcmA*-expressing plasmid (Michener *et al.*, 2014b). Populations from all but one of these initial ancestors showed substantial improvement in just 150 generations. Genome sequencing revealed no mutations on the *dcmA*-expressing plasmid, but rather, each genome had one mutation in a known or putative transporter. As it turns out, there were four separate genomic targets revealed that could lead to increased Cl^- export, which appears to have been the primary physiological challenge for these cells (more about repeatability in section 'Repeatability during adaptation to use of an introduced enzyme or metabolic pathway'). By making a construct that expressed *dcmA* and one of these identified exporters, *clcA*, he demonstrated across a wide swath of natural isolates that providing both the enzyme and a 'solution' to the primary physiological challenge – Cl^- export – strains could utilize DCM much better than with *dcmA* alone. Furthermore, when later testing chloromethane growth after having introduced the necessary '*cmu*' gene cluster (Vannelli *et al.*, 1999; Vuilleumier *et al.*, 2011), to our surprise, there was no correlation between the ability of these strains to use DCM and chloromethane, despite the fact that both lead to HCl production (Michener *et al.*, 2016).

Utilization of methylamine as a growth substrate by novel pathways

The third example of novel metabolic pathways involved two very different pathways of methylamine utilization. Another former graduate student, Dipti Nayak, initiated these projects due to curiosity as to why some strains of *M. extorquens* grow phenomenally faster on methylamine than others ($t_D \approx 4$ h versus $t_D \geq 24$ h). It turns out that the major difference came down to whether the cells were using the well-studied methylamine dehydrogenase system (the fast ones), or were using the *N*-methylglutamate (NMG) pathway (the slow ones; Nayak *et al.*, 2015). Besides having uncovered that the NMG pathway directs C_1 units through metabolism in a unique manner (Nayak and Marx, 2014b), and that all *Methylobacterium* genomes encode the NMG pathway, whereas only select few appear to have acquired methylamine dehydrogenase (Nayak *et al.*, 2015), she decided to test the ability to improve the methylamine use of strain in two ways. First, she tried introducing methylamine dehydrogenase (from *M. extorquens* AM1) into a strain without it (*M. extorquens* PA1). Remarkably, unlike all examples above, this immediately resulted in growth equivalent to that in *M. extorquens* AM1 without any further mutations (Nayak *et al.*, 2015)! Second, she tried to evolve strains of *M. extorquens* AM1 or PA1 that were left to utilize their NMG pathway (Nayak *et al.*, 2016). In this case, she observed only modest improvement. Why was adaptation so stunted? The beneficial mutations that Dipti identified were quite illuminating. Besides changes in the NMG pathway itself to increase its expression, she identified mutations that appeared to have relieved the challenges of cytoplasmic production of ammonium. These occurred in either a K^+/H^+ antiporter (KefB), or in a urea transporter (and urea was detected in the supernatant of this strain). Use of methylamine dehydrogenase obviates this challenge because it generates ammonium in the periplasm. This finding also hinted at a possible *advantage* to cytoplasmic production: to prevent ammonium loss when using methylamine as a nitrogen source. Indeed, she demonstrated that strains without the NMG pathway perform worse when using low concentrations of methylamine as a nitrogen source during growth on succinate. Thus, the outcome of experimental evolution provided the key hint as to selective pressures in nature to use different pathways for optimal use of methylamine as a carbon versus a nitrogen source. This distinction in which cellular need methylamine was fulfilling could thus maintain selection upon what would otherwise appear to be degenerate pathways.

Adaptation to ameliorate the negative effects of synonymous mutations

The final example in this section involves a study by a former postdoc, Deepa Agashe, to examine what selective pressures act upon synonymous codons, and how adaptation may proceed to overcome these challenges. Many C_1 pathways appear to have been acquired via horizontal gene transfer, and thus may initially be encoded with poor codons (e.g. Kalyuzhnaya *et al.*, 2005). As a target protein for this work we chose the small, highly expressed (1–2% total protein) enzyme FAE (formaldehyde-activating enzyme), which catalyses the condensation of formaldehyde with H_4MPT (Vorholt *et al.*, 2000). We designed seven versions of *fae* to be synthesized that only differed in their codon usage (Agashe *et al.*, 2013). These ranged from using the most frequent codon at all sites, to using the rarest codon at all sites. Remarkably, there were huge fitness differences between these versions, and they did not correlate with the proportion of frequent codons. The only commonality was that all versions led to insufficient FAE protein levels. Thus, although frequent codons are generally selected for across the genome, they may not always be advantageous in a particular gene.

Strains with these versions swapped into the chromosomal locus were used as ancestors for experimental evolution (Agashe *et al.*, 2016). Despite the fact that these variants contained up to 150 synonymous mutations, Deepa observed rapid adaptation in all lineages. Some of these beneficial mutations were in the promoter region of *fae*, but most were single SNPs in its coding sequence, including some synonymous mutations (more about repeatability in this system in section 'Repeatability during adaptation to ameliorate the negative effects of synonymous mutations'). This indicates that even single mutations can rapidly lead to improved usage of genes that have poor codon usage in their new host.

Surprises: unexpected lessons emerging from evolution experiments

Although all of the above experiments were initiated with clear goals to examine a particular question inspired by natural methylotrophs, one of the great benefits of experimental evolution is the rich set of unintended outcomes. Indeed, I would have to say that these collateral benefits have, in many cases, had a greater level of impact than the answers to the questions we intended to ask. To try to help maintain connection to the previous sections, within each theme below I have covered the model systems in the same order as they were described above.

Repeatability in adaptation

One of the most common outcomes to emerge from experimental evolution is the ability to ascertain whether – for a particular strain in the chosen selective environment – adaptation tends to proceed similarly across independent replicates, or rather differently. This question can be addressed at the level of the genetic basis of adaptation, or at the level of repeated, parallel changes in phenotype. Repeatability will be affected by the relative rates of different beneficial mutations, as well as by the magnitude of their selective coefficients. The latter is particularly important due to the fact that beneficial mutations tend not to be alone in a population: these populations are plenty big to have many beneficial mutations occurring and escaping drift simultaneously. Those with a greater selective benefit not only escape drift and 'establish' more easily, but also have an increased ability to outcompete the other beneficial mutations that have established. This process is known as 'clonal interference' (Gerrish and Lenski, 1998), and is in contrast to the traditional image of one beneficial mutation establishing and sweeping through a population is known as 'periodic selection' (for more detail, see 'Complex allele dynamics during adaptation', below).

Repeatability during evolution of trade-offs between C_1 and multi-C metabolism

Although we never broadly applied whole genome resequencing to uncover the genetic basis of adaptation from our trade-off experiment. We managed to uncover genetic repeatability in a couple of ways. The repeated occurrence of loss-of-function mutations in *ftfL* for succinate-evolved lineages that lost the ability to grow on methanol was already briefly described above in section 'Trade-offs between C1 and multi-C metabolism', and two other stories also emerged.

The first route to uncovering repeatability in adaptation originated in an odd way: from trying to uncover why my lab was initially unable to produce media that would lead to reliable growth. I had no troubles growing *M. extorquens* as a postdoc at Michigan State, but when I moved to Harvard in 2005, we were besieged with inconsistent growth. Our major hint came from one of the first beneficial mutations David Chou uncovered (Chou *et al.*, 2009). An insertion sequence (IS) element had transposed right upstream of a putative metal transporter, and kicked up expression of this locus. Upon examination of other populations, 30 of 32 that had evolved with methanol as a carbon source in their media had mutations at the same locus, but none of the eight populations that grew on succinate alone. Most remarkably, all of these mutations were caused by the same ISMex4 element, and it inserted into one of only two locations upstream of this gene. After a series of elegant experiments, David was able to demonstrate that this gene encoded a cobalt transporter, which we named *icuAB* (for increased cobalt uptake). Why was cobalt a problem, and only for the methanol-evolved populations? As it turns out, the recipe for the Vishniac trace metal mix that was used in our medium had changed compared with its original formulation in terms of strength and ratio of metals (Vishniac and Santer, 1957; Delaney *et al.*, 2013). Notably, the total metal concentration was now similar in strength as the EDTA chelator present. Owing to the light sensitivity of EDTA, this led to inconsistencies: David noted good growth on sunny days when the metal mix was a paler purple, and worse growth on cloudy days when the metal mix was more intensely coloured. Why did *icuAB*^{EVO} alleles only arise as beneficial mutations in medium containing methanol? Cobalt is required for B_{12}, and there are a couple of enzymatic steps unique to the ethylmalonyl-CoA pathway for glyoxylate regeneration in this organism that use B_{12} (Peyraud *et al.*, 2009). Mutants in this pathway have a defect on C_1 or C_2 compounds, which mirrored the scenarios that the medium was problematic, and were the media in which *icuAB*^{EVO} alleles were beneficial. Accordingly, just like for glyoxylate regeneration mutants, the addition of glyoxylate – the end product of this pathway – alleviated the problem. The laboratory of Julia Vorholt at ETH Zürich independently found the same cobalt problem in the medium, but their clue came from metabolomics. They noted accumulation of intermediates immediately upstream of the vitamin B_{12}-requiring reactions of the ethylmalonyl-CoA pathway (Kiefer *et al.*, 2009). Thankfully, the new minimal medium developed by Nigel Delaney and others avoids these issues and has become increasingly utilized in the field (Delaney *et al.*, 2013).

The second story of repeatability from these populations emerged in a different manner: use of comparative genomic hybridization to microarrays (remember those?) to identify potential gene amplification or deletions. Many other selection experiments have identified key selective targets due to amplifications that increased gene dosage, and thus expression (Dunham *et al.*, 2002; Maisnier-Patin and Roth, 2015). Thus, prior to today's ability to inexpensively sequence genomes, we decided to try it out. Miki Lee looked at a total of 44 isolates across 32 different populations that had evolved for 1500 generations, including the 16 populations described above in the trade-offs story (see 'Trade-offs between C_1 and multi-C metabolism'), and found much more than we could have guessed (Lee and Marx, 2012). About 80% of these populations experienced a series of similar deletions that removed 5–10% of the genome. This region was present on a 1.3 Mb megaplasmid unique to *M. extorquens* AM1 versus the other *M. extorquens* isolates (Vuilleumier *et al.*, 2009; Marx *et al.*, 2012). All of the deletions involved homologous recombination between pairs of matching IS elements, but the precise end points of these events were different between populations. Most importantly, Miki was able to generate clean deletions of these regions and demonstrated that they were selectively beneficial, but not in a manner that scaled with the length of the deletion. As such, it is the loss of particular gene products that was beneficial, rather than directly due to having a smaller genome to replicate.

Repeatability during adaptation to use of an introduced enzyme or metabolic pathway

Several stories of repeatability emerged from the multiple examples of evolving engineered ancestral strains that depended upon a novel, introduced metabolic capacity (section 'Adaptation requiring use of an introduced enzyme or metabolic pathway'). To see parallelism when selection acts upon a particular physiological challenge is not surprising, but the types and targets of adaptation were beyond what we could have expected.

Our richest information comes from the populations that evolved on methanol with the GSH-dependent pathway in place of the native H_4MPT-dependent one. Perhaps unsurprisingly, looking across populations, every winning lineage possessed mutations on the plasmid expressing the GSH-dependent pathway (Chou and Marx, 2012). David found that these were of three mutational classes: mutations that affected expression of the two introduced genes that encode the GSH-dependent pathway (*flhA* and *fghA*), mutations that affected the copy number of the plasmid, or integrations of the introduced plasmid into the host genome via incomplete transposition events. It was not a shock that all of these mutations, despite the molecular details being so different, all of these mutations affected levels of FlhA and FghA in the same direction, but we were rather surprised that they all led to *decreased* expression! As it turns out, the high level of FlhA and FghA arising from using the strong P_{mxaF} promoter on a multi-copy plasmid (Marx and Lidstrom, 2001) led to expression costs (Chou *et al.*, 2011). Although the expression phenotype was similar across lineages, the mutational paths to get there were quite distinct. Strangely, the one 'obvious' class of mutations – mutations to decrease the strength of the promoter – was never observed. More on that below in section 'Simple mathematical models to understand epistasis between beneficial mutations'.

Unlike the universal presence of beneficial mutations in the introduced plasmid expressing the GSH-dependent pathway in these strains, chromosomal mutational targets were much more variable between lineages (Carroll *et al.*, 2015). Sean Carroll examined the genome sequences for one strain from generation 600 for each of the eight populations. There was great variety in the total number of mutations present, from 4 to 18. The chromosomal locus most commonly mutated was *icuAB* described above in section 'Repeatability during evolution of trade-offs between C_1 and multi-C metabolism' (six of eight), followed by *gshA* mentioned above in section 'Adaptation requiring use of an introduced enzyme or metabolic pathway' (five of eight), and then seven other loci with mutations in two or three of the eight populations. This modest overlap of mutational targets is perhaps consistent with the relatively large spread in fitness values achieved by these lineages, ranging from 70% to 160%, and the corresponding maintenance of among-population variation in fitness through time (Lee and Marx, 2013).

Although the beneficial mutations may have been only modestly repeatable, analyses of gene expression and some currency metabolites were much more consistent across lineages. Sean took these same eight strains and analysed mRNA levels during exponential phase (Carroll and Marx, 2013). There were huge changes in gene expression from the ancestor to each of the eight evolved isolates examined, and these were remarkably parallel. How to rationalize this compared to the fairly different set of underlying mutations? Comparison back to wild-type, which grows quickly using the H_4MPT-dependent pathway was tremendously useful. Nearly all the genes that went up (such as stress responses) or down (metabolism and ribosomes) due to swapping the pathways in the original strain were those with a reversed pattern during adaptation. It is clear that most of these represent the indirect consequences of poor growth versus rapid growth, and this aspect changed in all lineages. Steady-state ratios of the concentrations of the pyridine nucleotides – NAD(P)(H) – also changed in this manner: increases in the ratio of NADPH/$NADP^+$ and NADH/NAD^+ occurred in the engineered ancestor compared with wild-type, and these were reversed during adaptation. Indeed, the exact ratios for each pair of nucleotides correlated quite well with the fitness of the strains.

This pattern of modest repeatability was also seen for the DCM adaptation by Josh Michener (Michener *et al.*, 2014b). Although Cl^- export appeared to be the central challenge in all cases, he found four distinct loci could ameliorate this challenge. Interestingly, the same locus was often observed as a beneficial mutation across different strains or species used as ancestors. Beneficial alleles in *secY* (subunit of protein secretion system) were found in *M. extorquens* AM1, *M. extorquens* BJ001 (formerly *M. populi*), and *M. nodulans*; *clcA* (Cl^-/H^+ antiporter) in *M. extorquens* PA1 and BJ001; and *edgA* (new locus named for evolved DCM growth) in *M. extorquens* PA1 and *M. nodulans*. This pattern seemed to imply that the same loci could solve this common problem regardless of the genomic background of the strain. Indeed, beneficial mutations from one strain could be introduced into another strain where none of the beneficial mutations were at that locus, and that allele was still beneficial in the new genomic context.

Perhaps the most exciting part of this study was Josh's examination of how *M. extorquens* DM4 first may have evolved originally to use *dcmA* in nature (Michener *et al.*, 2014b). The amino acid sequences for all four loci identified above to solve the Cl^- export problem were identical between the natural DCM utilizer (DM4), and the other *M. extorquens* strains (AM1, BJ001, CM4, PA1). There were, however, mutations unique to *M. extorquens* DM4 in the *clcA* promoter. Swapping just P_{clcA} between DM4 and the DCM-naive PA1 strain was sufficient to completely swap the phenotype! Furthermore, as there happens to be only one other strain of *M. extorquens* known to utilize DCM, Josh obtained this strain and sequenced P_{clcA}. A beneficial mutation (another IS insertion) was also found in this strain that increased *clcA* expression. Thus, not only was adaptation repeatable in the lab across different strains and species within a genus, we were able to identify the chromosomal locus that permitted effective DCM utilization in two independent natural isolates after they had acquired *dcmA* in the past.

Repeatability during adaptation to ameliorate the negative effects of synonymous mutations

Deepa Agashe's work on rapid adaptation to various synonymous re-codings of *fae* produced another remarkable pattern of repeatability (Agashe *et al.*, 2016). Although all of the strains initially grew poorly on methanol due to insufficient levels of FAE, they evolved in a manner that was highly consistent per strain, but very different between them. Replicate populations for three of the six variants always had beneficial mutations in the promoter of *fae* that increased gene expression, but never had coding mutations. On the other hand, for the other three synonymous variants the pattern was reversed: always coding mutations and never promoter mutations. For this last set, there were commonly repeated mutations to the exact same residue of FAE, such as three distinct mutations to Ile-12, two different synonymous mutations and one non-synonymous one. Furthermore, to briefly mention epistasis in this system, when 12 different beneficial coding mutations were moved to alternative synonymous ancestral versions, only two of these conferred benefit in the new context. This version-specific pattern of adaptation corroborated her earlier work (Agashe *et al.*, 2013) that suggested that each recoded version of *fae* was problematic to express in a unique way.

Complex allele dynamics during adaptation

Besides examination of the collection of alleles or phenotypes that have emerged in single evolved isolates, a complementary aspect of the evolutionary process is to uncover the trajectories of alleles in the populations. As I have commented upon before (Marx, 2013), these dynamics alone can give clues as to aspects such as multiple ecological niches. Recent work has greatly expanded the quantitative inferences that can be made in this regard (e.g. Lang *et al.*, 2013; Good *et al.*, 2017), but two simple stories have emerged from our work on adaptation of *M. extorquens* dependent upon the GSH-dependent pathway for formaldehyde oxidation.

The first story of allele dynamics built upon our desire to develop a simple, inexpensive method to follow allele frequencies in populations via amplicon sequencing. Before then, the standard method to follow allele dynamics was to isolate many (10–100) isolates per timepoint to interrogate, and then amplify each allele from each isolate at each timepoint, and submit these for standard Sanger sequencing. Although simple in design, this is rather laborious and expensive. A former postdoc, Lon Chubiz, worked with Miki Lee to develop a cheap method we named 'FREQ-Seq' (Chubiz *et al.*, 2012). The primary innovation was to recognize that we wanted to barcode individual samples, yet did not want to order tens or hundreds of barcoded primers for every locus in question. FREQ-Seq uses two rounds of PCR, beginning with a single set of allele-specific primers to amplify a locus in question from a mixed population timepoint (using primers with generic 5' extensions that match the 3' extensions of primers in step two). This product is then re-amplified in a mixture containing three primers: two universal primers (that are also the Illumina A and B adapters, eliminating the need for further sample prep) and a small amount of a long bridging primer that carries the sample-specific barcode. This worked amazingly well. The very first time we tested this in full, Lon sacrificed an opportunity to join the lab for a well-earned beer, and instead spent ≈3 h performing two rounds of PCR for each of three loci (*fghA*, *gshA*, *pntAB*; see 'Adaptation requiring use of an introduced enzyme or metabolic pathway') at 27 time-points of the population that CM1145 was isolated from (Chou *et al.*, 2011). These samples – as well as controls for each locus – were then put in to 3% of an Illumina control lane. Even at the modest sequencing depth available at the time, this resulted in an average coverage of 150,000-fold per timepoint per locus, with almost no PCR bias. What did we learn? First, it became clear that, of these three mutations, the *gshA*^{EVO} mutation occurred first, and then *fghA*^{EVO}, then *pntAB*^{EVO}. This matched the order of their selective benefits from largest to smallest (Chou *et al.*, 2011), which was on its own unsurprising. What was fascinating, however, was that the *gshA*^{EVO} had rapidly risen to ≈40% of the population by generation 108, and then fell to < 10% at generation 180, before rapidly rising again to near fixation by generation 210. This pattern is indicative of clonal interference: the *gshA*^{EVO} allele did not become less beneficial, but was temporarily passed by one (or more) lineages that were even more fit, before being able to rise to eventual fixation once coupled with the *fghA*^{EVO} allele. In this case we were lucky, and in the next story we figured out just what mutation(s) temporarily pushed *gshA*^{EVO} down in frequency.

The second window into allele dynamics arose due to a particular type of allele mentioned above in section 'Adaptation requiring use of an introduced enzyme or metabolic pathway': the beneficial integration of our introduced plasmid encoding the GSH-dependent pathway into the host genome (Chou and Marx, 2012). As bizarre of a genetic event as this would seem to be, it occurred multiple times in the small number of isolates David examined. These integrations were not due to traditional homologous recombination, but, rather, were caused by incomplete transposition of ISMex25 into our plasmid, pCM410, resulting in a cointegrate. Every one of these events that we obtained was an insertion into *trfA*, which encodes the essential replication initiation gene of this small IncP plasmid backbone (Marx and Lidstrom, 2001). Miki decided to follow-up on this, and devised a simple semi-quantitative PCR approach to identify such insertions. Her data were striking: not only could she follow the dynamics of these bizarre alleles in the populations where they had been previously identified, it turns out every population contained multiple (up to 17!) such insertions. This was possible to discern because distinct insertion sites or orientations of ISMex25 in *trfA* generated PCR products of different lengths. If these were beneficial (≈20% when tested alone) and so easy to produce, how could they have failed to become the winning lineage in all populations? In the five populations where these alleles did not fix, the whole set of alleles became detectable at the same time, rose in frequency together, and then fell in synchrony with each other. They were thus victims of clonal interference. In the population studied with FREQ-Seq and described above, we could clearly see that it was the 17 genotypes bearing one these insertion alleles (along with other mutations) that rose in frequency and peaked precisely when the *gshA*^{EVO}-containing genotype was at its lowest point, and then the insertion alleles fell in frequency together when the *gshA*^{EVO} rapidly rose towards eventual fixation. It should be noted that the complex dynamics of clonal interference are not just seen in laboratory-based experimental evolution, but also occur in natural situations such as persistent infections (Lieberman *et al.*, 2011; Yang *et al.*, 2011; Silva *et al.*, 2016; Levade *et al.*, 2017; Xue *et al.*, 2017).

Epistatic interactions between beneficial mutations

Once one has found the beneficial mutations that were targets of adaptation, and perhaps having uncovered their complex temporal dynamics, this opens the door to considering how the mutations affected each other along an adaptive trajectory. If the selective effect of an allele were independent of the genotype it arose upon, the order of beneficial mutations observed would solely be due to the balance between the likelihood of a mutation happening, and the probability a mutation of that (fixed) selective coefficient would escape drift and beat out other mutations it was clonally interfering with. It would also imply a perfectly smooth, single-peaked adaptive landscape. But if the fitness effect of a mutation is not fixed and depends upon the genetic background, then all bets are off. For a pair of mutations *i* and *j*, we typically consider epistasis (ε_{ij}) as the difference between the phenotype (here fitness, *W*) actually observed (W_{ij}) and the null expectation if there were no 'extra' effect of the two alleles having been combined. This null hypothesis is generally the product of the two individual effects ($W_i \times W_j$), consistent with the idea of independently acting allele having a constant proportional affect upon fitness in the absence of epistasis. Prior to taking on these studies, this type of effect had been examined for beneficial mutations within a single gene (e.g. Weinreich *et al.*, 2006), but had been little explored for mutations that had occurred across the genome. Having earlier 'snuck in' brief mentions of epistasis emerging from studies of metabolic trade-offs and selection upon codons, here I concentrate on three examples from work by David Chou on adaptation with the foreign, GSH-dependent pathway.

The first example of epistasis we uncovered involved the *icuAB* cobalt transporter we discovered (Chou *et al.*, 2009; see section 'Repeatability during evolution of trade-offs between C1 and multi-C metabolism'). As mentioned above, beneficial mutations in *icuAB* occurred in 30 of 32 populations examined that evolved using methanol as a substrate, and this included six of the eight populations with the GSH-dependent pathway. In the wild-type context, $icuAB^{EVO}$ conferred an 18% advantage during growth on methanol in the original (unintentionally) cobalt-limiting media. To David's great surprise, when he introduced $icuAB^{EVO}$ into the GSH-dependent ancestor, there was only a very slight increase in fitness (< 3%). If that was so, how could it have emerged repeatedly? He recognized that $icuAB^{EVO}$ likely was not the first allele to arise, however, and thus could have arisen on a fitter version of that strain. Having begun to re-construct strains bearing $fghA^{EVO}$, $gshA^{EVO}$, or $pntAB^{EVO}$, he introduced $icuAB^{EVO}$ into each of these somewhat faster strains. Plotting the selective coefficient of $icuAB^{EVO}$ against the baseline growth rate of the genotypes it was introduced into, there was a straight line with positive slope. This indicates that the beneficial effect of $icuAB^{EVO}$ was synergistic with the other mutations, and thus conferred a higher proportional benefit the fitter the strain background was. David then had the clever idea to test whether changes in the environment that would affect growth rate would generate a similar effect as epistasis. By simply growing wild-type at 16, 20, and 25 °C (rather than the standard 30 °C), he could generate a similar range of growth rates as the GSH-dependent strains grown at 30 °C. Indeed, the fitness effect of these manipulations plotted right on top of the epistasis data. This indicates that it really does not seem to matter what makes the cells grow more slowly, but simply doing so results in a consistent decrease in the fitness effect of $icuAB^{EVO}$. What underlies this commonality between epistasis (also known as genotype × genotype, or G × G interactions) and genotype × environment (i.e. G × E) interactions? This requires thinking about where cobalt 'goes' during cell growth. Unlike carbon, it is not breathed off, but simply becomes a part of the B_{12} pool. The only thing that makes the B_{12} concentration go down is dilution by cell growth. As such, the need for increased cobalt uptake becomes increasingly severe as the cell grows faster and faster.

The second example of epistasis in this system came from analysing a network of allele combinations from the best-studied evolved isolate, CM1145, whose genome was first sequenced (Chou *et al.*, 2011), and has been mentioned in multiple sections above. Although this strain contained nine mutations, many of them were either hard to reproduce (like a huge deletion of 10% of the genome, like what Miki had uncovered in other populations; Lee and Marx, 2012), of questionable effect (synonymous mutation or in a gene of unknown function), or clearly had to do with the cobalt-limitation in the medium (*icuAB*; Chou *et al.*, 2009). Thus, rather than labour to make all $2^9 = 512$ strains, David wisely chose to concentrate upon the loci that were directly related to the introduced GSH-dependent formaldehyde oxidation pathway (*fghA*, *gshA*, *pntAB*), and lump the rest of the six mutations as a single allele, the 'genetic background' (*GB*). The fitness value for all 16 combinations of these alleles was determined, and a clear pattern emerged. For three of the four alleles, the selective benefit of that allele steadily decreased with the fitness of the background it was introduced into. For example, the $gshA^{EVO}$ allele had a 51% benefit when present alone, but only conferred a 34% benefit if introduced into a strain already containing the three other evolved alleles. A similar trend was shown by my colleague Tim Cooper (now at Massey University, New Zealand) for the first mutational steps of one of the Lenski lineages of *E. coli* evolved to utilize glucose (Khan *et al.*, 2011). Tim and I coordinated our submissions, and these ended up being the first papers to suggest a generic trend for 'diminishing returns' epistasis between beneficial mutations. Mutations becoming less and less valuable when stacked on top of each other play a critical role in the well-established tendency of experimental evolution experiments to decelerate their rate of adaptation (Lenski *et al.*, 1991; Wiser *et al.*, 2013). This trend has also been observed in many other biological systems, suggesting one or more general principles at cause (Kvitek and Sherlock, 2011; Rokyta *et al.*, 2011; Kryazhimskiy *et al.*, 2014).

Our final example of epistasis came from an orthogonal approach that David took: combine beneficial mutations from separate evolved lineages that affected the same pathway (Chou *et al.*, 2014). He made a series of allele combinations where one mutation decreased expression of the GSH-dependent pathway due to lowering the expression per plasmid, and the second mutation was one that decreased plasmid copy number. His rationale was that these mutations should affect expression in independent ways, and thus be free from epistasis at the level of enzyme expression. This turned out to be the case; the expression level of the double mutant was well-predicted by the product of the individual mutation effects. When it came to fitness, however, $W_{ij} = W_i \times W_j$ failed miserably, with combinations of beneficial alleles consistently providing less benefit than expected. In some cases, two highly beneficial alleles were tremendously deleterious when combined for reasons described below.

Simple mathematical models to understand epistasis between beneficial mutations

Perhaps the aspects of all this work that I am the most excited about, and guides several of the current efforts in the lab, are efforts to propose mathematical models of biochemistry as hypotheses for how mutations should affect fitness, alone or together. My interest in this area was shaped tremendously by the pioneering work of Dan Dykhuizen, Tony Dean and Dan Hartl throughout the 1980s (e.g. Hartl *et al.*, 1985; Dean *et al.*, 1986; Dykhuizen *et al.*, 1987). They used theory from Metabolic Control Analysis (Kascer and Burns, 1973; Heinrich and Rapoport, 1974) to understand the fitness effects of mutations in *lacZ* or *lacY* upon growth of *E. coli* in lactose-limited chemostats. This was evolutionary systems biology long before the term was ever used (or was popular). Chemostats are great for the ability to focus selection largely upon a single limiting resource, but this is also their limitation; it is unclear how well biochemistry could predict growth rate in the absence of a single external limitation. On the other hand, my GSH-dependent *M. extorquens* has a single internal limitation, so I hoped there was a chance for this approach to work.

The first form of epistasis we modelled was the diminishing returns observed when combining alleles that arose in a single evolved isolate (Chou *et al.*, 2011). For this, we teamed up with my good friend and collaborator, Daniel Segrè (Boston University), and his former graduate student, Hsuan-Chao Chiu. David Chou had noted that the expression costs for synthesizing the two enzymes of the GSH-dependent pathway visually manifested as larger, morphologically abnormal cells. Mutations that directly reduced expression of the pathway (e.g. $fghA^{EVO}$) led to fewer such cells, however, even the $gshA^{EVO}$ and GB^{EVO} alleles led to fewer abnormal cells (but not $pntAB^{EVO}$). This suggested that the benefit of $gshA^{EVO}$ and GB^{EVO} might at least partially be due to reducing the protein expression costs of the two enzymes of the introduced GSH pathway (whereas the benefit of $pntAB^{EVO}$ is unrelated to expression costs). These three alleles that appeared to reduce expression costs – $fghA^{EVO}$, $gshA^{EVO}$ and GB^{EVO} – were also the three of the four that exhibited diminishing returns ($pntAB^{EVO}$ imparted a ≈ 10% benefit regardless of background). This led us to speculate a connection between expression costs and diminishing returns epistasis. We formulated a simple model in which we treated protein expression costs as a separate phenotype from everything else that affects growth rate. Then we imagined each mutation could affect one or both of these phenotypes and interact independently (i.e. no epistasis) upon that phenotype, yet fitness is the intrinsic growth rate minus expression costs. By using the proportion of morphological abnormalities as a proxy to partition the degree to which each single mutation affected benefits or costs, we were able to predict the fitness values of double, triple, and quadruple mutation combinations from the phenotypes of the single mutants with a surprising degree of success ($R^2 = 0.97$). The intuition for diminishing returns in this example of protein expression costs is easy to illustrate. As a thought experiment, imagine three mutations that each cut protein expression costs in half in distinct ways, and that these expression costs reduce fitness by 20%. The first of these beneficial mutations to occur will cut the cost from 20% to 10%, and will thus produce a fitness benefit of $(1.2 - 0.1)/(1.2 - 0.2) = 10\%$. The next mutation, however, would only reduce the expression cost a further 5%, and would have a smaller benefit of $(1.2 - 0.05)/(1.2 - 0.1) = 4.5\%$. The third mutation would have even less remaining effect, cutting the costs by 2.5%, thus a $(1.2 - 0.025)/(1.2 - 0.05) = 2.2\%$ fitness benefit. As developed in a follow-up theory paper led by Hsuan-Chao Chiu, continued proportional reductions to a single phenotype inherently leads to diminishing returns (Chiu *et al.*, 2012).

The second type of epistasis modelled was that of multiple mutations that reduced expression of the GSH-dependent pathway via independent molecular mechanisms (Chou *et al.*, 2014). As mentioned above in section 'Epistatic interactions between beneficial mutations', enzyme expression was well-predicted via a simple multiplicative model, but fitness was not. Rather than build a full metabolic control analysis model with many parameters that were never manipulated in this experiment, we generated a simplified algebraic expression for the 'control curve' that describes the consequences of changing activity of a single enzyme at a time upon the steady-state flux through a pathway (which in a case like this is proportional to fitness). As these control curves are inherently hyperbolic, they can be approximated in a manner analogous to a Michaelis–Menten relationship: the 'v_{max}' parameter is the maximal steady-state flux of the pathway given infinite activity of the enzyme in question, and the 'K_M'-like saturation parameter is the level of that enzyme needed to provide half-maximal steady-state flux through the pathway. Two more modifications to this logic were suggested by the data: linear costs of the enzymes, and an offset from the origin due to the apparent need for a threshold value of enzyme activity to avoid formaldehyde toxicity (and thus no growth at all). All of this led to single equation with five parameters. We fit these five parameters using 27 data points from strains with either inducible promoters driving known changes in enzyme levels, or single mutants. We then used this model to extrapolate to 17 mutational combinations not used in the fitting. Remarkably, despite the many simplifications involved, this model provided a fairly precise prediction of these fitness values ($R^2 = 0.98$). This demonstrates that the fitness landscape of enzyme expression is smooth and can be easily modelled if the data are available.

Most importantly, several aspects of the adaptation that had occurred were illuminated by the shape of the resulting fitness landscape. First, it became clear that the winning beneficial mutations affecting the levels of the two enzymes of the GSH-dependent pathway made it to the fitness peak in a single mutational step, rather than a long series of smaller mutations stacked on top of each other. Second, the lack of promoter mutations was now explained: although turning down transcription via the inducible promoter could be beneficial, the highest fitness benefit obtainable by doing so is ≈ 20%. The particular mutations that did occur generated benefit up to ≈ 40%, and thus would have generally outcompeted the lesser benefit solutions to this particular physiological challenge. Third, it was now clear why some mutations interacted epistatically so poorly with other mutations; these were the ones that sat closest to the 'cliff edge' of the fitness landscape, such that any further decrease in expression would lead to a massive decrease in growth.

Conclusions

Combining experimental evolution, microbial physiology, and mathematical biology has been an enjoyable and reasonably fruitful approach for my lab to address questions about *M. extorquens* that lie at the intersection of these fields. Although the original questions being asked with each experiment outlined here turned up answers, I strongly feel the 'collateral' findings that were not predicted from the start have been the most rewarding and broadly applicable to other systems. This balance between planned versus surprising results perhaps mirrors the tension of evolution itself: selection upon traits in a given environment is deterministic and (in principle) predictable, whereas the arrival and escape from loss due to drift of the very mutations that affect these traits is inherently stochastic.

Acknowledgements

CJM recognizes an NSF award (MCB-1714949) that supports work onv the experimental evolution of *Methylobacterium*.

References

Agashe, D., Martinez-Gomez, N.C., Drummond, D.A., and Marx, C.J. (2013). Good codons, bad transcript: large reductions in gene expression and fitness arising from synonymous mutations in a key enzyme. Mol. Biol. Evol. *30*, 549–560. https://doi.org/10.1093/molbev/mss273

Agashe, D., Sane, M., Phalnikar, K., Diwan, G.D., Habibullah, A., Martinez-Gomez, N.C., Sahasrabuddhe, V., Polachek, W., Wang, J., Chubiz, L.M., and Marx, C.J. (2016). Large-effect beneficial synonymous mutations mediate rapid and parallel adaptation in a bacterium. Mol. Biol. Evol. 33, 1542–1553. https://doi.org/10.1093/molbev/msw035

Bruger, E.L., and Marx, C.J. (2018). A decade of genome sequencing has revolutionized studies of experimental evolution. Curr. Opin. Microbiol. *45*, 149–155. https://doi.org/10.1016/j.mib.2018.03.002

Carroll, S.M., and Marx, C.J. (2013). Evolution after introduction of a novel metabolic pathway consistently leads to restoration of wild-type physiology. PLOS Genet. *9*, e1003427. https://doi.org/10.1371/journal.pgen.1003427

Carroll, S.M., Lee, M.C., and Marx, C.J. (2014). Sign epistasis limits evolutionary trade-offs at the confluence of single- and multi-carbon metabolism in *Methylobacterium extorquens* AM1. Evolution *68*, 760–771. https://doi.org/10.1111/evo.12301

Carroll, S.M., Chubiz, L.M., Agashe, D., and Marx, C.J. (2015). Parallel and divergent evolutionary solutions for the optimization of an engineered central metabolism in *Methylobacterium extorquens* AM1. Microorganisms 3, 152–174. https://doi.org/10.3390/microorganisms3020152

Chistoserdova, L.V., and Lidstrom, M.E. (1994). Genetics of the serine cycle in *Methylobacterium extorquens* AM1: identification of *sgaA* and *mtdA* and sequences of *sgaA*, *hprA*, and *mtdA*. J. Bacteriol. *176*, 1957–1968.

Chistoserdova, L., Vorholt, J.A., Thauer, R.K., and Lidstrom, M.E. (1998). C1 transfer enzymes and coenzymes linking methylotrophic bacteria and methanogenic Archaea. Science *281*, 99–102.

Chiu, H.C., Marx, C.J., and Segrè, D. (2012). Epistasis from functional dependence of fitness on underlying traits. Proc. Biol. Sci. *279*, 4156–4164. https://doi.org/10.1098/rspb.2012.1449

Chou, H.H., and Marx, C.J. (2012). Optimization of gene expression through divergent mutational paths. Cell Rep. *1*, 133–140. https://doi.org/10.1016/j.celrep.2011.12.003

Chou, H.H., Berthet, J., and Marx, C.J. (2009). Fast growth increases the selective advantage of a mutation arising recurrently during evolution under metal limitation. PLOS Genet. *5*, e1000652. https://doi.org/10.1371/journal.pgen.1000652

Chou, H.H., Chiu, H.C., Delaney, N.F., Segrè, D., and Marx, C.J. (2011). Diminishing returns epistasis among beneficial mutations decelerates adaptation. Science *332*, 1190–1192. https://doi.org/10.1126/science.1203799

Chou, H.H., Delaney, N.F., Draghi, J.A., and Marx, C.J. (2014). Mapping the fitness landscape of gene expression uncovers the cause of antagonism and sign epistasis between adaptive mutations. PLOS Genet. *10*, e1004149. https://doi.org/10.1371/journal.pgen.1004149

Chubiz, L.M., Lee, M.-C., Delaney, N.F., and Marx, C.J. (2012). FREQ-Seq: A rapid, cost-effective, sequencing-based method to determine allele frequencies directly from mixed populations. PLOS ONE. *7*, e47959. https://doi.org/10.1371/journal.pone.0047959

Crowther, G.J., Kosály, G., and Lidstrom, M.E. (2008). Formate as the main branch point for methylotrophic metabolism in *Methylobacterium extorquens* AM1. J. Bacteriol. *190*, 5057–5062. https://doi.org/10.1128/JB.00228-08

Dean, A.M., Dykhuizen, D.E., and Hartl, D.L. (1986). Fitness as a function of beta-galactosidase activity in *Escherichia coli*. Genet. Res. *48*, 1–8.

Delaney, N.F., Kaczmarek, M.E., Ward, L.M., Swanson, P.K., Lee, M.-C., and Marx, C.J. (2013). Development of an optimized medium, strain and high-throughput culturing methods for *Methylobacterium extorquens*. PLOS ONE *8*, e62957. https://doi.org/10.1371/journal.pone.0062957

Douglas, S.M., Chubiz, L.M., Harcombe, W.R., Ytreberg, F.M., and Marx, C.J. (2016). Parallel mutations result in a wide range of cooperation and community consequences in a two-species bacterial consortium.

PLOS ONE *11*, e0161837. https://doi.org/10.1371/journal.pone.0161837

Douglas, S.M., Chubiz, L.M., Harcombe, W.R., and Marx, C.J. (2017). Identification of the potentiating mutations and synergistic epistasis that enabled the evolution of inter-species cooperation. PLOS ONE *12*, e0174345. https://doi.org/10.1371/journal.pone.0174345

Dunham, M.J., Badrane, H., Ferea, T., Adams, J., Brown, P.O., Rosenzweig, F., and Botstein, D. (2002). Characteristic genome rearrangements in experimental evolution of *Saccharomyces cerevisiae*. Proc. Natl. Acad. Sci. U.S.A. *99*, 16144–16149. https://doi.org/10.1073/pnas.242624799

Dykhuizen, D.E., Dean, A.M., and Hartl, D.L. (1987). Metabolic flux and fitness. Genetics *115*, 25–31.

Gerrish, P.J., and Lenski, R.E. (1998). The fate of competing beneficial mutations in an asexual population. Genetica 102-103, 127–144.

Good, B.H., McDonald, M.J., Barrick, J.E., Lenski, R.E., and Desai, M.M. (2017). The dynamics of molecular evolution over 60,000 generations. Nature *551*, 45–50. https://doi.org/10.1038/nature24287

Harcombe, W.R., Riehl, W.J., Dukovski, I., Granger, B.R., Betts, A., Lang, A.H., Bonilla, G., Kar, A., Leiby, N., Mehta, P., et al. (2014). Metabolic resource allocation in individual microbes determines ecosystem interactions and spatial dynamics. Cell Rep. *7*, 1–12. https://doi.org/10.1016/j.cellrep.2014.03.070

Harcombe, W.R., Betts, A., Shapiro, J.W., and Marx, C.J. (2016). Adding biotic complexity alters the metabolic benefits of mutualism. Evolution *70*, 1871–1881. https://doi.org/10.1111/evo.12973

Harms, N., Ras, J., Reijnders, W.N., van Spanning, R.J., and Stouthamer, A.H. (1996). *S*-Formylglutathione hydrolase of *Paracoccus denitrificans* is homologous to human esterase D: a universal pathway for formaldehyde detoxification? J. Bacteriol. *178*, 6296–6299.

Hartl, D.L., Dykhuizen, D.E., and Dean, A.M. (1985). Limits of adaptation: the evolution of selective neutrality. Genetics *111*, 655–674.

Heinrich, R., and Rapoport, T.A. (1974). A linear steady-state treatment of enzymatic chains. General properties, control and effector strength. Eur. J. Biochem. *42*, 89–95.

Kalyuzhnaya, M.G., Korotkova, N., Crowther, G., Marx, C.J., Lidstrom, M.E., and Chistoserdova, L. (2005). Analysis of gene islands involved in methanopterin-linked C_1 transfer reactions reveals new functions and provides evolutionary insights. J. Bacteriol. *187*, 4607–4614. https://doi.org/10.1128/JB.187.13.4607-4614.2005

Kascer, H., and Burns, J.A. (1973). The control of flux. Symp. Soc. Exp. Biol. *27*, 65–104.

Kiefer, P., Buchhaupt, M., Christen, P., Kaup, B., Schrader, J., and Vorholt, J.A. (2009). Metabolite profiling uncovers plasmid-induced cobalt limitation under methylotrophic growth conditions. PLOS ONE. *4*, e7831. https://doi.org/10.1371/journal.pone.0007831

Khan, A.I., Dinh, D.M., Schneider, D., Lenski, R.E., and Cooper, T.F. (2011). Negative epistasis between beneficial mutations in an evolving bacterial population. Science *332*, 1193–1196. https://doi.org/10.1126/science.1203801

Knief, C., Frances, L., and Vorholt, J.A. (2010). Competitiveness of diverse *Methylobacterium* strains in the phyllosphere of *Arabidopsis thaliana* and identification of representative models, including *M. extorquens* PA1. Microb. Ecol. *60*, 440–452. https://doi.org/10.1007/s00248-010-9725-3

Kryazhimskiy, S., Rice, D.P., Jerison, E.R., and Desai, M.M. (2014). Microbial evolution. Global epistasis makes adaptation predictable despite sequence-level stochasticity. Science *344*, 1519–1522. https://doi.org/10.1126/science.1250939

Kvitek, D.J., and Sherlock, G. (2011). Reciprocal sign epistasis between frequently experimentally evolved adaptive mutations causes a rugged fitness landscape. PLOS Genet. *7*, e1002056. https://doi.org/10.1371/journal.pgen.1002056

Lang, G.I., Rice, D.P., Hickman, M.J., Sodergren, E., Weinstock, G.M., Botstein, D., and Desai, M.M. (2013). Pervasive genetic hitchhiking and clonal interference in forty evolving yeast populations. Nature *500*, 571–574. https://doi.org/10.1038/nature12344

Lee, M.-C., and Marx, C.J. (2012). Repeated, selection-driven genome reduction of accessory genes in experimental populations. PLOS Genet. *8*, e1002651. https://doi.org/10.1371/journal.pgen.1002651

Lee, M.-C., and Marx, C.J. (2013). Synchronous waves of failed soft sweeps in the laboratory: remarkably rampant clonal interference of alleles at a single locus. Genetics *193*, 943–952. https://doi.org/10.1534/genetics.112.148502

Lee, M.C., Chou, H.H., and Marx, C.J. (2009). Asymmetric, bimodal trade-offs during adaptation of *Methylobacterium* to distinct growth substrates. Evolution *63*, 2816–2830. https://doi.org/10.1111/j.1558-5646.2009.00757.x

Leiby, N., and Marx, C.J. (2014). Metabolic erosion primarily through mutation accumulation, and not tradeoffs, drives limited evolution of substrate specificity in *Escherichia coli*. PLOS Biol. *12*, e1001789. https://doi.org/10.1371/journal.pbio.1001789

Lenski, R.E. (2017). Experimental evolution and the dynamics of adaptation and genome evolution in microbial populations. ISME J. *11*, 2181–2194. https://doi.org/10.1038/ismej.2017.69

Lenski, R.E., Rose, M.R., Simpson, S.C., and Tadler, S.C. (1991). Long-term experimental evolution in *Escherichia coli*. I. Adaptation and divergence during 2,000 generations. Am. Nat. *138*, 1315–1341.

Levade, I., Terrat, Y., Leducq, J.B., Weil, A.A., Mayo-Smith, L.M., Chowdhury, F., Khan, A.I., Boncy, J., Buteau, J., Ivers, L.C., *et al.* (2017). *Vibrio cholerae* genomic diversity within and between patients. Microb. Genom. 3. https://doi.org/10.1099/mgen.0.000142

Lieberman, T.D., Michel, J.B., Aingaran, M., Potter-Bynoe, G., Roux, D., Davis, M.R. Jr., Skurnik, D., Leiby, N., LiPuma, J.J., Goldberg, J.B., *et al.* (2011). Parallel bacterial evolution within multiple patients identifies candidate pathogenicity genes. Nat. Genet. *43*, 1275–1280. https://doi.org/10.1038/ng.997

Luria, S.E., and Delbrück, M. (1943). Mutations of bacteria from virus sensitivity to virus resistance. Genetics *28*, 491–511.

Maisnier-Patin, S., and Roth, J.R. (2015). The origin of mutants under selection: how natural selection mimics mutagenesis (adaptive mutation). Cold Spring Harb

Perspect Biol. 7, a018176. https://doi.org/10.1101/cshperspect.a018176

Marx, C.J. (2008). Development of a broad-host-range *sacB*-based vector for unmarked allelic exchange. BMC Res. Notes *1*, 1. https://doi.org/10.1186/1756-0500-1-1

Marx, C.J. (2011). Evolution as an experimental tool in microbiology: 'Bacterium, improve thyself!' Environ. Microbiol. Rep. 3, 12–14.

Marx, C.J. (2012). Recovering from a bad start: rapid adaptation and tradeoffs to growth below a threshold density. BMC Evol. Biol. *12*, 109. https://doi.org/10.1186/1471-2148-12-109

Marx, C.J. (2013). Can you sequence ecology? Metagenomics of adaptive diversification. PLOS Biol. *11*, e1001487. https://doi.org/10.1371/journal.pbio.1001487

Marx, C.J., and Lidstrom, M.E. (2001). Development of improved versatile broad-host-range vectors for use in methylotrophs and other Gram-negative bacteria. Microbiology *147*, 2065–2075.

Marx, C.J., and Lidstrom, M.E. (2004). Development of an insertional vector system for *Methylobacterium extorquens* AM1 and generation of null mutants lacking *mtdA* and/or *fch*. Microbiology *150*, 9–19. https://doi.org/10.1099/mic.0.26587-0

Marx, C.J., O'Brien, B.N., Breezee, J., and Lidstrom, M.E. (2003a). Novel methylotrophy genes of *Methylobacterium extorquens* AM1 identified by using transposon mutagenesis including a putative dihydromethanopterin reductase. J. Bacteriol. *185*, 669–673. https://doi.org/10.1128/JB.185.2.669-673.2003

Marx, C.J., Chistoserdova, L., and Lidstrom, M.E. (2003b). The formaldehyde-detoxifying role of the tetrahydromethanopterin-linked pathway in *Methylobacterium extorquens* AM1. J. Bacteriol. *185*, 7160–7168. https://doi.org/10.1128/JB.185.24.7160-7168.2003

Marx, C.J., Laukel, M., Vorholt, J.A., and Lidstrom, M.E. (2003c). Purification of the formate-tetrahydrofolate ligase from *Methylobacterium extorquens* AM1 and demonstration of its requirement for methylotrophic growth. J. Bacteriol. *185*, 7169–7175. https://doi.org/10.1128/JB.185.24.7169-7175.2003

Marx, C.J., Van Dien, S.J., and Lidstrom, M.E. (2005). Flux analysis uncovers key role of functional redundancy in formaldehyde metabolism. PLOS Biol. 3, e16. https://doi.org/10.1371/journal.pbio.0030016

Marx, C.J., Bringel, F., Chistoserdova, L., Moulin, L., Farhan Ul Haque, M., Fleischman, D.E., Gruffaz, C., Jourand, P., Knief, C., Lee, M.C., *et al.* (2012). Complete genome sequences of six strains of the genus *Methylobacterium*. J. Bacteriol. *194*, 4746–4748. https://doi.org/10.1128/JB.01009-12

Michener, J.K., and Marx, C.J. (2015). After horizontal gene transfer, metabolic pathways may need further optimization. Microbe. *10*, 61–67.

Michener, J.K., Vuilleumier, S., Bringel, F., and Marx, C.J. (2014a). Phylogeny poorly predicts the utility of a challenging horizontally-transferred gene in *Methylobacterium* strains. J. Bacteriol. *196*, 2101–2107. https://doi.org/10.1128/JB.00034-14

Michener, J.K., Camargo Neves, A.A., Vuilleumier, S., Bringel, F., and Marx, C.J. (2014b). Effective use of a horizontally-transferred pathway for dichloromethane catabolism requires post-transfer refinement of the host. eLife. 3, e04279. https://doi.org/10.7554/eLife.04279

Michener, J.K., Vuilleumier, S., Bringel, F., and Marx, C.J. (2016). Transfer of a catabolic pathway for chloromethane in *Methylobacterium* strains highlights different limitations for growth with chloromethane or with dichloromethane. Front. Microbiol. 7, 1116. https://doi.org/10.3389/fmicb.2016.01116

Muller, E.E., Bringel, F., and Vuilleumier, S. (2011). Dichloromethane-degrading bacteria in the genomic age. Res. Microbiol. *162*, 869–876. https://doi.org/10.1016/j.resmic.2011.01.008

Nayak, D.D., and Marx, C.J. (2014a). Genetic and phenotypic comparison of facultative methylotrophy between *Methylobacterium extorquens* strains PA1 and AM1. PLOS ONE. *9*, e107887. https://doi.org/10.1371/journal.pone.0107887

Nayak, D.D., and Marx, C.J. (2014b). Methylamine utilization via the *N*-methylglutamate pathway in *Methylobacterium extorquens* PA1 involves a novel flow of carbon through C_1 assimilation and dissimilation pathways. J. Bacteriol. *196*, 4130–4139. https://doi.org/10.1128/JB.02026-14

Nayak, D.D., and Marx, C.J. (2015). Experimental horizontal gene transfer of methylamine dehydrogenase mimics prevalent exchange in nature and overcomes the methylamine growth constraints posed by the sub-optimal *N*-methylglutamate pathway. Microorganisms 3, 60–79. https://doi.org/10.3390/microorganisms3010060

Nayak, D.D., Agashe, D., Lee, M.-C., and Marx, C.J. (2016). Selection maintains apparently degenerate metabolic pathways due to tradeoffs in using methylamine for carbon versus nitrogen. Curr. Biol. *26*, 1416–1426. https://doi.org/10.1016/j.cub.2016.04.029

Peel, D., and Quayle, J.R. (1961). Microbial growth on C1 compounds. I. Isolation and characterization of *Pseudomonas* AM 1. Biochem. J. *81*, 465–469.

Peyraud, R., Kiefer, P., Christen, P., Massou, S., Portais, J.C., and Vorholt, J.A. (2009). Demonstration of the ethylmalonyl CoA pathway by using ^{13}C metabolomics. Proc. Natl. Acad. Sci. U.S.A. *106*, 4846–4851. https://doi.org/10.1073/pnas.0810932106

Ras, J., Van Ophem, P.W., Reijnders, W.N., Van Spanning, R.J., Duine, J.A., Stouthamer, A.H., and Harms, N. (1995). Isolation, sequencing, and mutagenesis of the gene encoding NAD- and glutathione-dependent formaldehyde dehydrogenase (GD-FALDH) from *Paracoccus denitrificans*, in which GD-FALDH is essential for methylotrophic growth. J. Bacteriol. *177*, 247–251.

Rokyta, D.R., Joyce, P., Caudle, S.B., Miller, C., Beisel, C.J., and Wichman, H.A. (2011). Epistasis between beneficial mutations and the phenotype-to-fitness Map for a ssDNA virus. PLOS Genet. 7, e1002075. https://doi.org/10.1371/journal.pgen.1002075

Silva, I.N., Santos, P.M., Santos, M.R., Zlosnik, J.E., Speert, D.P., Buskirk, S.W., Bruger, E.L., Waters, C.M., Cooper, V.S., and Moreira, L.M. (2016). Long-term evolution of *Burkholderia multivorans* during a chronic cystic fibrosis infection reveals shifting forces of selection. mSystems *1*, e00029-16. https://doi.org/10.1128/mSystems.00029-16

Van Dien, S.J., Marx, C.J., O'Brien, B.N., and Lidstrom, M.E. (2003). Genetic characterization of the carotenoid biosynthetic pathway in *Methylobacterium extorquens* AM1 and isolation of a colorless mutant. Appl. Environ. Microbiol. *69*, 7563–7566.

Vannelli, T., Messmer, M., Studer, A., Vuilleumier, S., and Leisinger, T. (1999). A corrinoid-dependent catabolic pathway for growth of a *Methylobacterium* strain with chloromethane. Proc. Natl. Acad. Sci. U.S.A. *96*, 4615–4620.

Vasi, F., Travisano, M., and Lenski, R.E. (1994). Long-term experimental evolution in *Escherichia coli*. II. Changes in life-history traits during adaptation to a seasonal environment. Am. Nat. *144*, 432–456.

Vishniac, W., and Santer, M. (1957). The thiobacilli. Bacteriol. Rev. *21*, 195–213.

Vorholt, J.A., Marx, C.J., Lidstrom, M.E., and Thauer, R.K. (2000). Novel formaldehyde-activating enzyme in *Methylobacterium extorquens* AM1 required for growth on methanol. J. Bacteriol. *182*, 6645–6650.

Vuilleumier, S., Chistoserdova, L., Lee, M.C., Bringel, F., Lajus, A., Zhou, Y., Gourion, B., Barbe, V., Chang, J., Cruveiller, S., *et al.* (2009). *Methylobacterium* genome sequences: a reference blueprint to investigate microbial metabolism of C1 compounds from natural and industrial sources. PLOS ONE *4*, e5584. https://doi.org/10.1371/journal.pone.0005584

Vuilleumier, S., Nadalig, T., Ul Haque, M.F., Magdelenat, G., Lajus, A., Roselli, S., Muller, E.E., Gruffaz, C., Barbe, V., Médigue, C., *et al.* (2011). Complete genome sequence of the chloromethane-degrading *Hyphomicrobium* sp. strain MC1. J. Bacteriol. *193*, 5035–5036. https://doi.org/10.1128/JB.05627-11

Weinreich, D.M., Delaney, N.F., Depristo, M.A., and Hartl, D.L. (2006). Darwinian evolution can follow only very few mutational paths to fitter proteins. Science *312*, 111–114. https://doi.org/10.1126/science.1123539

Wiser, M.J., Ribeck, N., and Lenski, R.E. (2013). Long-term dynamics of adaptation in asexual populations. Science *342*, 1364–1367. https://doi.org/10.1126/science.1243357

Xue, K.S., Stevens-Ayers, T., Campbell, A.P., Englund, J.A., Pergam, S.A., Boeckh, M., and Bloom, J.D. (2017). Parallel evolution of influenza across multiple spatiotemporal scales. eLife *6*, e26875. https://doi.org/10.7554/eLife.26875

Yang, L., Jelsbak, L., Marvig, R.L., Damkiær, S., Workman, C.T., Rau, M.H., Hansen, S.K., Folkesson, A., Johansen, H.K., Ciofu, O., *et al.* (2011). Evolutionary dynamics of bacteria in a human host environment. Proc. Natl. Acad. Sci. U.S.A. *108*, 7481–7486. https://doi.org/10.1073/pnas.1018249108

Index

D

E

F

G

H

I

L

M

N

O

P

Current Books of Interest

www.ingramcontent.com/pod-product-compliance
Lightning Source LLC
LaVergne TN
LVHW071340210926
848486LV00003B/199
* 9 7 8 1 9 1 2 5 3 0 0 4 5 *